Elementary Algebra
Equations and Graphs

Elementary Algebra
Equations and Graphs

Table of Contents

Chapter 1 Variables and Equations

Lesson 1.1   Variables

- Use letters to represent variables
- Make a table of values for an expression
- Find a pattern and write a rule
- Plot points from a table of values
- Read values from a graph

Lesson 1.2   Algebraic Expressions

- Write expressions
- Evaluate expressions
- Plot points from a table of values
- Use formulas to write and evaluate expressions

Lesson 1.3   Equations and Graphs

- Graph equations
- Solve equations graphically

Lesson 1.4   Solving Equations

- Solve linear equations with one operation
- Solve problems using formulas
- Write an equation to solve a problem

Lesson 1.5   Order of Operations

- Use the associative laws for addition and multiplication
- Follow the order of operations to simplify expressions

1.1 Variables
What Is a Variable?

> A **variable** is a numerical quantity that changes over time or in different situations.

We can show the values of a variable in a table or a graph.

◼ Example 1   Life expectancy is the average age to which people will live. The table at right shows that your life expectancy depends on the year of your birth. Thus, life expectancy is a variable.

a. To what age could people born in 1940 expect to live?

b. In what year did people's life expectancy reach 70 years of age?

Solutions a. People born in 1940 lived to 63 years of age on average.

b. Life expectancy reached 70 years of age in 1960.

Year Born	Life Expectancy
1900	49
1910	51
1920	58
1930	59
1940	63
1950	68
1960	70
1970	71
1980	74
1990	75

◼ Example 2 The graph at right shows the average annual salaries of NFL football players, starting in 1940. The average salary of NFL football players is a variable.

a. Use the graph to estimate the average annual salary of NFL football players in 1975.

b. By how much did salaries increase from 1975 to 1980?

Solutions a.   In 1975, the average NFL player's salary was $40,000.

b. In 1980, the average salary was $80,000, so salaries had increased from $40,000 to $80,000, or by

$$\$80,000 - \$40,000 = \$40,000$$

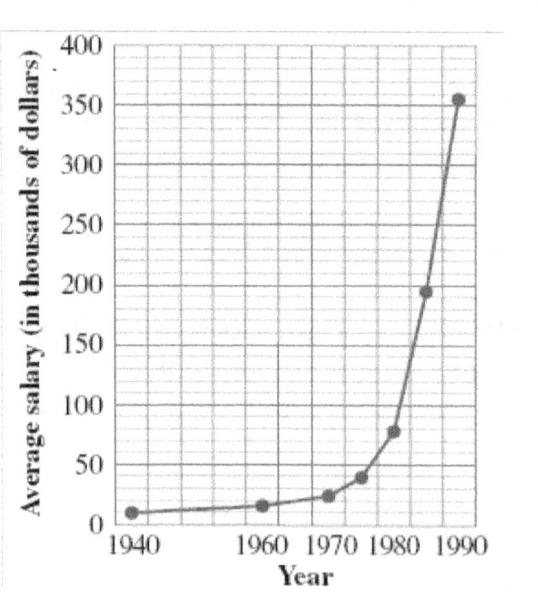

RQ1.   A numerical quantity that changes over time or in different situations is called a _____ .

RQ2.   We can show the values of a variable in a _____ or a _____ .

Look Closer: By displaying the values of a variable in a table or a graph, we sometimes see trends or patterns in those values. For example, we see that life expectancy has been increasing slowly over time, and that salaries of NFL football players have been increasing ever more rapidly since 1940.

Using Letters to Represent Variables

Imagine that you are traveling by airplane to another city. The table below shows the altitude of the airplane, in thousands of feet, at ten-minute intervals after take-off. The two quantities shown in the table, time and altitude, are both variables.

Time (min)	0	10	20	30	40	50	60	70	80	90	100	110
Altitude (1000 ft)	0	8	20	21	23	28	29	30	25	13	5	0

We often use a letter as a kind of short-hand to represent a variable. For this example, we use the following letters:

t stands for the **time** elapsed after take-off,
h stands for the plane's **altitude** at that time

We can get a better feel for the variables by plotting them on a graph. In the figure at right, the values of t are displayed on the horizontal scale, or **axis**, and the values of h are shown on the vertical axis. The graph shows how the values of h are related to the values of t.

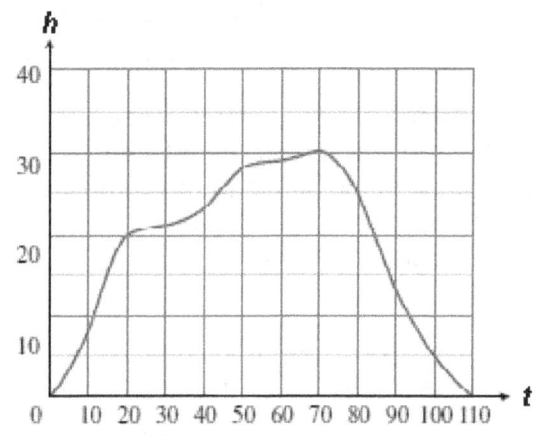

Example 3a. What is the value of h when $t = 85$? What do these values mean in this situation?
b. When is the plane descending?

Solutions a.   The graph shows that h is approximately 20 when $t = 85$. This means that at 85 minutes into the flight, the plane's altitude is about 20,000 feet.
b. The plane is descending when its altitude is decreasing. On the graph, h begins to decrease at $t = 70$ and continues decreasing until $t = 110$. Thus, the plane is descending from the 70th minute until the end of the flight at 110 minutes.

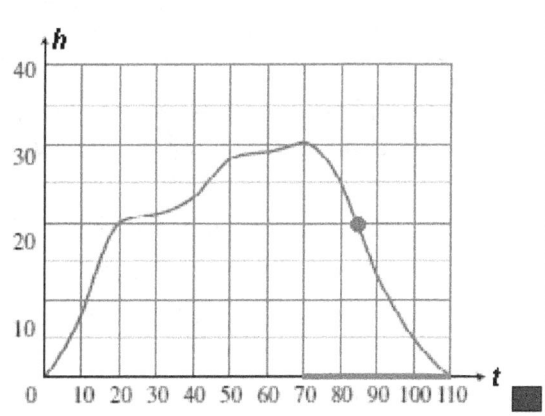

RQ3.   We often use _____ as a shorthand to represent variable quantities.
RQ4.   Which variable is displayed on the vertical axis in Example 3?

Look Ahead: It would be useful if we could discover a formula or rule to help us *predict* the values of an interesting variable.

Connections Between Variables

Sometimes there is a simple mathematical relationship between the values of two variables.

■ Example 4   Fernando plans to share an apartment with three other students next year. The table shows his share of the rent for apartments of various prices.

Rent	280	300	340	360	400	460	500
Fernando's share	70	75	85	90	100	?	?

a. Fill in the blanks in the table. Describe in words how you found the unknown values.

Divide the rent by 4

b. Write a mathematical sentence that gives Fernando's share of the total rent.

Fernando's share = **Rent** ÷ 4

c. The total rent and Fernando's share are variables. Let R stand for the total rent, and let F stand for Fernando's share. Using these letters, write a formula for Fernando's share of the rent.

$$F = R \div 4$$

d. Plot the points from the table and connect them with a smooth curve. Extend your line so that it reaches across the entire grid, not just the points that you plotted.

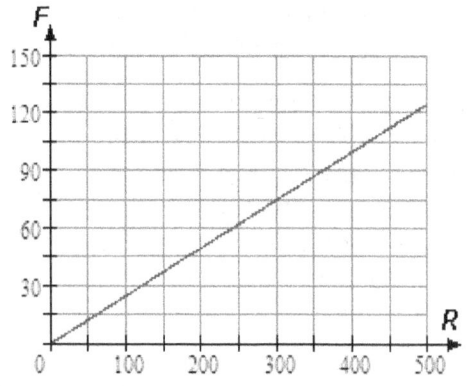

A formula relating two variables is a type of **equation**, and an equation is just a statement that two quantities are equal.

By studying the values in a table, we may be able to find a relationship between the values, and then write an equation relating the two variables.

■ Example 5   Write an equation that relates the two variables in the table.

x	5	10	15	20	25
y	2	7	12	17	22

Solution The two variables are x and y. Each column of the table shows a related pair of values. Notice that 2 is 3 less than 5, 7 is 3 less than 10, and so on. That is:

$$2 = 5 - 3, \quad 7 = 10 - 3,$$
$$12 = 15 - 3, \quad 17 = 20 - 3,$$

and so on. In each column, we can find y by subtracting 3 from x. Thus, $y = x - 3$. ■

RQ5.   A statement that two quantities are equal is called an _____.

Skills Warm-Up

▪ Choose the correct arithmetic operation (addition, subtraction, multiplication, or division) and explain how to answer the question.

1. An air conditioner keeps the inside temperature 16° cooler than the outside temperature. If the outside temperature is 90°, how can you find the inside temperature?

2. Tom's recipe for punch calls for three times as much fruit juice as soda. If he has half a gallon of soda, how can he find the amount of fruit juice he needs?

3. A clothesline should be 2 feet longer than the distance between the supporting poles so that it can be tied at each end. If the poles are 20 feet apart, how can you find the length of the clothesline?

4. The weight of a bridge is supported equally by 8 pillars. If the bridge weighs one million tons, how can you find the weight each pillar must support?

5. Katrin has 4 hours to complete both her math and geography homework. If her math assignment takes $2\frac{1}{2}$ hours, how can she calculate how long she has for geography?

6. The cost of leasing a compact car is 63% the cost of a luxury car. If the lease on the luxury car is $500 per month, how can you find the cost of leasing the compact car?

Answers: 1. Subtract 16 from the outside temperature. **2.** Multiply the amount of soda by 3. **3.** Add 2 feet to the distance between the poles. **4.** Divide the weight of the bridge by 8. **5.** Subtract the time for math from 4 hours. **6.** Multiply the cost of the luxury car by 0.63.

Homework 1.1

Skills Practice

■ For Problems 1-4,
   a. Find the pattern and fill in the table.
   b. Write an equation for the second variable in terms of the first variable.

1.

m	g
2	5
3	6
5	8
10	13
12	
16	
18	
m	

2.

t	w
0	20
2	18
4	16
5	15
6	
10	
12	
t	

3.

b	x
0	0
2	1
4	2
5	2.5
6	
8	
9	
b	

4.

z	3	6	8	12	15	18	20	z
r	2	4	$\frac{16}{3}$	8				

Hint: What fraction can you multiply z by to get r?

■ In Problems 5-6, make your own table. Choose values for the first variable, and use the rule to find the values of the second variable.

5. $W = 1.2 \times n$

n					
W					

6. $M = \dfrac{3}{2} \times x$

x					
M					

■ For Problems 7-8, fill in the tables. What do you notice? Explain.

7a. $y = \dfrac{x}{8}$

x	y
4	
8	
10	
16	

b. $y = \dfrac{1}{8} \times x$

x	y
4	
8	
10	
16	

c. $y = 0.125 \times x$

x	y
4	
8	
10	
16	

8a. $y = \dfrac{3 \times x}{5}$

x	y
5	
10	
12	
1	

b. $y = \dfrac{3}{5} \times x$

x	y
5	
10	
12	
1	

c. $y = 0.6 \times x$

x	y
5	
10	
12	
1	

Applications

9. The temperature in Sunnyvale is usually 15° hotter than it is in Ridgecrest, which is at a higher elevation. Fill in the table.

Temperature in Ridgecrest	70	75	82	86	90	R
Calculation						
Temperature in Sunnyvale						

a. Explain in words how to find the temperature in Sunnyvale.

b. Write your explanation as a mathematical sentence:

Temp in Sunnyvale =

c. Let R stand for the temperature in Ridgecrest and S for the temperature in Sunnyvale. Write an equation for S in terms of R.

d. Plot the points from the table and connect them with a smooth curve.

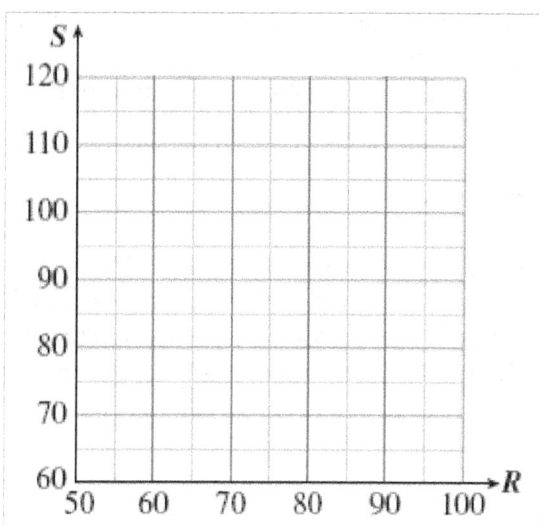

10. Jerome is driving from his home in White Falls to his parents' home in Castle Heights, 200 miles away. Fill in the table.

Miles driven	40	60	90	120	140	170	d
Calculation							
Miles remaining							

a. Explain in words how to find the number of miles Jerome has left to drive.

b. Write your explanation as a mathematical sentence:

Miles remaining =

c. Let d stand for the number of miles Jerome has driven and r for the number of miles that remain. Write an equation for r in terms of d.

d. Plot the points from the table and connect them with a smooth curve.

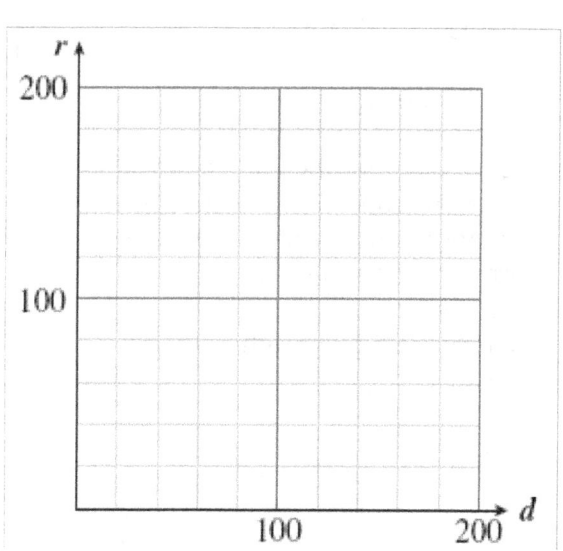

11. Milton goes to a restaurant with two friends, and they agree to split the bill equally. Fill in the table.

Total bill	15	30	45	60	75	81	b
Calculation							
Milton's share							

a. Explain in words how to find Milton's share of the bill.

b. Write your explanation as a mathematical sentence:

Milton's share =

c. Let b stand for the bill and s for Milton's share. Write an equation for s in terms of b.

d. Plot the points from the table and connect them with a smooth curve.

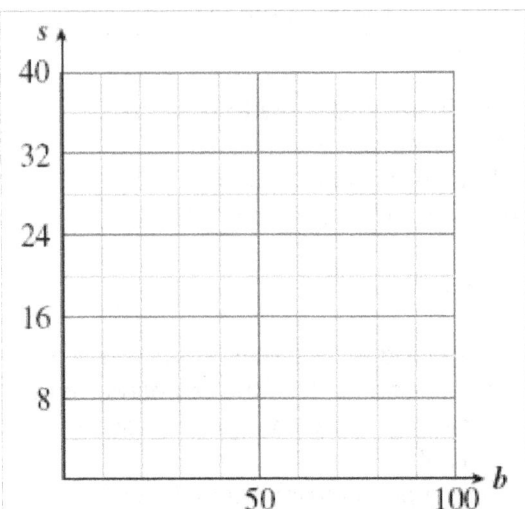

12. Nutrition experts tell us that no more than 30% of the calories we consume should come from fat. Fill in the table with the fat calories allowed daily for various calorie levels.

Total calories	1000	1500	2000	2500	3000	C
Calculation						
Fat calories						

a. Explain in words how to find the number of fat calories allowed.

b. Write your explanation as a mathematical sentence:

Fat calories =

c. Let C stand for the total number of calories and F for the number of fat calories. Write an equation for F in terms of C.

d. Plot the points from the table and connect them with a smooth curve.

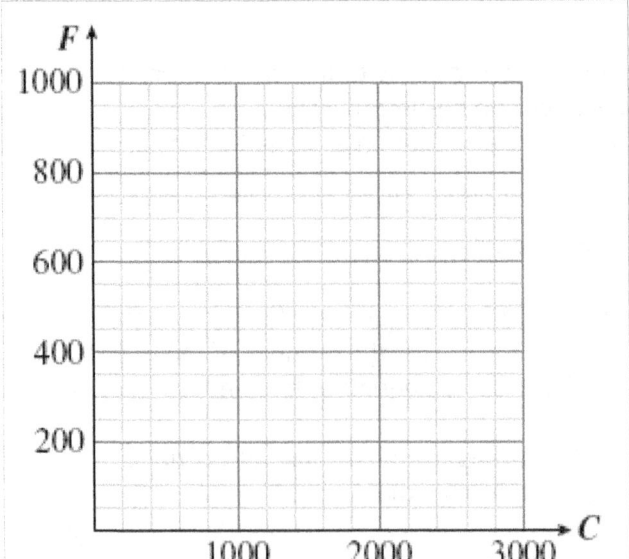

■ Use the graphs to answer the questions in Problems 13 and 14. You may have to estimate some of your answers.

13. Suppose you invest $2000 in a retirement account that pays 8% interest compounded continuously. The amount of money in the account each year is shown by the graph at left below.
   a.   What variable is displayed on the horizontal axis? The vertical axis?
   b.   How much money will be in the account 5 years from now?
   c.   When will the amount of money in the account exceed $6000?
   d.   How much will the account grow between the second and third years?
   e.   How much will the account grow between the twelfth and thirteenth years?

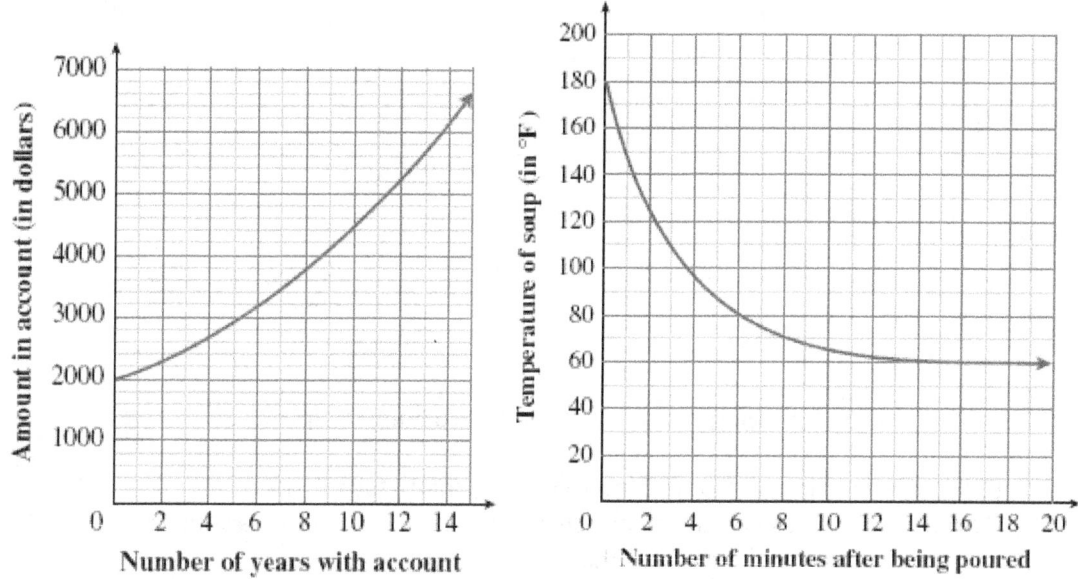

14. Wendy brings a Thermos of soup for her lunch on a hike in the mountains. When she pours the soup into a bowl, it begins to cool off. The graph above right shows the temperature of the soup after it is served.
   a.   What variable is displayed on the horizontal axis? The vertical axis?
   b.   How long does it take the soup to cool below 100°?
   c.   What is the temperature of the soup after 8 minutes?
   d.   How many degrees does the soup cool in the first 2 minutes?
   e.   After a while, the temperature of the soup levels off at the same as the outside temperature. What was the outside temperature that day?

15. A cantaloupe is dropped from a tall building. Its speed increases until it hits the ground. Which of the four graphs below best represents the speed of the cantaloupe? Explain why your choice is the best one.

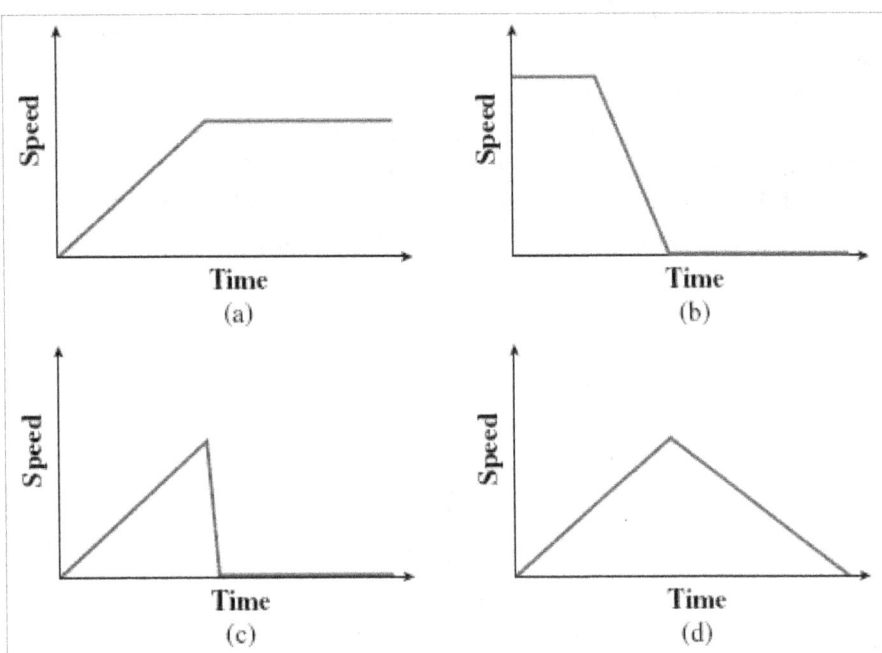

16. a. Match each of the following stories with the appropriate graph.

    I. Delbert walks directly from home to school, then stays there.

    II. Delbert walks towards school, but stops at coffee shop for a cappuccino, then he goes on to school and stays there.

    III. Delbert walks toward school, but decides to return home, where he stays.

    b. Write your own story for the remaining graph.

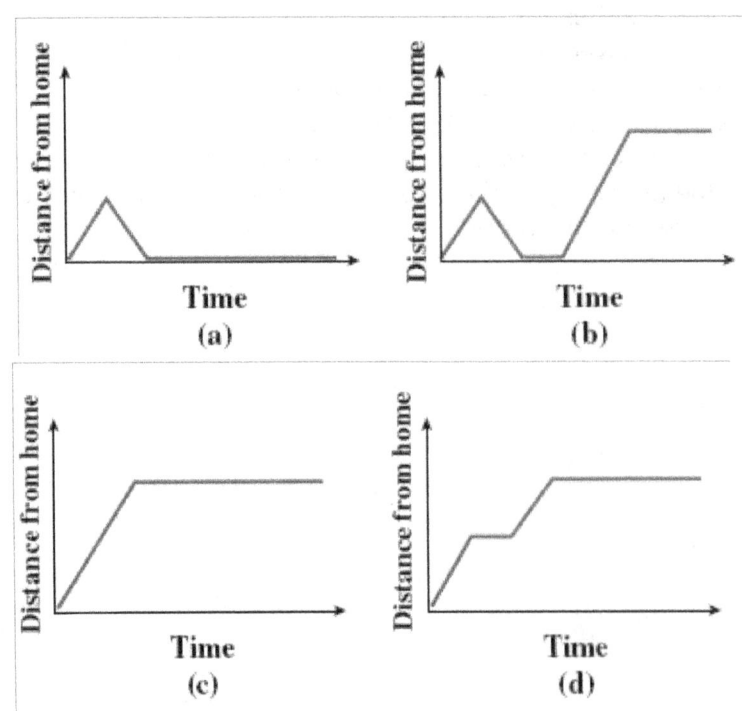

17. Plot each pair of values on the grid provided. Connect the points with a smooth curve. Which graph appears to be a straight line?

a.

t	0	1	2	5	6	8
v	16	14	12	6	4	0

b.

w	0	1	2	4	6	10
Q	12	8	6	4	3	2

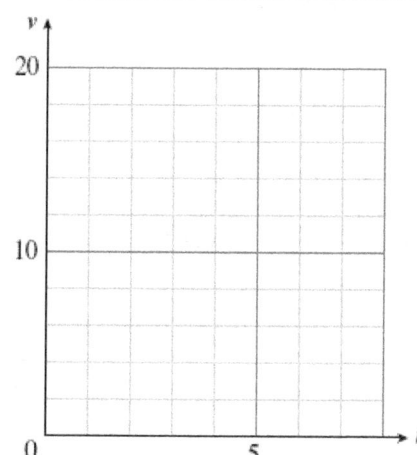
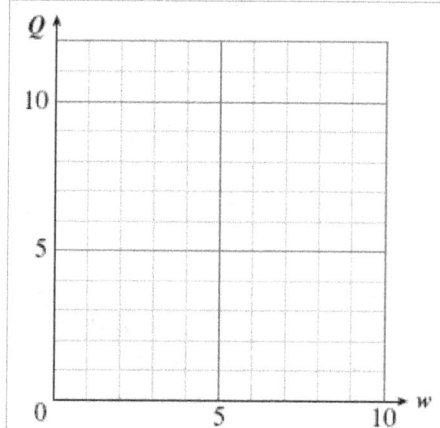

18. a. Read the graph at right to fill in the table.

x	0	10	30	40	60	70
y						

b. Find a formula expressing y in terms of x.

For Problems 19-20,
  a. Write an equation for the second variable in terms of the first variable.
  b. Fill in the table. Plot the points and connect them with a smooth curve.

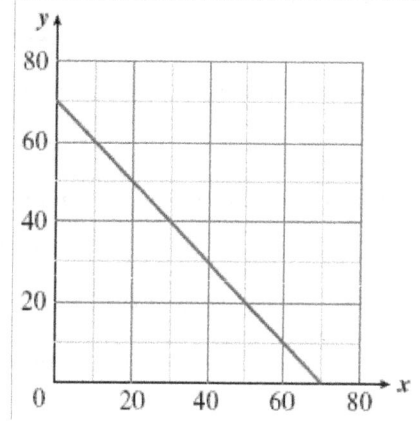

19.

x	0	0.2	0.3	0.5	0.6	0.7
y	2	1.8	1.7			

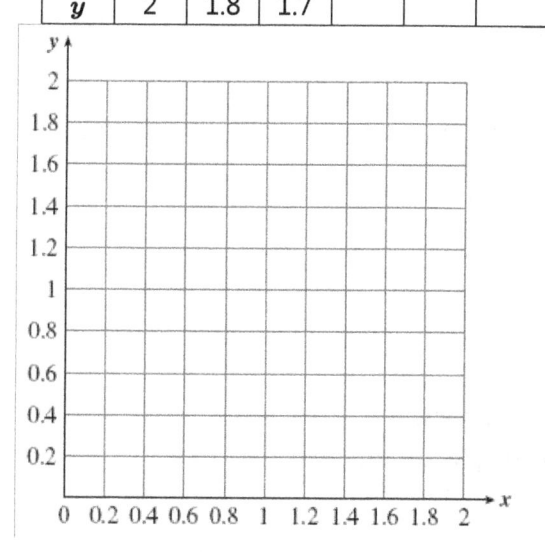

20.

x	1	2	3	4	6	8
y	120	60	40			

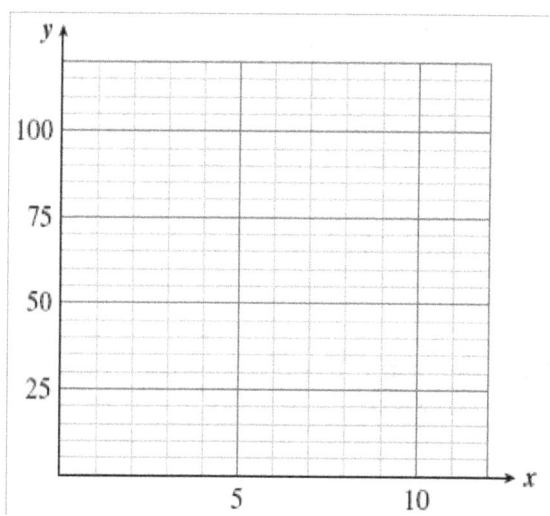

1.2 Algebraic Expressions
Writing an Algebraic Expression

> An **algebraic expression**, or simply an **expression**, is any meaningful combination of numbers, variables, and operation symbols.

For example,

$$4 \times g, \quad 3 \times c + p, \quad \text{and} \quad \frac{n-7}{w}$$

are algebraic expressions. An important part of algebra involves translating word phrases into algebraic expressions.

■ **Example 1**   Write an algebraic expression for each quantity.
a. 30% of the money invested in stocks
b. The cost of the dinner split three ways

Solutions a. The amount invested is unknown, so we choose a variable to represent it.

Amount invested in stocks: s

Next, we identify the operation described: "30% of" means 0.30 times:

The expression is $0.30\,s$

b. The cost of the dinner is unknown, so we choose a variable to represent it.

Cost of dinner: C

Next, we identify the operation described: "Split" means divided:

The expression is $\dfrac{C}{3}$ ■

RQ1.   A meaningful combination of numbers, variables and operation symbols is called an _____.

Sums

When we add two numbers a and b, the result is called the **sum** of a and b. We call the numbers a and b the **terms** of the sum.

■ **Example 2**   Write sums to represent the following phrases.
a. Six more than x **b.** Fifteen greater than r

Solutions a. $6 + x$ or $x + 6$ **b.** $15 + r$ or $r + 15$ ■

The terms can be added in either order. We call this fact the **commutative law** for addition.

> **Commutative Law for Addition**
>   If a and b are any numbers, then
>
> $$a + b = b + a$$

Products

When we multiply two numbers a and b, the result is called the **product** of a and b. The numbers a and b are the **factors** of the product.

Look Closer: In arithmetic we use the symbol $\times$ to denote multiplication. However, in algebra, $\times$ may be confused with the variable x. So, instead, we use a dot or parentheses for multiplication, like this:

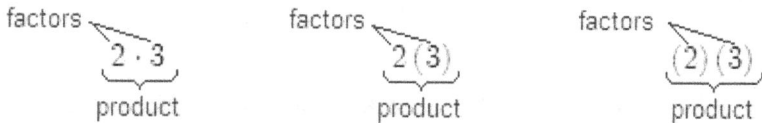

If one of the factors is a variable, we can just write the factors side by side. For example,

ab means   the product of a times b
$3x$ means   the product of 3 times x

> ■ **Example 3**   Write products to represent the following phrases.
> **a.**   Ten times z **b.** Two-thirds of y **c.**   The product of x and y
>
> **Solutions a.** $10z$ **b.** $\dfrac{2}{3}y$ **c.** xy or yx ■

Look Closer: In Example 3a, $10z = 10 \cdot z = z \cdot 10$, because two numbers can be multiplied in either order to give the same answer. This is the **commutative law** for multiplication.

> **Commutative Law for Multiplication**
>   If a and b are any numbers, then
>
> $$a \cdot b = b \cdot a$$

In algebra, we usually write products with the numeral first. Thus, we write $10z$ for "10 times z".

> **RQ2.**   When we add two numbers, the result is called the _____ and the two
>   numbers are called the _____ .
> **RQ3.**   When we multiply two numbers, the result is called the _____ and the
>   two numbers are called the _____ .

Differences

When we subtract b from a the result is called the **difference** of a and b. As with addition, a and b are called **terms**.

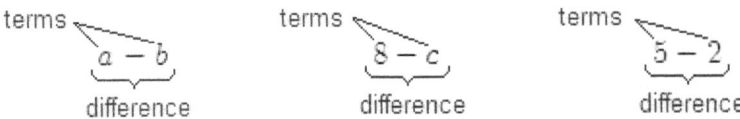

> **Caution!**   The difference $5 - 2$ is *not* the same as $2 - 5$. When we translate "a subtracted from b," the order of the terms is important.
>
> $$b - a \quad \text{means} \quad a \text{ subtracted from } b$$
> $$x - 7 \quad \text{means} \quad 7 \text{ subtracted from } x$$
>
> The operation of subtraction is *not* commutative.

Quotients

When we divide a by b the result is called the **quotient** of a and b. We call a the **dividend** and b the **divisor**. In algebra we indicate division by using the division symbol, $\div$, or a fraction bar.

division symbol

dividend $\rightarrow$ $a \div b$ $\leftarrow$ divisor

quotient

dividend $\rightarrow$ $\dfrac{a}{b}$ $\leftarrow$ quotient

fraction bar $\rightarrow$ $\dfrac{a}{b}$ divisor

Look Closer: The operation of division is **not** commutative. The order of the numbers in a quotient makes a difference. For example, $12 \div 3 = 4$ but $3 \div 12 = \frac{1}{4}$.

Example 4   Write an algebraic expression for each phrase.

a. z subtracted from 13   b. 25 divided by R

Solutions a. $13 - z$ b. $\dfrac{25}{R}$

> **RQ4.**   Which operations obey the **commutative laws**? What do these laws say?
>
> **RQ5.**   When we subtract two numbers, the result is called the _____ .
>
> **RQ6.**   When we divide two numbers, the result is called the _____ .

Evaluating an Algebraic Expression

An algebraic expression is a pattern or rule for different versions of the situation it describes. We can replace the variable by specific numbers to fit a particular situation.

> Substituting a specific value into an expression and calculating the result is called **evaluating** the expression.

Skills Warm-Up

▣ Recall the formulas for the **area** and **perimeter** of a rectangle:

$$A = lw \quad \text{and} \quad P = 2l + 2w$$

1. Delbert's living room is 20 feet long and 12 feet wide. How much oak baseboard does he need to border the floor? (Don't worry about doorways.)
2. How much wood parquet tiling must he buy to cover the floor?

▣ Which of the following phrases describe a perimeter, and which describe an area?

3. The distance you jog around the shore of a small lake
4. The amount of Astroturf needed for the new football field
5. The amount of grated cheese needed to cover a pizza
6. The number of tulip bulbs needed to border a patio

▣ Find the perimeter and area of each figure. All angles are right angles.

7.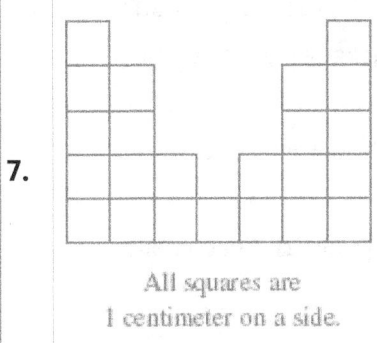

All squares are
1 centimeter on a side.

8.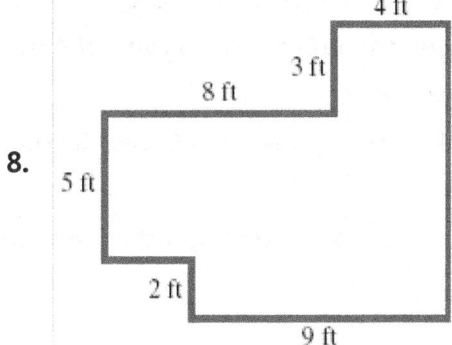

Answers: **1.** 64 ft   **2.** 240 sq ft   **3.** perimeter   **4.** area   **5.** area   **6.** perimeter
7. 32 cm; 23 sq cm   **8.** 44 ft; 90 sq ft

Homework 1.2

Skills Practice

■ For Problems 1-10, write the word phrase as an algebraic expression.

1.   Product of 4 and y

2.   Twice b

3.   115% of g

4. t decreased by 5

5.   The quotient of 7 and w

6.   4 divided into B

7.   20 more than T

8.   16 reduced by p

9.   The ratio of 15 to M

10.   The difference of R and 3.5

■ For Problems 11-14, choose the correct algebraic expression.

$$n+6 \qquad n-6 \qquad 6-n$$
$$6n \qquad \frac{n}{6} \qquad \frac{6}{n}$$

11.   Rashad is 6 years younger than Shelley. If Shelley is n years old, how old is Rashad?

12.   Each package of sodas contains 6 cans. If Antoine bought n packages of sodas, how many cans did he buy?

13.   Lizette and Patrick together own 6 cats. If Lizette owns n cats, how many cats does Patrick own?

14.   Mitra divided 6 cupcakes among n children. How much of a cupcake did each child get?

■ For Problems 15-17, write an algebraic expression for the area or perimeter of the figure. Include units in your answers. The dimensions are given in inches.

15.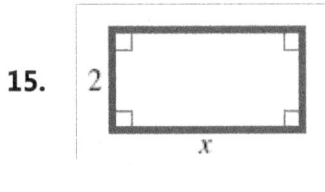

area =

16.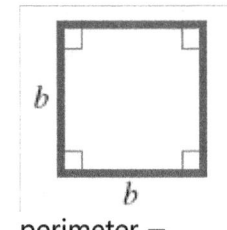

perimeter =

17.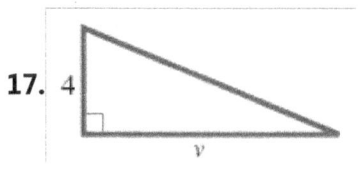

area =

18.   a.   Which of these expressions says "one-sixth of x" ?

$$\frac{x}{6}, \qquad \frac{6}{x}, \qquad \frac{1}{6}x, \qquad \frac{1}{6x}$$

b.   Which of these expressions says "6% of x" ?

$$0.6x, \qquad 0.06x, \qquad \frac{6}{100}x, \qquad \frac{1}{6}x$$

19. Francine is saving up to buy a car, and deposits $\frac{1}{5}$ of her spending money each month into a savings account.
   a. If m stands for Francine's spending money, write an expression for the amount she saves.
   b. Evaluate the expression to complete the table.

Spending money (dollars)	m	20	25	60	80
Amount saved (dollars)					

20. Delbert wants to enlarge his class photograph. Its height is $\frac{1}{4}$ of its width. The enlarged photo should have the same shape as the original.
   a. If W stands for the width of the enlargement, write an expression for its height.
   b. Evaluate the expression to complete the table.

Width of photo (inches)	W	6	10	24	30
Height of photo (inches)					

21. If the governor vetoes a bill passed by the State Assembly, $\frac{2}{3}$ of the members present must vote for the bill in order to overturn the veto.
   a. If p stand for the number of Assembly members present, write an expression for the number of votes needed to overturn a veto.
   b. Evaluate the expression to complete the table.

Members present	p	90	96	120	129
Votes needed					

22. Marla's investment club buys some stock. She will get $\frac{3}{5}$ of the dividends.
   a. If D stands for the stock dividends, write an expression for Marla's share.
   b. Evaluate the expression to complete the table.

Stock dividend	D	40	85	115	170
Marla's share					

Applications

For Problems 23-30, choose a variable for the unknown quantity and translate the phrase into an algebraic expression. Use the three steps you learned in the Lesson.

23.   The product of the ticket price and $15
24.   Three times the cost of a light bulb
25.   Three-fifths of the savings account balance
26.   The price of the pizza divided by 6
27.   The weight of the copper in ounces divided by 16
28.   9% of the school buses
29. $16 less than the cost of the vaccine
30.   The total cost of 32 identical computers

■ For Problems 31-32, write algebraic expressions in terms of x.

31. a. Daniel and Lara together made $480. If Daniel made x dollars, how much did Lara make?

  b. Alix spent $500 on tuition and books. If she spent x dollars on books, how much was her tuition?

  c. Thirty children signed up for summer camp. If x boys signed up, how many girls signed up?

32. a. Rona spent $15 less than her sister on shoes. If Rona's sister spent x dollars, how much did Rona spend?

  b. Phoenix had 12 fewer rain days than Boston last year. If Boston had x rain days, how many rain days did Phoenix have?

  c. Jared scored 18 points lower on his second test than he scored on his first test. If he scored x points on the first test, what was his score on the second test?

■ For Problems 33-36, name the variable and write an algebraic expression.

33. Eggnog is 70% milk. Write an expression for the amount of milk in a container of eggnog.

34. Errol has saved $1200 for his vacation this year. Write an expression for the average amount he can spend on each day of his vacation.

35. Garth received 432 fewer votes than his opponent in the election. Write an expression for the number of votes Garth received.

36. The cost of the conference was $2000 over budget. Write an expression for the cost of the conference.

■ For Problems 37-40, use the formulas in this Lesson.

37. a. Write an equation for the distance d traveled in t hours by a small plane flying at 180 miles per hour.

  b. How far will the plane fly in 2 hours? In $3\frac{1}{2}$ hours? In half a day?

38. a. A certain pesticide contains 0.02% by volume of a dangerous chemical by volume. Write an equation for the amount of chemical C that enters the environment in terms of the number of gallons g of pesticide used.

  b. How much of the chemical enters the environment if 400 gallons of the pesticide are used? 5000 gallons? 50,000 gallons?

39. a. BioTech budgets 8.5% of its revenue for research. Write an equation for the research budget B in terms of BioTech's revenue R.

  b. What is the research budget if BioTech's revenue is $100,000? $500,000? $2,000,000?

40. a. Hugo's Auto Shop paid $4000 in expenses this month. Write an equation for their profit P in terms of their revenue R.

  b. What was their profit if their revenue was $10,000? $6500? $2500?

1.3 Equations and Graphs

Anatomy of a Graph

We use equations to express the relationship between two variabes. A graph is a way of visualizing an equation.

> A graph has two **axes**, horizontal and vertical, and the values for the variables are displayed along the axes. The first, or **input,** variable is displayed on the horizontal axis. The second, or **output,** variable is displayed on the vertical axis.

The graph itself shows how the two variables are related.

■ Example 1 The graph at right shows the relationship between Delbert's age, D, and Francine's age, F. Write an equation for Francine's age in terms of Delbert's age.

Solution In this graph, D is the input variable and F is the output variable. We make a table of values by reading points on the graph.

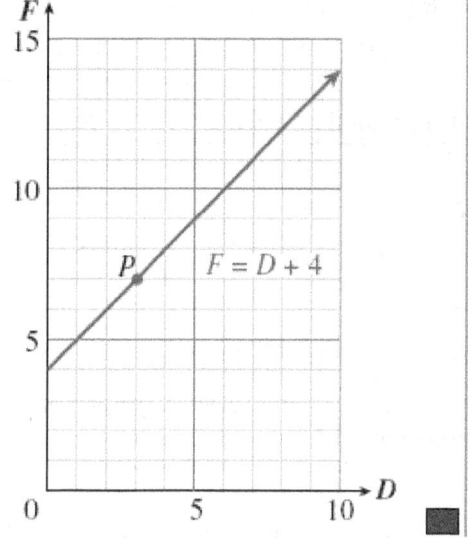

$F = D + 4$

D	0	2	3	4	6	9
F	4	6	7	8	10	13

From the table, we see that the value of F is always 4 more than the value of D. Thus, Francine is exactly four years older than Delbert, or $F = D + 4$. ■

RQ1. A _____ is a way of visualizing an algebraic equation.

RQ2. The values of the variables are displayed on the _____ .

RQ3. The input variable is located on the _____ axis.

Each point on a graph has two **coordinates,** which designate the position of the point. For example, the point labeled P in Example 1 has horizontal coordinate 3 and vertical coordinate 7.

> In Example 1 above, we write the coordinates of point P inside parentheses as an **ordered pair**: $(3, 7)$. The order of the coordinates makes a difference. *We always list the horizontal coordinate first, then the vertical coordinate.*

RQ4. The position of a point on the graph is given by its _____ .

RQ5. The notation (x, y) is called _____ .

RQ6. In an ordered pair, we always list the _____ coordinate first.

Look Closer: How does the graph illustrate the equation $F = D + 4$? The coordinates of each point on the graph are values for D and F that make the equation true. The coordinates of the point P, namely $D = 3$ and $F = 7$, represent the fact that when Delbert was 3 years old, Francine was 7 years old. If we substitute these values into our equation we get

$$F = D + 4$$
$$7 = 3 + 4$$

which is true.

> An ordered pair that makes an equation true is called a **solution** of the equation. Each point on the graph represents a solution of the equation.

> **Exercise a.**   Locate on the graph each of the ordered pairs listed in the table above, and make a dot there. Label each point with its coordinates
> **b.** By substituting its coordinates into the equation, verify that each point you labeled in part (a) represents a solution of the equation $F = D + 4$.

> **RQ7.**   An ordered pair whose coordinates make the equation true is called a
> _____ of the equation.

Graphing an Equation

The simplest way to draw a graph for an equation is to make a table of values and plot the points.

> **■ Example 2** Graph the equation $y = 8 - x$.
>
> **Solution**   In this equation, x is the input variable and y is the output variable. We choose several values for x and use the equation to find the corresponding values for y.
>
>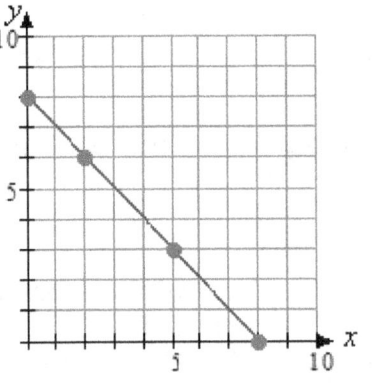
>
> | x | Calculation | y | (x, y) |
> |---|---|---|---|
> | 0 | $y = 8 - 0 = 8$ | 8 | $(0, 8)$ |
> | 2 | $y = 8 - 2 = 6$ | 6 | $(2, 6)$ |
> | 5 | $y = 8 - 5 = 3$ | 3 | $(5, 3)$ |
> | 8 | $y = 8 - 8 = 0$ | 0 | $(8, 0)$ |
>
> Each ordered pair (x, y) represents a point on the graph of the equation. We plot the points on the grid and connect them with a smooth curve, as shown at right.

Choosing Scales for the Axes

If the variables in an equation have very large (or very small) values, we must choose scales for the axes that fit these values.

■ **Example 3**   Graph the equation $y = 250 + x$.

Solution   To graph this equation, we choose multiples of 50 for the x-values.

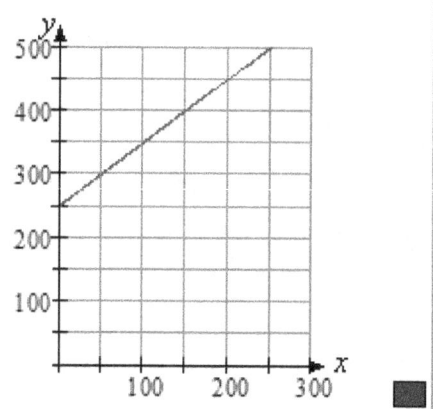

x	Calculation	y	(x, y)
0	$y = 250 + 0 = 250$	250	$(0, 250)$
50	$y = 250 + 50 = 300$	300	$(50, 300)$
100	$y = 250 + 100 = 350$	350	$(100, 350)$
200	$y = 250 + 200 = 450$	450	$(200, 450)$

The largest y-value in the table is 450, so we scale the axis to a little larger than 450, say, 500. We plot the ordered pairs to obtain the graph shown at right.

Look Closer: How do we know what scales to use on the axes? For most graphs, it is best to have between ten and twenty tick marks on each axis, or the graph will be hard to read. We choose intervals of convenient size for the particular problem, such as 5, 10, 25, 100, or 1000. It is not necessary to use the same scale on both axes.   ●

Caution!   It is important that the scales on the axes be evenly spaced. Each tick mark must represent the same interval. The scales on the graph at right are incorrectly labeled.

Incorrect! →

Solving Equations with Graphs

In Example 1 we showed a graph of the equation

$$F = D + 4$$

which gives Francine's age, F, in terms of Delbert's age, D. We can use the graph to answer two types of questions about the equation $F = D + 4$:

> **1.**   Given a value of D, find the corresponding value of F.
> **2.**   Given a value of F, find the corresponding value of D.
>
> The first of these tasks is another way of **evaluating** the algebraic expression $D + 4$, and the second task is another way of **solving** an equation.

■ **Example 4a.**   Use the graph of the equation $F = D + 4$ to evaluate the expression $D + 4$ for $D = 7$. Verify your answer algebraically.
b. Use the graph of the equation $F = D + 4$ to solve the equation $13 = D + 4$. Verify your answer algebraically.

Solutions a.   We locate $D = 7$ on the horizontal axis, as shown at left below. Then we move vertically to point A on the graph with D-coordinate 7. Finally, we move horizontally from point A to the vertical axis to find the F-coordinate. The coordinates of point A are $(7, 11)$, which tells us that when $D = 7$, $F = 11$. To verify the answer algebraically, we **evaluate** the expression for $D = 7$: we substitute 7 for D into the equation:

$$F = D + 4 = 7 + 4 = 11$$

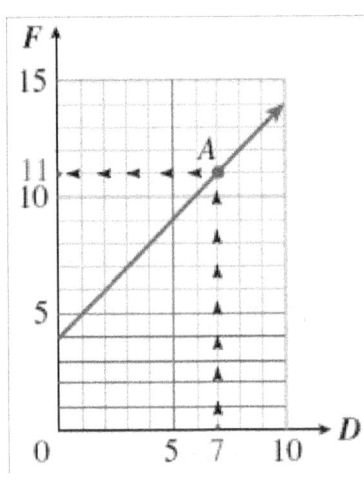

a.

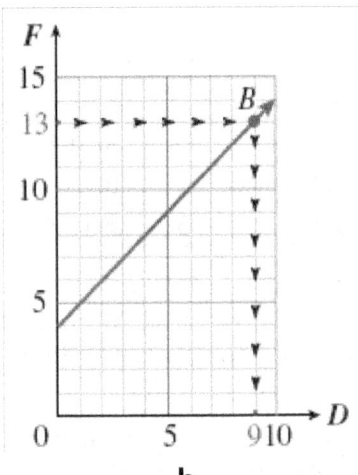

b.

b. We locate $F = 13$ on the vertical axis. We move horizontally to point B on the graph, with F-coordinate 13. From point B, we move vertically to the horizontal axis to find the D-coordinate. The coordinates of point B are $(9, 13)$, which tells us that when $F = 13$, $D = 9$. To verify the answer algebraically, we **solve** the equation when $F = 13$: we subtract 4 from both sides of the equation:

$$13 = D + 4 \qquad \text{Subtract 4 from both sides.}$$
$$\underline{-4 \qquad -4}$$
$$9 = D$$

Example 5   Here is a graph of

$$y = 1.5x$$

Use the graph to solve the equation $1.5x = 3.75$.

Solution By comparing the equation of the graph with the equation we want to solve, we see that y has been replaced by 3.75.

$$\boldsymbol{y} = 1.5x$$
$$\downarrow$$
$$\boldsymbol{3.75} = 1.5x$$

We locate 3.75 on the y-axis, then find the point on the graph with y-coordinate 3.75. The x-coordinate of this point is 2.5, so the ordered pair $(2.5,\ 3.75)$ is a solution of the equation $y = 1.5x$, and $x = 2.5$ is the solution of the equation $1.5x = 3.75$.

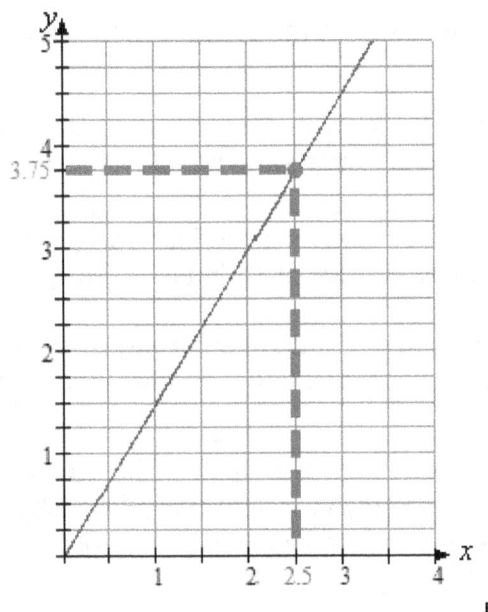

RQ8.   The graph of an equation is a picture of its _____ .
RQ9.   To solve an equation using a graph, we start on the _____ axis.

Skills Warm-Up

In Exercises 1 and 2, what value corresponds to each labeled point?

1.

2.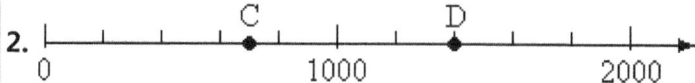

3. **a.** Label the axis.

   b. On the axis in part (a), plot 1400 and 8350.

4. **a.** Label the axis below with 16 tick marks (not counting zero), the highest being 800.

   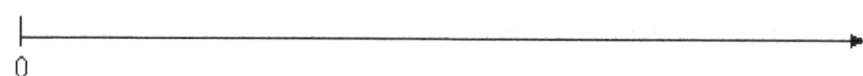

   b. On the axis in part (a), plot 132 and 614.

5. **a.** Label an axis in increments of 40,000 from 0 to 600,000.

   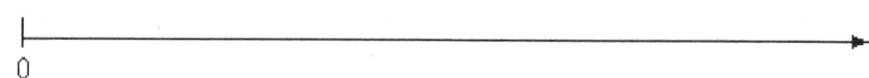

   b. On your axis, plot 250,000 and 472,600.

6. **a.** Label an axis in increments of 0.5 from 0 to 5.

   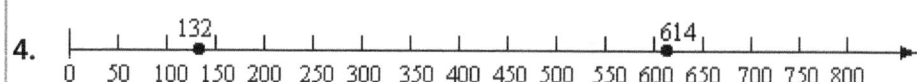

   b. On your axis, plot 1.3 and 3.77.

Answers: 1. A : 350; B : 825   **2.** 700; 1400

3.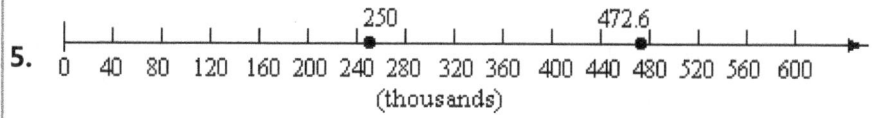

4.

5.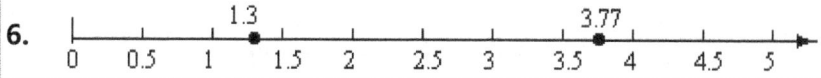
   (thousands)

6.

Homework 1.3

Skills Practice

▪ For Problems 1-2, decide whether the ordered pairs are solutions of the equation whose graph is shown.

1. **a.** $(6.5, 3)$ **b.** $(0, 3.5)$
    c. $(8, 2)$ **d.** $(4.5, 1)$

2. **a.** $(2, 6)$ **b.** $(4, 2)$
    c. $(10, 0)$ **d.** $(11, 7)$

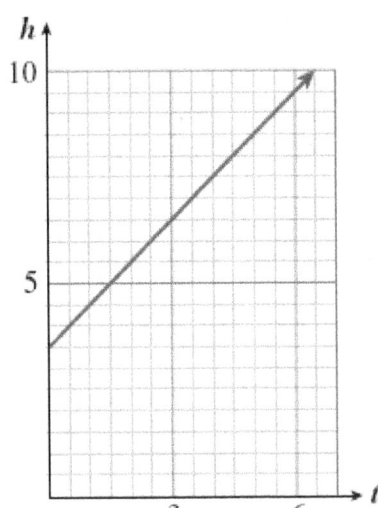   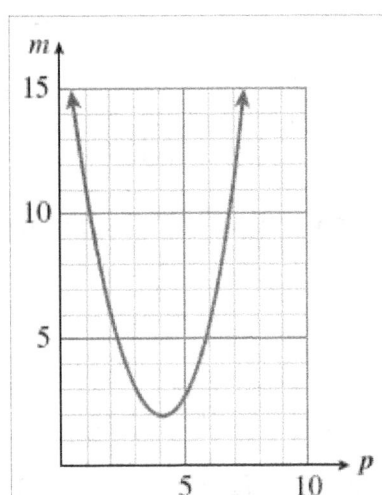

▪ For Problems 3-6, decide whether the ordered pairs are solutions of the given equation.

3. $y = \dfrac{3}{4}x$

    a. $(8, 6)$ **b.** $(12, 16)$

    c. $(2, 3)$ **d.** $(6, \dfrac{9}{2})$

4. $y = \dfrac{x}{2.5}$

    a. $(1, 2.5)$ **b.** $(25, 10)$

    c. $(5, 2)$ **d.** $(8, 20)$

5. $w = z - 1.8$

    a. $(10, 8.8)$ **b.** $(6, 7.8)$

    c. $(2, \frac{1}{5})$ **d.** $(9.2, 7.4)$

6. $w = 120 - z$

    a. $(0, 120)$ **b.** $(65, 55)$

    c. $(150, 30)$ **d.** $(9.6, 2.4)$

▪ For Problems 7-8, state the interval that each grid line represents on the horizontal and vertical axes.

7.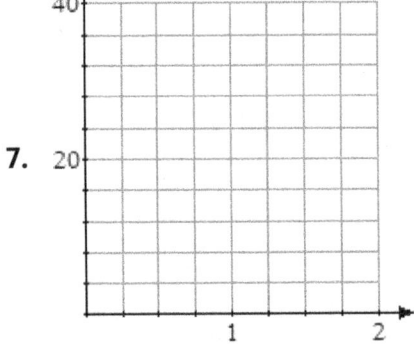

8.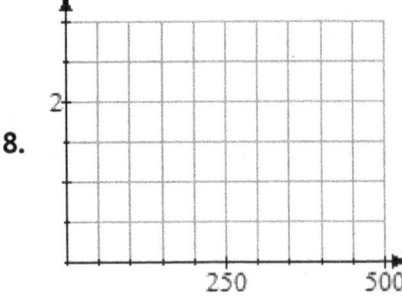

For the graphs in Problems 9-10:
  a. Fill in the table with the correct coordinates. Choose your own points to complete the table.
  b. Look for a pattern in your table, and write an equation for the second variable in terms of the first variable.

9.

x		2	5		10		
y	16			10			

10.

x	0	1			7		
y			16	20			

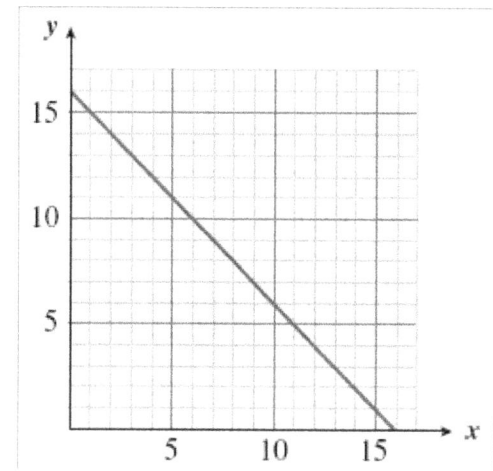

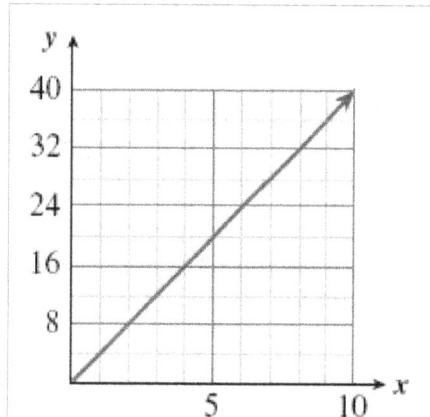

For Problems 11-14, choose the equation that describes the graph. (A graph might have more than one equation.)

a. $y = 150 - x$	**b.** $y = 15 - x$	**c.** $y = 0.2x$	**d.** $y = \dfrac{x}{60}$
e. $y = \dfrac{x}{5}$	**f.** $y = 15 + x$	**g.** $y = \dfrac{60}{x}$	**h.** $y = 5x$

11.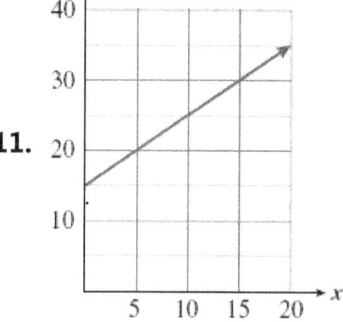

12.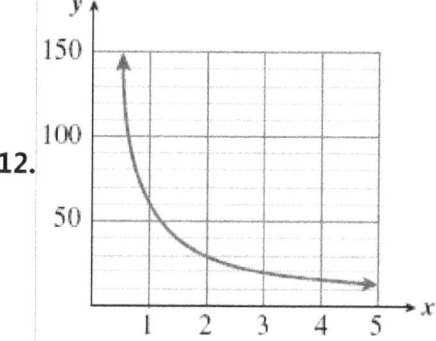

13.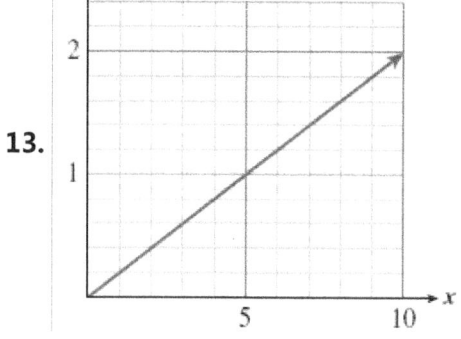

14.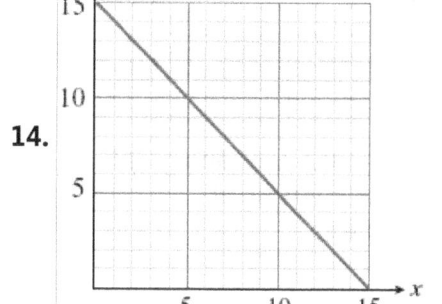

■ For Problems 22-25, use the formulas in Section 1.2.

22. a. Suppose your father loans you $2000 to be repaid with 4% annual interest when you finish school. Write an equation for the amount of interest I you will owe after t years.

  b. How much interest will you owe if you finish school in 2 years? In 4 years? In 7 years? Make a table showing your answers as ordered pairs.

  c. Use your table to choose appropriate scales for the axes, then plot points and sketch a graph of the equation.

23. a. The Earth Alliance made $6000 in revenue from selling tickets to Earth Day, an event for children. Write an equation for its profit P in terms of its costs.

  b. What was their profit if their costs were $800? $1000? $2500? Make a table showing your answers as ordered pairs.

  c. Use your table to choose appropriate scales for the axes, then plot points and sketch a graph of the equation.

24. a. Edgar's great-aunt plans to put $5000 in a trust for Edgar until he turns 21, three years from now. She has a choice of several different accounts. Write an equation for the amount of interest I the money will earn in 3 years if the account pays interest rate r.

  b. How much interest will the $5000 earn in an account that pays 8% interest? $10\frac{1}{4}$% interest? 12% interest? Make a table showing your answers as ordered pairs.

  c. Use your table to choose appropriate scales for the axes, then plot points and sketch a graph of the equation.

25. a. Delbert keeps track of the total number of points he earns on homework assignments, each of which is worth 30 points. At the end of the semester he has 540 points. Write an equation for Delbert's average homework score S in terms of the number of assignments, n.

  b. What is Delbert's average score if there were 20 assignments? 25 assignments? 30 assignments? Make a table showing your answers as ordered pairs.

  c. Use your table to choose appropriate scales for the axes, then plot points and sketch a graph of the equation.

1.4 Solving Equations

What is an Equation?

> An **equation** is a statement that two expressions are equal. It may involve one or more variables.

> **Example 1**   Liz makes $6 an hour as a tutor in the Math Lab. Her weekly earnings, w, depend on the number of hours she works, h. The equation relating these two variables is
>
> $$w = 6h$$
>
> If we know the value of h, we can evaluate the expression $6h$ to find Liz's earnings. For example, if $h = 7$, then
>
> $$w = 6(7) = 42$$
>
> Liz makes $42 for 7 hours of work.

In Example 1 we used evaluation to find the value of w when we knew the value of h. What if we know the value of w and want to find h? Suppose Liz earned $54 last week. How many hours did she work? To answer this question we substitute 54 for w, so that the equation becomes

$$54 = 6h$$

Now we have an equation in just one variable. We would like to find the value of h that makes this equation true.

> A value of the variable that makes an equation true is called a **solution** of the equation, and the process of finding this value is called **solving** the equation.

Trial and Error

Sometimes we can solve an equation by trial and error. By trying different values of h, you will quickly see that the solution of the equation $54 = 6h$ is 9. We can use a table to help us solve an equation by trial and error.

> **Exercise** One of Aunt Esther's Chocolate Dream cookies contains 42 calories, so d cookies contain c calories, and $c = 42d$. If Albert consumed 546 calories, how many cookies did he eat? Use trial and error to help you solve the equation $546 = 42d$.
>
> | d | 5 | 6 | 7 | 8 | 9 | 10 | 11 | 12 | 13 | 14 | 15 |
> |---|---|---|---|---|---|---|---|---|---|---|---|
> | c | | | | | | | | | | | |

Look Ahead: Trial and error can be time-consuming, especially for complicated equations. In this section we investigate some algebraic methods for solving equations. •

> **RQ1.**   What is an equation?
> **RQ2.**   What is a solution of an equation?

Inverse Operations

Instead of relying on trial and error to solve equations, we will use **inverse operations**. Try the Exercise below.

Exercise a.   Choose any number for x: $x =$ _____ .

Multiply your number by 5: $5x =$ _____.

Divide the result by 5: $\dfrac{5x}{5} =$ _____.

Did you end up with your original number?
Try multiplying and dividing your number by another number besides 5. Do you still end up with your original number?

Multiplication and division are opposite or **inverse operations**, because each operation undoes the effects of the other.

Exercise b.   Choose any number for x: $x =$ _____.

Add 4 to your number: $x + 4 =$ _____.

Subtract 4 from the result: $x + 4 - 4 =$ _____.

Did you end up with your original number?
Try adding and subtracting another number besides 4. Do you still end up with your original number?

Addition and subtraction are opposite or **inverse operations**, because each operation undoes the effects of the other.

We see that the opposite operation for multiplication is division, and the opposite operation for addition is subtraction. Now we can solve the equation in Example 1.

■ **Example 2**   Solve the equation $54 = 6h$.

Solution We will transform this equation into a new one of the form $h = k$ (or $k = h$), where the number k is the solution. To do this, we must isolate the variable h on one side of the equals sign. Because h is multiplied by 6 in this equation, we divide both sides by 6, like this:

$$\frac{54}{6} = \frac{6h}{6} \qquad \textbf{Divide both sides by 6.}$$
$$9 = h$$

The solution is 9. Liz worked for 9 hours.   ■

In Example 2, we transformed the equation into a simpler one (namely, $h = 9$) which tells us the solution. We can use the following operations to transform an equation:

To solve an equation:
1. We can add or subtract the same number from both sides.
2. We can multiply or divide both sides by the same number, as long as that number is not zero.

Look Closer: Our method for solving equations works because of the **properties of equality**, one for each of the four arithmetic operations. They are listed below.

Addition and Subtraction Properties of Equality
    If the same quantity is added to or subtracted from both sides of an equation, the solution is unchanged. In symbols,

$$\text{If } \ a = b, \ \text{ then } \ a + c = b + c \ \text{ and } \ a - c = b - c$$

Multiplication and Division Properties of Equality
    If both sides of an equation are multiplied or divided by the same nonzero quantity, the solution is unchanged. In symbols,

$$\text{If } \ a = b, \ \text{ then } \ ac = bc \ \text{ and } \ \frac{a}{c} = \frac{b}{c}, \ c \neq 0$$

    Applying any one of these properties to an equation produces an **equivalent equation**: one with the same solutions as the original equation.

Solving Equations

    Now we can solve equations algebraically. We'll start with the simple Example from Section 1.3 about Delbert's and Francine's ages.

■ **Example 3**   Francine is exactly four years older than Delbert, so an equation relating their ages is $F = D + 4$. How old is Delbert when Francine is 19?

Solution We must solve the equation

$$19 = D + 4$$

We see that 4 has been added to the variable, D. To isolate D on one side of the equals sign, we subtract 4 from both sides of the equation, like this:

$$\begin{array}{r} 19 = D + 4 \\ \underline{-4 \quad\quad -4} \\ 15 = D \end{array}$$ Note that $D + 4 - 4 = D$.

The solution to this equation is 15. You can check that substituting 15 for D does make the equation true.

Check: $19 = D + 4$ Substitute 15 for D.
          $19 = 15 + 4$ True: the solution checks.   ■

    The last example is easy enough to solve without algebra, but the method will help us with harder problems. Here is a general strategy for solving such equations:

To solve an equation algebraically:
Step 1   Ask yourself what operation has been performed on the variable.
Step 2   Perform the opposite operation on *both* sides of the equation in order to isolate the variable.

RQ3.   What is the opposite operation for subtraction? For multiplication?
RQ4.   What are equivalent equations?

Problem Solving with Equations

Some problems can be solved by applying one of our familiar formulas.

■ **Example 4**   A saving's account at Al's Bank earns 5% annual interest. Jan's account earned $42.50 in interest last year. What was her balance at the beginning of the year?

Solution First, we choose the appropriate formula. This is a problem about interest, so we use the interest formula: $I = Prt$
Next, we list the values of the variables. (Which variable is unknown?)

$$I = 42.50 \qquad r = 0.05$$
$$P = \text{unknown} \qquad t = 1$$

We substitute the values into the formula: $42.50 = P(0.05)(1)$
Then we solve the equation:

$$\frac{42.50}{0.05} = \frac{P(0.05)}{0.05} \qquad \textbf{Divide both sides by 0.05.}$$
$$850 = P$$

Finally, we answer the question in a sentence:

Jan's account balance was $850 at the beginning of the year.   ■

Many practical problems can be solved by first writing an equation that describes or **models** the problem, and then solving the equation. We will use three steps to solve applied problems.

Steps for solving an applied problem:
Step 1   Identify the unknown quantity and choose a variable to represent it.
Step 2   Find some quantity that can be described in two different ways, and write an equation using the variable to model the problem situation.
Step 3   Solve the equation and answer the question in the problem.

■ **Example 5**   Jerry needs an additional $35 for airfare to New York. The ticket to New York costs $293. How much money does Jerry have?

Solution We follow the three steps.
Step 1   The amount of money Jerry has is unknown.

Amount of money Jerry has: m

Step 2   The airfare to New York can be described in two different ways:

$$\underset{\substack{\text{amount} \\ \text{Jerry has}}}{m} + \underset{\substack{\text{amount} \\ \text{he needs}}}{35} = \underset{\text{airfare}}{293}$$

Step 3   We solve the equation and answer the question.

$$m + 35 = 293 \quad \textbf{Subtract 35 from both sides.}$$
$$\underline{-35 \quad -35}$$
$$m = 258$$

Jerry has $258.

Caution! In Example 5, each side of the equation represents the airfare to New York, **not** the amount of money Jerry has, which is what we want to find. This is typical of applied problems: the equation does not usually represent the unknown quantity itself, but some related quantity.

Look Ahead: Although you may be able to solve the problems in this lesson with arithmetic, *the important thing is to learn to write the algebraic equation, or model,* for the problem. This skill will help you to solve problems that are too difficult to solve with arithmetic alone.

> **RQ5.**   What is the first step in solving an applied problem?
> **RQ6.**   True or false: One side of the equation that models a problem will be the unknown variable.

Skills Warm-Up

Choose the correct algebraic expression.

$$m + 15 \qquad m - 15 \qquad 15 - m$$

1. Carol weighs 15 pounds less than Garth. If Garth weighs m pounds, how much does Carol weigh?
2. Amber and Beryl together planted 15 trees. If Amber planted m trees, how many trees did Beryl plant?
3. Fred earned $15 more this week than last week. If he earned m dollars last week, how much did he earn this week?
4. Meg bicycled 15 miles farther than Kwan. If Kwan rode m miles, how far did Meg ride?
5. There are 15 children in Amy's swim class. If there are m girls, how many are boys?
6. The sale price of a sweater is $15 less than the regular price. If the regular price is m dollars, what is the sale price?

Choose the correct algebraic expression.

$$12p \qquad \frac{12}{p} \qquad \frac{p}{12}$$

7. Julian earns $12 an hour. If he works for p hours, how much will he make?
8. Farmer Brown collected p eggs this morning. How many dozen is that?
9. Melissa bought 12 colored markers. If their total cost was p dollars, how much did each marker cost?

10. Rosalind is baby-sitting for p children. If she brings 12 puzzles, how many will each child get?

11. Hector has to read 12 chapters in his history text. If he has p days to complete the assignment, how many chapters should he read per day?

12. Roma swims 12 laps per day. After p days, how many laps has she swum?

Answers: 1. $m - 15$ **2.** $15 - m$ **3.** $m + 15$ **4.** $m + 15$ **5.** $15 - m$ **6.** $m - 15$
7. $12p$ **8.** $\dfrac{p}{12}$ **9.** $\dfrac{p}{12}$ **10.** $\dfrac{12}{p}$ **11.** $\dfrac{12}{p}$ **12.** $12p$

Homework 1.4

Skills Practice

■ For Problems 1-2, fill in the table for each equation. Explain how you found the unknown values.

1. $q = 9 + t$

t	2	4			21	
q			15	18		39

2. $p = 5n$

n	0	2		5		
p			20		35	55

3. Decide whether the given value for the variable is a solution of the equation.

    a. $x - 4 = 6;\ \ x = 10$ **b.** $4y = 28;\ \ y = 24$ **c.** $\dfrac{0}{z} = 0;\ \ z = 19$

■ For Problems 4-12, solve, and check your solution.

4. $x - 3 = 11$ **5.** $10.6 = 7.8 + y$ **6.** $3y = 108$

7. $42 = 3.5b$ **8.** $2.6 = \dfrac{a}{1.5}$ **9.** $x - 4 = 0$

10. $34x = 212$ **11.** $6z = 20$ **12.** $9 = k + 9$

■ For Problems 13-16,
    a. Choose the appropriate formula to write an equation.
    b. Solve the equation and answer the question.

13. Clive loaned his brother some money to buy a new truck, and his brother agreed to repay the loan in 1 year with 3% interest. Clive earned $75 interest on the loan. How much did Clive loan his brother?

14. Andy's average homework score on eight assignments was 38.25. How many homework points did Andy earn altogether?

15. How long will it take a cyclist traveling at 13 miles an hour to cover 234 miles?

16. A roll of carpet contains 400 square feet of carpet. If the roll is 16 feet wide, how long is the piece of carpet?

■ For Problems 17-22, choose the appropriate equation.

$x + 7 = 26$	$7x = 26$	$\dfrac{x}{7} = 26$
$x - 7 = 26$	$\dfrac{x}{26} = 7$	$\dfrac{26}{x} = 7$

17. Sarah drove 7 miles farther to her high school reunion than Jenni drove. If Sarah drove 26 miles, how far did Jenni drive?

18. Lurline and Rozik live 26 miles apart. They meet at a theme park between their homes. If Lurline drove 7 miles to the park, how far did Rozik drive?

19. Doris is training for a triathlon. This week she averaged 26 miles per day on her bicycle. If she rode every day, what was her total mileage?

20. Glynnis jogged the same route every day this week for a total of 26 miles. how long is her route?

21. Astrid divided her supply of colored pencils among the 26 children in her class, and each child got 7 pencils. How many pencils does she have?

22. Ariel lost 7 of the beads on her necklace, and now there are 26. How many were there originally?

Applications
■ Follow the steps to solve Problems 23-26.

23. Lupe spent $24 at the Craft Fair. She now has $39 left. How much did she have before the Craft Fair?
   a. What are we asked to find? Choose a variable to represent it.
   b. Find two ways to express the amount of money Lupe had after the Craft Fair, and write an equation.
   c. Solve the equation and answer the question in the problem.

24. Danny weighs 32 pounds more than Brenda. If Danny weighs 157 pounds, how much does Brenda weigh?
   a. What are we asked to find? Choose a variable to represent it.
   b. Find two ways to express Danny's weight, and write an equation.
   c. Solve the equation and answer the question in the problem.

25. Miranda worked 20 hours this week and made $136. What is Miranda's hourly wage?
   a. What are we asked to find? Choose a variable to represent it.
   b. Find two ways to express Miranda's total earnings, and write an equation.
   c. Solve the equation and answer the question in the problem.

26. Struggling Students Gardening Service splits their profit equally among their eight members. If each member made $64 last week, what was the total profit?
   a. What are we asked to find? Choose a variable to represent it.
   b. Find two ways to express each member's share, and write an equation.
   c. Solve the equation and answer the question in the problem.

■ Write algebraic equations to solve Problems 27-29. Follow the three steps in the Lesson.

27. Martha paid $26 less for a suit at a discount store than her mother paid at a boutique for the same suit. If Martha paid $89 for the suit, how much did her mother pay?

28. Emily spends 40% of her monthly income on rent. If her rent is $360 a month, how much does Emily make?

29. After she wrote a check for $2378, Avril's bank account shows a balance of $1978. How much money was in Avril's account before the check cleared?

In Problems 30-32, we compare evaluating an expression and solving an equation.

30. A used car costs $3400 less than the new version of the same model.
   a. Choose two variables and write an equation for the cost of the used car in terms of the cost of the new car.
   b. If the new car costs $14,500, how much does the used car cost?
   c. If the used car costs $9200, how much does the new car cost?

31. The Dodgers won 60% of their games last season.
   a. Choose two variables and write an equation for the number of games the Dodgers won in terms of the number of games they played.
   b. If the Dodgers played 120 games, how many did they win?
   c. If the Dodgers won 96 games, how many did they play?

32. Sunshine Industries manufactures beach umbrellas. Their profit on each umbrella is 18% of the selling price.
   a. Choose two variables and write an equation for the profit on each umbrella in terms of its selling price.
   b. If a beach umbrella sells for $60, what is the profit?
   c. If the profit on one umbrella is $7.20, what is the selling price?

1.5 Order of Operations

Addition and Multiplication

If you add together three or more numbers, such as

$$2 + 5 + 8$$

it doesn't matter which addition you do first; you will get the same answer either way.

■ Example 1a.   A sum of three or more terms can be added in any order. In the sums below, the parentheses tell us which part of the expression to simplify first.

$$\begin{array}{ccc} (2+5)+8 & \text{and} & 2+(5+8) \\ = 7+8 = 15 & & = 2+13 = 15 \end{array}$$

b. Similarly, a product of three factors can be multiplied in any order. Thus

$$\begin{array}{ccc} (3 \cdot 2) \cdot 4 & \text{and} & 3 \cdot (2 \cdot 4) \\ = 6 \cdot 4 = 24 & & = 3 \cdot 8 = 24 \end{array} \qquad ■$$

These two facts illustrate the **associative laws** for addition and multiplication.

Associative Law for Addition
   If a, b, and c are any numbers, then

$$(a + b) + c = a + (b + c)$$

Associative Law for Multiplication
   If a, b, and c are any numbers, then

$$(a \cdot b) \cdot c = a \cdot (b \cdot c)$$

Subtraction and Division

What about a string of subtractions or a string of divisions, such as

$$20 - 8 - 5 \qquad \text{or} \qquad 36 \div 6 \div 2$$

In these calculations, we get different answers, depending on which operations we perform first, as you can see in Example 2.

■ Example 2a.   Subtraction:

$$(20 - 8) - 5 = 12 - 5 = 7 \qquad \text{but} \qquad 20 - (8 - 5) = 20 - 3 = 17$$

b. Division:

$$(36 \div 6) \div 2 = 6 \div 2 = 3 \qquad \text{but} \qquad 36 \div (6 \div 2) = 36 \div 3 = 12 \qquad ■$$

The associative laws do **not** hold for subtraction or division.

So, if there are no parentheses in the expression, how do we know which operations to perform first?

1. In a string of additions and subtractions, we perform the operations **in order from left to right**.
2. Similarly, we perform multiplications and divisions **in order from left to right**.

Example 3 Simplify each expression.

a. $20 - 8 - 5$

b. $36 \div 6 \div 2$

Solutions Perform the operations in order from left to right.

a. $20 - 8 - 5 = 12 - 5 = 17$

b. $36 \div 6 \div 2 = 6 \div 2 = 3$

RQ1. What do the associative laws tell us?

RQ2. Which two operations are not associative?

RQ3. In what order should we perform a string of additions and subtractions?

Combined Operations

How should we simplify the expression $4 + 6 \cdot 2$?

If we do the addition first, we get

$$(4 + 6) \cdot 2 = 10 \cdot 2 = 20$$

If we do the multiplication first, we get

$$4 + (6 \cdot 2) = 4 + 12 = 16$$

Which one is correct? In order to avoid confusion, we make the following agreement.

Always perform multiplications and divisions before additions and subtractions.

Example 4 The correct way to simplify the expression $4 + 6 \cdot 2$ is

$$4 + 6 \cdot 2 \qquad \text{Multiply first.}$$
$$= 4 + 12 \qquad \text{Then add.}$$
$$= 16$$

Look Closer: In longer expressions, it can be helpful to group the expression into its terms before beginning. **Terms** are expressions separated by addition or subtraction symbols. We simplify each term before combining them. ●

Example 5 Simplify $6 + 2 \cdot 5 - 12 \div 3 \cdot 2$.

Solution We start by underlining each term of the expression separately. Then we simplify each term.

$$\underline{6} + \underline{2 \cdot 5} - \underline{12 \div 3 \cdot 2} \qquad \begin{array}{l}\text{Perform multiplications and} \\ \text{divisions from left to right.}\end{array}$$
$$= 6 + 10 - 8 \qquad \begin{array}{l}\text{Perform additions and} \\ \text{subtractions from left to right.}\end{array}$$
$$= 16 - 8 = 8$$

Caution!   **1.** In Example 5, we do **not** start by adding $6 + 2$. We must perform multiplications before additions.

2. We simplify $12 \div 3 \cdot 2$ by performing the operations **from left to right.**

$$12 \div 3 \cdot 2 = 4 \cdot 2 = 8$$

Parentheses

What if the addition should come first in a particular calculation? In that case, we use parentheses to enclose the sum, like this:

$$(4 + 6) \cdot 2$$

> **Perform any operations inside parentheses first.**

■ **Example 6**   Simplify.

a. $2 + 5(7 - 3)$ Subtract inside parentheses.

$\qquad = 2 + 5(4)$ Multiply.

$\qquad = 2 + 20 = 22$ Add.

b. $6(10 - 2 \cdot 4) \div 4$ Multiply inside parentheses.

$\qquad = 6(10 - 8) \div 4$ Subtract inside parentheses.

$\qquad = 6(2) \div 4$ Multiply and divide in order from left to right.

$\qquad = 12 \div 4 = 3$ ■

RQ4.   Is it true that multiplications should be performed before divisions? Why or why not?

RQ5.   What do parentheses tell us?

RQ6.   If there are no parentheses, in what order should we perform multiplications and divisions?

Fraction Bars

Like parentheses, a fraction bar is a **grouping device.**

> **Expressions that appear above or below a fraction bar**
> **should be simplified first.**

■ **Example 7**   Simplify $\dfrac{24 + 6}{12 + 6}$

Solution   We begin by computing the sums above and below the fraction bar. Then we can reduce the fraction. Thus,

$$\frac{24 + 6}{12 + 6} = \frac{30}{18} = \frac{5}{3}$$

■

Caution!   Do not be tempted to "cancel" the terms in Example 7. For example, it would be incorrect to write

$$\frac{24+6}{12+6} = \frac{\cancel{2}4+6}{\cancel{1}2+6} = \frac{2+6}{1+6} \quad \leftarrow \text{Incorrect!}$$

According to the order of operations, we must simplify the numerator and denominator first, before dividing.

RQ7. Besides parentheses, what other symbol serves as a grouping device?
RQ8. When adding fractions, which should you do first: reduce each fraction, or find an LCD?

Summary

Combining all our guidelines for simplifying expressions, we state the rules for the order of operations.

Order of Operations
1. First, perform any operations that appear inside parentheses, or above or below a fraction bar.
2. Next, perform all multiplications and divisions in order from left to right.
3. Finally, perform all additions and subtractions in order from left to right.

Algebraic Expressions

The order of operations applies to variables as well as constants.   For example, the expression

$$5 + 2x$$

tells us to multiply x by 2, then add 5 to the result. Thus, to evaluate the expression for $x = 8$, we write

For $x = 8$, $5 + 2x = 5 + 2(8)$ **Multiply first.**
$$= 5 + 16 = 21$$

In Example 8, note how parentheses change the meaning of an expression.

Example 8   Write algebraic expressions for each of the following phrases. Then evaluate each phrase for $x = 2$ and $y = 6$.
a. Three times the sum of x and y **b.** The sum of $3x$ and y

Solutions a. $3(x + y)$ **b.** $3x + y$
$$3(2 + 6) = 3(8) = 24$$ $$3(2) + 6 = 6 + 6 = 12$$

Skills Warm-Up

Perform the operations on fractions as indicated.

1a. $\dfrac{5}{4} + \dfrac{3}{4}$ **b.** $\dfrac{5}{4} - \dfrac{3}{4}$ **c.** $\dfrac{5}{4} \cdot \dfrac{3}{4}$ **d.** $\dfrac{5}{4} \div \dfrac{3}{4}$

2a. $\dfrac{1}{2} + \dfrac{1}{6}$ **b.** $\dfrac{1}{2} - \dfrac{1}{6}$ **c.** $\dfrac{1}{2} \cdot \dfrac{1}{6}$ **d.** $\dfrac{1}{2} \div \dfrac{1}{6}$

3a. $\dfrac{2}{3} + \dfrac{1}{4}$ **b.** $\dfrac{2}{3} - \dfrac{1}{4}$ **c.** $\dfrac{2}{3} \cdot \dfrac{1}{4}$ **d.** $\dfrac{2}{3} \div \dfrac{1}{4}$

Answers: 1a. 2 **b.** $\dfrac{1}{2}$ **c.** $\dfrac{15}{16}$ **d.** $\dfrac{5}{3}$ **2a.** $\dfrac{2}{3}$ **b.** $\dfrac{1}{3}$ **c.** $\dfrac{1}{12}$ **d.** 3

3a. $\dfrac{11}{12}$ **b.** $\dfrac{5}{12}$ **c.** $\dfrac{1}{6}$ **d.** $\dfrac{8}{3}$

Homework 1.5

Skills Practice

■ For Problems 1-12, simplify each expression by following the order of operations.

1. $2 + 4(3)$

2. $15 - \frac{3}{4}(16)$

3. $6 \div \frac{1}{4} \cdot 3$

4. $3 + 3(2 + 3)$

5. $\frac{1}{3} \cdot 12 - 3\left(\frac{5}{6}\right)$

6. $2 + 3 \cdot 8 - 6 + 3$

7. $\frac{3(8)}{12} - \frac{6 + 4}{5}$

8. $28 + 6 \div 2 - 2(5 + 3 \cdot 2)$

9. $3[3(3 + 2) - 8] - 17$

10. $\frac{3(3) + 5}{6 - 2(2)} + \frac{2(5) - 4}{9 - 4 - 2}$

11. $7[15 - 24 \div 2] - 9[5(4 + 2) - 4(6 + 1)]$

12. $20 - 4(9 - 3 \cdot 2) + 8 - 5 \cdot 3$

■ For Problems 13-16, use your calculator to simplify each expression. Round your answers to three decimal places if necessary.

13. $\frac{6.4 + 3.5}{3.6(3.2)}$

14. $\frac{14.6 - 6.8}{7.3 + 3.4}$

15. $\frac{26.2 - 9.1}{8.4 \div 7.7} + 5.1(6.9 - 1.6)$

16. $\frac{1728(847 - 603)}{216(98 - 38)} + 6(876 - 514)$

■ For Problems 17-22, evaluate each pair of expressions mentally.

17a. $8 + 2 \cdot 5$

b. $(8 + 2) \cdot 5$

18a. $\frac{24}{2 + 6}$

b. $\frac{24}{2} + 6$

19a. $(9 - 4) - 3$

b. $9 - (4 - 3)$

20a. $6 \cdot 8 - 6$

b. $6(8 - 6)$

21a. $\frac{36}{6(3)}$

b. $\frac{36}{6}(3)$

22a. $(30 - 5)(5)$

b. $30 - (5)(5)$

■ For Problems 23-24, fill in the table to evaluate the expression in two steps.

23. $5z - 3$

z	$5z$	$5z - 3$
2		
4		
5		

24. $2(12 + Q)$

Q	$12 + Q$	$2(12 + Q)$
0		
4		
8		

■ For Problems 25-30, evaluate for the given values.

25. $2y + x$ for $x = 8$ and $y = 9$

26. $4a + 3b$ for $a = 8$ and $b = 7$

27. $\frac{a}{b} - \frac{b}{a}$ for $a = 8$ and $b = 6$

28. $\frac{24 - 2x}{2 + y} - \frac{4x + 1}{3y}$ for $x = 4$ and $y = 6$

29. $\frac{a}{1 - r}$ for $a = 6$ and $r = 0.2$

30. $mx + b$ for $m = \dfrac{3}{5},\ x = \dfrac{2}{3},\ $ and $b = \dfrac{9}{10}$

Applications

31. Consider the expression $20 - 2 \cdot 8 + 1.$ How would you change the expression if you wanted:
   a. The subtraction performed first?
   b. The addition performed first?
   c. The multiplication performed first?

32. Write an algebraic expression for the following instructions:
   "Multiply the sum of 5 and 7 by 4, then divide by the difference of 10 and 8."

33. Write two algebraic expressions for "the difference of 20 and 8, divided by the sum of 6 ane 4 and 10":
   a. Using a fraction bar
   b. Using a division symbol

34. Without performing the calculations, write down the steps you would use to simplify the expression: $825 - 32(12) \div 4 + 2$

■   For problems 35-38, choose the correct algebraic expression for each English phrase.

$\dfrac{m}{12} - 3$	$\dfrac{m - 3}{12}$
$\dfrac{12}{m - 3}$	$3 - \dfrac{12}{m}$

35. Twelve divided by 3 less than m.
36. Three less than the quotient of m divided by 12.
37. The quotient of 3 less than m divided by 12.
38. Subtract from 3 the quotient of 12 and m.

39. Find and correct the error in the calculation:

$$(5 + 4) - 3(8 - 3 \cdot 2)$$
$$= 9 - 3(8 - 6)$$
$$= 6(2) = 12$$

40. Think of the region shown at right as a rectangle with a smaller rectangle removed, and write an expression for its area. Then simplify your expression to find the area. (The measurements given are in centimeters.)

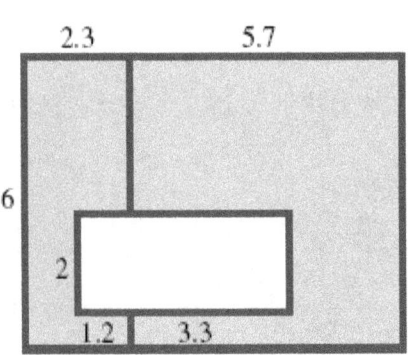

Chapter 1 Summary and Review

Lesson 1.1 Variables

- A **variable** is a numerical quantity that changes over time or in different situations.
- We can use a table, a graph, or an equation to compare the values of two related variables.
- A formula relating two variables is a type of **equation**, and an equation is just a statement that two quantities are equal.

Lesson 1.2 Algebraic Expressions

- An **algebraic expression** is any meaningful combination of numbers, variables, and operation symbols.

Commutative Laws

    Addition: If a and b are any numbers, then

$$a + b = b + a$$

    Multiplication: If a and b are any numbers, then

$$a \cdot b = b \cdot a$$

- Subtraction and division are not commutative; the order of the numbers matters in a difference or a quotient.

To write an algebraic expression:

Step 1	Identify the unknown quantity and write a short phrase to describe it.
Step 2	Choose a variable to represent the unknown quantity.
Step 3	Use mathematical symbols to represent the relationship.

- The process of substituting a specific value for the variable into an expression is called **evaluating** the expression.
- An **equation** is a mathematical statement that two algebraic expressions are equal. A **formula** is an equation that relates two or more variables.

Some Useful Algebraic Formulas

1. distance $=$ rate $\times$ time $d = rt$
2. profit $=$ revenue $-$ cost $P = R - C$
3. interest $=$ principal $\times$ interest rate $\times$ time $I = Prt$
4. part $=$ percentage rate $\times$ whole $P = rW$
5. average score $= \dfrac{\text{sum of scores}}{\text{number of scores}}$ $A = \dfrac{S}{n}$

Lesson 1.3 Equations and Graphs

- A graph is a way of visualizing an algebraic equation in two variables. The first or **input** variable is displayed on the horizontal axis and the second or **output** variable is displayed on the vertical axis.
- Each point on a graph has two **coordinates,** which designate the position of the point.

Review Problems

1. Martha can travel 22 miles on a gallon of gasoline. Fill in the table with the distance Martha can drive on various amounts of gasoline.

Gallons of gas	3	5	8	10	11	12
Calculation						
Miles driven						

    a. Explain in words how to find the distance Martha can drive.

    b. Write your explanation as a mathematical sentence:

        Miles driven =

    c. Let g stand for the number of gallons of gas in Martha's car and m for the number of miles she can drive. Write an equation for m in terms of g.

    d. Use the grid below left to plot the values from your table. Connect the data points with a smooth curve.

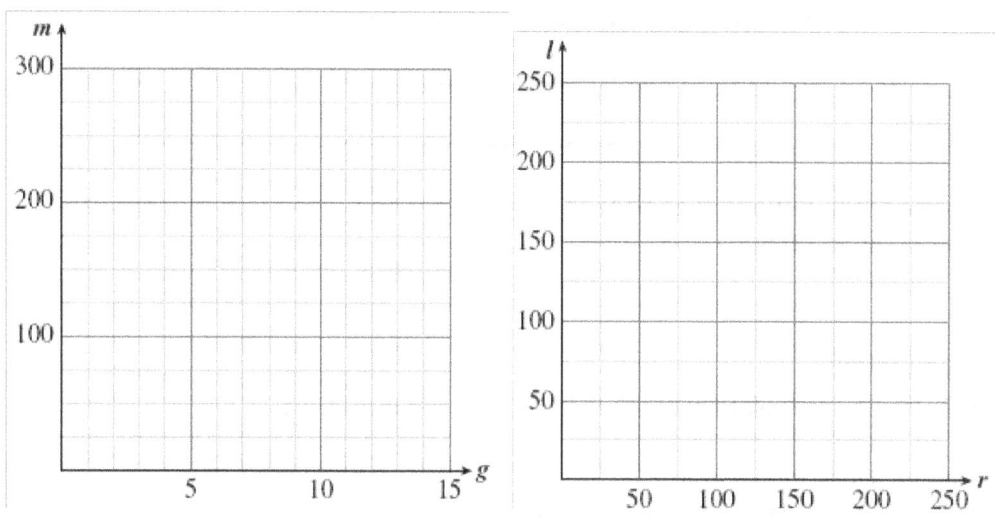

2. Mariel has a 200-page history assignment to read. Fill in the table with the number of pages Mariel has left to read after completing a certain number of pages

Pages read	20	50	85	110	135	180
Calculation						
Pages left						

    a. Explain in words how to find the number of pages left to read.

    b. Write your explanation as a mathematical sentence:

        Pages left =

    c. Let r stand for the number of pages Mariel has already read, and l for the number of pages she has left to read. Write an equation for l in terms of r.

    d. Use the grid above right to plot the values from your table. Connect the data points with a smooth curve.

3. Find the pattern and fill in the table. Then write an equation for the second variable in terms of the first.

x	0.5	1.0	1.5	2.0	4.0	6.0	7.5
y	0.125	0.25	0.375	0.5			

4. Use the graph to fill in the table. Then write an equation for the second variable in terms of the first.

x	y

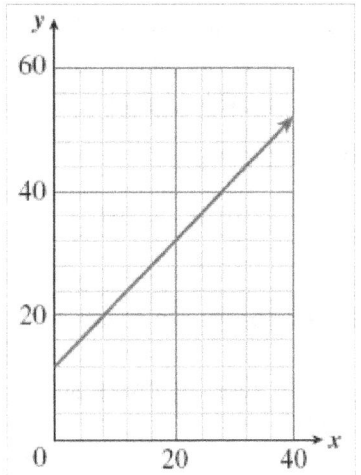

■ For Problems 5-12, write an algebraic expression for the phrase.

5. Five greater than z

6. 28% of t

7. Sixty dollars less than first-class air fare

8. The quotient of the volume of the sphere and 6

9. Six degrees hotter than yesterday's temperature

10. The cost of the gasoline divided three ways

11. 8% of the total bill

12. Five inches less than the height of the triangle

13. a. Write an expression for the distance traveled in t seconds by a car moving at 88 feet per second.
   b. How far will the car travel in 5 seconds? In half a second? In half a minute?

14. a. Write an expression for the distance you can travel in 3 hours at a speed of r miles per hour.
   b. How far can you travel in 3 hours on a bicycle at a speed of 6 miles per hour? On a motorboat at a speed of 20 miles per hour?

15. a. Write an expression for the amount of interest earned in t years by $500 deposited in an account that pays 7% annual interest.
   b. How much interest will the account earn after 1 year? After 2 years? After 5 years?

16. a. Write an expression for the amount of interest earned in two years by $1500 deposited in an account that pays interest rate r.
   b. How much interest will the money earn at $6\frac{1}{2}$% interest? At 8.3% interest?

17. Rachel has saved $60 to help pay for her books this semester. Let b stand for the price of her books, and n for the amount she still needs.
   a. Write an equation for n in terms of b.
   b. Fill in the table.

b	100	120	150	180	200
n					

   c. Choose appropriate scales for the axes and graph your equation.

■ For Problems 39-42, simplify by following the order of operations.

39. $6 + 18 \div 6 - 3 \cdot 2$

40. $\dfrac{2}{3} + \dfrac{1}{3} \cdot 4$

41. $36 \div (9 - 3 \cdot 2) \cdot 2 - 24 \div 4 \cdot 3$

42. $\dfrac{1.2(7.7)}{4.3 - 2.3}$

■ For Problems 43-36, write an algebraic expression for the phrase.

43. Four pages longer than half the length of the first draft
44. $100 short of twice last year's price
45. Three times the sum of the length and 5.6
46. 73% of the difference of Robin's weight and 100 pounds

■ For Problems 47-50, evaluate for the given values.

47. $mx + b$ for $m = \dfrac{1}{2},\ x = 3,\ b = \dfrac{5}{2}$

48. $\dfrac{1}{m} - \dfrac{1}{n}$ for $m = 4,\ n = 6$

49. $\dfrac{3w + z}{z}$ for $w = 8,\ z = 6$

50. $2(l + w)$ for $l = \dfrac{1}{3},\ w = \dfrac{1}{6}$

51. The total amount of money in an account is given by $P(1 + rt)$, where P is the initial investment, r is the interest rate, and t is the number of years the money has been in the account. How much money is in an account that earns 10% interest if $1000 was invested 3 years ago?

52. If an object is thrown downwards with initial speed v, then after t seconds it will fall a distance given in feet by $t(16t + v)$. How far will a penny fall in 2 seconds if it is thrown at a speed of 10 feet per second?

Chapter 2 Linear Equations

Lesson 2.1   Signed Numbers

- Perform arithmetic operations on signed numbers
- Write a subtraction fact as an equivalent addition fact
- Simplify quotients involving zero
- Use operations on signed numbers in applications

Lesson 2.2   Algebraic Expressions

- Write an expression with two operations
- Evaluate expressions using the order of operations
- Apply the order of operations to expressions involving signed numbers

Lesson 2.3   Graphs of Linear Equations

- Plot points on a Cartesian coordinate system
- Graph a linear equation in four quadrants
- Graph a linear model
- Use a graph to evaluate an expression

Lesson 2.4   Linear Equations and Inequalities

- Solve equations with two or more operations
- Write an equation to model an applied problem
- Solve an inequality

Lesson 2.5   Like Terms

- Decide whether two expressions are equivalent
- Combine like terms
- Write and simplify expressions using like terms
- Write an equation to model a situation
- Solve equations that involve like terms

2.1 Signed Numbers

Types of Numbers

Numbers greater than zero are called **positive numbers**, and numbers less than zero are **negative numbers**. We use a **number line** to illustrate relationships among numbers.

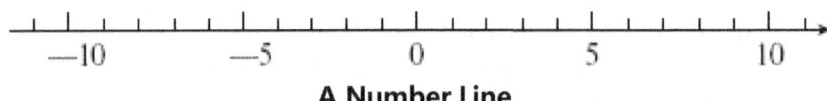

A Number Line

To the right of zero we mark the numbers 1, 2, 3, 4, … at evenly spaced intervals. These numbers are called the **natural** or **counting numbers**. The **whole numbers** include the natural numbers and zero: 0, 1, 2, 3, … .

> The natural numbers, zero, and the negatives of the natural numbers are called the **integers**:
>
> $$\cdots, -3, -2, -1, 0, 1, 2, 3, \cdots$$

Fractions, such as $\frac{2}{3}$, $-\frac{5}{4}$, and 3.6, lie between the integers on the number line.

> **RQ1.**   What are integers?

As we move from left to right on a number line, the numbers increase. In the figure below, the graph of −6 lies to the left of the graph of −2. Therefore, −6 is less than −2, or, equivalently, −2 is greater than −6.

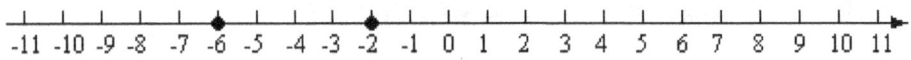

> We use special symbols to indicate order:
>
> <   means   **is less than**
> >   means   **is greater than**

For example,

$$-6 < -2 \quad \text{means} \quad -6 \text{ is less than } -2$$
$$-2 > -6 \quad \text{means} \quad -2 \text{ is greater than } -6$$

The small ends of the symbols < and > always point to the smaller number.

■ **Example 1a.**   Which is the lower altitude, −81 feet or −94 feet?
b. Express the relationship using one of the order symbols.

Solutions a.   Negative altitudes correspond to feet below sea level, and 94 feet is farther below sea level than 81 feet. Therefore, −91 feet is the lower altitude.
b. $-94 < -81$, or $-81 > -94$.   ■

Example 5   Simplify $7 - 9 + 12 - 6$

Solution We add the signed numbers 7 and −9 and 12 and −6, in order from left to right.

$$7 - 9 + 12 - 6 = -2 + 12 - 6 \qquad \text{The sum of 7 and } -9 \text{ is } -2;$$
$$= 10 - 6 = 4 \qquad \text{the sum of } -2 \text{ and 12 is 10.}$$

Caution!   The following calculation is incorrect:

$$16 - 9 - 4 + 3 = 16 - 5 + 3 = 11 + 3 = 14 \qquad \leftarrow \textbf{Incorrect!}$$

We must perform additions and subtractions in order from left to right:

$$16 - 9 - 4 + 3 = 7 - 4 + 3 = 3 + 3 = 6$$

Multiplying Signed Numbers

Case 1: The product of numbers with opposite signs

We can think of multiplication as repeated addition. For example,

$$3(2) \qquad \text{means} \qquad 2 + 2 + 2 \qquad \text{The sum of three 2's}$$

Similarly,

$$3(-2) \qquad \text{means} \qquad -2 + (-2) + (-2) \qquad \text{The sum of three } -2\text{'s}$$

which is −6, so $3(-2) = -6$. Think in terms of money: if you owe three different people 2 dollars each, then you are 6 dollars in debt.

Because multiplication is commutative, it is also true that $-2(3) = -6$. This example illustrates the following fact.

> **The product of a positive number and a negative number is a negative number.**

Example 6   You can use your calculator to verify each of the following products.
a. $(-4)(7) = -28$ **b.** $5(-2.6) = -13$

Case 2: The product of numbers with the same sign

You know that the product of two positive numbers is positive. It is also true that the product of two negative numbers is positive.

To understand why the product of two negative numbers is positive, you might think of canceling, or subtracting, several debts of the same amount. Or, you could observe the pattern in the lists of products below. As we move down the list, the product increases by equal amounts.

$$
\begin{array}{ll}
3(-2) = -6 & 3(-5) = -15 \\
2(-2) = -4 & 2(-5) = -10 \\
1(-2) = -2 & 1(-5) = -5 \qquad \leftarrow \text{Positive times negative is negative.} \\
0(-2) = 0 & 0(-5) = 0 \\
-1(-2) = 2 & -1(-5) = 5 \qquad \leftarrow \text{Negative times negative is positive.} \\
-2(-2) = 4 & -2(-5) = 10 \\
-3(-2) = 6 & -3(-5) = 15
\end{array}
$$

Based on the pattern in these calculations, the following rule seems reasonable.

> **The product of two negative numbers is a positive number.**

Example 7   You can use your calculator to verify the following products.

a. $(-6)(-3) = 18$ **b.** $(-2.1)(-3.4) = 7.14$

Now we have a pair of rules about the products of signed numbers.

> **Products of Signed Numbers**
> 1. The product of two numbers with opposite signs is a negative number.
> 2. The product of two numbers with the same sign is a positive number.

Dividing Signed Numbers

Division is the inverse operation for multiplication. For example, we know that

$$\frac{12}{3} = \boxed{4} \qquad \text{because} \qquad 3 \cdot \boxed{4} = 12$$

Thus, every division fact can be converted into an equivalent multiplication fact. For example,

$$\text{The division fact} \quad \frac{400}{16} = 25 \quad \text{is equivalent to} \quad 16 \cdot 25 = 400$$

Note that the *numerator* in the division fact becomes the *product* in the multiplication fact. The same relationship between multiplication and division holds for negative numbers as well.

Example 8   Rewrite each division fact as a multiplication fact.

a. $\dfrac{-144}{64} = -2.25$ **b.** $-36 \div \dfrac{-3}{8} = 96$

Solutions The numerator in each division fact becomes the product in the multiplication fact.

a. $-144 = -2.25(64)$ **b.** $-36 = \dfrac{-3}{8}(96)$

Example 8 illustrates the rules for division of signed numbers.

> **Quotients of Signed Numbers**
> 1. The quotient of two numbers with opposite signs is a negative number.
> 2. The quotient of two numbers with the same sign is a positive number.

RQ6.   Which of these is a product?

$$3 + (-4) \qquad -3 - 4 \qquad -3(-4) \qquad 3 - (-4)$$

RQ7.   Rewrite the division fact $\dfrac{48}{D} = Q$ as a multiplication fact.

Remember that a fraction can be thought of as a division. For example, $\frac{3}{4}$ means $3 \div 4$, or 0.75. What about negative fractions? Because the quotient of two numbers with unlike signs is negative, the fractions $\frac{-3}{4}$ and $\frac{3}{-4}$ both represent the negative number $-\frac{3}{4}$. On the other hand, $\frac{-3}{-4}$ is equal to the positive number $\frac{3}{4}$.

For many of the calculations in algebra, it is more convenient to write a negative fraction with the minus sign in the numerator, like this: $\frac{-3}{4}$.

A negative fraction is written in **standard form** when the minus sign is in the numerator: $\frac{-a}{b}$.

$$\frac{-a}{b} = \frac{a}{-b} = -\frac{a}{b}$$

The standard form for a positive fraction has positive numerator and denominator.

$$\frac{a}{b} = \frac{-a}{-b}$$

Quotients Involving Zero

What is the meaning of an expression such as $\frac{0}{2}$? We can rewrite this quotient as an equivalent multiplication fact.

$$\frac{0}{2} = \boxed{?} \qquad \text{is equivalent to} \qquad 2 \cdot \boxed{?} = 0$$

What number can we substitute for the question mark to make $2 \cdot \boxed{?} = 0$ a true statement? The only solution for this equation is 0. Thus, replacing the question marks by 0, we have

$$\frac{0}{2} = 0 \qquad \text{because} \qquad 2 \cdot 0 = 0$$

In general, **0 divided by any (non-zero) number is 0.**

Now consider a quotient with zero in the denominator, such as $\frac{2}{0}$.

$$\frac{2}{0} = \boxed{?} \qquad \text{is equivalent to} \qquad 0 \cdot \boxed{?} = 2$$

What number can we substitute for the question mark to make $0 \cdot \boxed{?} = 2$ a true statement? This equation has no solution, because 0 times any number is 0, not 2. Thus, there is no answer for the division problem $\frac{2}{0}$. We say that the quotient $\frac{2}{0}$ is **undefined**. In general, **we cannot divide any number by zero.**

Quotients Involving Zero

If a is any nonzero number, then

$$\frac{0}{a} = 0 \qquad \text{and} \qquad \frac{a}{0} \text{ is undefined.}$$

■ **Example 9a.** $6 \div 0$ is undefined.   **b.** $0 \div 8 = 0$ ■

RQ8.   Which of these is undefined? $\dfrac{-6}{0}$ or $\dfrac{0}{-6}$

Skills Warm-Up

1. Graph the numbers on the number line: $0,\ -3,\ 3,\ \dfrac{1}{2},\ \dfrac{-5}{3},\ 5\dfrac{1}{4}$

2. Graph the numbers on the number line: $-4.5,\ -2,\ -1.4,\ 0.6,\ 2$

■ Replace the comma in each pair by the proper symbol, $<,\ >,$ or $=$.

3. a. $0, -4$ **b.** $-5, -9$ **c.** $-2\frac{5}{8}, -2\frac{1}{8}$ **d.** $13.6, 13.66$
4. a. $3, -7$ **b.** $-1, -6$ **c.** $-3, -3\frac{3}{4}$ **d.** $-18.4, -19.6$

Answers: 3.a. $>$ **b.** $>$ **c.** $<$ **d.** $<$ **4.a.** $>$ **b.** $>$ **c.** $>$ **d.** $>$

Homework 2.1

Skills Practice

■ For Problems 1-4, add.

1. a. $5 + (-3)$ **b.** $-5 + 3$ **c.** $-5 + (-3)$

2. a. $-15 + (-20)$ **b.** $15 + (-20)$ **c.** $-15 + 20$

3. a. $-47 + 22$ **b.** $6.8 + (-2.7)$ **c.** $-\dfrac{5}{6} + \dfrac{2}{3}$

4. a. $-13 + (-36)$ **b.** $-2.5 + 4.9$ **c.** $-\dfrac{3}{4} + \left(-\dfrac{3}{8}\right)$

5. Rewrite each subtraction problem as an addition problem, and give the answer.

    a. $4 - 8$ **b.** $3 - (-9)$ **c.** $-8 - (-6)$ **d.** $-6 - 5$

■ For Problems 6-8, add or subtract as indicated.

6. a. $12 + (-6)$ **b.** $6 - (-4)$ **c.** $-2 - 8$ **d.** $-7 + 9$

7. a. $-14 - (-3)$ **b.** $-5 + (-4)$ **c.** $-6 - (-6)$ **d.** $-4 - 4$

8. a. $24 - (-10)$ **b.** $18 + (-12)$ **c.** $-16 + 14$ **d.** $-25 - (-15)$

■ For Problems 9-10, compute each sum or difference in parts (a) and (b), and decide which is easier. Then decide how to simplify the expression in part (c).

9. a. $15 - (+5)$ **b.** $15 + (-5)$ **c.** $15 - 5$

10. a. $-6 - (+2)$ **b.** $-6 + (-2)$ **c.** $-6 - 2$

■ For Problems 11-12, compute mentally.

11. a. $2 - 7$ **b.** $-2 - 7$ **c.** $-2 + 7$ **d.** $2 - (-7)$ **e.** $-2 - (-7)$

12. a. $13 - 5$ **b.** $-13 - 5$ **c.** $-13 + 5$ **d.** $13 - (-5)$ **e.** $-13 - (-5)$

■ For Problems 13-14, add.

13. a. $-5 + 3 + (-4)$ **b.** $-4 + (-7) + (-7)$

14. a. $6 + (-14) + 12 + (-17)$ **b.** $-35 + (-5) + 28 + (-21) + 13 + (-14)$

15. Simplify by adding the signed numbers.

    a. $2 + 5 - 8 - 1$ **b.** $-23 + 28 - 14 + 21$ **c.** $-34 - 52 + 68 - 21$

16. Simplify by following the rules for addition and subtraction.

    a. $-6 + 5 + (-3) - (-8)$ **b.** $-11 - 2 - (-4) - (-3)$ **c.** $-14 - (-16) - 4 + (-7)$

■ For problems 17-20, multiply or divide.

17. a. $(-8)(-4)$ **b.** $\dfrac{12}{-4}$ **c.** $-20 \div (-5)$

18. a. $-6(-1)(3)$ **b.** $\dfrac{-8}{0}$ **c.** $(-5)(0)(6)$

19. a. $0\left(\dfrac{-7}{15}\right)$ **b.** $(-2)(-2)(-2)$ **c.** $-30 \div 0$

20. a. $\left(\dfrac{1}{2}\right)\left(\dfrac{-3}{4}\right)$ **b.** $-0.1(-26)$ **c.** $\dfrac{-1}{2} \div \left(\dfrac{-3}{4}\right)$

For Problems 21-22, perform the indicated operations.

21. a. $-12 - 4$
   b. $-12(-4)$
   c. $-12 - (-4)$
   d. $-12 + (-4)$
   e. $-12 \div 4$

22. a. $-3 - 3$
   b. $-3(-3)$
   c. $-3 - (-3)$
   d. $-3 + (-3)$
   e. $-3 \div (-3)$

23. a. Find two numbers whose sum equals zero.
   b. Find two numbers whose difference equals zero.
   c. Find two numbers whose product equals zero.
   d. Find two numbers whose quotient equals zero.

24. a. Find two numbers whose product is 1.
   b. Find two numbers whose product is –1.
   c. Find two numbers whose quotient is 1.
   d. Find two numbers whose quotient is –1.

For Problems 25-30, solve.

25. $x - 9 = -4$ **26.** $-9z = 12$ **27.** $\dfrac{-a}{4} = 8$

28. $9 - x = 3$ **29.** $t + 5 = -8$ **30.** $-6b = -27$

31. a. What is the solution of the equation $-x = -3$?
   b. What is the solution of the equation $-x = 6$?

32. Find and correct the error in the following calculation:

$$18 - 6 + 2 - 5 = 18 - 8 + 5 = 15 \quad \textbf{(Incorrect!)}$$

Applications

For problems 33-40, write and simplify an expression using negative numbers to answer the question.

33. Thelma's failing company is worth –$1000. She decides to merge with Louise's company, whose net worth is $1500. What is the combined worth of the two companies?

34. Emily is at an elevation of –87 feet when she begins climbing a mountain. What is her elevation after ascending 127 feet?

35. The temperature at noon was 2° and dropped 8° over the next 12 hours. What was the temperature at midnight?

36. Rocky lost $450 betting on the horses this afternoon and then dropped $245 at a poker game in the evening. What was the net change in his financial status for the day?

37. A tourist in California can travel from Death Valley, at an elevation of −282 feet, to Mt. Whitney, at 14,494 feet. What is his net change in elevation?

38. Despite being overdrawn by $24.20, Nelson writes a check for $11.20. What is his new balance?

39. Whitney is climbing down a sheer cliff. She has pitons spaced vertically every 6 meters. What is her net elevation change after descending to the eighth piton from the top?

40. The temperature on the ice planet Hoth dropped 115° in just 4 hours. What was the average temperature change per hour?

■ Choose the correct equation to model Problems 41-44.

$-12 + n = -5$	$-12 + n = 5$
$n + 12 = -5$	$n + 12 = 5$

41. Last night the temperature rose 12° and this morning the temperature is 5°. What was the temperature yesterday?

42. Marta's score was −12, but after the last hand her score rose to −5. How many points did she make on the last hand?

43. Bradley rode the elevator down 12 floors and emerged five floors below ground. What floor did he start on?

44. Orrin was $12 in debt, but he did some yard work this weekend and after paying his debt he has $5. How much did he make?

2.2 Expressions and Equations

Writing Algebraic Expressions

When we write an algebraic expression, we use the same operations on a variable that we would use to calculate with a specific number. Writing down the expression for a specific numerical value can help us write an algebraic expression.

■ Example 1   Alida keeps $100 in cash from her weekly paycheck, and deposits 40% of the remainder in her savings account. If Alida's paycheck is p, write an expression for the amount she deposits in savings.

Solution   How would we calculate Alida's deposit if we knew her paycheck? Suppose Alida's paycheck is $500. First she subtracts $100 from that amount to get $500 - 100$, and then she takes 40% of the remainder for savings:

$$0.40\,(500 - 100) \quad \textbf{Perform operations inside parentheses first.}$$

If her paycheck is p dollars, we perform the same operations on p instead of on 500. The expression is thus

$$0.40\,(p - 100)$$ ■

RQ1.   We write an expression with variables using the same operations we would use to calculate with a _____ .

RQ2.   In Example 1, how does Alida calculate the amount to deposit in savings? How do we write that in an algebraic expression?

Evaluating Algebraic Expressions

When we evaluate an algebraic expression, we must follow the order of operations.

■ Example 2   Four students bought concert tickets with their $20 student-discount coupons.
a. If the regular price of a ticket is t dollars, write an expression for the total amount the four students paid.
b. How much did the students pay if the regular price of a ticket is $38?

Solutions a.   The discount price of one ticket is $t - 20$ dollars. So four tickets cost

$$4(t - 20) \text{ dollars}$$

b. We evaluate the expression in part (a) for $t = 38$.

$$4(t - 20) = 4(38 - 20) \quad \textbf{Simplify inside parentheses first.}$$
$$= 4(18) = 72$$

The students paid $72 for the tickets.   ■

RQ3.   Explain why we need parentheses in Example 2.

RQ4.   How should we write "3 times the sum of x and 12"?

Negative Numbers

The order of operations applies to signed numbers. Operations inside parentheses or other grouping devices should be performed first.

Example 3   Simplify $4 - 3 - [-6 + (-5) - (-2)]$

Solution Perform the operations inside brackets first. Simplify each step by rewriting subtractions as equivalent additions.

$$4 - 3 - [-6 + (-5) - (-2)]$$
$$= 4 - 3 - [-6 - 5 + 2]$$
$$= 4 - 3 - [-9]$$
$$= 4 - 3 + 9 = 10$$

$-5 - (-2) = -5 + 2$

$-6 - 5 = -11; \ -11 + 2 = -9$

Rewrite as an addition.

Add from left to right.   ■

RQ5.   Which operation is performed first in the expression $6 - 4x$?

RQ6.   What is the first step in evaluating the expression $2(18 - x)$?

Caution!   When using negative numbers, we must be careful to distinguish between products and sums.   In Example 4, note how the parentheses and minus signs are used in each expression.

Example 4 Simplify each expression.

a. $3(-8)$ **b.** $3 - (-8)$ **c.** $3 - 8$ **d.** $-3 - 8$

Solutions a.   This expression is a product: $3(-8) = -24$

b. This is a subtraction. We follow the rule for subtraction by changing the sign of the second number and then adding: $3 - (-8) = 3 + 8 = 11$

c.   This is an addition; the negative sign in front of 8 tells us that we are adding -8 to 3. Thus, $3 - 8 = 3 + (-8) = -5$

d. This is also an addition: $-3 - 8 = -3 + (-8) = -11$ ■

RQ7.   What is wrong with this calculation: $4 - 3 + 9 = 4 - 12 = -8$

Look Closer: When we evaluate an algebraic expression at a negative number, we enclose the negative numbers in parentheses. This will help prevent us from confusing multiplication with subtraction.   •

Example 5   Evaluate $2x - 3xy$ for $x = -5$ and $y = -2$.

Solution We substitute -5 for x and -2 for y, then follow the order of operations.

$$2x - 3xy = 2(-5) - 3(-5)(-2)$$ **Do multiplications first.**
$$= -10 - (-15)(-2)$$
$$= -10 - 30 = -40$$ ■

RQ8.   When we evaluate an expression at a negative number, we should _____ the negative number in _____ .

Skills Warm-Up

Choose the correct algebraic expression for each of the following situations.

$$n + 12 \qquad 12n \qquad \frac{n}{12}$$

$$\frac{12}{n} \qquad 12 - n \qquad n - 12$$

1. Helen bought n packages of tulip bulbs. If each package contains 12 bulbs, how many bulbs did she buy?
2. Henry bought a package of n gladiolus bulbs, then bought 12 loose bulbs. How many bulbs did he buy?
3. Together Karen and Dave sold 12 tickets to the spring concert. If Karen sold n tickets, how many did Dave sell?
4. Together Karl and Diana collected n used books for the book sale. If Karl collected 12 books, how many did Diana collect?
5. Greta made n dollars last week. If she worked for 12 hours, how much did she make per hour?
6. Gert jogged for 12 minutes. If she jogged n miles, how many minutes does it take her to jog 1 mile?

Answers: 1. $12n$ **2.** $n + 12$ **3.** $12 - n$ **4.** $n - 12$ **5.** $\dfrac{n}{12}$ **6.** $\dfrac{12}{n}$

Homework 2.2

Skills Practice

1. The perimeter of a rectangle of length l and width w is given by $P = 2l + 2w$. Find the perimeter of a rectangular meeting hall with dimensions 8.5 meters by 6.4 meters.

2. The area of a trapezoid with bases B and b and height h is given by $A = \frac{1}{2}(B + b)h$. Find the area of a trapezoid whose bases are 9 centimeters and 7 centimeters and whose height is 3 centimeters.

3. If a company sells n items at a cost of c dollars each and sells then at a price of p dollars each, the company's profit is given by $P = n(p - c)$. Find the profit earned by a manufacturer of bicycle equipment by selling 300 bicycle helmets that cost \$32 each to produce and sell for \$50 apiece.

4. The temperature in degrees Celsius (°C) is given by $C = \dfrac{5}{9}(F - 32)$, where F stands for the temperature in degrees Fahrenheit (°F). Find normal body temperature in degrees Celsius if normal temperature is 98.6° Fahrenheit.

◼ In Problems 5-8, simplify by following the order of operations.

5. $-18 - [8 - 12 - (-4)]$

6. $3 - (-6 + 2) + (-1 - 4)$

7. $-7 + [-8 - (-2)] - [6 + (-4)]$

8. $0 - [5 - (-1)] + [-6 - 3]$

9. Find each product.
   a. $(-2)(3)(4)$
   b. $(-2)(-3)(4)$
   c. $(-2)(-3)(-4)$
   d. $(-2)(-3)(4)(2)$
   e. $(-2)(-3)(-4)(2)$
   f. $(-2)(-3)(-4)(-2)$

10. Use your results from Problem 9 to complete the statements:
    a. The product of an odd number of negative numbers is _____ .
    b. The product of an even number of negative numbers is _____ .

◼ In Problems 11-14, use the order of operations to simplify each expression.

11a. $-2(-3) - 4$

b. $5(-4) - 3(-6)$

12a. $(-4 - 3)(-4 + 3)$

b. $-3(8) - 6(-2) - 5(2)$

13a. $\dfrac{15}{-3} - \dfrac{4 - 8}{8 - 12}$

b. $\dfrac{4 - 2(-5)}{-4 + 3(-1)}$

14a. $12.6 - 0.32(0.25)(4.2 - 8.7)$

b. $(5.8 - 2.6)(-2.5)(-0.6 + 3)$

15. Simplify each expression.
    a. $-3(-4)(-5)$
    b. $-3(-4) - 5$
    c. $-3(-4 - 5)$
    d. $-3 - (-4 - 5)$
    e. $-3 - (-4)(-5)$
    f. $(-3 - 4)(-5)$

16. Simplify each expression.
    a. $24 \div 6 - 2$
    b. $24 \div (-6 - 2)$
    c. $24 \div (-6) - 2$
    d. $24 - 6 \div (-2)$
    e. $24 \div (-6) \div (-2)$
    f. $24 \div (-6 \div 2)$

■ In Problems 17-22, evaluate for the given values.

17. $15 - x - y$ for $x = -6,\ y = 8$

18. $p - (4 - m)$ for $p = -2,\ m = -6$

19. $12x - 3xy$ for $x = -3,\ y = 2$

20. $\dfrac{y - 3}{x - 4}$ for $x = -9,\ y = 2$

21. $\dfrac{1}{2}t(t - 1)$ for $t = \dfrac{2}{3}$

22. $\dfrac{x - m}{s}$ for $x = 4,\ m = \dfrac{9}{4},\ s = \dfrac{3}{2}$

23. a. Evaluate each expression for $x = 5$. What do you notice?

     (i) $\dfrac{-3}{4}x$ **(ii)** $\dfrac{-3x}{4}$ **(iii)** $-0.75\,x$

    b. Does $\dfrac{-8}{5}x = \dfrac{-8}{5x}$? Support your answer with examples.

24. Which of the following are equivalent to $-\dfrac{4}{9}x$?

     a. $\dfrac{4}{9}(-x)$ **b.** $\dfrac{-4x}{-9}$ **c.** $\dfrac{-4}{9x}$ **d.** $\dfrac{-4x}{9}$ **e.** $\dfrac{-4}{9}(-x)$

■ Write an algebraic expression for area or perimeter.

25.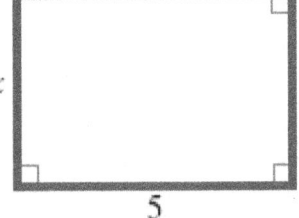

Perimeter =

26.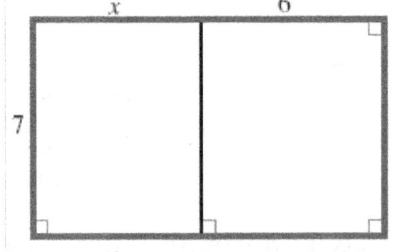

Area =

Applications

27. Salewa saved $5000 to live on while going to school full time. She spends $200 per week on living expenses.

    a. How much of Salewa's savings will be left after 3 weeks? After 12 weeks?

    b. Describe in words how you calculated your answers to part (a).

    c. Fill in the table below.

Number of weeks	2	4	5	6	10	15	20
Calculation							
Savings left							

    d. Write an algebraic expression for Salewa's savings after w weeks.

    e. Write an equation that gives Salewa's savings, S, in terms of w.

28. To calculate how much state income tax she owes, Francine subtracts $2000 from her income, and then takes 12% of the result.

    a. What is Francine's state income tax if her income is $8000? If her income is $10,000?

    b. Describe in words how you calculated your answers to part (a).

    c. Fill in the table below.

Income	5000	7000	12,000	15,000	20,000	24,000	30,000
Calculation							
State Tax							

    d. Write an algebraic expression for Francine's state income tax if her income is I.

    e. Write an equation that gives Francine's state income tax, T, in terms of I.

▧ For Problems 29-32,

    a. Choose a variable for the unknown quantity and write an algebraic expression.

    b. Evaluate the expression for the given values.

29. a. Three inches less than twice the width

    b. The width is 13 inches.

30. a. Twenty dollars more than 40% of the principal

    b. The principal is $500.

31. a. $8 times the number of children's tickets subtracted from 150

    b. There are 83 children's tickets

32. a. One-third of $50 less than the profit

    b. The profit is $500

▧ In Problems 33-40, write an algebraic expression.

33. The oven temperature started at 65° and is rising at 30° per minute. Write an expression for the oven temperature after t minutes.

34. Luisa's parents have agreed to pay her tuition ($800 per year) plus half her annual living expenses while she is in school. Write an expression for the amount her parents will pay if Luisa's annual expenses are a dollars.

35. Mildred canned 80 pints of tomatoes. She kept some for herslf and divided the rest equally among her four daughters. If Mildred kept M pints, write an expression for the number of pints she gave each daughter.

36. Moira's income is $50 more than one-third of her mother's income. Write an expression for Moira's income if her mother's income is I.

37. Otis buys 200 pounds of dog food and uses 15 pounds a week for his dog Ralph. Write an expression for the amount of dog food Otis has left after w weeks.

38. Renee receives $600 for appearing in a cola commercial, plus a residual of $80 each time the commercial is aired. Write an expression for Renee's earnings if the commercial plays t times.

39. Digby bought scuba diving gear by making a $50 down payment and arranging to pay the balance in ten equal installments. If the total cost of the gear is C dollars, write an expression for the amount of each installment.

40. Each passengers and three crew members on a small airplane is allowed 35 pounds of luggage. Write an expression for the weight of the luggage if there are x passengers.

■ In Problems 41-42,
   a. Write an algebraic expression.
   b. Make a table for the expression showing at least two positive values for the variable and two negative values.

41.   The ratio of a number to 4 more than the number.
42.   The sum of a and 5, times the difference of a and 5.

2.3 Graphs of Linear Equations

Graphing Equations

> An equation of the form $y = ax + b$, where a and b are constants, is called a **linear equation** because its graph is a straight line.

We can graph a linear equation by evaluating the expression $ax + b$ at several values of x and then plotting points.

■ Example 1 At 6 am the temperature was 50°, and it has been rising by 4° every hour.
a. Write an equation for the temperature, T, after h hours.
b. Graph your equation.

Solutions a.   We set $h = 0$ at 6 am. At that time, $T = 50$. Because the temperature is rising by 4° every hour, we multiply the number of hours by 4 and add the result to 50 to get the new temperature. Thus, our equation is $T = 50 + 4h$.
b. Step 1 We make a table of values for the equation. We choose some values for h, as shown in the first column of the table below. For each value of h, we evaluate the formula for T, and we record these values in the second column of the table.

h	T		Ordered pairs
0	50	$T = 50 + 4(\mathbf{0}) = 50 + 0 = 50$	$(0, 50)$
1	54	$T = 50 + 4(\mathbf{1}) = 50 + 4 = 54$	$(1, 54)$
3	62	$T = 50 + 4(\mathbf{3}) = 50 + 12 = 62$	$(3, 62)$
5	70	$T = 50 + 4(\mathbf{5}) = 50 + 20 = 70$	$(5, 70)$

Step 2 We use the values in the table to help us choose scales for the axes. For this graph, the x-values are less than 10, but the largest y-value is 70. We use increments of 1 on the horizontal or h-axis and increments of 10 on the vertical or T-axis.

Step 3 Each ordered pair gives us a point on the graph. We plot the points on the grid and connect them with a smooth curve. All the points should lie on a straight line.

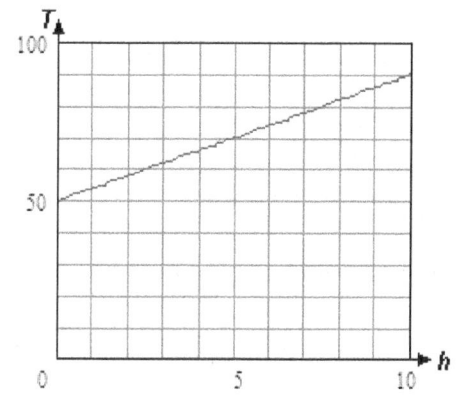

RQ1.   A linear equation has the form _____ .
RQ2.   We use _____ to help us choose scales for the axes.

Cartesian Coordinate System

Many graphs include negative values as well as positive values. To make such a graph, we construct a **Cartesian coordinate system**, named after the French mathematician and philosopher René Descartes.

We draw two perpendicular number lines for the horizontal and vertical axes, as shown below left. The two axes divide the plane into four **quadrants**, or regions, as shown at right. We often use x for the input variable and y for the output variable, so the horizontal axis is called the **x-axis** and the vertical axis is the **y-axis**. The axes intersect at the zero point of each number line. The coordinates of this intersection point, called the **origin**, are $(0,0)$.

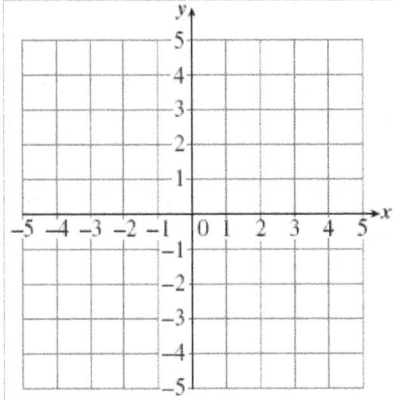

RQ3. What are the coordinates of the origin?

RQ4. True or False: The x-coordinate of any point on the y-axis is zero.

■ Example 2   The figure at right shows how to plot points on a Cartesian grid. Points in the upper right quadrant have both coordinates positive. Points with negative x-coordinates are plotted to the left of the y-axis, and points with negative y-coordinates are plotted below the x-axis. For example, the point $(-2, 1)$ is plotted 2 units to the left of the y-axis and 1 unit above the x-axis.

In which quadrant are both coordinates negative?

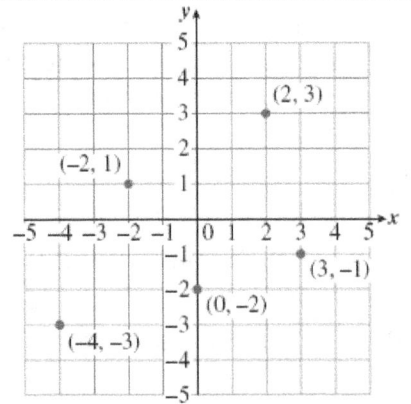

RQ5. True or False: The notation $(5, 3)$ represents two points, one on the x-axis and one on the y-axis.

RQ6. True or False: If both coordinates of a point are negative, the point is located in the lower left quadrant.

Now we can graph an equation that includes negative values.

■ Example 3   The temperature in Nome was $-12°F$ at noon. It has been rising at a rate of $2°F$ per hour all day.

a. Write an equation for the temperature, T, after h hours.

b. Fill in the table and graph your equation. (Negative values of h represent hours before noon.)

h	-3	-2	-1	0	1	2	3
T							

Use your graph to answer the following questions:

c. What was the temperature 8 hours before noon?

d. When will the temperature reach $-4°$F?

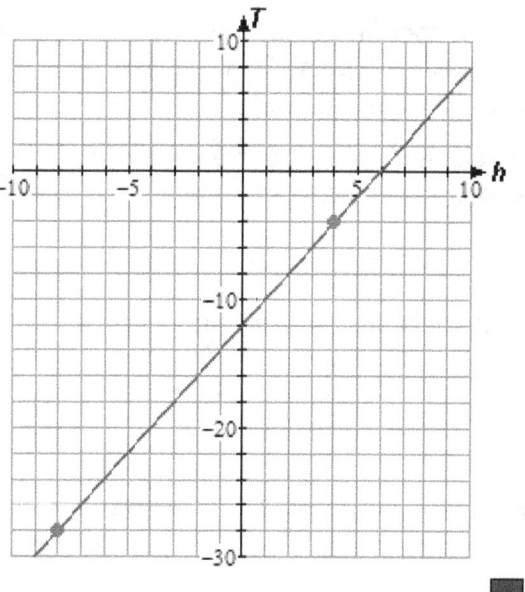

Solutions a. The temperature at h hours after noon is given by $T = -12 + 2h$.
b. We evaluate the expression for T at each of the h-values in the table.

h	-3	-2	-1	0	1	2	3
T	-18	-16	-14	-12	-10	-8	-6

c. We find the point on the graph with h-coordinate -8. Its T-coordinate is -28, so the temperature was $-28°$.
d. We find the point on the graph with T- coordinate -4. Its h-coordinate is 4, so it will be $-4°$ at 4 pm.

Look Closer: We can also answer parts (c) and (d) of Example 3 using algebra. For part (c), we evaluate the expression for T, namely $-12 + 2h$, at $h = -8$. For part (d), we solve the equation $T = -12 + 2h$ with $T = -4$. Make sure that you understand both methods, graphical and algebraic. •

> **Graphing Tip #1**
> If the coefficient of x is a fraction, we can make our work easier by choosing multiples of the denominator for the x-values. That way we won't have to work with fractions to find the y-values. In Example 4, the coefficient of x is $\frac{-2}{3}$, so we choose multiples of 3 for the x-values.

■ **Example 4a.** Graph $y = \dfrac{-2}{3}x - 4$

b. Use the graph to solve the equation $\dfrac{-2}{3}x - 4 = -2$.

Solutions a. We choose x-values that are multiples of 3 and make a table of values.

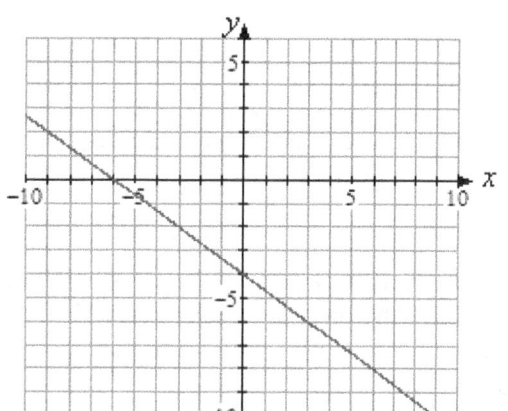

x	y	
-9	2	$y = -\frac{2}{3}(-9) - 4 = 2$
-6	0	$y = -\frac{2}{3}(-6) - 4 = 0$
0	-4	$y = -\frac{2}{3}(0) - 4 = -4$
3	-6	$y = -\frac{2}{3}(3) - 4 = -6$
6	-8	$y = -\frac{2}{3}(6) - 4 = -8$

Then we plot the points and connect them with a straight line.
b. We see that y has been replaced by -2 in the equation of the graph. So we look for the point on the graph that has y-coordinate -2. This point, labeled P on the graph

at right, has x-coordinate -3. Because it lies on the graph, the point $P(-3, -2)$ is a solution of the equation $y = \dfrac{-2}{3}x - 4$.

When we substitute $x = -3$ and $y = -2$ into the equation, we get a true statement:

$$-2 = \frac{-2}{3}(-3) - 4$$

But this statement also tells us that -3 is a solution of the equation $-2 = \dfrac{-2}{3}x - 4$.

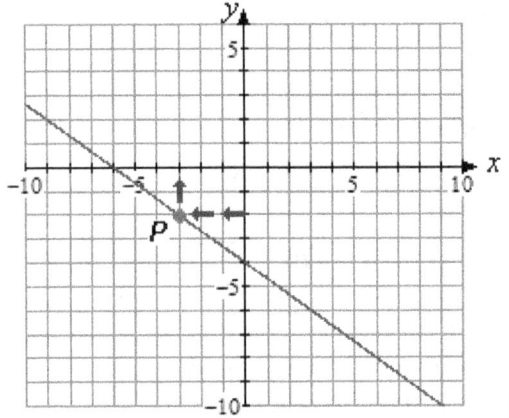

RQ7.   What is the first step in graphing an equation?

RQ8.   Which x-values would you choose to graph $y = \frac{5}{4}x - 3$?

Graphing Tip #2

    When you graph a linear equation, you should extend the line far enough in both directions so that it will cross both the x-axis and the y-axis. The points where the graph crosses the axes are important for applications.

Look Closer: The graph of an equation is a picture of the solutions of the equation. Each point on the graph represents a solution. If a point lies on the graph, its coordinates make the equation true.

    For example, the point $(-3, -2)$ lies on the graph in Example 4. When we substitute $x = -3$ and $y = -2$ into the equation $y = \dfrac{-2}{3}x - 4$, we get a true statement:

$$y = \frac{-2}{3}x - 4 \qquad \textbf{Substitute } x = -3 \textbf{ and } y = -2.$$

$$-2 = \frac{-2}{3}(-3) - 4$$

$$-2 = 2 - 4 \qquad \textbf{True.} \qquad \bullet$$

Skills Warm-Up

    Choose the correct algebraic expression for each situation.

$$2t + 12 \qquad\qquad 12 - 2t$$
$$2(t + 12) \qquad\qquad 2t - 12$$

1. Janine's history book has 12 chapters. If she studies 2 chapters a week, how many chapters will she have left after t weeks?

2. Arturo is 12 years older than twice the age of his nephew. If Arturo's nephew is t years old, how old is Arturo?

3. Rick made 12 fewer than twice as many phone calls as his roommate made this month. If Rick's roommate made t phone calls, how many calls did Rick make?

4. Every winter, the Civic Society knits mittens for the children of the county orphanage. This year there are 12 more children than last year. If there were t children last year, how many mittens will they need this year?

Answers: 1. $12 - 2t$ **2.** $2t + 12$ **3.** $2t - 12$ **4.** $2(t + 12)$

Homework 2.3

Skills Practice

In Problems 1-2, the figure shown is the graph of an equation. Decide which of the given points are solutions of the equation.

1. **a.** $(2, 0)$ **b.** $(3, 2)$ **2.** **a.** $(4, -2)$ **b.** $(4, 4)$
     c. $(-1, -2)$ **d.** $(-3, -6)$ **c.** $(0, 4)$ **d.** $(-1, -4)$

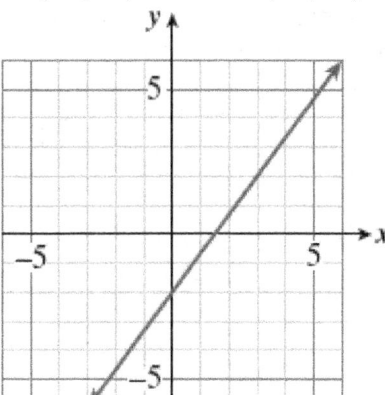   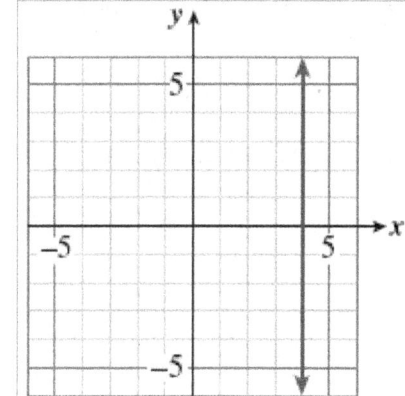

For Problems 3-5, give the coordinates of the labeled points.

3. 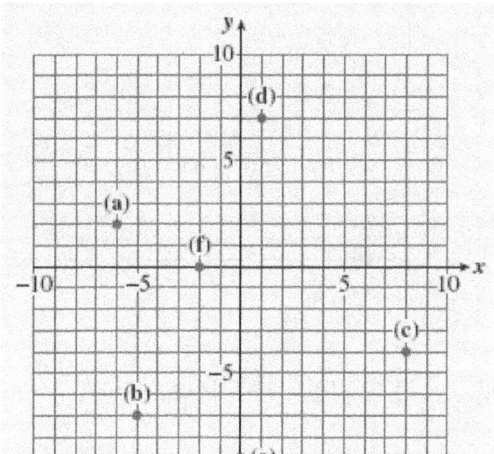   **4.**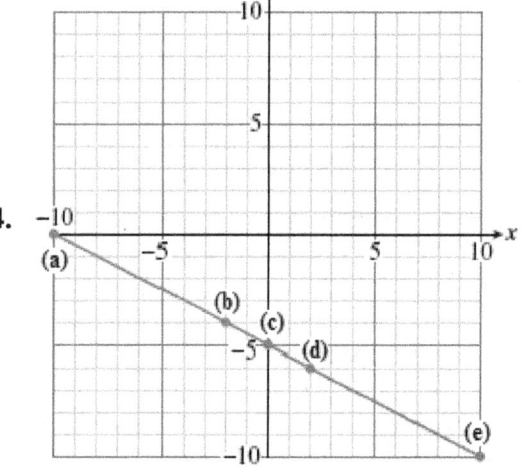

5.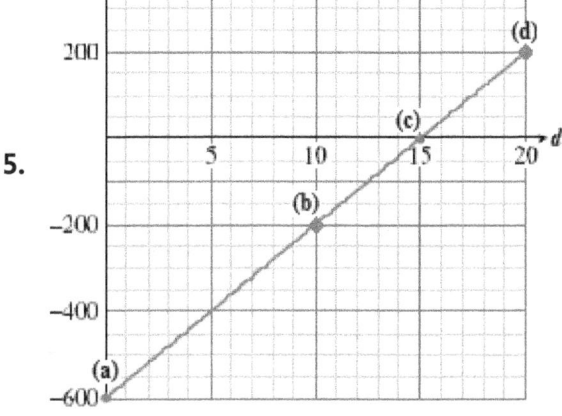

Applications

◼ In Problems 6-7,
   a. Write an equation relating the variables.
   b. Complete the table of values.
   c. Graph your equation on the grid.

6. Greta's math notebook has 100 pages, and she uses on average 6 pages per day for notes and homework. How many pages, P, will she have left after d days?

d	0	2	5	10	15
P					

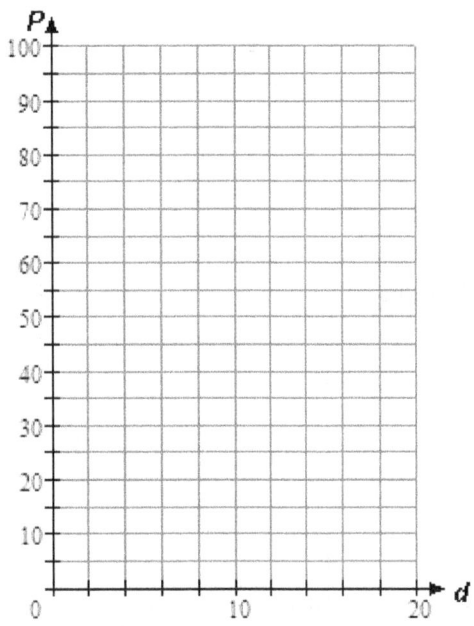

7. Delbert shares a house with four roommates. He pays $200 rent per month, plus his share of the utilities. If the utilities cost U dollars, how much money, M, does Delbert owe?

U	20	40	80	100	200
M					

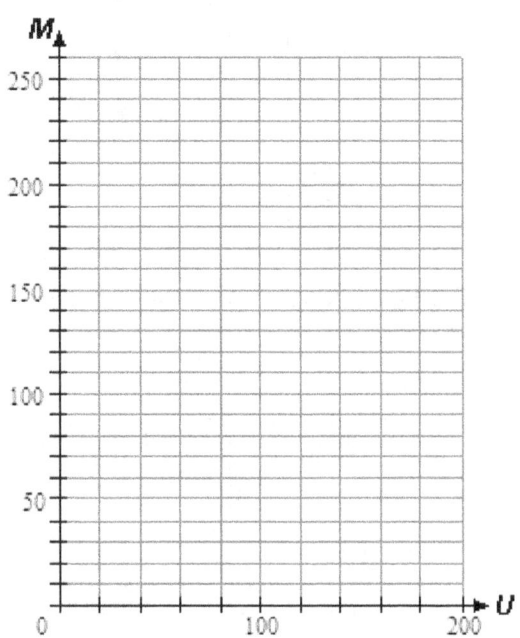

8. Darren inherited some money and has been spending it at the rate of $100 per day. Right now he has $2000 left.
   a. Write an equation for Darren's balance, B, after d days.
   b. Fill in the table. Negative values of d represent days in the past.

d	−15	−5	0	5	10	15	20
B							

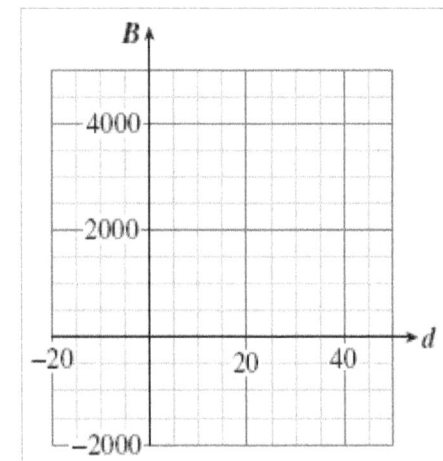

   c. Graph your equation, using the values in the table.
      Use your graph to answer the following questions:
   d. What was Darren's balance 15 days ago?
   e. When will Darren's balance reach $500?

9. Gregory purchased stereo equipment on a monthly installment plan. After m months, Gregory still owes a balance of B dollars.
   a. What was the price of the stereo equipment?
   b. How much does Gregory pay each month?
   c. How much does Gregory owe after 6 months?
   d. How many monthly payments will Gregory make?
   e. Write an algebraic equation for B in terms of m.

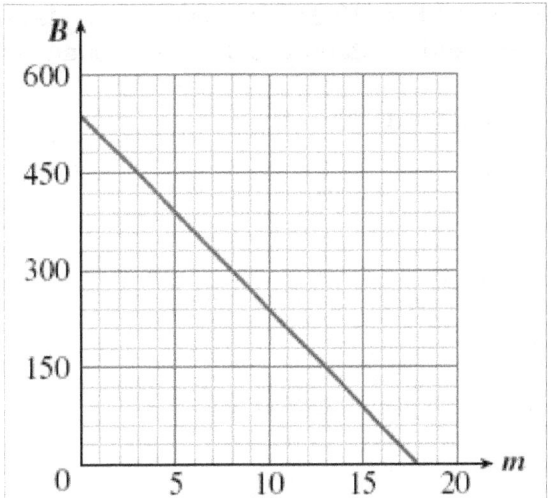

For part (c) of Problems 10-11, write an equation and use your graph to solve it.

10. On a 100-point test, Lori loses 5 points for each wrong answer.
    a. Write an equation for Lori's score, s, if she gives x wrong answers.
    b. Complete the table of values and graph your equation.

x	2	5	6	12
s				

   c. If Lori's score is 65, how many wrong answers did she give?

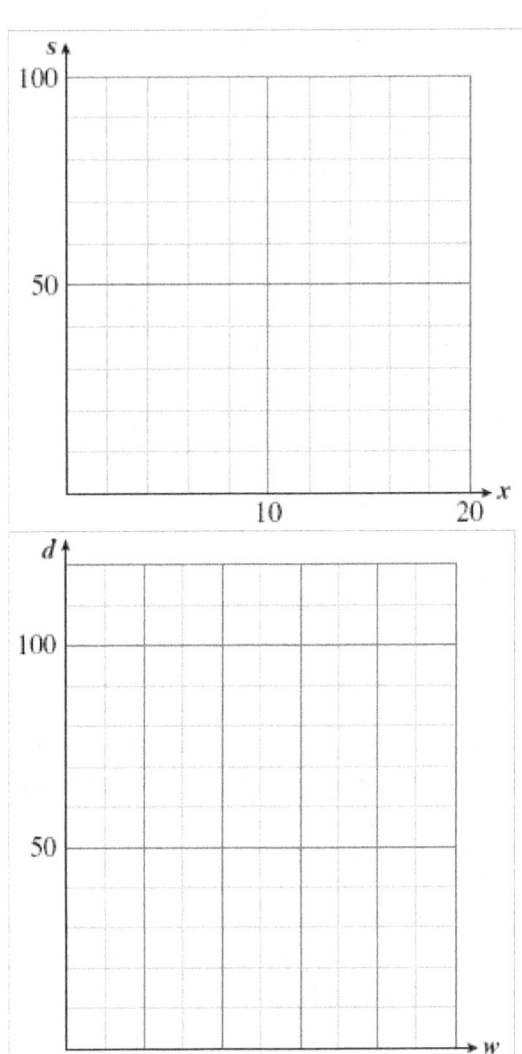

11. The water in Silver Pond is 10 feet deep, but the water level is dropping at a rate of $\frac{1}{2}$ inch per week.
    a. Write an equation for the depth, d, of the pond after w weeks.
    b. Complete the table of values and graph your equation.

w	24	36	84	120
d				

   c. How long will it take until the depth of the pond is 8 feet? (**Hint:** Convert all the units to inches.)

▮ In Problems 12-15, choose four x-values and make a table of values, then graph the equation. Extend your line far enough that it crosses both axes.

12. $y = x + 3$

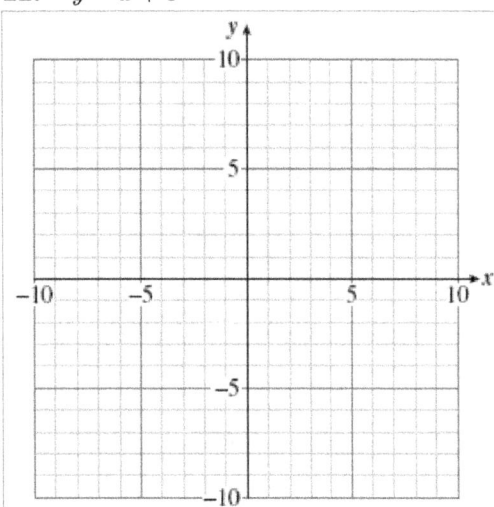

13. $y = 2x + 1$

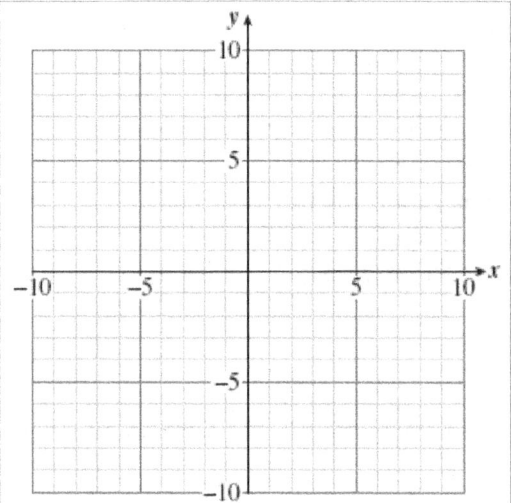

14. $y = -\dfrac{1}{2}x - 5$

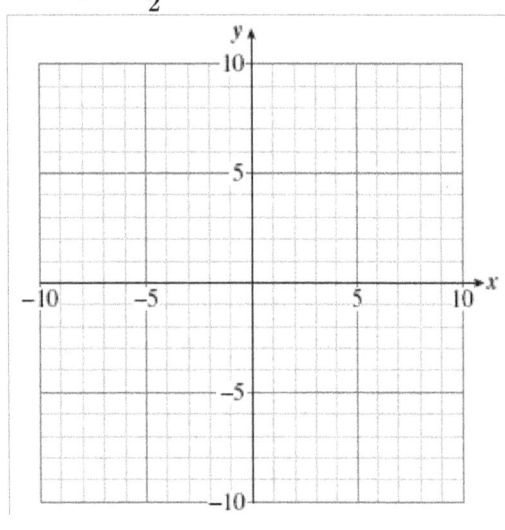

15. $y = \dfrac{5}{4}x - 4$

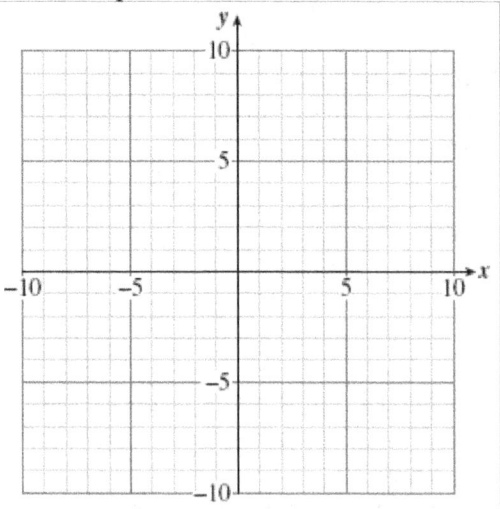

16. The graph of $y = 2x + 6$ is shown at right.
   a. Use the graph to evaluate the expression $2x + 6$ for $x = -5$.
   b. Find a point on the graph with $y = -4$. What is the x-value of the point?
   c. Verify that the coordinates of your point in part (b) satisfy the equation of the graph.
   d. Use the graph to find an x-value that produces a y-value of 8.
   e. Find two points on the graph for which $y > -4$. What are their x-values?

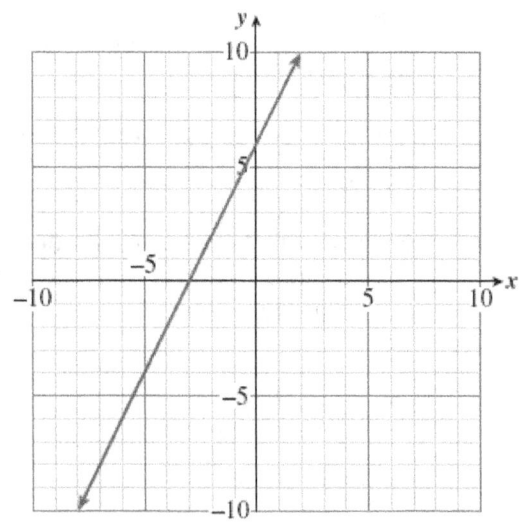

■ In Problems 17-18, use the graph to solve the equations. Show your solutions on the graph. Then check your solutions algebraically.

17.

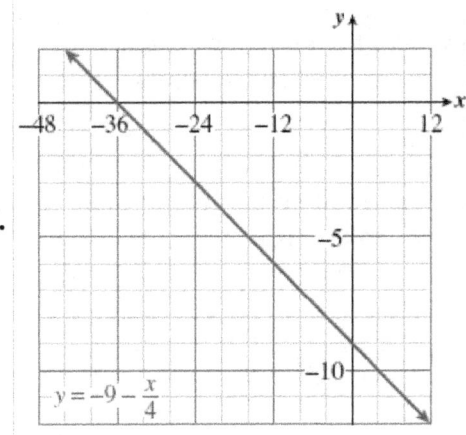

18.

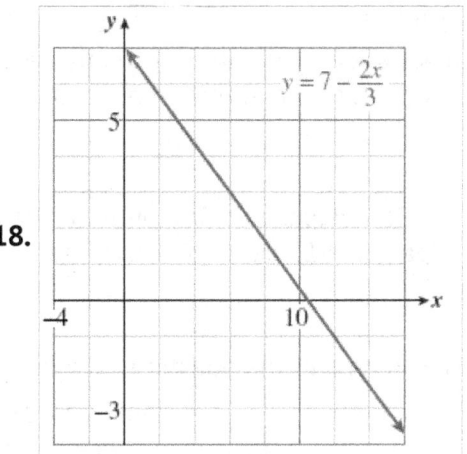

a. $-9 - \dfrac{x}{4} = -5$

b. $-9 - \dfrac{x}{4} = 1$

a. $7 - \dfrac{2x}{3} = 3$

b. $7 - \dfrac{2x}{3} = -1$

■ In Problems 19-20, use the graph to estimate the solution of the equation. Then use a calculator to help you verify the solution algebraically. Do you think your estimate was too big or too small?

19. $37.21 - 8.4t = 24.61$

20. $-26.4 = -3.65 + 9.1x$

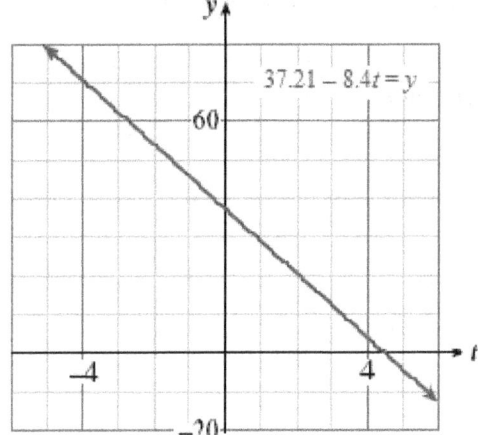

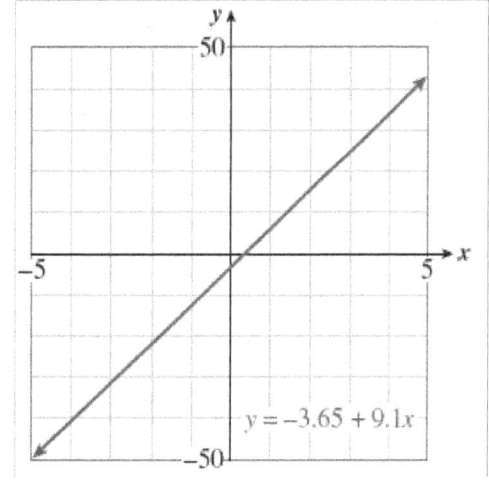

2.4 Linear Equations and Inequalities

Solving Equations

We solve an equation by isolating the variable on one side of the equation. If an equation involves two or more operations, we must undo those operations in reverse order.

■ Example 1   The expression $T = 50 + 15u$ describes college tuition consisting of a $50 registration fee plus $15 per unit. How many units can you take if you have $290 saved for tuition?

Solution   We can answer the question by solving the equation

$$290 = 50 + 15u$$

where we have substituted $290 for the tuition, T. Think about the expression $50 + 15u$. How would you evaluate this expression if you were given a value for u? Following the order of operations, you would

1. first multiply by 15
2. then add 50

In order to solve the equation, we must *reverse these two steps* to undo the operations and isolate the variable. We first subtract 50 from both sides of the equation:

$$
\begin{array}{rl}
290 = & 50 + 15u \qquad \textbf{Subtract 50 from both sides.} \\
\underline{-\,50} & \underline{-\,50} \\
240 = & 15u
\end{array}
$$

This isolates the term that contains the variable, $15u$. Then we divide both sides of the equation by 15.

$$\frac{240}{15} = \frac{15u}{15} \qquad \textbf{Divide both sides by 15.}$$
$$16 = u$$

You can enroll in 16 units. We can check the solution by substituting 16 for u in the original equation.

Check: $50 + 15u = 290$
$$50 + 15(16) = 50 + 240 = 290 \quad \textbf{True.}$$

Because a true statement results, the solution checks.   ■

Look Closer: In Example 1, notice how we reversed the operations used in the equation.

Operations performed on u	Steps for solution
1. Multiply by 15	1. Subtract 50
2. Add 50	2. Divide by 15

Strategy for solving equations.

To solve an equation that involves two or more operations, we undo those operations *in reverse order.*

■ Example 2   Solve $8 - 3x = -10$

Solution   The left side of the equation has two terms: 8 and $-3x$. We want to isolate the term containing the variable, so we subtract 8 from both sides.

$$\begin{array}{rl} 8 - 3x = -10 & \text{\textbf{Subtract 8 from both sides.}} \\ \underline{-8 \qquad -8} & \\ -3x = -18 & \end{array}$$

Next, we divide both sides by -3 to get

$$\begin{array}{cl} \dfrac{-3x}{-3} = \dfrac{-18}{-3} & \text{\textbf{Divide both sides by} } -\text{\textbf{3.}} \\ x = 6 & \end{array}$$

Thus, the solution is 6.

Check: $8 - 3(6) = 8 - 18 = -10$ ■

Caution!   In Example 1, the term $-3x$ means "-3 times x," so we **divide** both sides by -3. Do not try to add 3 to both sides.   ■

RQ1.   If an equation involves more than one operation, how must we undo those operations?

RQ2.   Delbert says he solved the equation $-5x = 15$ by adding 5 to both sides. What is wrong with his method?

Applied Problems

Problem solving often involves signed numbers, either in the equation that models the problem or in its solution, or both.

■ Example 3   The trout population in Clear Lake is decreasing by approximately 60 fish per year, and this year there are about 430 trout in the lake. If the population drops to 100, the Park Service will have to restock the lake.

a. Write an equation for the population P of trout x years from now.

b. When will the Park Service have to restock the lake?

c. Graph your equation for P, and illustrate your answer to part (b) on the graph.

Solutions a.   The population starts this year at 430, and decreases by 60 for each following year. Thus, $P = 430 - 60x$.

b. We would like to find the value of x when $P = 100$. We substitute 100 for P, and solve the equation for x.

$$\begin{array}{rl} 100 = 430 - 60x & \text{\textbf{Subtract 430 from both sides.}} \\ \underline{-430 -430} & \\ -330 = -60x & \end{array}$$

$$\begin{array}{cl} \dfrac{-330}{-60} = \dfrac{-60x}{-60} & \text{\textbf{Divide both sides by} } -\text{\textbf{60.}} \\ 5.5 = x & \end{array}$$

The Park Service will have to restock the lake in five and a half years, if the population continues to decline at the current rate.

c. The figure shows the graph of the equation $P = 430 - 60x$. To solve

$$100 = 430 - 60x$$

we locate the point on the graph with $P = 100$, and read its x-coordinate, at about 5.5.

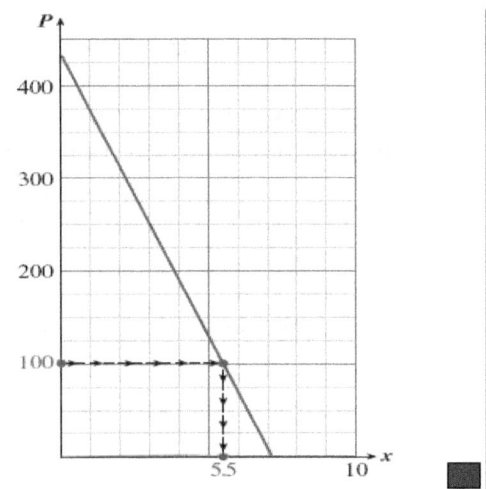

Solving Inequalities

> A statement that uses one of the symbols $<$ or $>$ is called an **inequality**.

Examples of inequalities are

$$-1 > -3 \quad \text{and} \quad x < 2$$

Unlike the equations we have studied, which have at most one solution, an inequality can have infinitely many solutions. A solution of an equation or inequality is said to **satisfy** the equation or inequality.

■ **Example 4**   Solve the inequality $x < 2$.

Solution The solutions include $1, 0, -1, -2$, and all the other negative integers, as well as fractions less than 2, such as $1\frac{3}{5}, \frac{2}{3}$, and $-\frac{17}{8}$. All of these values satisfy the inequality $x < 2$. In fact, *all* the numbers to the left of 2 on the number line are solutions of $x < 2$. Because we cannot list all these solutions, we often graph them on a number line as shown below.

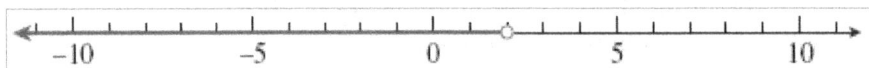

The open circle at 2 shows that 2 itself is not a solution (because 2 is not less than 2.) ■

Look Closer: An inequality that uses the symbol for less than, $<$, or greater than, $>$, is called a **strict inequality**. A **nonstrict inequality** uses one of the following symbols.

> $\geq$ means **greater than or equal to**
> $\leq$ means **less than or equal to**

For example, the graph of all solutions to the inequality

$$x \geq -2$$

is shown on the next page. We use a solid dot at -2 to show that $x = -2$ is included in the solutions.

RQ3. A solution of an inequality is said to _____ the inequality.
RQ4. What is the difference between a strict and a nonstrict inequality?

The rules for solving inequalities are very similar to the rules for solving equations, with one important difference. In the Activities we will develop the following strategies for solving inequalities.

> **To Solve an Inequality:**
> 1. We can add or subtract the same quantity on both sides.
> 2. We can multiply or divide both sides by the same positive number.
> 3. If we multiply or divide both sides by a **negative** number, we must **reverse** the direction of the inequality.

■ **Example 5**   Solve $-3x + 1 > 7$, and graph the solutions on a number line.

Solution We isolate x on one side of the inequality.

$$-3x + 1 - 1 > 7 - 1 \qquad \text{Subtract 1 from both sides.}$$
$$-3x > 6$$
$$\frac{-3x}{-3} < \frac{6}{-3} \qquad \text{Divide both sides by } -3; \text{ reverse the direction of the inequality.}$$
$$x < -2$$

The graph of the solutions is shown below.

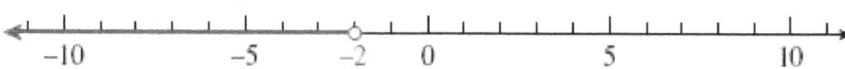

■

Compound Inequalities

An inequality in which the variable expression is bounded from above and from below is called a **compound inequality**. For example,

$$-3 < 2x - 5 \le 6$$

is a compound inequality. To solve a compound inequality, we must perform the steps needed to isolate x on all three sides of the inequality.

■ **Example 6**   Solve $-3 < 2x - 5 \le 6$

Solution To solve for x, we first add 5 on each side of the inequality symbols.

$$-3 < 2x - 5 \le 6$$
$$\underline{+5 \qquad +5 \quad +5}$$
$$2 < 2x \quad \le 11$$

Next, to solve $2 < 2x \le 11$, we divide each side by 2.

$$\frac{2}{2} < \frac{2x}{2} \le \frac{11}{2}$$

$$1 < x \le \frac{11}{2}$$

The solution consists of all numbers greater than 1 but less than or equal to $\frac{11}{2}$. The graph of the solutions is shown below.

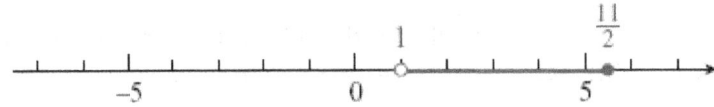

RQ5. When do we need to reverse the direction of an inequality?
RQ6. What is a compound inequality?

Skills Warm-Up

▪ Solve each equation. Try to do so mentally (without using pencil and paper.)

1. $\frac{u}{3} = 6$ **2.** $7 = 3 + s$ **3.** $a - \frac{1}{3} = \frac{2}{3}$

4. $20 = 5m$ **5.** $\frac{1}{4}p = 8$ **6.** $7t = 5$

▪ Fill in the tables. Then analyze the order of operations in your calculations:

Table 1		
n	$3n$	$3n - 5$
2		
5		
		7
		22

Table 2		
m	$\frac{m}{4}$	$\frac{m}{4} + 1$
8		
12		
		6
		2

7. Consider the equation $3n - 5 = p$. Look at Table 1 to help you answer the questions:
   a. Let $n = 2$. Explain how to find p in two steps.
   b. Let $p = 7$. Explain how to find n in two steps.

8. Consider the equation $\frac{m}{4} + 1 = h$. Look at Table 2 to help you answer the questions:
   a. Let $m = 8$. Explain how to find h in two steps.
   b. Let $h = 6$. Explain how to find m in two steps.

9. a. If you put on socks and then put on shoes, what operations are needed to reverse the process?
   b. You leave home and bicycle north for 3 miles and then east for 2 miles. Suddenly you notice that you have dropped your wallet. How should you retrace your steps?

Answers: 1. 18 **2.** 4 **3.** 1 **4.** 4 **5.** 32 **6.** $\frac{5}{7}$

Homework 2.4

Skills Practice

▪ Fill in the table in Problems 1-2. Use two steps for each row: Fill in the middle column first. Then use the table to help you answer the questions.

1.

x	$2x$	$2x+4$
3		
6		
		14
		20

Consider the equation $y = 2x + 4$.
a. Let $x = 3$. Explain how to find y in two steps.
b. Let $y = 14$. Explain how to find x in two steps.

2.

q	$q-3$	$5(q-3)$
3		
		10
4		
		20

Consider the equation $5(q - 3) = R$.
a. Let $q = 3$. Explain how to find R in two steps.
b. Let $R = 10$. Explain how to find q in two steps.

▪ For Problems 3-19, solve.

3. $6x - 13 = 5$

4. $\dfrac{2a}{5} = 8$

5. $\dfrac{x}{4} + 2 = 3$

6. $6x + 5 = 5$

7. $24 = 4(p - 7)$

8. $0 = \dfrac{5z}{7}$

9. $\dfrac{k+4}{5} = 9$

10. $\dfrac{2x}{3} - 5 = 7$

11. $7 = \dfrac{4b - 3}{3}$

12. $3c - 7 = -13$

13. $-5 = -2 - 3t$

14. $-3(2h - 6) = -12$

15. $1 - \dfrac{b}{3} = -5$

16. $\dfrac{3y}{5} + 2 = -4$

17. $\dfrac{5x}{2} + 10 = 0$

18. $11.8w - 37.8 = 120.32$

19. $9.7(2.6 + v) = 58.2$

▪ In Problems 20-28,
a. Solve each inequality algebraically.
b. Graph your solutions on a number line.
c. Give at least one value of the variable that is a solution, and one value that is not a solution.

20. $x + 10 \leq -5$

21. $-3y < 15$

22. $\dfrac{x}{3} \leq 4$

23. $2x + 3 > 7$

24. $-3x + 2 \leq 11$

25. $-3 > \dfrac{2x}{3} + 1$

26. $-3 \leq 3x \leq 12$

27. $23 > 9 - 2b \geq 13$

28. $-8 \leq \dfrac{5w + 3}{4} < -3$

Applications

29. Delbert is thinking of a number. If he multiplies the number by 5 and then subtracts 6, the result is 29. What is the number?

■ In Problems 30-31, find the error in the "solution," then write a corrected solution.

30. $6 - 3x = -12$
$-3x = -6$
$x = 2$

31. $-2 + \dfrac{2}{3}x = -4$
$-2 + 2x = -12$
$2x = -10$
$x = -5$

■ In Problems 32-33, use the graphs to answer the questions.

32. Francine's new puppy weighed 3.8 pounds when she brought it home, and it should gain approximately 0.6 pound per week. The graph shows the puppy's weight, W, after t weeks.

a. Estimate the puppy's weight after 8 weeks.
b. Estimate how long it will take for the puppy to reach 26 pounds.
c. Write an algebraic equation for W in terms of t.
d. Write and solve an equation to verify your answer to part (b).

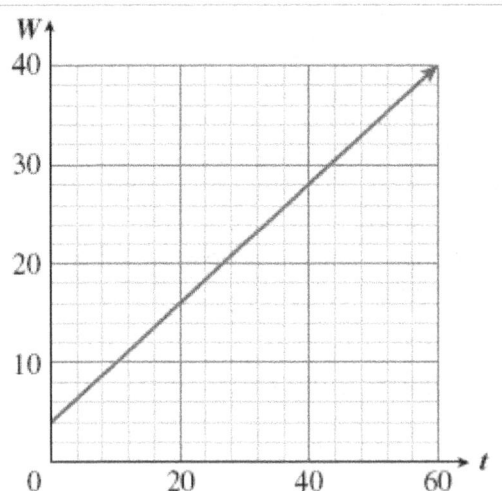

33. Marvin's telephone company charges $13 a month plus $0.15 per minute. The graph shows Marvin's phone bill, P, if he talks for m minutes.

a. Estimate Marvin's phone bill if he talks for 45 minutes.
b. Estimate the number of minutes Marvin talked if his phone bill is $25.
c. Write an algebraic equation for P in terms of m.
d. Write and solve an equation to verify your answer to part (b).

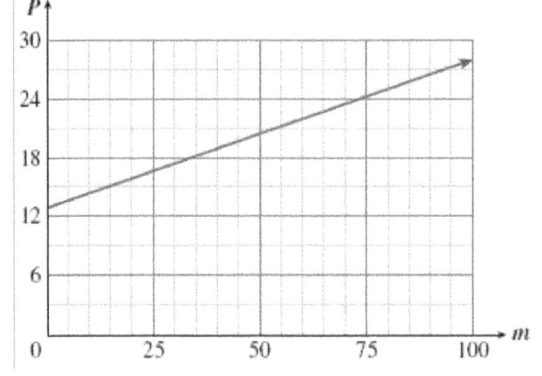

■ In Problems 34-38, solve in three steps:
a. Choose a variable for the unknown quantity.
b. Write an equation that involves your variable.
c. Solve your equation, and answer the question.

34. Simona bought a car for $10,200. She paid $1200 down and will pay the rest in 36 monthly installments. How much is each installment?
(**Hint:** Write an equation about the total amount Simona pays.)

35. Every school day, each member of the cross-country team runs a certain distance d, which is assigned on the basis of ability, plus a half mile around the track. Greta runs 22.5 miles per week. What is her assigned distance?
(**Hint:** Write an equation about the total distance Greta runs in one school week.)

36. The Tree People planted new trees in an area that was burned by brush fires. 60% of the seeds sprouted, but gophers ate 38 of the new sprouts. That left 112 new saplings. How many seeds were planted?
(**Hint:** Write an equation about the number of seeds that survived.)

37. During a four-day warming trend, the temperature rose from $-6°$ to $26°$. What was the average change in temperature per day?

38. Eric is on a diet to reduce his current weight of 196 pounds to 162 pounds. If he loses 4 pounds per week, how long will it take him to reach his desired weight?

■ Use one of the formulas below to solve Problems 39-40.

$$\text{Perimeter of a rectangle} \quad P = 2l + 2w$$
$$\text{Area of a triangle} \quad A = \frac{bh}{2}$$

39. A farmer has 500 yards of fencing material to enclose a rectangular pasture. He would like the pasture to be 75 yards wide. How long will it be?

40. A triangular sail requires 12 square meters of fabric. If the base of the sail measures 4 meters, how tall is the sail?

■ In Problems 41-42, write and solve an equation to find the value of x.

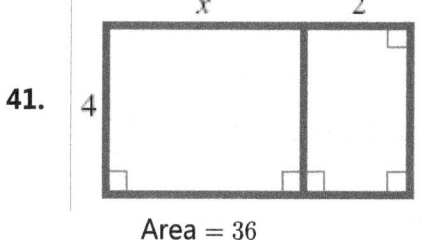

41. Area $= 36$

42. Area $= 20$

2.5 Like Terms

Equivalent Expressions

When we solve an equation such as

$$2 + 3x = 17$$

we find a value for the variable that makes the equation true. The solution to this equation is 5, because the two expressions $2 + 3x$ and 17 are equal if $x = 5$. If we use any other value for x, say $x = 9$ or $x = -1$, then $2 + 3x$ does not equal 17.

It is possible for two algebraic expressions to be equal for **all** values of the variable. For instance, the expression $2x + 3x$ and $5x$ will always give the same value when the same number is substituted for x in each expression. For example,

$$\text{if } x = 3, \qquad \text{then} \qquad 2x + 3x = 2(3) + 3(3)$$
$$= 6 + 9 = 15$$
$$\text{and} \qquad 5x = 5(3) = 15$$

Both expressions equal 15 when $x = 3$.

Exercise   Verify that $2x + 3x$ and $5x$ are equal when $x = -4$.

$$2x + 3x = \underline{\hspace{3cm}}$$
$$\text{and} \qquad 5x = \underline{\hspace{3cm}}$$

> Two algebraic expressions are **equivalent** if they name the same number for all values of the variable.

Thus, $2x + 3x$ and $5x$ are equivalent expressions, and we can replace one by the other if it will simplify our work.

Caution!   Although two expressions may name the same number for *some* values of the variable, they are not necessarily equivalent. For example, to show that the expressions $2 + 3x$ and $5x$ are *not* equivalent, we only have to find a single value of x for which they are not equal. If we evaluate each expression for $x = 6$, we find:

$$2 + 3x = 2 + 3(6) = 20$$
$$5x = 5(6) = 30$$

Because they are not equal for $x = 6$, the expressions $2 + 3x$ and $5x$ are not equivalent. ■

RQ1.   When are two expressions equivalent?

Like Terms

We saw above that $2x + 3x$ and $5x$ are equivalent expressions. This makes sense when we consider what these expressions mean: $2x$ means two x's added together, and $3x$ means three x's added together. Then

$$2x + 3x = (x + x) + (x + x + x) = 5x$$

We can add two x's to three x's and get five x's, just as we can add two pencils to three pencils, or two dollars to three dollars. In algebra this is called **combining like terms**.

Like terms are any terms that are exactly alike in their variable factors.

Example 1a. $2x$ and $3x$ are like terms

$-4a$ and $7a$ are like terms

because their variable factors are identical.
b.

$2x$ and $3y$ are not like terms

$2x$ and 3 are not like terms

because their variable factors are different.

The numerical factor in a term is called the **numerical coefficient**, or just the **coefficient** of the term.

Example 2a.   In the expression $3xy$, the number 3 is the numerical coefficient.
b. In a term such as xy or b, the numerical coefficient is 1.
c. If a variable is preceded by a negative sign, the coefficient of the term is -1. For example,

$-x$ means $-1 \cdot x$

$-a$ means $-1 \cdot a$

Adding and Subtracting Like Terms

To add like terms:
    Add the numerical coefficients of the terms. Do not change the variable factors of the terms.

Example 3   Add like terms.
a. $5n + 3n$ **b.** $-4y - 3y$ **c.** $-6st + 9st$

Solutions   In each calculation, we do not change the variables we are adding; only the coefficient changes.
a. $5n + 3n = (5 + 3)n = 8n$ **Add the coefficients, 5 and 3.**
b. $-4y - 3y = (-4 - 3)y = -7y$ **Add the coefficients, -4 and -3.**
c. $-6st + 9st = (-6 + 9)st = 3st$ **Add the coefficients, -6 and 9.**

Look Ahead: In Example 3, we replaced each expression by a simpler but equivalent expression. In other words, $-7y$ is equivalent to $-4y - 3y$, and $3st$ is equivalent to $-6st + 9st$.

Replacing an expression by a simpler equivalent one is called **simplifying** the expression.

We subtract like terms in the same way that we add like terms.

To subtract like terms:
   Subtract the numerical coefficients of the terms. Do not change the variable factors of the terms.

■ **Example 4** $6a - (-8a) = [6 - (-8)]a = 14a$ **Subtract the coefficients.**   ■

RQ2. What are like terms?
RQ3. How do we add or subtract like terms?

Caution!   We cannot add or subtract unlike terms. Thus,

$$2x + 3y \qquad \text{cannot be simplified}$$
$$-6x + 4xy \qquad \text{cannot be simplified}$$
$$6 - 2x \qquad \text{cannot be simplified} \qquad ■$$

Many expressions contain both like and unlike terms. In such expressions we can combine only the like terms.

■ **Example 5**   Simplify $-4a + 2 - 5 + 8a + 5a$

Solution We combine all the a-terms, and all the constant terms, like this:

$$-4a + 8a + 5a = 9a \qquad \text{and} \qquad 2 - 5 = -3$$

so

$$-4a + 2 - 5 + 8a + 5a = 9a - 3 \qquad ■$$

Caution!   In Example 5, $9a - 3 \neq 6a$, because $9a$ and -3 are not like terms.   ■

Removing Parentheses

   When we add signed numbers, we can remove parentheses that follow an addition symbol. For instance,

$$5 + (-3) = 5 - 3 = 2$$

We can also remove parentheses when we add like terms. Thus,

$$5x + (-3x) = 5x - 3x = 2x$$

Removing parentheses that follow a plus sign:
   Parentheses preceded by a plus sign may be omitted, and each term within parentheses keeps its original sign.

Example 6   Simplify $(2x + 3) + (5x - 7)$

Solution Each set of parentheses is preceded by a plus sign, so we may remove the parentheses.

$$(2x + 3) + (5x - 7) = 2x + 3 + 5x - 7 \qquad \textbf{Combine like terms.}$$
$$= 7x - 4$$

We must be careful when removing parentheses preceded by a **minus** sign. In the expression

$$(5x - 2) - (7x - 4)$$

the minus sign preceding $(7x - 4)$ applies to *each term* inside the parentheses. Therefore, when we remove the parentheses we must change the sign of $7x$ from $+$ to $-$ *and* we must change the sign of -4 from $-$ to $+$. Thus,

$$(5x - 2) - (7x - 4) = 5x - 2 - 7x + 4 \qquad \textbf{Change each sign inside parentheses.}$$
$$= 5x - 7x - 2 + 4 \qquad \textbf{Combine like terms.}$$
$$= -2x + 2$$

> **Removing parentheses that follow a minus sign:**
>   If an expression inside parentheses is preceded by a minus sign, we **change the sign of each term within parentheses** and then omit the parentheses.

Example 7   Simplify $-5b - (3 - 2b) + (4b - 6)$

Solution Before combining like terms, we must remove the parentheses. We change the signs of any terms inside parentheses preceded by a minus sign; we do *not* change the signs of terms inside parentheses preceded by a plus sign.

$$-5b - (3 - 2b) + (4b - 6) \;=\; -5b - 3 + 2b + 4b - 6$$

signs changed

signs not changed

$$= -5b + 2b + 4b - 3 - 6$$
$$= b - 9$$

RQ4.   We can remove parentheses that follow _____.

RQ5.   What must you do if you remove parentheses that follow a minus sign?

Solving Equations and Inequalities

We can now solve equations in which two or more terms contain the variable.

Example 8   Solve the equation $2x + 7 = 4x - 3$

Solution We first get all terms containing the variable on one side of the equation. For this example, we will subtract $2x$ from both sides of the equation to get

$$2x + 7 = 4x - 3 \qquad \text{Subtract } 2x \text{ from both sides.}$$
$$\underline{-2x \qquad -2x}$$
$$7 = 2x - 3$$

Now the equation looks like ones we already know how to solve, and we proceed as usual to isolate the variable.

$$7 = 2x - 3 \qquad \text{Add 3 to both sides.}$$
$$\underline{+3 \qquad +3}$$
$$10 = 2x \qquad \text{Divide both sides by 2.}$$
$$\frac{10}{2} = \frac{2x}{2}$$
$$5 = x$$

The solution is 5.

Check: Does $2(\mathbf{5}) + 7 = 4(\mathbf{5}) - 3$?   True.   ■

If one side of an equation or inequality contains like terms, we should combine them before beginning to solve.

■ **Example 9**   Solve $2x - 4x + 14 < 3 + 5x - 10$

Solution Combine like terms on each side of the inequality, to get

$$\mathbf{2x - 4x + 14 < 3 + 5x - 10} \qquad 2x - 4x = -2x \text{ and } 3 - 10 = -7$$
$$-2x + 14 < 5x - 7$$

Now continue solving as usual: We want to get all the terms containing x on one side of the inequality, and all the constant terms on the other side.

$$-2x + 14 < 5x - 7 \qquad \text{Subtract } 5x \text{ from both sides.}$$
$$\underline{-5x \qquad -5x}$$
$$-7x + 14 < -7$$
$$\underline{-14 \qquad -14} \qquad \text{Subtract 14 from both sides.}$$
$$-7x < -21$$
$$\frac{-7x}{-7} > \frac{-21}{-7} \qquad \begin{array}{l}\text{Divide both sides by } -7; \\ \text{reverse the direction of the inequality.}\end{array}$$
$$x > 3$$

The solution is all x-values greater than 3. You can check the solution by substituting one x-value greater than 3 and one value less than 3 into the original inequality, for instance:

Check: $x = 4:$ $2(\mathbf{4}) - 4(\mathbf{4}) + 14 < 3 + 5(\mathbf{4}) - 10$?   **True:** $6 < 13$
$\phantom{\text{Check:}} x = 2:$ $2(\mathbf{2}) - 4(\mathbf{2}) + 14 < 3 + 5(\mathbf{2}) - 10$?   **False:** $10 \not< 3$ ■

RQ6.   What should you do if one side of an equation or inequality contains like terms?

Here are our guidelines for solving linear equations.

Steps for Solving Linear Equations
1. Combine like terms on each side of the equation.
2. By adding or subtracting the same quantity on both sides of the equation, get all the variable terms on one side and all the constant terms on the other.
3. Divide both sides by the coefficient of the variable to obtain an equation of the form $x = a$.

Skills Warm-Up

Write algebraic expressions.

1. Getaway Tours offers a Caribbean cruise for $2000 per person if 12 people sign up. For each additional person who signs up, the price per person is reduced by $60. How much will you pay for a cruise if p additional people sign up?

2. Anita figures her taxes by subtracting a $1200 deductible from her income, then taking 6% of the result. What is her tax bill if her income is I?

3. The area of a trapezoid is one-half the product of its height h and the sum of the two bases, a and b.

4. The ratio of a number k to 5 less than k.

Evaluate.

5. $3n(n - k)(n + k)$ for $n = -4$ and $k = -6$

6. $(t - 1)(2t + 3)$ for $t = \dfrac{2}{3}$

Answers: **1.** $2000 - 60p$ **2.** $0.06(I - 1200)$ **3.** $\dfrac{1}{2}h(a + b)$

4. $\dfrac{k}{k - 5}$ **5.** 240 **6.** $\dfrac{-13}{9}$

Homework 2.5

Skills Practice

■ In Problems 1-4, add or subtract like terms.

1. $-6x + 2x$

2. $-7.6a - 5.2a$

3. $3t - 4t + 2t$

4. $3bc - (-4bc) - 8bc$

■ In Problems 5-6, combine like terms.

5. $3 + 4y - (-8y) - 7$

6. $-2st - 2 + 5s - 6st - (-4s)$

■ In Problems 7-10, simplify and combine like terms.

7. $4x + (5x - 2)$

8. $22y - 34 - (16y - 24)$

9. $6a - 5 - 2a - (2a - 5)$

10. $(-2 - 3b) - (2b - 4)$

■ In Problems 11-16, solve each equation or inequality.

11. $4m - 3 = 2m + 5$

12. $15 - 9t = 33 - 5t$

13. $-6s = 3s$

14. $3x + 5 > 2x + 3$

15. $-8g + 35 = 9g - 13 + g$

16. $-15y + 5 - 2y - 4 \geq -12y + 21$

17. $-2 - 3w + 5w \leq 4w - 44$

18. $-5p - 4 + 3p = 2 + 7p + 3$

19. a. Evaluate the expression $-3y + 2 + 7y - 6y - 4y - 8$ for $y = 2.5$.
  b. Simplify the expression in part (a).
  c. Evaluate your answer to part (b) for $y = 2.5$. Check that you got the same answer for part (a).

20. Evaluate the expression for the given values of the variables.
  $$4 - 3a + 6ab - (-8a) - 10 + 9ab + 2a \quad \text{for } a = -2, \ b = 5$$

Applications

■ In Problems 21-22, choose a value for the variable and show that the following pairs of expressions are **not** equivalent.

21. $2 + 7x$ and $9x$

22. $-(a - 3)$ and $-a - 3$

■ Write and simplify an algebraic expression for the perimeter of the figure in Problems 23-24.

23.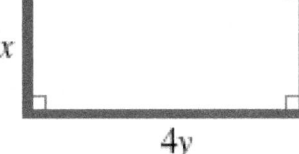

24.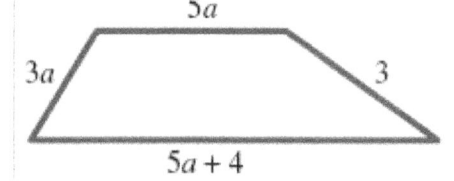

■ For Problems 25-29, write an algebraic expression or an equation to answer each question.

25. There are 8 more cats than dogs at the county shelter. Write expressions in terms of the number of dogs, d.
   a. The number of cats
   b. The total number of cats and dogs
   c. If there are 40 dogs and cats at the shelter, how many dogs are there?

26. For every smoker in the restaurant there are three nonsmokers. Write expressions in terms of x, the number of smokers.
   a. The number of nonsmokers
   b. The total number of people in the restaurant
   c. If there are 36 people in the restaurant, how many are smokers?

27. A tortoise and a hare are having a race. After t seconds, the tortoise has traveled $0.2t$ feet and the hare has traveled $10t$ feet.
   a. Express the distance between them in terms of t.
   b. How far apart are the tortoise and the hare after 10 seconds?
   c. When will they be 147 feet apart?

28. Delbert bought a high definition TV at a price of x dollars. The sales tax in his state is 8%.
   a. Write an expression for the tax on Delbert's TV.
   b. Write an expression for the price Delbert paid, including tax.
   c. Delbert paid $928.80 for his TV. What was the price before tax?

29. The cost of producing m stereos is $251m + 1355$ dollars, and each stereo sells for $847.
   a. What profit is made from producing and selling m stereos?
   b. How many stereos must be produced to make a profit of $16,525?

■ For Problems 30-34, write and solve an equation.

30. Last summer Kevin made 8 batches of barbecue sauce using one bottle of catsup. This summer he made only 5 batches and had 12 ounces of catsup left over. How much catsup does Kevin use in a batch of barbecue sauce?

31. Arch has two wooden beams of equal length. He makes 3 posts for mailboxes with one beam, and has 33 inches left. He makes another mailbox post with the other beam, and has 131 inches left. How long is each mailbox post?

32. Quentin bought 12 cans of cat food and a half-gallon of milk that cost $1.80. Tilda bought 8 cans of cat food and a $5 bag of kibble. They paid the same price at the check-out stand. How much does one can of cat food cost?

33. If Sarah drinks three glasses of milk a day she will exceed her minimum daily requirement for calcium by 70 milligrams. If she drinks only two glasses of milk she will still need another 220 milligrams of calcium. How many milligrams of calcium does a glass of milk contain?

34. On her last history test, Parisa answered 9 short-answer questions correctly and earned 15 points on the essay. Seana answered 10 short-answer questions correctly and got 23 points on the essay. Seana's score on the test was 14 points higher than Parisa's. How many points was each short-answer question worth?

Chapter 2 Summary and Review

Section 2.1 Signed Numbers

- The natural numbers, zero, and the negatives of the natural numbers are called the **integers.**

- We use special symbols to indicate order:

$<$ means   **is less than**

$>$ means   **is greater than**

- The **absolute value** of a number is the unsigned part of the number. It is never negative.

Rules for Operations on Integers

To add two integers:
1. To add two numbers with the same sign, add their absolute values. The sum has the same sign as the numbers.
2. To add two numbers with opposite signs, subtract their absolute values. The sum has the same sign as the number with the larger absolute value.

To subtract b from a:
1. Change the sign of b.
2. Change the subtraction to addition.
3. Proceed as in addition.

To multiply or divide two integers:
1. The product or quotient of two numbers with opposite signs is a negative number.
2. The product or quotient of two numbers with the same sign is a positive number.

Quotients Involving Zero
If a is any nonzero number, then

$$\frac{0}{a} = 0 \qquad \text{and} \qquad \frac{a}{0} \text{ is undefined}$$

Section 2.2 Expressions and Equations

- When we evaluate an algebraic expression, we follow the order of operations.

- When we evaluate an algebraic expression at a negative number, we enclose the negative numbers in parentheses.

Section 2.3 Graphs of Linear Equations

- An equation of the form $y = ax + b$, where a and b are constants, is called a **linear equation** because its graph is a straight line.

- We can graph a linear equation by evaluating the expression $ax + b$ at several values of x and then plotting points.

- We use a **Cartesian coordinate system** to make a graph that includes negative values of the variables.

- The graph of an equation is a picture of the solutions of the equation. Each point on the graph represents a solution.

Section 2.4 Linear Equations and Inequalities

- To solve an equation that involves two or more operations, we undo those operations *in reverse order*.

- A statement that uses one of the symbols $<$ or $>$ is called an **inequality**.

Rules for Solving Inequalities
1. We can add or subtract the same quantity on both sides.
2. We can multiply or divide both sides by the same positive number.
3. If we multiply or divide both sides by a **negative** number, we must **reverse** the direction of the inequality.

- Each point on the graph of an equation represents a solution of the equation. To solve the equation for a particular y-value, we locate the corresponding point on the graph.

- To solve a **compound inequality**, we perform the steps needed to isolate x on all three sides of the inequality.

Section 2.5 Like Terms

- Two algebraic expressions are **equivalent** if they name the same number for *all* values of the variable.

- **Like terms** are any terms that are *exactly alike* in their variable factors. By adding or subtracting like terms we can replace one algebraic expression by a shorter or simpler one.

- The numerical factor in a term is called the **numerical coefficient**, or just the **coefficient** of the term.

- Replacing an expression by a simpler equivalent one is called **simplifying** the expression.

To add or subtract like terms:
   Add or subtract the numerical coefficients of the terms. Do not change the variable factors of the terms.

Removing parentheses
1. Parentheses following a plus sign may be omitted; each term within parentheses keeps its original sign.
2. To remove parentheses that follow a minus sign, we change the sign of each term within parentheses and then omit the parentheses and the minus sign.

Steps for Solving Linear Equations
1. Combine like terms on each side of the equation.
2. By adding or subtracting the same quantity on both sides of the equation, get all the variable terms on one side and all the constant terms on the other.
3. Divide both sides by the coefficient of the variable to obtain an equation of the form $x = a$.

Review Questions

■ Use complete sentences to answer the questions in Problems 1-10.

1. Explain the terms natural numbers, whole numbers, and integers.
2. A classmate claims that the opposite of a number and the absolute value of a number are the same, and uses $x = -3$ as an example. Do you agree? Give examples of your own.
3. You have probably heard people say that "two negatives make a positive." For which of the four arithmetic operations is this statement
   a. always true?
   b. always false?
   c. sometimes true and sometimes false?
   Make up an example for each case.
4. In a Cartesian coordinate system, the axes divide the plane into four _____ . The point $(0, 0)$ is called the _____ .
5. Explain how to add or subtract like terms.
6. What is wrong with this statement? $8x - (2x - 3) = 6x - 3$
7. When solving an inequality, we must remember to _____ if we _____ both sides by _____ .
8. Explain how to use a graph to solve an equation.
9. Give an example of a quotient that is undefined.
10. Explain why $-\dfrac{a}{b}$ is not equal to $\dfrac{-a}{-b}$.

Review Problems

■ For Problems 1-6, choose the appropriate equation.

a. $5x - 8 = 30$	b. $\dfrac{x}{5} - 30 = 8$	c. $\dfrac{x - 30}{5} = 8$
d. $5x + 8 = 30$	e. $\dfrac{x}{5} - 8 = 30$	f. $\dfrac{x + 30}{5} = 8$

1. Ilciar has earned a total of 30 points on the first four quizzes in his biology class. What must he earn on the fifth quiz to end up with an average of 8?
2. Jocelyn ordered five exotic plants from a nursery. She paid a total of $30, including an $8 shipping fee. How much did she pay for each plant?
3. The five members of the chess team pitched in to buy new equipment. They used $30 from their treasury, and each member donated $8. How much did the new equipment cost?
4. Hemman bought 5 tapes on sale, and he cashed in a gift certificate for $8. He then owed the clerk $30. How much was each tape?
5. Nirusha and four other people won the office baseball pool. After spending $30 of her share, Nirusha had $8 left. What was the total amount in the pool?
6. One-fifth of the members at Sportslife Health Club signed up for a yoga class. Eight of them dropped out, leaving 30 in the class. How many members are in the Club?

■ For Problems 7-10, solve.

7. $3x - 4 = 1$
8. $1.2 + 0.4z = 3.2$
9. $\dfrac{7v}{8} - 3 = 4$
10. $13 = \dfrac{2}{7}x + 13$

◼ For Problems 11-12, write an equation relating the variables, make a table of values, and graph your equation on the grid provided.

11. A new computer station for a graphics design firm costs $2000 and depreciates in value $200 every year. Write and graph an equation that gives the value, V, of the station after t years. Use the grid below left.

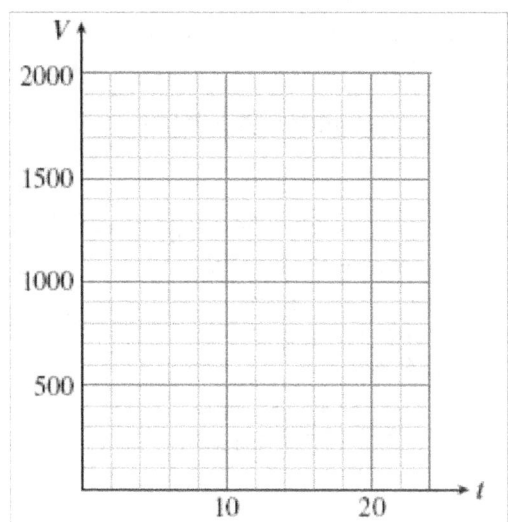
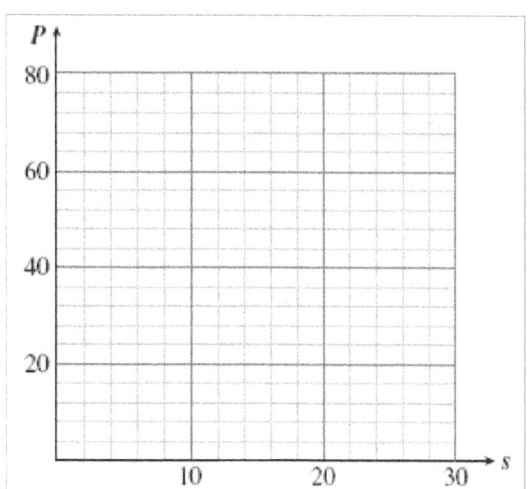

12. Ranwa teaches aerobic dance. Her gym pays her $5 for an hour-long class, plus $2 for each paying student in the class. Write and graph an equation that gives Ranwa's pay, P, if there are s students in the class. Use the grid above right.

◼ For Problems 13-14, write and solve an equation to answer the question.

13. Customers at Sunny Orchard can pick their own fruit. There is a $5.75 entrance fee, and the fruit costs $4.50 a bushel. Bahn spent $19.25 at the orchard. How many bushels of fruit did he pick?

14. The perimeter of a rectangle is the sum of twice its width and twice its length. If 150 meters of fence enclose a rectangular yard of length 45 meters, what is the width of the yard?

15. If $x = -3$, evaluate each expression.
   a. $-x$ **b.** $x - 3$ **c.** $3 - x$ **d.** $-(-x)$

16. Graph each set of numbers on a separate number line.
   a. 1.4, $2\frac{1}{2}$, $-2\frac{1}{3}$, -3.5 **b.** $-15, -5, 10, 25$

◼ For Problems 17-20, replace the comma by the proper symbol; $>$, $<$, or $=$.

17. $-2, -3$ **18.** $-2.02, -2.1$
19. $2 - (-5), -7$ **20.** $-6\left(\frac{-1}{3}\right), -2$

◼ For Problems 21-28, simplify.

21. $28 - 14 - 9 + 15$ **22.** $11 - 14 + (-24) - (-18)$
23. $12 - [6 - (-2) - 5]$ **24.** $-2 + [-3 - (-14) + 6]$
25. $5 - (-4)3 - 7(-2)$ **26.** $5 - (-4)(3 - 7)(-2)$

27. $\dfrac{6(-3) - 8}{-4(-3 - 5)}$

28. $\dfrac{6(-2) - 8(-9)}{5(2 - 7)}$

For Problems 29-32, evaluate.

29. $2 - ab - 3a$, for $a = -5$, $b = -4$

30. $(8 - 6xy)xy$, for $x = -2$, $y = 2$

31. $\dfrac{-3 - y}{4 - x}$, for $x = -1$, $y = 2$

32. $\dfrac{5}{9}(F - 32) + 273$, for $F = -22$

33. If the overnight low in Lone Pine was −4°F, what was the temperature after it had warmed by 10°F?

34. The winter temperature in the city is typically 3°F warmer than in the adjoining suburb. If the temperature in the suburb is −7°F, what is the temperature in the city?

35. Jordan's clothing company is worth 280,000 dollars and Asher's is worth −180,000 dollars. How much more is Jordan's company worth than Asher's?

36. A certain arsenic compound has a melting point of −8.5°F. If the melting point is reduced by 1.1°F, what is the new melting point?

37. In his first football game, Bo rushes for three consecutive losses of 4 yards each. What is his total net yardage for the three plays?

38. In testing a military aircraft's handling at low altitude, the pilot runs a series of flights over a set course, each at an altitude 150 feet lower than the previous flight. What is the net change in altitude between the first and the sixth flights?

For Problems 39-42, find three solutions for the inequality.

39. $x - 2 < 5$

40. $4x \geq -12$

41. $-9 \leq -3x$

42. $-15 > 5 + x$

43. The temperature in Maple Grove was 18°F at noon, and it has been dropping ever since at a rate of 3°F per hour.

   a. Fill in the table. T stands for the temperature h hours after noon. Negative values of h represent hours before noon.

h	−4	−2	0	1	3	5	8
T							

   b. Write an equation for the temperature, T, after h hours.
   c. Graph your equation on the grid.
   d. What was the temperature at 10 am?
   e. When will the temperature reach −15°F?
   f. How much did the temperature drop between 3 pm and 9 pm?

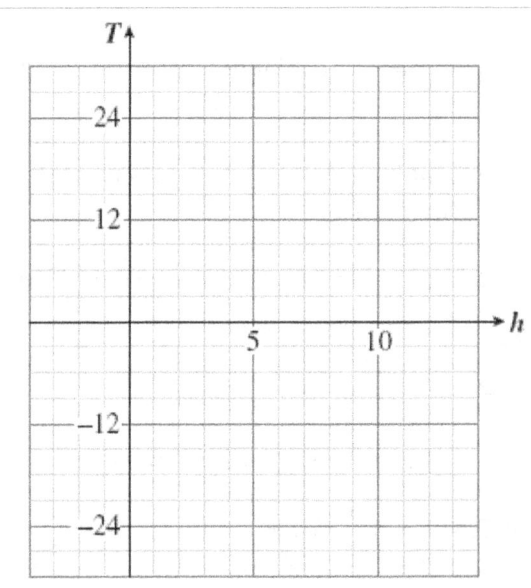

44. Graph each pair of points on the grid, then find the distance between them.

  a. $A(-2, -4)$ and $B(-7, -4)$

  b. $P(6, -2)$ and $Q(6, 8)$

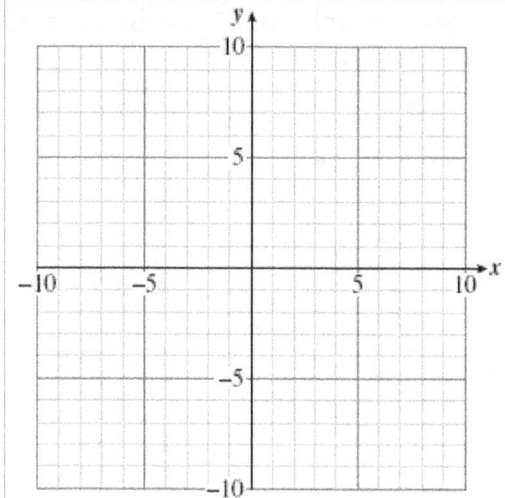

For Problems 45-46, make a table of values and graph each equation.

45. $y = -2x + 7$

46. $y = \dfrac{4}{3}x - 2$

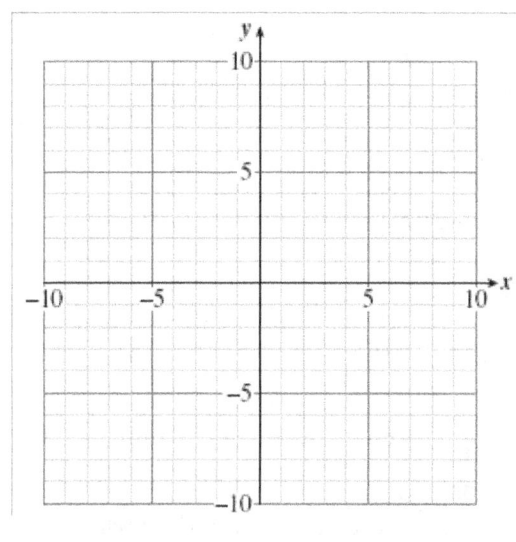

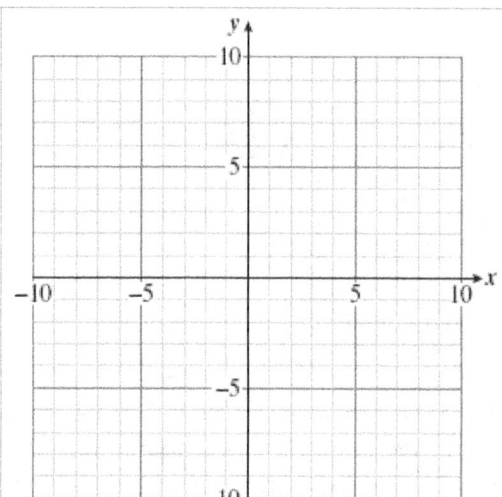

For Problems 47-48, simplify the expression.

47. $(4m + 2n) - (2m - 5n)$

48. $(-5c - 6) + (-11c + 15)$

For problems 49-54, solve.

49. $4z - 6 = -10$

50. $3 - 5x = -17$

51. $-1 = \dfrac{5w}{3} + 4$

52. $4 - \dfrac{2z}{5} = 8$

53. $3h - 2 = 5h + 10$

54. $7 - 9w = w + 7$

For Problems 55-56, use the graph to solve the equation. Estimate your solutions if necessary.

55. a. $-3x + 9 = 24$
   b. $-3x + 9 = 3$

56. a. $24x - 1800 = -1250$
   b. $24x - 1800 = 500$

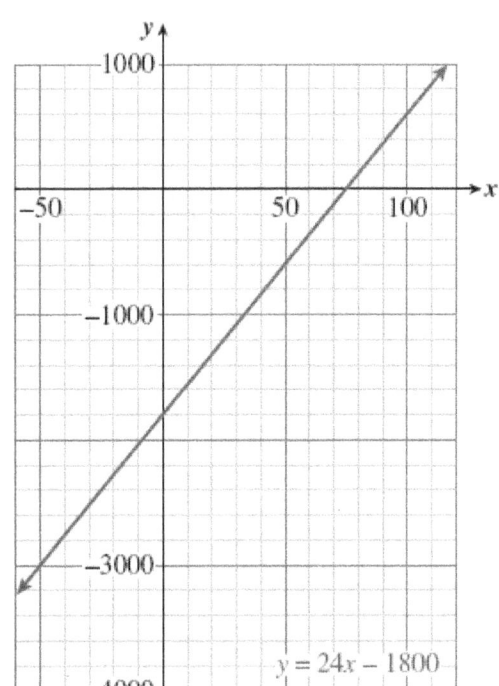

57. The length of a rectangle is 3 times its width.
   a. If the width of a rectangle is x, what is its length?
   b. Express the perimeter of the rectangle in terms of x.
   c. Suppose the perimeter of the rectangle is 48 centimeters. Find the dimensions of the rectangle.

58. In a city council election the winner received 132 votes more than her opponent.
   a. If the winner received y votes, how many did her opponent receive?
   b. Write an expression for the total number of votes cast for the two candidates.
   c. If 12,822 votes were cast, how many did each candidate receive?

For Problems 59-64, solve the inequality algebraically, and graph the solutions on a number line.

59. $2 - 3z \le -7$

60. $\dfrac{t}{-3} - 1.7 > 2.8$

61. $3k - 13 < 5 + 6k$

62. $12a - 28 < -18 + 2a$

63. $-9 < 5 - 2n \le -1$

64. $15 \ge -6 + 3m \ge -6$

Chapter 3 Graphs of Linear Equations

Lesson 3.1   Intercepts

- Find the intercepts of a line from its equation
- Graph a line by the intercept method
- Interpret the intercepts in context

Lesson 3.2   Ratio and Proportion

- Compare two quantities with a ratio or a rate
- Solve a proportion
- Write a proportion to model a problem
- Recognize proportional variables

Lesson 3.3   Slope

- Compute the slope of a line
- Interpret the slope as a rate of change

Lesson 3.4   Slope-Intercept Form

- Write an equation in slope-intercept form
- Identify the slope and y-intercept of a line from its equation
- Graph a linear equation by the slope-intercept method
- Interpret the slope and y-intercept in context

Lesson 3.5   Properties of Lines

- Find equations for parallel or perpendicular lines
- Find equations for horizontal or vertical lines
- Use the two-point formula to compute slope

3.1 Intercepts

Intercepts of a Line

> The **intercepts** of a line are the points where the graph crosses the axes.

It is easy to recognize the intercepts of a line on a good graph. Here is a graph of the line

$$y = \frac{1}{2}x + 4$$

We can see that its x-intercept is the point $(-8, 0)$, and its y-intercept is $(0, 4)$.

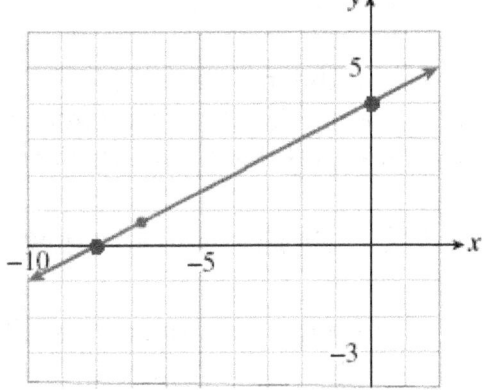

If we have an equation for a line we can find the intercepts of its graph algebraically. In Example 1 we'll use algebra to find the intercepts of this line.

Look Closer: Look at the intercepts on the graph above. Because the y-intercept of a graph lies on the y-axis, its x-coordinate must be zero. And because the x-intercept lies on the x-axis, its y-coordinate must be zero.   •

■ **Example 1**   Use algebra to find the intercepts of the graph of $y = \frac{1}{2}x + 4$.

Solution To find the y-intercept, we substitute 0 for x in the equation and solve for y.

$$y = \frac{1}{2}(0) + 4 = 4$$

The y-intercept is $(0, 4)$, as we saw above. To find the x-intercept we substitute 0 for y and solve the equation for x.

$$0 = \frac{1}{2}x + 4 \qquad \text{Subtract 4 from both sides.}$$
$$-4 = \frac{1}{2}x \qquad \text{Multiply both sides by 2.}$$
$$-8 = x$$

The x-intercept is $(-8, 0)$, as expected.   ■

To find the x-intercept of a graph:
    Substitute 0 for y in the equation and solve for x.

To find the y-intercept of a graph:
    Substitute 0 for x in the equation and solve for y.

RQ1.   What are the intercepts of a line?
RQ2.   How do we find the x-intercept of the graph of an equation?

The Intercept Method of Graphing

We can use the x- and y-intercepts to graph a linear equation quickly. Instead of choosing several different values of x to find points on the graph, we find the two intercepts and fill in the short table shown at right. By finding the missing values for x and y we are finding the intercepts of the graph.

x	y
0	
	0

■ **Example 2**   Graph the equation $3x + 2y = 7$ by the intercept method.

Solution   First, we find the x- and y-intercepts of the graph.

To find the y-intercept, we substitute 0 for x and solve for y.
The y-intercept is the point $(0, 3\frac{1}{2})$.

$$3(0) + 2y = 7 \qquad \text{Simplify the left side.}$$
$$2y = 7 \qquad \text{Divide both sides by 2.}$$
$$y = \frac{7}{2} = 3\frac{1}{2}$$

To find the x-intercept, we substitute 0 for y and solve for x.
The x-intercept is the point $(2\frac{1}{3}, 0)$.

$$3x + 2(0) = 7 \qquad \text{Simplify the left side.}$$
$$3x = 7 \qquad \text{Divide both sides by 3.}$$
$$x = \frac{7}{3} = 2\frac{1}{3}$$

Here is a table showing the two intercepts. We plot the intercepts and connect them with a straight line to obtain the graph below.

x	y
0	$3\frac{1}{2}$
$2\frac{1}{3}$	0

It is a good idea to find a third point as a check. We choose $x = 1$ and solve for y.

$$3(1) + 2y = 7 \qquad \text{Subtract 3 from both sides.}$$
$$2y = 4 \qquad \text{Divide by 2.}$$
$$y = 2$$

You can check that the point $(1, 2)$ lies on the graph.   ■

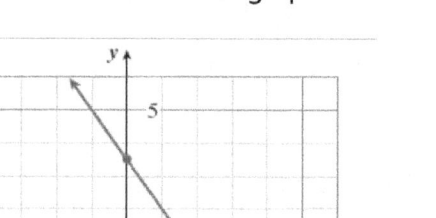

The **intercept method** for graphing a line is often faster and more efficient than making a table and plotting points.

> **To Graph a Linear Equation Using the Intercept Method:**
> 1. Find the x- and y-intercepts of the graph.
> 2. Draw the line through the two intercepts.
> 3. Find a third point on the graph as a check.
>   (Choose any convenient value for x and solve for y.)

RQ3. How do we graph a line by the intercept method?
RQ4. Make a table that you can use with the intercept method.
RQ5. What is the y-intercept of the line $3x + 2y = 7$?

Interpreting the Intercepts

The intercepts of a graph give us valuable information about a problem. They often represent the starting or ending values for a particular variable.

> ■ **Example 3**   The temperature in Nome was $-12°$ at noon and has been rising at a rate of $2°$ per hour all day. Find the intercepts of the graph. What do they tell us about the temperature in Nome?
>
> **Solution**   An equation for the temperature T at time h, where h represents the number of hours from noon, is
>
> $$T = -12 + 2h$$
>
> The graph of the equation is shown below. To find the T-intercept, we set $h = 0$ and solve for T.
>
> $$T = -12 + 2(0) = -12$$
>
> The T-intercept is $(0, -12)$. This point tells us that when $h = 0$, $T = -12$, or the temperature at noon was $-12°$. To find the h-intercept, we set $T = 0$ and solve for h.
>
>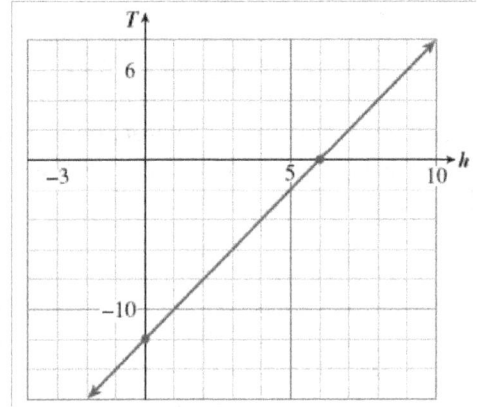
>
> $\quad 0 = -12 + 2h$ **Add 12 to both sides.**
> $\quad 12 = 2h$ **Divide by 2.**
> $\quad\ \ 6 = h$
>
> The h-intercept is the point $(6, 0)$. This point tells us that when $h = 6$, $T = 0$, or the temperature will reach zero degrees at six hours after noon, or 6 pm.   ■

> **RQ6.**   The intercepts of a graph often represent the _____ or _____ values for a particular variable.
>
> **RQ7.**   What does the h-intercept tell us in Example 3?

Skills Warm-Up

■ Choose the correct algebraic expression for each of the following situations.

$$5x - 8 = 30 \qquad \frac{x}{5} - 30 = 8 \qquad \frac{x - 30}{5} = 8$$

$$5x + 8 = 30 \qquad \frac{x}{5} + 30 = 8 \qquad \frac{x + 30}{5} = 8$$

1. Ilciar has earned a total of 30 points on the first four quizzes in his biology class. What must he earn on the fifth quiz to end up with an average of 8?
2. Jocelyn ordered five exotic plants from a nursery. She paid a total of $30, including an $8 shipping fee. How much did she pay for each plant?
3. To buy new equipment, the five members of the chess club used $30 from the treasury, and each member donated $8. How much did the equipment cost?

4. Hemman bought 5 tapes on sale, and he cashed in a gift certificate for $8. He then owed the clerk $30. How much was each tape?

5. Nirusha and four other people won the office baseball pool. After spending $30 of her share, Nirusha had $8 left. What was the total amount in the pool?

Answers: 1. $\dfrac{x+30}{5} = 8$ **2.** $5x + 8 = 30$ **3.** $\dfrac{x-30}{5} = 8$ **4.** $5x - 8 = 30$

5. $\dfrac{x}{5} - 30 = 8$

Homework 3.1

Skills Practice

For Problems 1-6,
   a. Find the x- and y-intercepts of each line.
   b. Use the intercept method to graph the line.

1. $2x + 4y = 8$

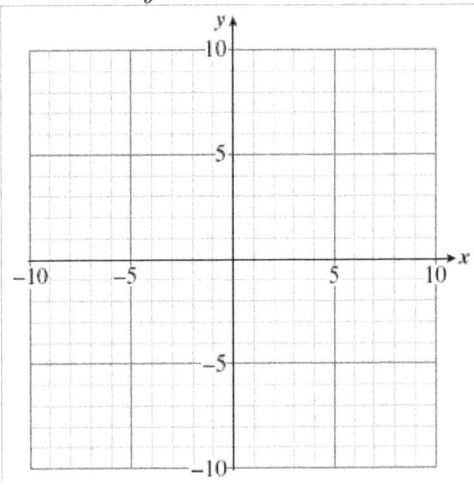

2. $x + 2y + 10 = 0$

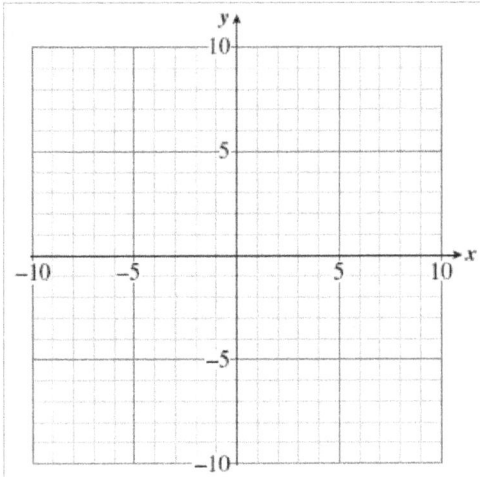

3. $2x = 14 + 7y$

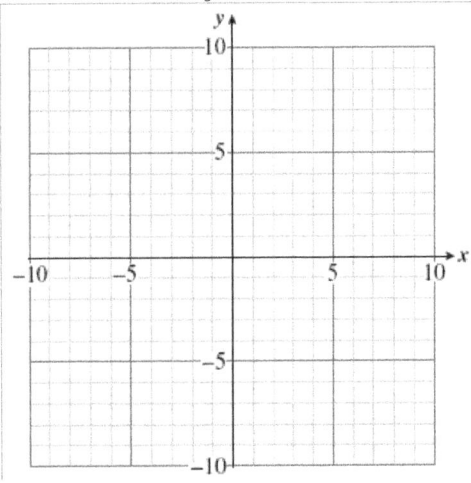

4. $y = -4x + 8$

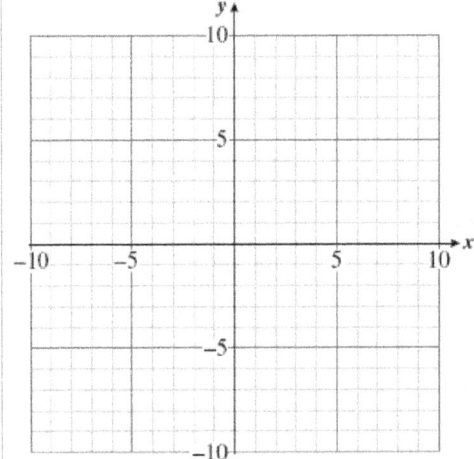

5. $\dfrac{x}{20} + \dfrac{y}{30} = 1$

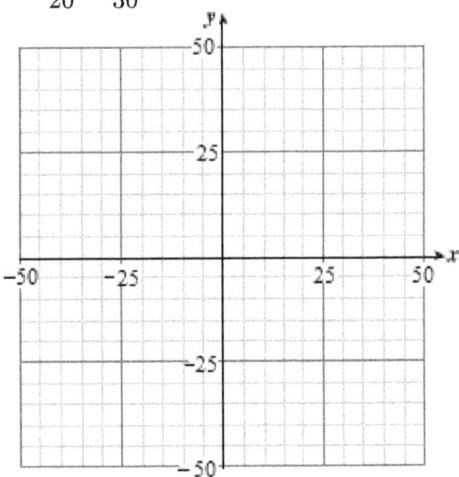

6. $3x - 2y = 120$

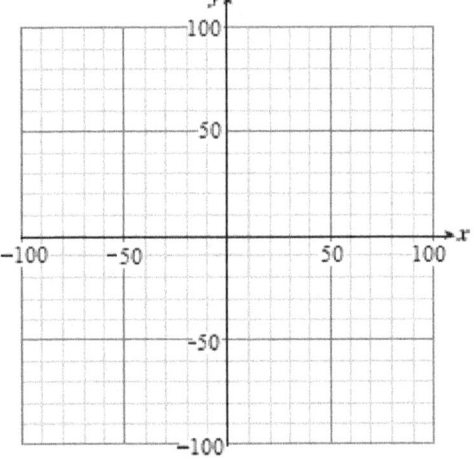

■ For Problems 7-12, match each equation with its graph. (More than one equation may describe the same graph.)

7. $2x + 3y = 12$

8. $2x - 3y = 12$

9. $3x - 2y = 12$

10. $-3y - 2x = 12$

11. $\dfrac{x}{6} - \dfrac{y}{4} = 1$

12. $\dfrac{x}{4} - \dfrac{y}{6} = 1$

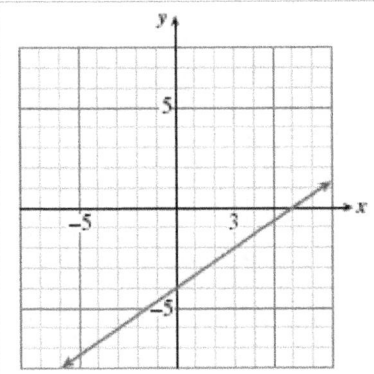

a.

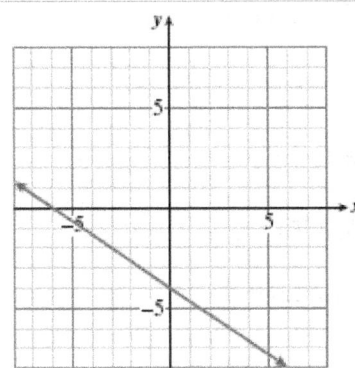

b.

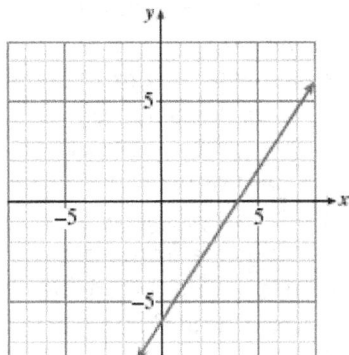

c.

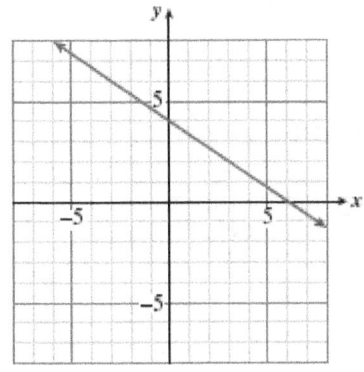

d.

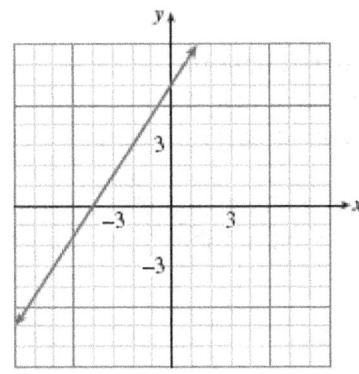

e.

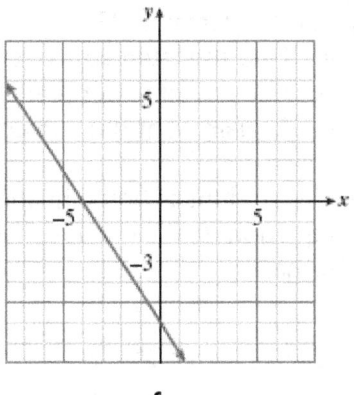

f.

■ For Problems 13-18, solve each pair of equations. In part (b) of each problem, your answer will involve the constant k.

13. a. $-2x = 6$
   b. $-2x = k$

14. a. $x + 4 = 7$
   b. $x + k = 7$

15. a. $x - 5 = 9$
   b. $x - 5 = k$

16. a. $2x + 3 = 8$
   b. $2x + k = 8$

17. a. $15 - 4x = 3$
   b. $15 - 4x = k$

18. a. $9 + 3x = -1$
   b. $9 + kx = -1$

Applications

◼ Use the graph to answer the questions.

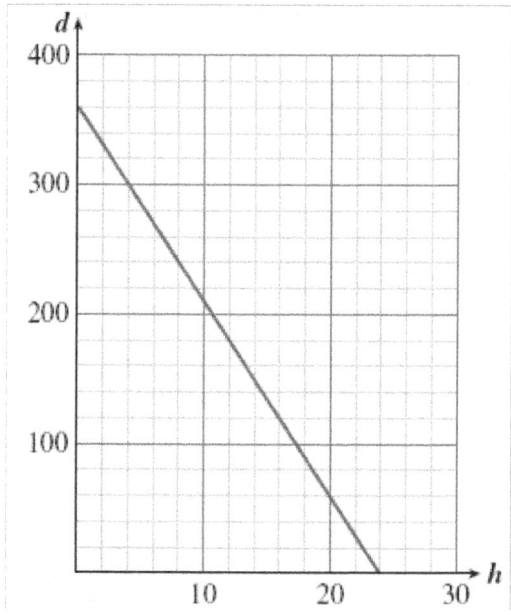

19. During spring break, Francine took the train to San Francisco and then bicycled home. The graph below left shows Francine's distance d from home, in miles, after cycling for h hours.
    a. How far is it from San Francisco to Francine's home?
    b. How many hours did Francine cycle to get home?
    c. After cycling for 12 hours, how far is Francine from home?
    d. How far did Francine cycle in the first four hours?

◼ For Problems 20-22,
    a. Find the intercepts of each linear equation.
    b. Use the intercept method to graph the line.
    c. Explain what the intercepts mean in terms of the problem situation.

20. The amount of home heating oil (in gallons) in the Olsons' tank is given by the equation $G = 200 - 15w$, where w is the number of weeks since they turned on the furnace.

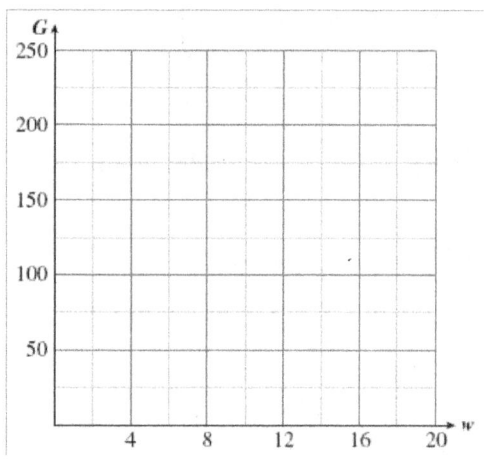

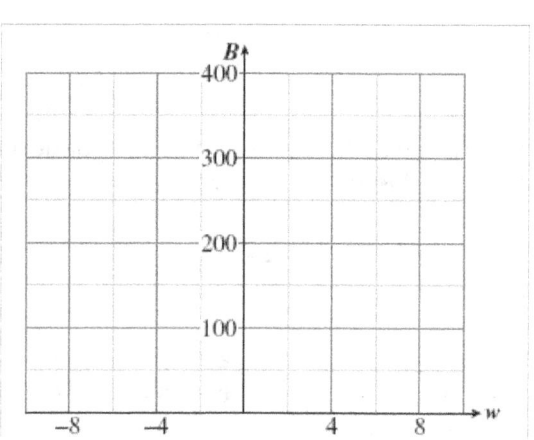

21. Dana joined a savings plan some weeks ago. Her bank balance is growing each week according to the formula $B = 225 + 25w$, where $w = 0$ represents this week.

22. Delbert bought some equipment and went into the dog-grooming business. His profit is increasing according to the equation $P = -600 + 40d$, where d is the number of dogs he has groomed.

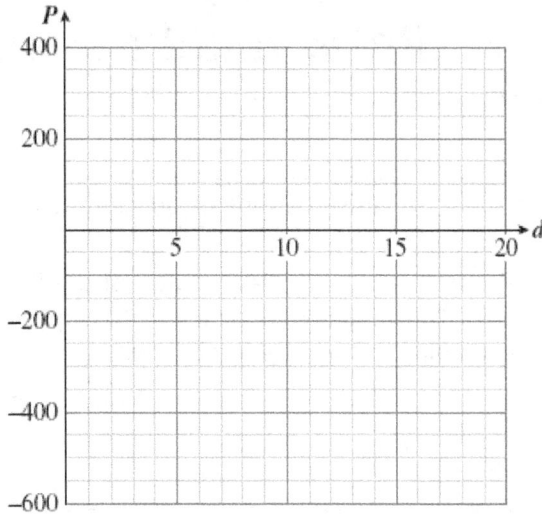

23. a. At what point does the graph of $2x - 3y = 25$ cross the x-axis?

b. At what point does the graph of $1.4x + 3.6y = -18$ cross the y-axis?

24. The x-intercept of a line is positive and its y-intercept is negative. Is the line increasing or decreasing? Sketch a possible example of such a line.

3.2 Ratio and Proportion

Ratios

> A **ratio** is a type of quotient used to compare two numerical quantities.

For example, suppose the ratio of pupils to computers at a local elementary school is 10 to 3. This means that for every 10 pupils in the school there are 3 computers. We often display a ratio as a fraction, like this:

$$\frac{\text{Number of pupils}}{\text{Number of computers}} = \frac{10}{3}$$

> **Caution!**   This ratio does *not* mean that there are exactly 10 pupils and 3 computers at the school. There could be 40 pupils and 12 computers or 200 pupils and 60 computers (or many other combinations) because
>
> $$\frac{40}{12} = \frac{200}{60} = \frac{10}{3}$$ ■

> ■ **Example 1**   In a survey of 100 of his classmates, Greg found that 68 students supported a strict gun control law, while 32 students did not. What is the ratio of survey respondents who support gun control to those opposed?
>
> **Solution** The ratio of those supporting gun control to those opposed is 68 to 32, or $\frac{68}{32}$. This ratio reduces to $\frac{17}{8}$, or 17 to 8. ■

> **Caution!**   When we write a ratio, it is important to notice which number is the numerator, and which is the denominator. When we write "the ratio of a to b" as a fraction, a is the numerator and b is the denominator. ■

In Example 1, we could also compute the ratio of opponents to supporters of the gun control law. In that case, the answer would be $\frac{32}{68}$ or $\frac{8}{17}$, the reciprocal of the answer to Example 1.

> **Ratio**
>   A **ratio** is a type of quotient used to compare two numerical quantities. The **ratio of a to b** is written $\frac{a}{b}$.

Look Closer: A ratio can be expressed as a decimal instead of a common fraction. Because $\frac{17}{8}$ is equal to 2.125, we can write the ratio in Example 1 as 2.125 to 1. We usually express a ratio as a decimal fraction if the numbers being compared are decimal numbers. ●

Example 2   A small circuit board measures 4.2 centimeters long by 2.4 centimeters wide. What is the ratio of its length to its width?

Solution The ratio of length to width is $\dfrac{4.2 \text{ cm}}{2.4 \text{ cm}}$, or 1.75, as you can verify with your calculator. (Can you express this ratio as a common fraction?)

RQ1. What is a ratio?
RQ2. When do we express a ratio as a decimal fraction?

Rates

A **rate** is a ratio that compares two quantities with different units.

You are already familiar with many rates. If you say that your average speed was 50 miles per hour, or that apples cost 89 cents per pound, you are using rates. Because *per* usually indicates division,

$$50 \text{ miles per hour} \qquad \text{means} \qquad \frac{50 \text{ miles}}{1 \text{ hour}}$$

$$89 \text{ cents per pound} \qquad \text{means} \qquad \frac{89 \text{ cents}}{1 \text{ pound}}$$

Example 3   Sarah traveled 390 miles on 15 gallons of gas. Express her rate of fuel consumption as a ratio, and then as a rate.

Solution   As a ratio, Sarah's rate of fuel consumption was $\dfrac{390 \text{ miles}}{15 \text{ gallons}}$. We simplify by dividing the denominator into the numerator, to get $\dfrac{26 \text{ miles}}{1 \text{ gallon}}$, or 26 miles per gallon.

Proportions

A **proportion** is a statement that two ratios are equal.

Thus, a proportion is an equation in which both sides are ratios. Here are some proportions:

$$\frac{6}{9} = \frac{8}{12} \qquad \text{and} \qquad \frac{3.2}{8} = \frac{1}{2.5}$$

A proportion may involve variables. How can we solve a proportion such as this one?

$$\frac{7}{5} = \frac{x}{6}$$

Because x is divided by 6, we could multiply both sides of the equation by 6 to clear the fraction. But we can clear both fractions at the same time if we multiply by the LCD (Lowest Common Denominator) of the two fractions. In this case, the LCD is $5 \cdot 6 = 30$. Multiplying both sides by 30 gives us

$$30\left(\frac{7}{5}\right) = \left(\frac{x}{6}\right)30$$
$$42 = 5x$$

Now we can divide both sides by 5 to get the answer,

$$x = \frac{42}{5} = 8.4$$

Look Ahead: There is a short-cut we can use that avoids calculating an LCD. We can obtain the equation $42 = 5x$ by *cross-multiplying.*

$$\frac{7}{5} = \frac{x}{6}$$
$$42 = 5x$$

Then we complete the solution as before.

Property of Proportions
   We can clear the fractions from a proportion by **cross-multiplying.**

   If $\frac{a}{b} = \frac{c}{d}$,   then $ad = bc$.

■ **Example 4**   Solve $\frac{2.4}{1.5} = \frac{8.4}{x}$

Solution We apply the property of proportions and cross-multiply to get

$$2.4x = 1.5(8.4) \qquad \text{\textbf{Divide both sides by 2.4.}}$$
$$x = \frac{1.5(8.4)}{2.4} \qquad \text{\textbf{Simplify the right side.}}$$
$$x = 5.25$$

We can check the solution by substituting $x = 5.25$ into the original proportion.   ■

Caution!   Cross-mutiplying works **only** for solving proportions! Do not try to use cross-multiplying on other types of equations, or for other operations on fractions. In particular, do not use cross-multiplying when multiplying fractions or adding fractions.   ■

RQ3.   What is a rate?
RQ4.   What is a proportion?
RQ5.   What is the short-cut for solving a proportion called?

Proportional Variables

Suppose that grape juice costs 80 cents per quart, and complete the table.

Number of Quarts	Total Price (cents)	$\dfrac{\text{Total Price}}{\text{Number of Quarts}}$
1	80	$\dfrac{80}{1} = 80$
2	160	$\dfrac{160}{2} =$
3	240	
4	320	

Did you find that the ratio $\dfrac{\text{total price}}{\text{number of quarts}}$, or *price per quart*, is the same for each row of the table? This agrees with common sense: The *price per quart* of grape juice is the same no matter how many quarts you buy.

> Two variables are **proportional** if their ratio is always the same.

■ Example 5 One quart of grape juice costs 80 cents. The price of grape juice is proportional to the number of quarts you buy. How much grape juice can you buy for $10.00?

Solution Let x represent the number of quarts you can buy for $10.00, or 1000 cents. Because the variables *total price* and *number of quarts* are proportional, we know that their ratio is constant. Thus, the ratio

$$\dfrac{1000 \text{ cents}}{x \text{ quarts}} \quad \text{is equal to the ratio} \quad \dfrac{80 \text{ cents}}{1 \text{ quart}}$$

$$\dfrac{1000}{x} = \dfrac{80}{1}$$

We solve the proportion by cross-multiplying to get

$$1000\,(1) = 80x \qquad \textbf{Divide both sides by 80.}$$

$$x = \dfrac{1000}{80} = 12.5$$

You can buy 12.5 quarts of grape juice for $10.00. (Or, if you can only buy quart bottles, then you can buy 12 quarts and have 40 cents left over.) ■

Caution! When writing a proportion, we must be careful that *both* ratios have the *same units* in their numerators, and the same units in their denominators. In the example above, it would *not* be correct to equate $\dfrac{1000}{x}$ and $\dfrac{1}{80}$, because the ratios do not have the same units. ■

For the grape juice example, the ratio $\dfrac{\text{total price}}{\text{number of quarts}}$ was always equal to 80. The number 80 is called the **constant of proportionality**.

Look Closer: If we let P stand for the price of the grape juice and q stand for the number of quarts purchased, we have the equation

$$\frac{P}{q} = 80$$

or $P = 80q$

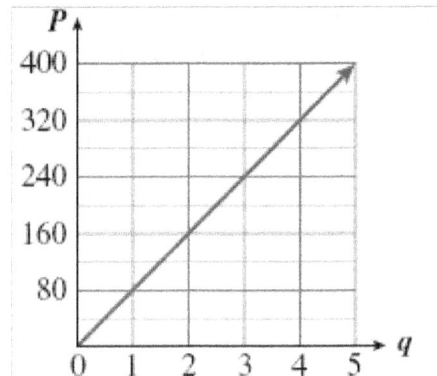

The graph of this equation is a straight line that passes through the origin, as shown at right.

Proportional Variables
If two variables are proportional, they are related by the equation

$$y = kx$$

where k is the **constant of proportionality**.

RQ6.   What are proportional variables?
RQ7.   If x and y are proportional, what equation do they satisfy?

Skills Warm-Up
Convert each common fraction to a decimal, and each decimal to a common fraction.

1. $\dfrac{3}{8}$ **2.** $\dfrac{7}{5}$ **3.** $\dfrac{4}{3}$ **4.** $\dfrac{5}{12}$

5. 0.04   **6.** $0.\overline{6}$ **7.** 1.875   **8.** 2.2

Answers: 1. 0.375   **2.** 1.4   **3.** $1.\overline{3}$ **4.** $0.41\overline{6}$ **5.** $\dfrac{1}{25}$ **6.** $\dfrac{2}{3}$ **7.** $\dfrac{15}{8}$ **8.** $\dfrac{11}{5}$

Homework 3.2

Skills Practice

■ For Problems 1-4, write ratios or rates.

1. In a survey of 300 employees at a large company, 125 used public transportation to commute to work. What is the ratio of employees who use public transportation to those who do not?

2. Erica tutored in the math lab for 6.5 hours last week and made $56.68. What is Erica's rate of pay in dollars per hour?

3. The instructions for mixing a rose fertilizer call for $\frac{3}{4}$ cup of potash and $1\frac{3}{8}$ cups of nitrogen. What is the ratio of nitrogen to potash?

4. An orange contains 0.14 milligram of thiamin and 0.6 milligram of niacin. What is the ratio of niacin to thiamin?

5. Which of the following expressions are proportions?

   a. $\dfrac{3}{x} = \dfrac{8}{15}$
   b. $\dfrac{v}{9} - \dfrac{12}{5}$
   c. $\dfrac{w}{6} + \dfrac{2}{5} = \dfrac{3}{2}$
   d. $\dfrac{2.6}{m} = \dfrac{m}{1.6}$

■ For Problems 6-9, solve the proportion.

6. $\dfrac{x}{16} = \dfrac{9}{24}$
7. $\dfrac{182}{65} = \dfrac{21}{w}$
8. $\dfrac{7}{b} = \dfrac{5}{9}$
9. $\dfrac{7/3}{3} = \dfrac{z}{9/2}$

■ For Problems 10-12, decide whether the two variables are proportional.

10.

Time	Distance
1	45
2	90
4	180
5	225

11.

Length	Area
3	9
4	16
8	64
10	100

12.

Rate	Time
20	40
40	20
50	16
80	10

■ For Problems 13-16, decide whether the two quantities are proportional.

13. A 9-inch pizza costs $6, and a 12-inch pizza costs $10.
14. A commuter train travels 10 miles in 20 minutes, and 15 miles in 30 minutes.
15. It takes $4\frac{1}{2}$ cups of flour to make two loaves of bread, and 18 cups to make eight loaves.
16. An 18-foot sailboat sleeps four, and a 32-foot sailboat sleeps six.

Applications

■ For Problems 17-22,
   a. Write a proportion.
   b. Solve the proportion and answer the question.

17. If 3 pounds of coffee makes 225 cups, how many pounds of coffee is needed to make 3000 cups of coffee?

18.   Gunther's car uses 32 liters of gas to travel 184 kilometers. How many liters will he need to travel 575 kilometers?

19.   On a map of Fairfield County, $\frac{3}{4}$ centimeter represents a distance of 10 kilometers. If Eastlake and Kenwood are 6 centimeters apart on the map, what is the actual distance between the two towns?

20.   In a survey of 1200 voters, 863 favored the construction of a light rail system. If 8000 people vote, about how many will vote for the light rail system?

21.   If 1 inch equals 2.54 centimeters, what is the length in inches of a wire 35 centimeters long?

22.   A cinnamon bread recipe calls for $1\frac{1}{4}$ tablespoons of cinnamon and 5 cups of flour. How much cinnamon would be needed with 8 cups of flour?

▨ For Problems 23-24,
   a.   Find the constant of proportionality and write an equation relating the variables.
   b.   Make a table of values and graph the equation.
   c.   Use your equation to answer the question.

23.   Megan's wages, w, are proportional to the time, h, she works. Megan made $60 for 8 hours work as a lab assistant. How long will it take her to make $500?

h	w

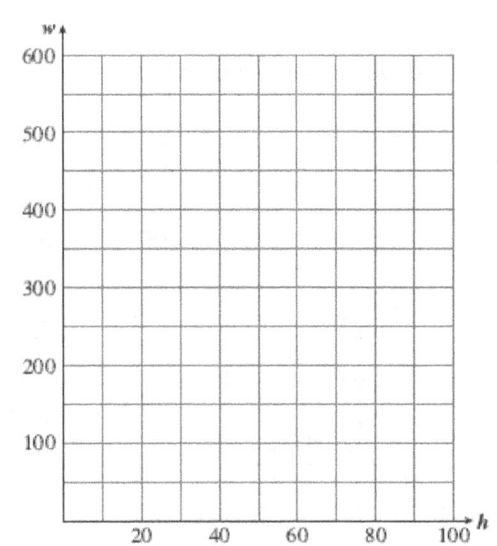

24.   The amount of pure copper, c, produced by Copperfield Mine is proportional to the amount of ore extracted, g. A new lode produced 24 grams of copper from 800 grams of ore. How many grams of copper will one kilogram of ore yield?

g	c

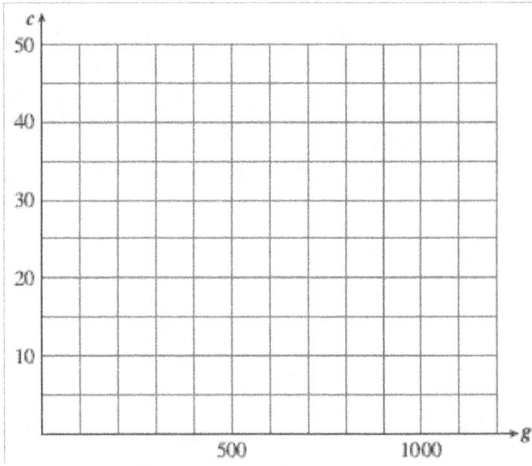

25. Trish's real-estate commission is proportional to the amount of the sale. She made $3200 commission on a sale of $80,000.
    a. Write an equation relating Trish's sales and her commission.
    b. What would her commission be on the sale of a $200,000 property?

26. Everett can bicycle 16 miles in 2 hours and 24 miles in 3 hours. The distance d that Everett travels is proportional to the time t that he bicycles.
    a. Write an equation for d in terms of t.
    b. Graph your equation.
    c. Complete the tables, and use them to answer the question below.

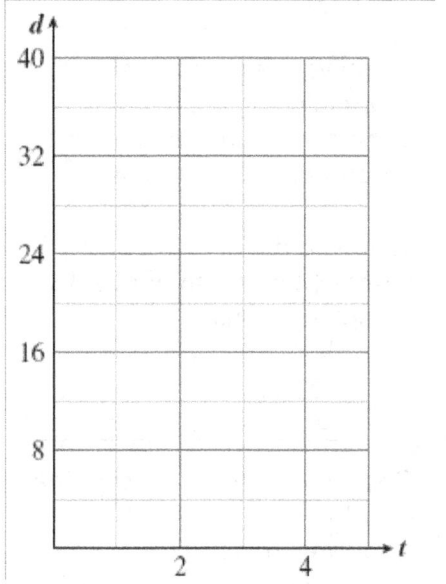

t	d		t	d
2			4	
3			6	
5			10	

What happens to the distance Everett travels when he doubles the time he bicycles?

27. The sales tax on $12 is 72 cents, and the sales tax on $15 is 90 cents. The tax T is proportional to the price P.
    a. Write an equation for T in terms of P.
    b. Graph your equation.
    c. Complete the tables, and use them to answer the question below.

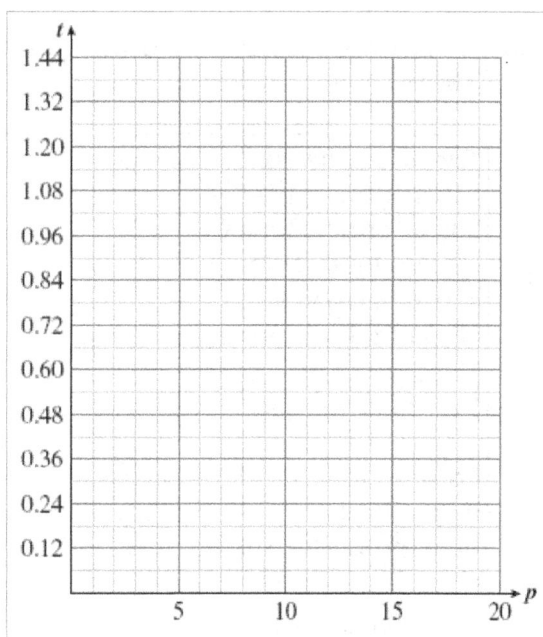

P	T		P	T
12			6	
20			10	
30			15	

What happens to the sales tax when the price is cut in half?

28. The cost of tuition at Walden College is given by the formula $T = 500 + 40u,$ where u is the number of units you take.
    a. Fill in the tables.

Units	Tuition		Units	Tuition
3			6	
5			10	
8			16	

    b. Does doubling the number of units you take double your tuition?
    c. Is T proportional to u?

29. a. State a formula for the circumference of a circle in terms of its radius.

b. Complete the table.

r	2	5	7	10
C				

c. Graph your equation, using the values in your table.

d. Is the circumference of a circle proportional to its radius?

30. a. State a formula for the area of a square in terms of the length of its side.

b. Complete the table.

s	2	5	6	8
A				

c. Graph your equation, using the values in your table.

d. Is the area of a square proportional to the length of its side?

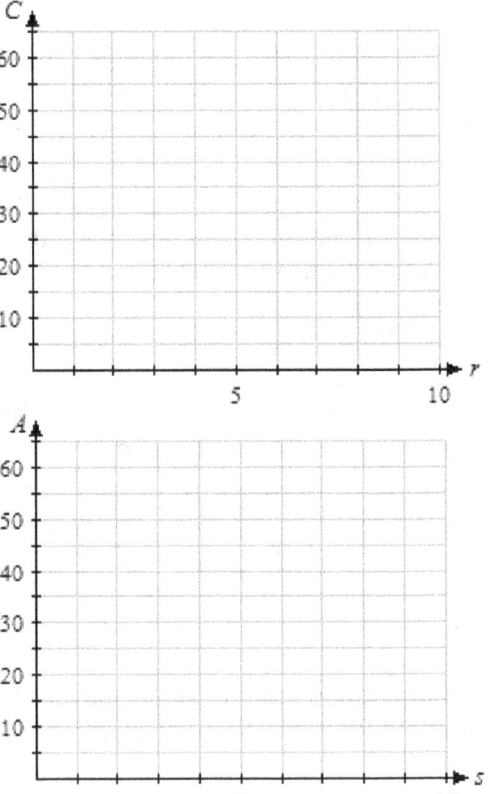

31. The following recipe for Cadet Mess Sloppy Joes comes from the West Point Officers Wives' Club Cookbook.

For 1000 servings:
310 lb. ground beef
100 lb. ground turkey
$5\frac{1}{2}$ cans mushrooms
4 lb. green peppers, sliced
18 cans tomato puree
12 cans tomato catsup
1 qt. cider vinegar

2 cans diced red peppers
20 lb. onions, diced
3 lb. chili powder
1 lb. garlic powder
$3\frac{1}{2}$ lb. sugar
1 qt. light flavor glow
28 oz. salt and pepper

Cook off turkey and beef in oven; place in pots. Braise off chopped peppers and diced onions (turkey will absorb liquid). Add stock to turkey to moisten well before adding other products. Stir together. Heat till piping hot. Serve on rolls.

---

a. What is the ratio of tomato puree to tomato catsup? How many servings of sloppy joes can you make if you have only 5 cans of tomato catsup? How many cans of tomato puree will you need?

b. How much chili powder (in ounces) will you need to serve sloppy joes to 50 people? How many pounds of ground beef?

c. The cookbook also lists the ingredients needed for 4500 servings. How many cans of mushrooms are required for 4500 servings? How many pounds of onions?

3.3 Slope

Rate of Change

> A **rate of change** is a type of ratio that measures how one variable changes with respect to another.

■ **Example 1** In order to fire a particular kind of pottery, the pieces must first be cured by raising the temperature slowly and evenly. Sonia checks the temperature in the drying oven at ten-minute intervals, and records the following data.

Time, x	0	10	20	30	40	50	60
Temperature, y	70	74	78	82	86	90	94

The heat in the oven should not increase any faster than 0.5 degree per minute. Is the temperature in the oven within the safe limits?

Solution A graph of the data is shown at right. Sonia calculates the rate at which the temperature is rising by finding the following ratio:

$$\frac{\text{change in temperature}}{\text{change in time}}$$

For example, over the first 10 minutes, the temperature rises from 70 degrees to 74 degrees, so

$$\frac{\text{change in temperature}}{\text{change in time}} = \frac{4 \text{ degrees}}{10 \text{ minutes}}$$

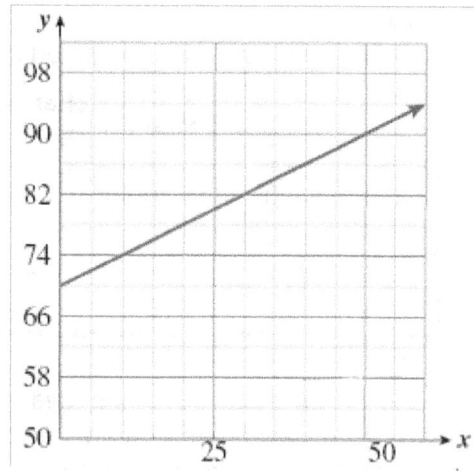

or 0.4 degree per minute. This is less than the maximum rate recommended for curing the pottery. ■

You can check that over each ten-minute interval the temperature again rises by four degrees, so it appears that the oven is heating up at an acceptable rate.

> **RQ1.** What is a rate of change?
> **RQ2.** What are the units of the rate of change in Example 1?

Slope

We introduce some new notation to use when calculating a rate of change.

> The Greek letter Δ ("delta") is used in mathematics to indicate **change**.

In Example 1, we used the variable x to represent time and y to represent the temperature, so we denote the ratio $\dfrac{\text{change in temperature}}{\text{change in time}}$ by $\dfrac{\Delta y}{\Delta x}$. With this notation, we calculate the rate of change of temperature between the data points $(20, 78)$ and $(50, 90)$ as follows:

$$\frac{\Delta y}{\Delta x} = \frac{12 \text{ degrees}}{30 \text{ minutes}} = 0.4 \text{ degrees per minute}$$

We can illustrate the rate of change on a graph of the data, as shown at right. We move from the point $(20, 78)$ to the point $(50, 90)$ by moving horizontally a distance of $\Delta x = 30$ and then vertically a distance of $\Delta y = 12$.

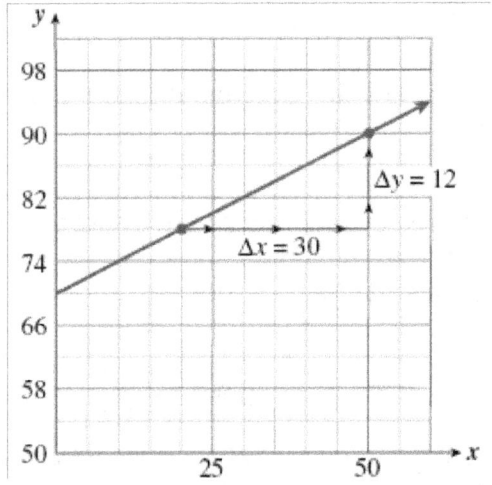

The rate of change of one variable with respect to another is so important in applications that the ratio $\dfrac{\Delta y}{\Delta x}$ is given a name; it is called **slope**, and is usually denoted by the letter m.

Slope

The **slope** of a line is defined by the ratio

$$\text{slope} = \frac{\text{change in } y\text{-coordinate}}{\text{change in } x\text{-coordinate}}$$

as we move from one point to another on the line. In symbols,

$$m = \frac{\Delta y}{\Delta x}$$

Look Closer: The slope of a line measures how fast the y-coordinate changes as we increase the x-coordinate of points on the line. More specifically, when we move one unit in the x-direction, how many units should we move in the y-direction to get back to the line?   ●

Example 2   Use the points A and B to compute the slope of the line shown.

Solution   The point A has coordinates $(1, 3)$, and B has coordinates $(5, 6)$. As we move along the line from $A(1, 3)$ to $B(5, 6)$, the y-coordinate changes by 3 units, and the x-coordinate changes by 4 units. The slope of the line is thus

$$\frac{\Delta y}{\Delta x} = \frac{\text{change in } y\text{-coordinate}}{\text{change in } x\text{-coordinate}} = \frac{3}{4}$$

The slope tells us that if we start at any point on the line and move 1 unit in the x-direction, we must move $\frac{3}{4}$ unit in the y-direction to return to the line.

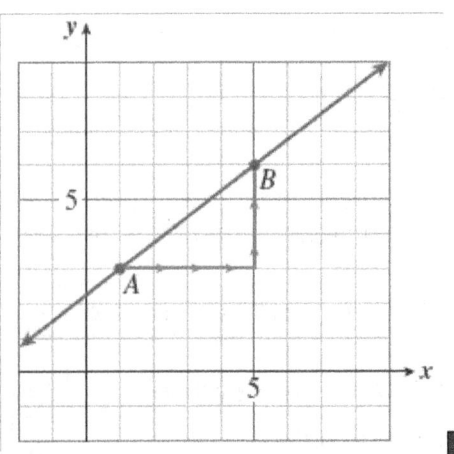

Caution!   Note the difference between the statements $y = 3$ and $\Delta y = 3$; they are not the same.   When we discuss a graph, $y = 3$ means that the y-coordinate of a particular point is 3, but $\Delta y = 3$ means that the y-coordinate changes by 3 units when we move from one point to another. ◼

> **RQ3.**   What does Δ mean in mathematics?
> **RQ4.**   How do we indicate Δx on a graph?
> **RQ5.**   What is the name of the ratio $\dfrac{\Delta y}{\Delta x}$, and what letter is used to represent it?

Meaning of Slope

In Example 1, we graphed the temperature of a pottery oven over time. We calculated the slope of the graph as

$$\frac{\Delta y}{\Delta x} = 0.4 \text{ degrees per minute}$$

The slope gives us the rate of change of the temperature with respect to time: the temperature is increasing at a rate of 0.4 degrees per minute.

> The **slope** of a line measures the **rate of change** of y with respect to x.

In different situations, this rate might be interpreted as a rate of growth or a speed. The slope of a graph can give us valuable information about the variables involved.

◼ **Example 3**   The graph shows the distance traveled by a driver for a cross-country trucking firm in terms of the number of hours she has been on the road.

a. Compute the slope of the graph.

b. What is the meaning of the slope for this problem?

Solutions a.   Choose any two points on the line, say $G(2, 100)$ and $H(4, 200)$ shown in the figure. As we move from G to H we find

$$m = \frac{\Delta D}{\Delta t} = \frac{100}{2} = 50$$

The slope of the line is 50.

b. The best way to understand the slope is to include units in the calculation.

$$\frac{\Delta D}{\Delta t} \text{ means } \frac{\text{change in distance}}{\text{change in time}} \quad \text{or} \quad \frac{\Delta D}{\Delta t} = \frac{100 \text{ miles}}{2 \text{ hours}} = 50 \text{ miles per hour}$$

The slope represents the trucker's average speed or velocity. ◼

Caution!   In Example 3, we refer to a point by a capital letter and the coordinates of the point, like this: $H(4, 200)$. This means that $t = 4$ and $D = 200$ at the point H. Do not confuse the coordinates (t, D) of a particular point with the values of Δt and ΔD obtained by moving from one point to a second point. ◼

> **RQ6.**   What does the slope of a line measure?
> **RQ7.**   What does the slope of the line measure in Example 3?

Geometrical Meaning of Slope

Suppose we graph two lines with positive slope on the same coordinate system. If we move along the lines from left to right, then the line with the larger slope will be steeper. This makes sense if we think of the slope as a rate of change: The line whose y-coordinate is increasing faster with respect to x is the steeper line.

Look Closer: You can verify the slope given for each line in Figure (a) by computing $\frac{\Delta y}{\Delta x}$. For each unit you increase in the x-direction, the steepest line increases 2 units in the y-direction, the middle line increases 1 unit in the y-direction, and the flattest line increases only $\frac{1}{3}$ unit.

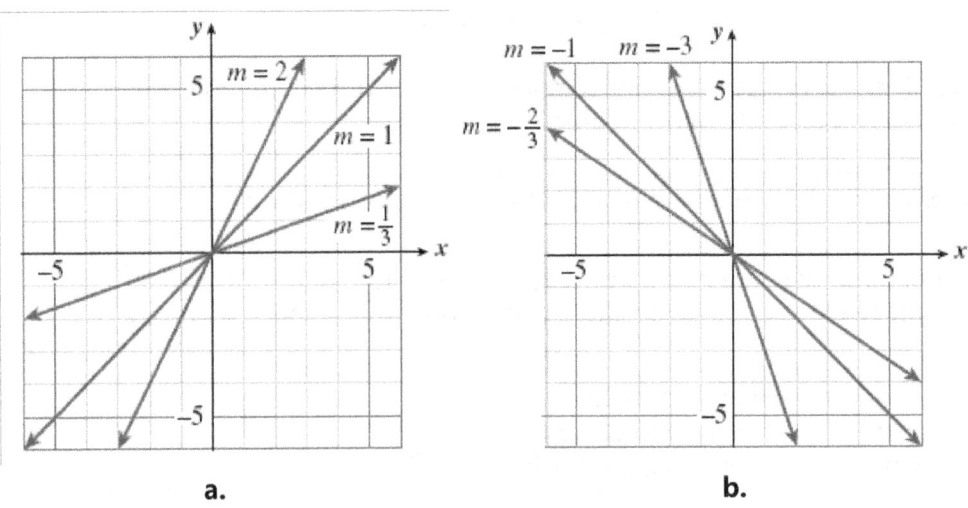

a.   b.

Figure (b) shows several lines with negative slopes. These lines slant downwards or decrease as we move from left to right. The more negative the slope, the more sharply the line decreases. For both increasing and decreasing graphs, the larger the absolute value of the slope, the steeper the graph.

Caution!   Slopes measure the relative steepness of two lines only if they are graphed on axes with the same scales. Changing the scale on either the x-axis or the y-axis can greatly alter the appearance of a graph.

> **RQ8.**   What sort of lines have negative slopes?

Skills Warm-Up

▓ Write a rate for each of the following situations, including units.

1. Zack's average speed, if he drove 426 miles in 9 hours.
2. Zelda's average speed, if she ran 6.6 miles in 55 minutes.
3. The rate at which water flows through a pipe, if a 400-gallon storage tank fills in 20 minutes.
4. A baby whale's rate of growth, if it gains 3000 pounds in its first 40 days of life.
5. Earnest's rate of pay, if he earns $344 for a 40-hour week.
6. Meg's rate of pay, if she charges $90 to type a 40-page paper.

Answers: **1.** $47.\overline{3}$ miles per hour   **2.** 0.12 miles per minute   **3.** 20 gallons per minute
4. 75 pounds per day   **5.** 8.60 dollars per hour   **6.** 2.25 dollars per page

Homework 3.3

Skills Practice

For Problems 1-2, find the slope of each line segment.

1a. 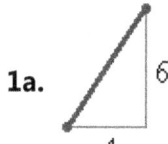   b. 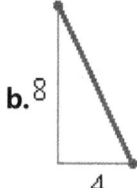   2a.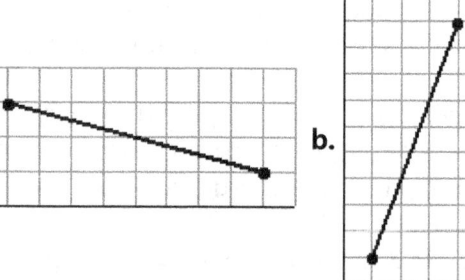

3. Choose two points from the table and compute the slope of the line.

x	0	2	6	8
y	−30	0	60	90

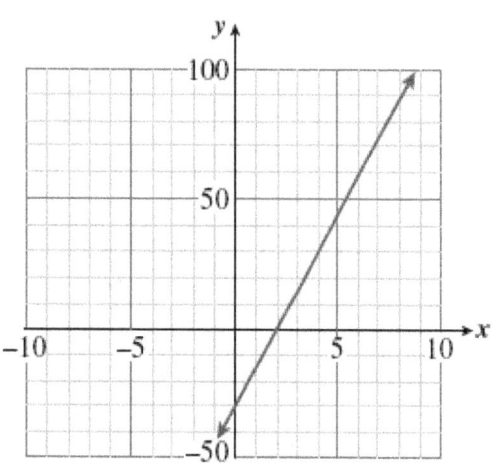

4. Graph the line and compute its slope:
$y = -12x + 32$

x	−2	0	3	4
y				

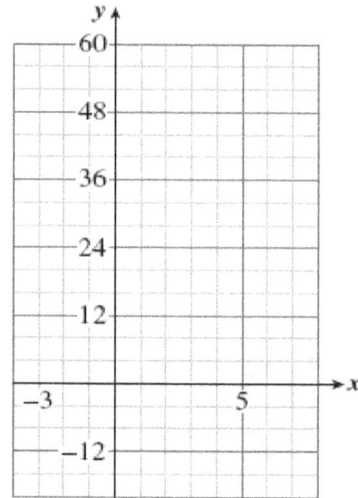

For Problems 5-6, find the slope of the line. Illustrate Δx and Δy on the graph.

5. $x + 2y = 6$

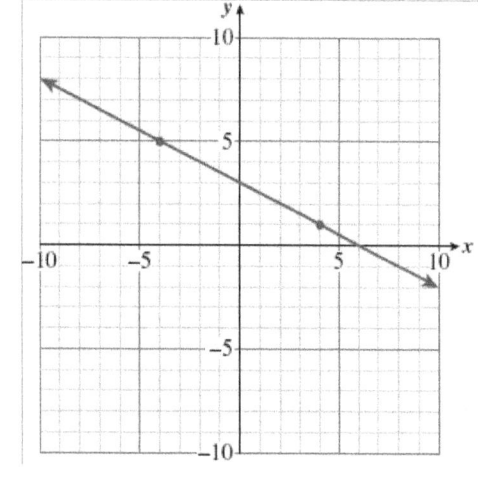

6. $3x - 2y = 0$

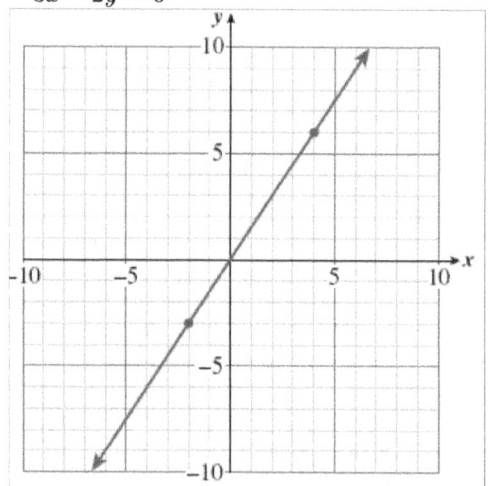

For Problems 7-10,
   a. Find the intercepts of each line.
   b. Graph the line on the grid provided. Use the intercept method.
   c. Use the intercepts to calculate the slope of the line.
   d. Calculate the slope again using the suggested points on the line.

7. $2x + 3y = 12$
        $(-3, 6)$ and $(3, 2)$

8. $5x - 2y = 10$
        $(-2, -10)$ and $(4, 5)$

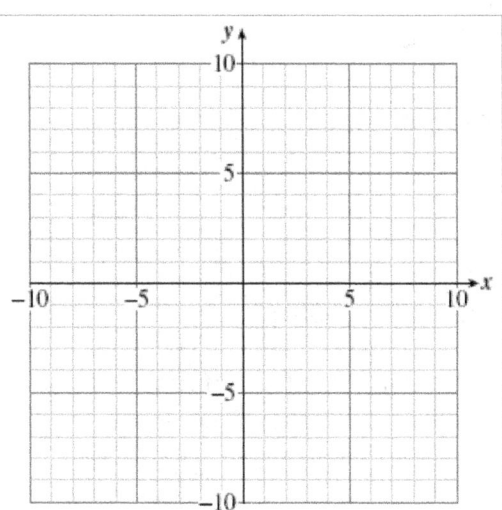
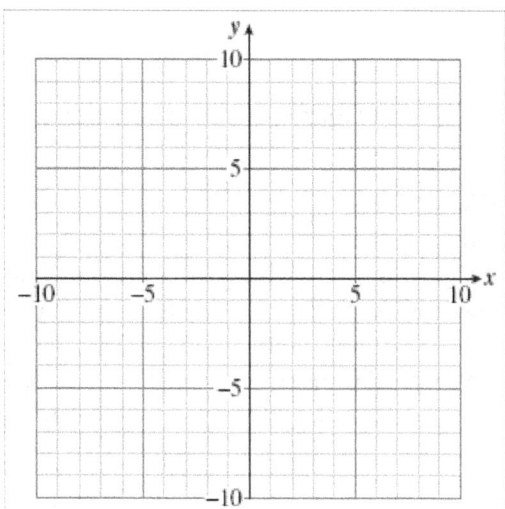

9. $x + y = 5$
        $(-3, 8)$ and $(8, -3)$

10. $x - 2y = 4$
        $(6, 1)$ and $(-4, -4)$

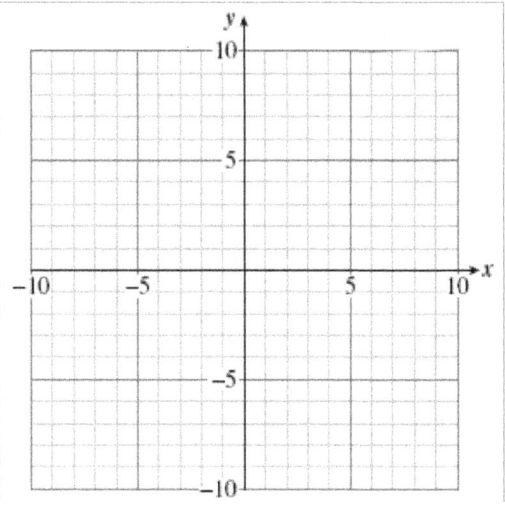
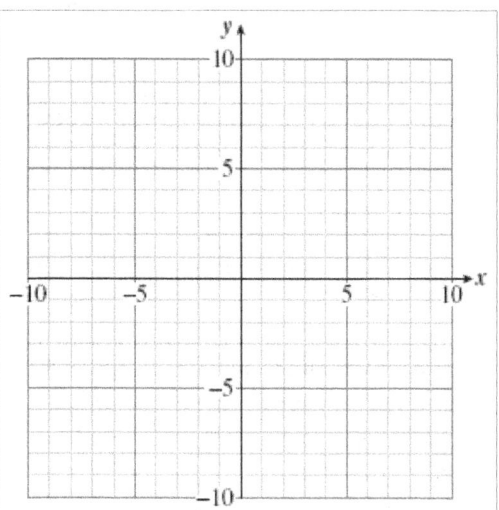

11.   A line contains the points $(0, 0)$ and $(3, 2)$. What is its slope?
12.   A line contains the points $(0, 0)$ and $(-30, 50)$. What is its slope?
13.   Which line is steeper: one with slope $\frac{3}{5}$ or one with slope $\frac{5}{3}$?
14.   Which line is decreasing: one with slope $\frac{1}{4}$ or one with slope -2?

15. The line shown in Figure (a) below has slope $\frac{5}{2}$. If $\Delta x = 7$, find Δy.

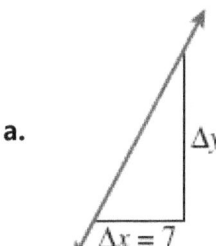

a.

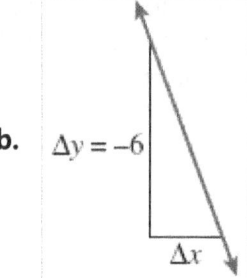

b.

16.   The line shown in Figure (b) above has slope -4. If $\Delta y = -6$, find Δx.

Applications

■ For Problems 17-18,
  a. Compute the slope of the graph, including units.
  b. Interpret the slope as a rate; what does it tell you about the problem?

17. Audrey can drive 150 miles on 6 gallons of gas, and 225 miles on 9 gallons of gas. Write an equation for the distance, d, that Audrey can drive on g gallons of gas.

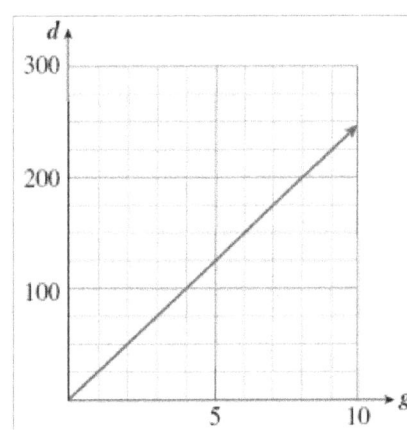
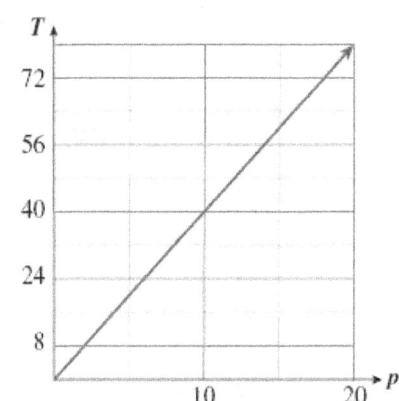

18. The sales tax on a $15 purchase is 60 cents, and 80 cents on a $20 purchase. Write an equation for the tax T on a purchase of p dollars.

19. Lynette is saving money for the down payment on a new car. The figure at right shows the amount A she has saved, in dollars, w weeks after the first of the year.
  a. How much does Lynette save each week?
  b. Give the coordinates (w, A) of any two points on the graph. Use those coordinates to compute the slope of the graph, $\dfrac{\Delta A}{\Delta w}$.
  c. What are the units of the slope? What does the slope tell you about the problem?

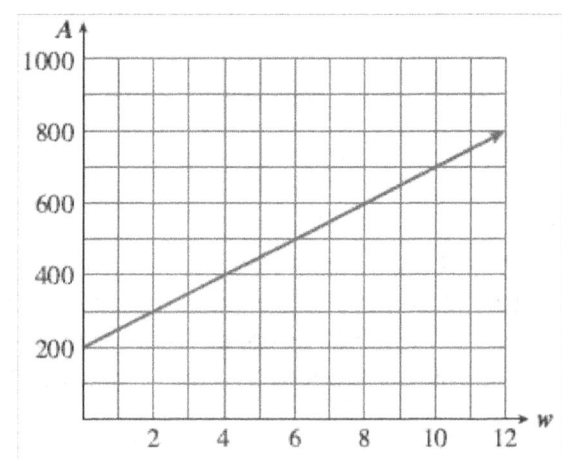

20. Jason is raising a rabbit for the county fair. The figure at right shows the rabbit's weight W when it was t weeks old.

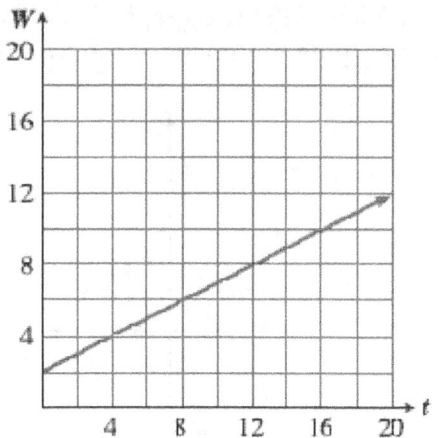

  a. How much did the rabbit's weight increase from the fourth week to the twelfth week? From the second week to the eighth week?

  b. Compute the rabbit's rate of growth, including units.

  c. Illustrate the rate of growth, $\dfrac{\Delta W}{\Delta t}$, on the graph.

For Problems 21-22, draw and label a sketch for the situation. Use the definition of slope to answer the questions.

21. A sign on the highway says "6% grade, next 3 miles." This means that the slope of the road ahead is $\frac{6}{100}$. How much will you climb in elevation (in feet) over the next 3 miles?

22. A wheelchair ramp must have a slope of 0.125. If the ramp must reach a door whose base is 2 feet off the ground, how far from the building should the base of the ramp be placed?

23. a. Calculate the slope of the line in Figure (a) below.
  b. Explain why $\Delta y = 0$ for any two points on the line.
  c. Explain why the slope of any horizontal line is zero.

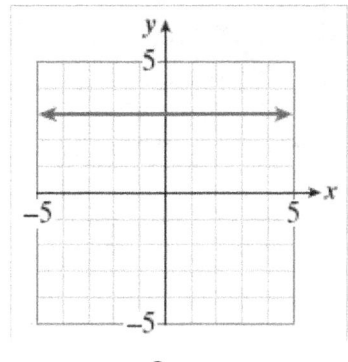

a.

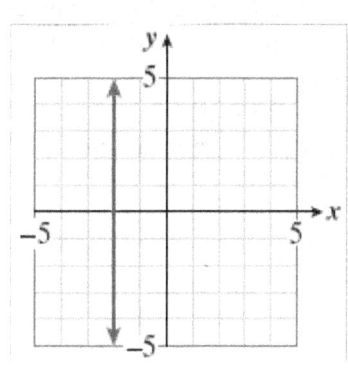

b.

24. a. Calculate the slope of the line in Figure (b) above.
  b. Explain why $\Delta x = 0$ for any two points on the line.
  c. Explain why the slope of any vertical line is undefined.

25. a. Which of the two graphs in the figure below appears steeper?
b. Compute the slopes of the two graphs. Which has the greater slope?

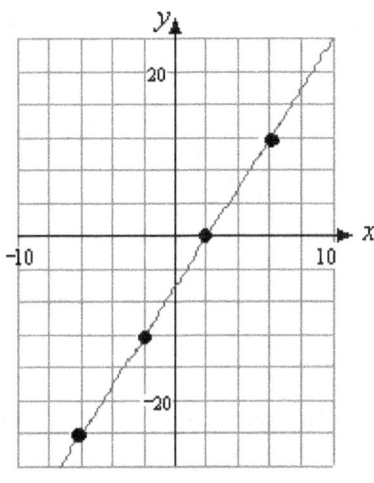

I

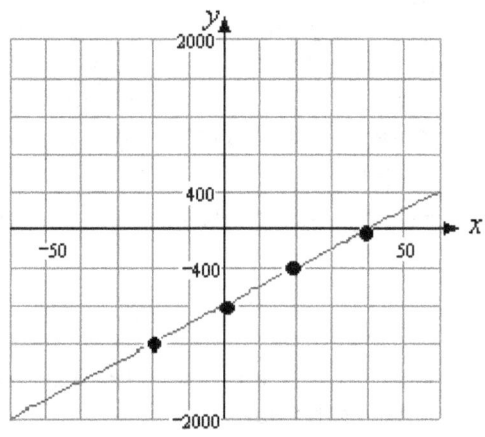

II

26. Kira buys granola in bulk at the health food store. There are two standard containers customers can use. The size of each container in ounces and its price in dollars are shown in the graph.

a. Read the coordinates of the two points shown on the graph.

b. Calculate the slope of the graph, including units. What does the slope tell us about the granola?

c. Extend the graph to include 25 ounces of granola. How much taller must you make the vertical axis?

d. Extend the graph to include $9.00 worth of granola. How far must you extend the horizontal axis?

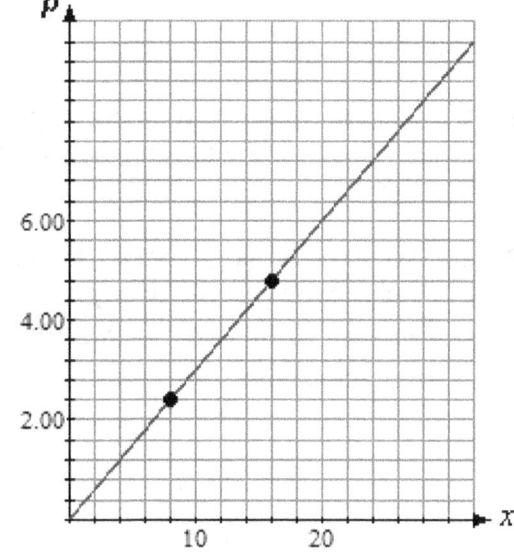

3.4 Slope-Intercept Form

Writing a Linear Equation

In Lesson 3.3, we plotted data for the temperature inside an oven used to cure pottery.

Time, x	0	10	20	30	40	50	60
Temperature, y	70	74	78	82	86	90	94

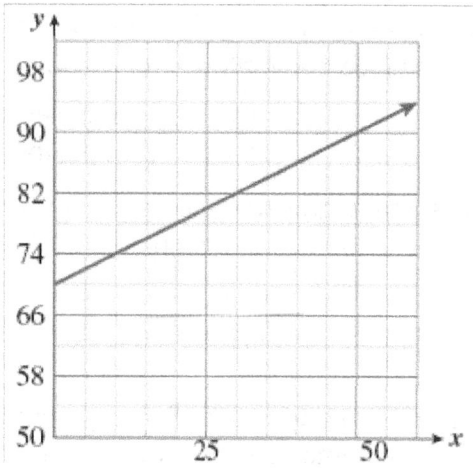

We computed the slope of the graph, shown in the figure. The slope is

$$\frac{\text{change in temperature}}{\text{change in time}} = \frac{\Delta y}{\Delta x} = \frac{4 \text{ degrees}}{10 \text{ minutes}}$$

or $m = 0.4$ degrees per minute.

You can see from either the table or the graph that the y-intercept is the point $(0, 70)$. This means that the initial temperature inside the drying oven was 70 degrees.

Look Ahead: If we know two pieces of information about a line: its slope and its initial value or y- intercept, we can write its equation.

■ **Example 1**   Write an equation for the temperature, H, inside the pottery drying oven t minutes after the oven is turned on.

Solution   The initial temperature in the oven was 70 degrees, so $H = 70$ when $t = 0$. The temperature rose at a rate of 0.4 degrees per minute, so we add 0.4 degrees to H for each minute that passes. After t minutes, we have added $0.4t$ degrees, giving us a temperature of

$$H = 70 + 0.4t$$

We can also write the equation as $H = 0.4t + 70$. ■

RQ1.   What does the constant term in an equation tell us about the graph?
RQ2.   What does the coefficient of the input variable tell us?

From Example 1, we see that the coefficients of a linear equation tell us something about its graph.

---

1. The constant term tells us the vertical intercept of the graph.
2. The coefficient of the input variable tells us the slope of the graph.

Slope-Intercept Form
     A linear equation written in the form

$$y = mx + b$$

is said to be in **slope-intercept** form. The coefficient m is the **slope** of the graph, and b is the y-**intercept**.

> **RQ3.**   What does the y-intercept tell us in Example 1?
> **RQ4.**   What is the slope-intercept form of an equation?

Slope-Intercept Method of Graphing

A linear equation has the form $Ax + By = C$, and its graph is a straight line. We have already studied two methods for graphing linear equations:

> 1. Make a table of values and plot points
> 2. Find and plot the intercepts (the intercept method)

There is a third graphing method that makes use of the slope of the line. We can use the slope-intercept form to sketch a graph quickly, without having to plot a lot of points.

■ **Example 2**   Graph the equation $y = \dfrac{3}{4}x - 2$

Solution   The slope of the line is $\frac{3}{4}$ and its y-intercept is the point $(0, -2)$. We begin by plotting the y-intercept, as shown in the figure. Next, we use the slope to find another point on the line. The slope,

$$m = \frac{\Delta y}{\Delta x} = \frac{3}{4}$$

gives the ratio of the change in y-coordinate to the change in x-coordinate as we move from any point on the line to another. Thus, starting at the point $(0, -2)$, we move:

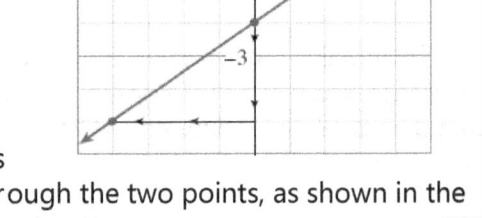

> **3 units up** (the positive y-direction),
> then **4 units right** (the positive x-direction)

to locate another point on the line. The coordinates of this new point are $(4, 1)$. Finally, we draw a line through the two points, as shown in the figure. ■

Look Closer: To improve the accuracy of the graph in Example 2, we can find a third point on the line by writing the slope in an equivalent form. We change the sign of both numerator and denominator of the slope to get

$$m = \frac{\Delta y}{\Delta x} = \frac{-3}{-4}$$

Starting again from the y-intercept $(0, -2)$, we now move 3 units down and 4 units left, and find the point $(-4, -5)$ on the graph.

> **RQ5.**   Name two methods of graphing that we have already studied.
> **RQ6.**   When using the slope-intercept method, what is the first point we plot?
> **RQ7.**   How do we use the slope to find a second point on the line?

The **slope-intercept method** can be used to graph any non-vertical line.

> **To Graph a Line Using the Slope-Intercept Method**
> 1. Write the equation in the form $y = mx + b$.
> 2. Plot the y-intercept, $(0, b)$.
> 3. Write the slope as a fraction, $m = \dfrac{\Delta y}{\Delta x}$.
> 4. Use the slope to find a second point on the graph:
>   Starting at the y-intercept, move Δy units in the y-direction,
>   then Δx units in the x-direction.
> 5. Find a third point by moving $-\Delta y$ units in the y-direction,
>   then $-\Delta x$ units in the x-direction, starting from the y-intercept.
> 6. Draw a line through the three plotted points.

Finding the Slope-Intercept Form

Not all linear equations appear in slope-intercept form. However, we can write the equation of any non-vertical line in slope-intercept form by solving the equation for y in terms of x.

■ Example 3 Find the slope and y-intercept of the graph of $3x - 4y = 8$

Solution To write the equation in slope-intercept form, we solve for y in terms of x.

$$3x - 4y = 8 \qquad \text{Subtract } 3x \text{ from both sides.}$$
$$-4y = -3x + 8 \qquad \text{Divide both sides by } -4.$$
$$\frac{-4y}{-4} = \frac{-3x + 8}{-4} \qquad \text{Divide each term of the right side by } -4.$$
$$y = \frac{-3x}{-4} + \frac{8}{-4} \qquad \text{Simplify each quotient.}$$
$$y = \frac{3}{4}x - 2$$

The equation is now in slope-intercept form, with $m = \frac{3}{4}$ and $b = -2$. Thus, the slope of the graph is $\frac{3}{4}$ and the y-intercept is the point $(0, -2)$. ■

Caution!   Do not confuse solving for y with finding the y-intercept. In Example 3, we do **not** set $x = 0$ before solving for y. When we find the y-intercept, we are looking for a specific point, namely, the point with x-coordinate zero, so we replace x by 0. When we "solve for y," we are writing the equation in another form, so both variables, x and y, still appear in the equation. ■

RQ8. How do we put an equation into slope-intercept form?

Skills Warm-Up

■ Solve for the indicated variable.

1. $2q + p = 10$ for p 2. $2l + 2w = 18$ for l
3. $3a + 9 = -6b$ for a 4. $2c = 2d + 22$ for d
5. $5r - 4s = 24$ for s 6. $2m = 11 - 3n$ for n

Answers: **1.** $10 - 2q$ **2.** $9 - w$ **3.** $-3 - 2b$ **4.** $c - 11$ **5.** $\dfrac{5r - 24}{4}$ **6.** $\dfrac{11 - 2m}{3}$

Homework 3.4

Skills Practice

■ For Problems 1-4,
   a. Write the equation in slope-intercept form.
   b. State the slope and y-intercept of the graph.

1. $y = 3x + 4$ **2.** $6x + 3y = 5$
3. $2x - 3y = 6$ **4.** $5x = 4y$

■ For Problems 5-8,
   a. Find the slope and the y-intercept of the line.
   b. Write an equation for the line.

5.

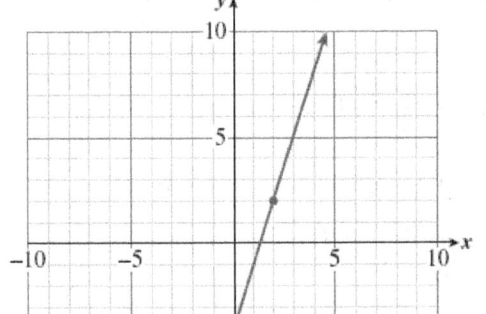

6.

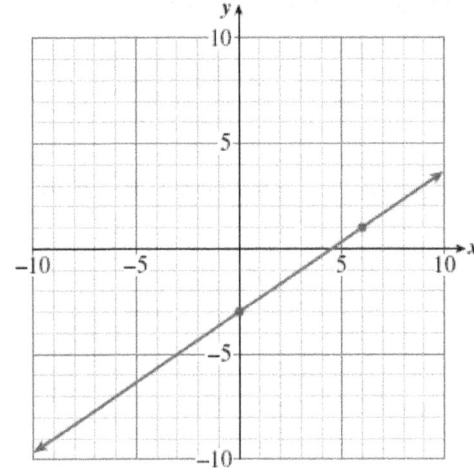

7.

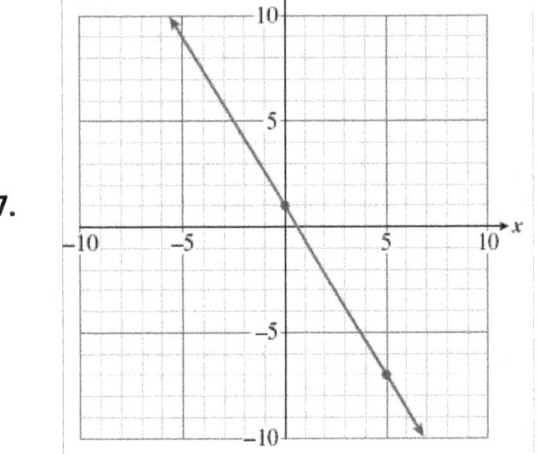

8.

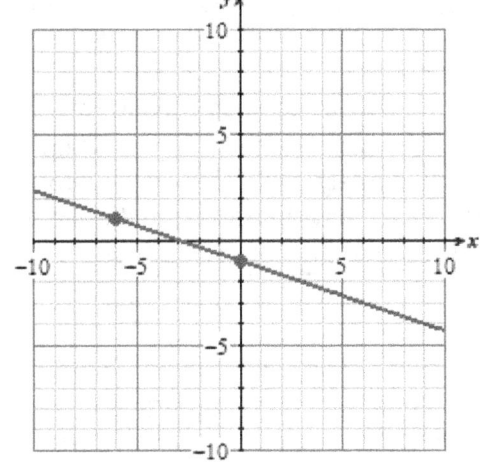

■ For Problems 9-12,
   a. Fill in the y-values in the tables and graph the lines.
   b. Choose two points on each line and compute its slope.
   c. What is the y-intercept of each line?

9. I. $y = 2x - 6$

x	-1	0	1	2	3
y					

II. $y = 2x + 1$

x	-1	0	1	2	3
y					

III. $y = 2x + 3$

x	-1	0	1	2	3
y					

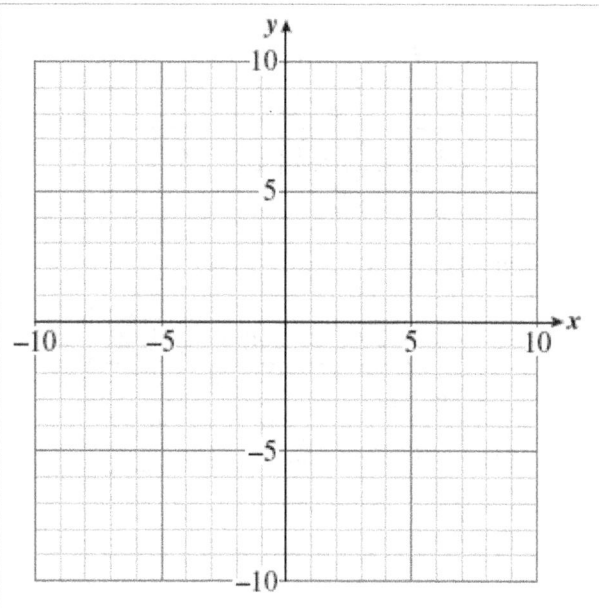

All three lines have the same _____.

10. I. $y = \dfrac{-3}{2}x - 4$

x	-6	-4	-2	0	2
y					

II. $y = \dfrac{-3}{2}x + 2$

x	-6	-4	-2	0	2
y					

III. $y = \dfrac{-3}{2}x + 6$

x	-6	-4	-2	0	2
y					

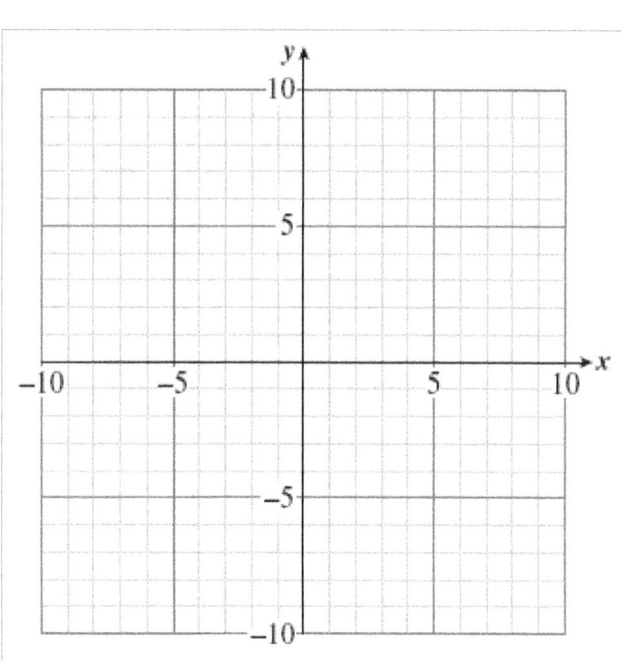

All three lines have the same _____.

11. I. $y = \frac{1}{4}x + 2$

x	-4	-2	0	2	4
y					

II. $y = \frac{1}{2}x + 2$

x	-4	-2	0	2	4
y					

III. $y = x + 2$

x	-4	-2	0	2	4
y					

All three lines have the same _____.

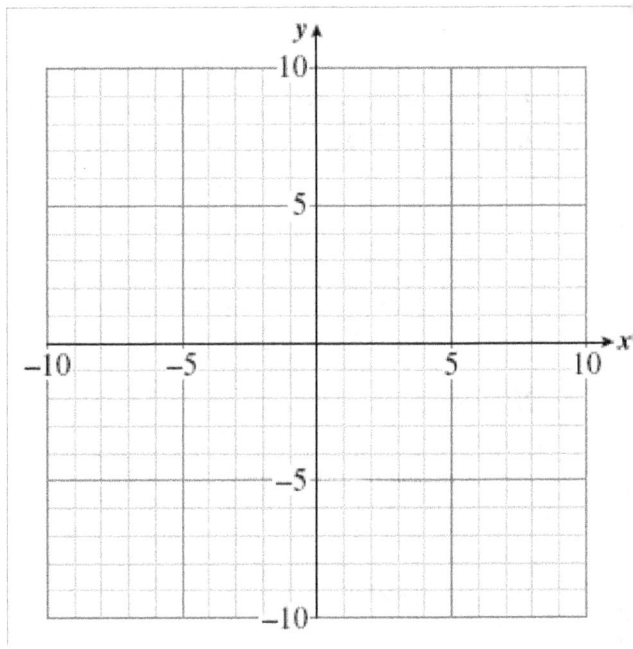

12. I. $y = -3x - 2$

x	-6	-3	0	3	6
y					

II. $y = -2x - 2$

x	-6	-3	0	3	6
y					

III. $y = \frac{-5}{3}x - 2$

x	-6	-3	0	3	6
y					

All three lines have the same _____.

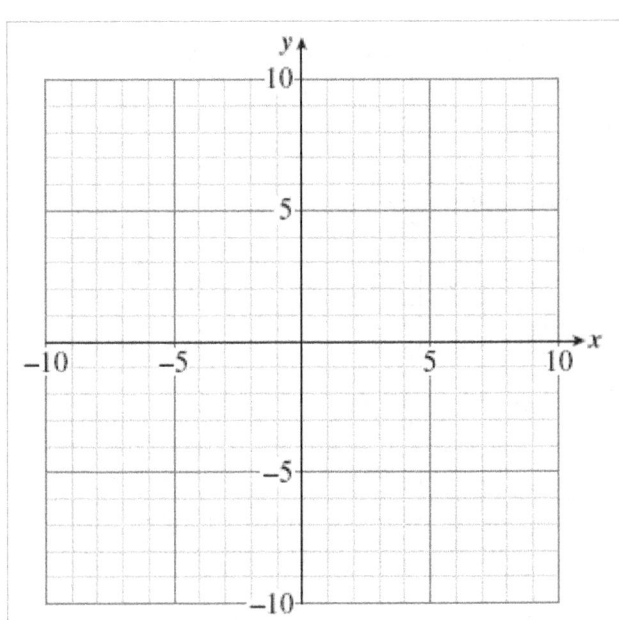

For Problems 13-14,
   a. Find the intercepts of the graph and graph the line.
   b. Compute the slope of the line.
   c. Put the equation in slope-intercept form.

13. $3x + 4y = 12$

x	0	
y		0

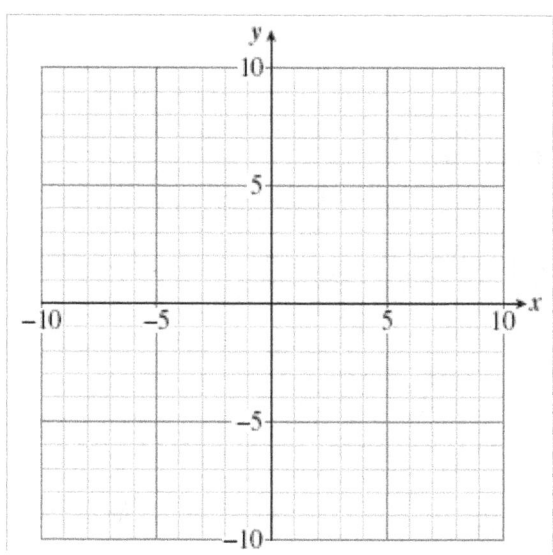

14. $y + 3x - 8 = 0$

x	0	
y		0

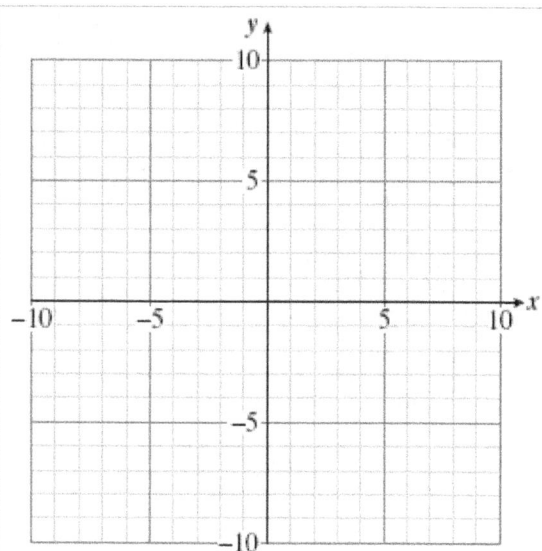

In Problems 15-16,
   a. Put the equation in slope-intercept form.
   b. What is the y-intercept of each line? What is its slope?
   c. Use the slope to find two more points on the line.
   d. Graph the line.

15. $3x - 5y = 0$

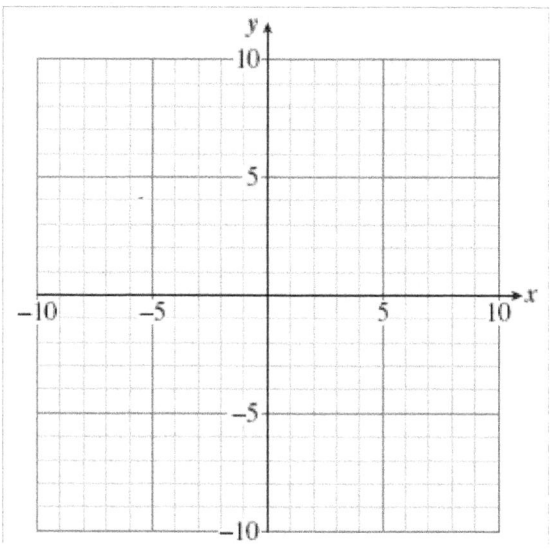

16. $5x + 4y = 0$

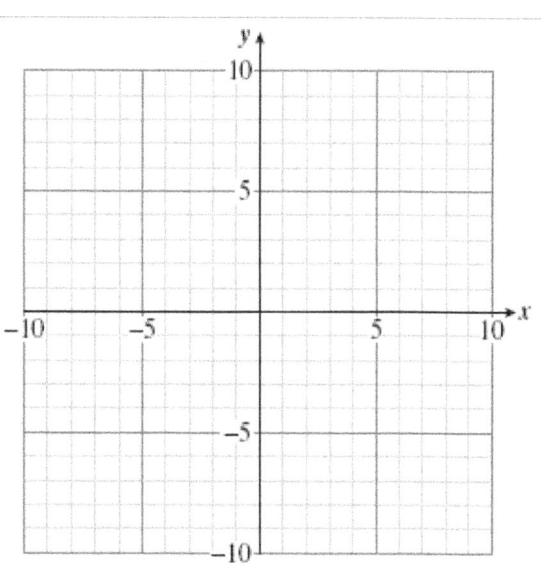

Applications

■ For Problems 17-20, graph the equation by using the slope-intercept method.

17. $y = 3 - x$

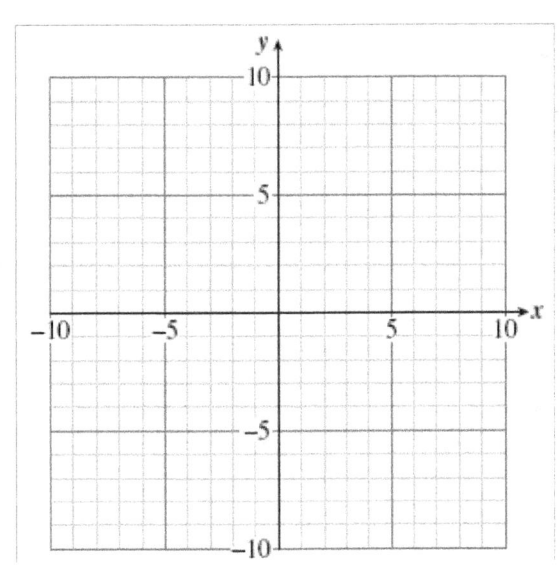

18. $y = 3x - 1$

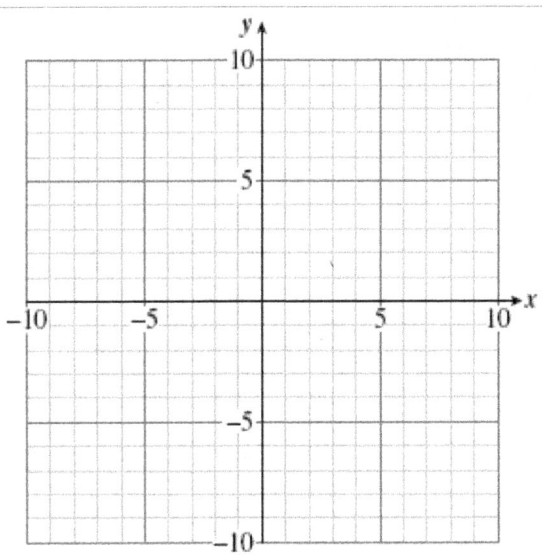

19. $y = \dfrac{3}{4}x + 2$

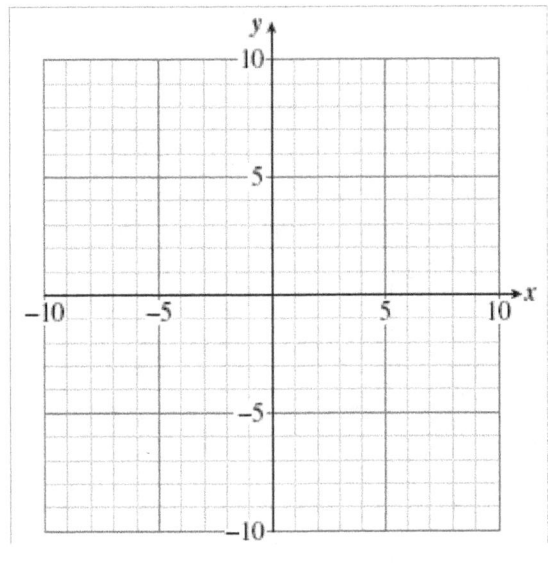

20. $y = -2 - \dfrac{4}{3}x$

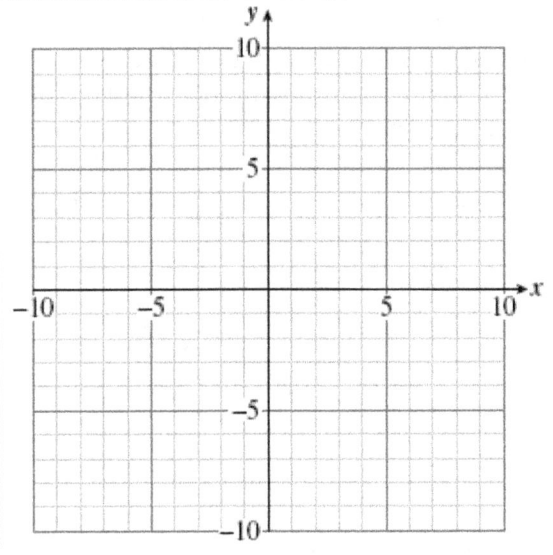

21. Robin opened a yogurt smoothie shop near campus. The graph shows Robin's profit P after selling s smoothies.

  a. What is the P-intercept of the line?
  b. Calculate the slope of the line.
  c. Write an equation for the line in slope-intercept form.
  d. What do the slope and the P-intercept tell us about the problem?

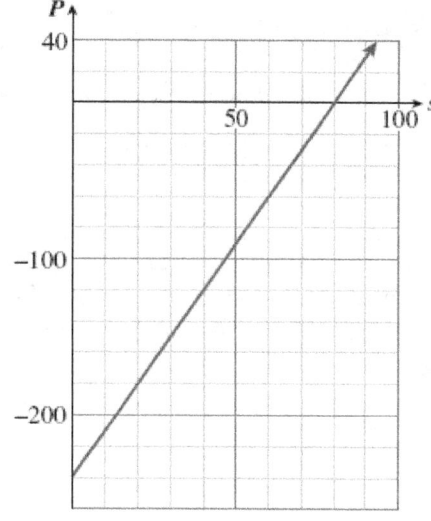

For Problems 22-24,
  a. Graph the line by the slope-intercept method.
  b. Explain what the slope and the vertical intercept tell us about the problem.

22. Serda's score on her driving test is computed by the equation $S = 120 - 4n$, where n is the number of wrong answers she gives.

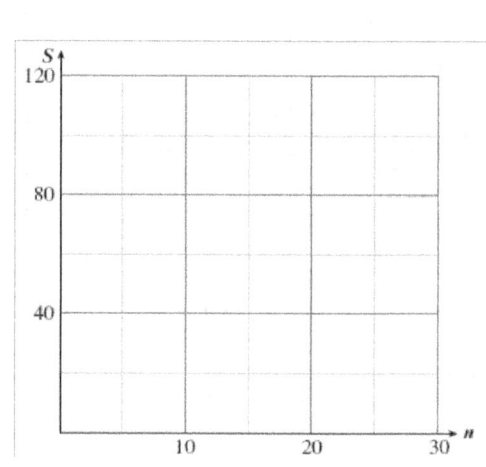

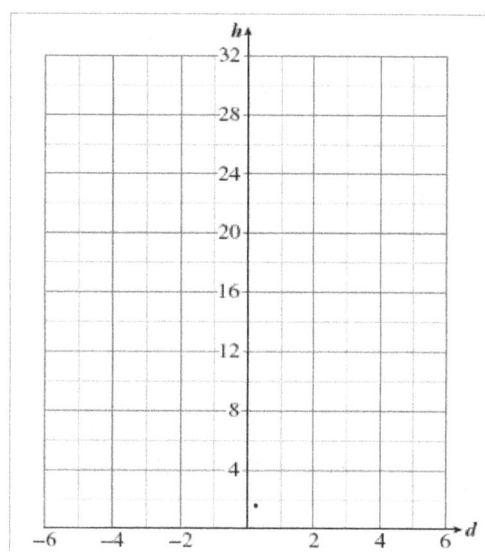

23. Greg is monitoring the growth of a new variety of string beans. The height of the vine each day is given in inches by $h = 18 + 3d$, where $d = 0$ represents today.

24. Cliff's score was negative at the end of the first round of College Quiz, but in Double Quiz his score improved according the equation $S = -400 + 20q$, where q is the number of questions he answered.

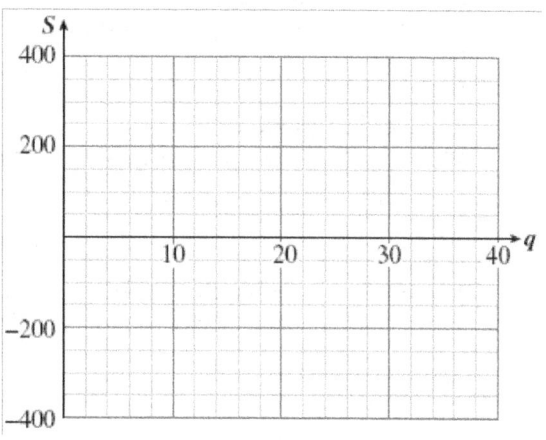

3.5 Properties of Lines

Parallel and Perpendicular Lines

Lines that lie in the same plane but never intersect are called **parallel lines**. It is easy to understand that parallel lines have the same slope.

Exercise Calculate the slopes of the two parallel lines in the figure. (We denote the slope of the first line by m_1 and the slope of the second line by m_2.)

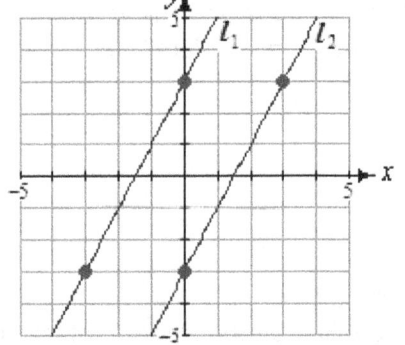

$$m_1 = \frac{\Delta y}{\Delta x} =$$

$$m_2 = \frac{\Delta y}{\Delta x} =$$

Lines that intersect at right angles are called **perpendicular lines**. It is a little harder to see the relationship between the slopes of perpendicular lines.

Exercise Calculate the slopes of the two perpendicular lines in the figure.

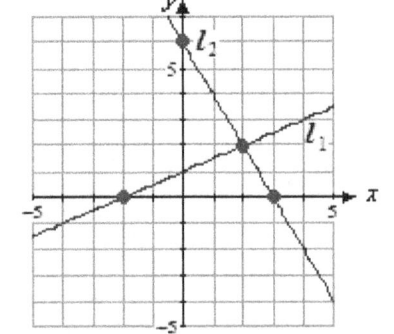

$$m_1 = \frac{\Delta y}{\Delta x} =$$

$$m_2 = \frac{\Delta y}{\Delta x} =$$

In the Exercise above, note that the product of m_1 and m_2 is -1, that is,

$$m_1 m_2 = \frac{1}{2}(-2) = -1$$

This relationship holds for any pair of perpendicular lines.

Parallel and Perpendicular Lines

1. Two lines are **parallel** if their slopes are equal, that is, if

$$m_1 = m_2$$

   or if both lines are vertical.

2. Two lines are **perpendicular** if the product of their slopes is -1, that is, if

$$m_1 m_2 = -1$$

   or if one of the lines is horizontal and one is vertical.

■ **Example 1**   Decide whether the lines

$$2x + 3y = 6 \qquad \text{and} \qquad 3x - 2y = 6$$

are parallel, perpendicular, or neither.

Solution We could graph the lines, but we can't be sure from a graph if the lines are exactly parallel or exactly perpendicular. A more accurate way to answer the question is to find the slope of each line. To do this we write each equation in slope-intercept form, that is, we solve for y.

$$2x + 3y = 6 \qquad\qquad\qquad\qquad 3x - 2y = 6$$
$$3y = -2x + 6 \qquad\qquad\qquad\qquad -2y = -3x + 6$$
$$y = \frac{-2}{3}x + 2 \qquad\qquad\qquad\qquad y = \frac{3}{2}x - 3$$

The slope of the first line is $m_1 = \frac{-2}{3}$, and the slope of the second line is $m_2 = \frac{3}{2}$. The slopes are not equal, so the lines are not parallel. However, the product of the slopes is

$$m_1 m_2 = \left(\frac{-2}{3}\right)\left(\frac{2}{3}\right) = -1$$

so the lines are perpendicular. ■

Look Closer: Another way to state the condition for perpendicular lines is

$$m_2 = \frac{-1}{m_1}$$

Because of this relationship, we often say that the slope of one perpendicular line is the **negative reciprocal** of the other.   ●

Equations for Horizontal and Vertical Lines

In Lesson 3.3 we learned that the slope of a horizontal line is zero. What does this tell us about the equation of a horizontal line that passes through a particular point?

■ **Example 2**   Find the equation of the horizontal line that passes through $(5, 3)$.

Solution Looking at the graph of the line shown at right, we see that the y-coordinate of every point on the line is 3. In particular, the y-intercept of the line is the point $(0, 3)$, so $b = 3$. As we noted above, the slope of the line is $m = 0$. Substituting these values into the slope-intercept form gives us the equation $y = 0 \cdot x + 3$, or just $y = 3$. The fact that x does

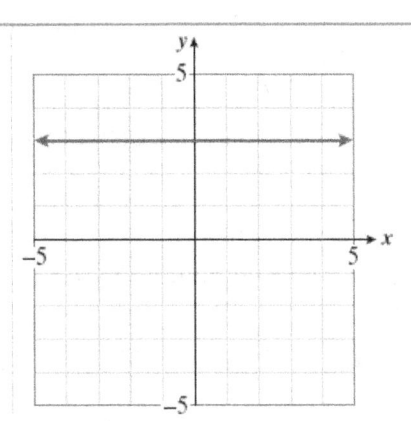

not appear in the equation means that $y = 3$ for every point on the line, no matter what the value of x is. ■

Look Closer: What about the equation of a vertical line? The slope of a vertical line is undefined; a vertical line does not have a slope. We cannot use the slope-intercept form to write the equation of a vertical line. However, we can use what we learned about horizontal lines in Example 2.

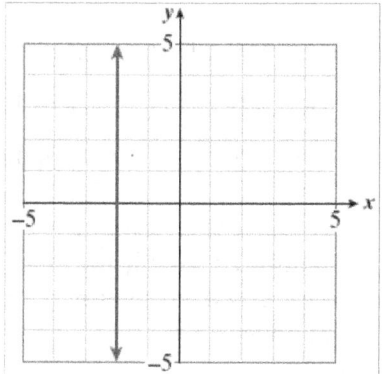

Look at the graph of the vertical line at right. Every point on the line has x-coordinate -2, no matter what the y-coordinate is. An equation for this line is $x = -2$. The value of x does not depend upon y; it is constant, so y does not appear in the equation. ●

The y-intercept of the horizontal line $y = 3$ is $(0, 3)$; it has no x-intercept. The x-intercept of the vertical line $x = -2$ is $(-2, 0)$; it has no y-intercept.

Horizontal and Vertical Lines
1. The equation of the **horizontal line** passing through $(0, b)$ is
$$y = b$$
2. The equation of the **vertical line** passing through $(a, 0)$ is
$$x = a$$

Caution!   The equation for a line is not the same thing as the slope of the line! The slope of every horizontal line is zero, but the equation of a horizontal line has the form $y = b$, where b is the y-coordinate of every point on the line. Similarly, the equation of a vertical line has the form $x = a$, but the slope of a vertical line is undefined ■

RQ5.   Give an example of an equation of a vertical line.
RQ6.   Give an example of an equation of a horizontal line.

Distance Between Points

It is easy to compute the distance between two points that lie on the same horizontal or vertical line. We subtract the smaller coordinate from the larger one.

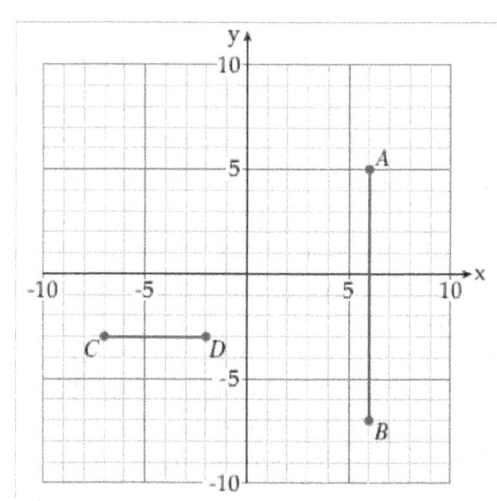

■ **Example 3**   The distance between the points $A(6, 5)$ and $B(6, -7)$ in the figure is
$$\Delta y = 5 - (-7) = 12$$
and the distance between points $C(-7, -3)$ and $D(-2, -3)$ is
$$\Delta x = -2 - (-7) = 5$$ ■

Note that the distance between two points is always a positive number.

Look Closer: When we compute slope, the direction in which we move on the line makes a difference. We will call this the **directed distance**, and it can be either positive or negative. If we move from C to D, we have moved in the positive x-direction, so the directed distance is positive:

$$\Delta x = -2 - (-7) = 5$$

If we move from D to C, we have moved in the negative x-direction, so the directed distance is negative:

$$\Delta x = -7 - (-2) = -5 \qquad\qquad\qquad \bullet$$

> To find the **directed distance** between two points on a number line, we subtract the initial coordinate from the final coordinate.

> **RQ7.**   How do we find the distance between two points that lie on the same vertical line?
>
> **RQ8.**   A directed distance can be either _____ or _____.

Look Ahead: So far we have computed the slope of a line by finding Δy and Δx on the graph. Now we use directed distance develop a formula for slope. $\bullet$

Example 4   Compute the slope of the line segment joining P and R in two ways:
a. Find Δy and Δx using the graph.
b. Find Δy and Δx using coordinates.

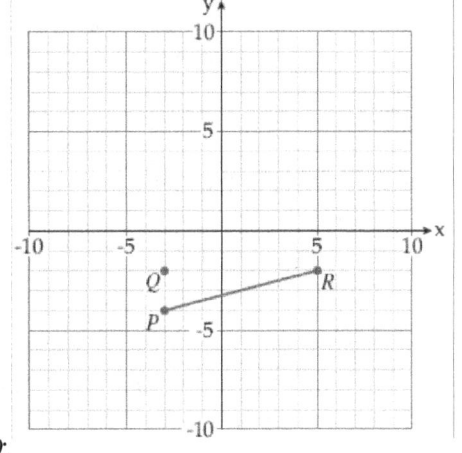

Solutions a.   As we move from P to Q, we move up 2 squares on the graph, so $\Delta y = 2$. As we move from Q to R, we move 8 squares to the right, so $\Delta x = 8$. Thus, the slope of the line is

$$m = \frac{\Delta y}{\Delta x} = \frac{2}{8} = \frac{1}{4}$$

b.   First we write down the coordinates of P and Q:

$$P(-3, -4) \quad \text{and} \quad Q(-3, -2)$$

and compute the directed distance from P to Q:

$$\Delta y = -2 - (-4) = 2$$
$$\text{final} - \text{initial}$$

Then we write down the coordinates of Q and R:

$$Q(-3, -2) \quad \text{and} \quad R(5, -2)$$

and compute the directed distance from Q to R:

$$\Delta x = 5 - (-3) = 8$$
$$\text{final} - \text{initial}$$

We get the same value for the slope as in part (a), $m = \dfrac{\Delta y}{\Delta x} = \dfrac{2}{8} = \dfrac{1}{4}$.   ■

Subscript Notation

   When computing the slope of a line joining two points, we must be careful to subtract their x-coordinates and their y-coordinates in the same order, final minus initial. To distinguish between the coordinates of different points, we use a new notation called **subscripts**.

> We denote the coordinates of a particular point by (x_1, y_1) and the coordinates of a second point by (x_2, y_2).

■ **Example 5**   Compute the slope between points D and F in the figure.

Solution We'll call D the first point and F the second point. Then their coordinates are

$$D: \quad x_1 = -6 \quad \text{and} \quad y_1 = -2$$
$$F: \quad x_2 = 9 \quad \text{and} \quad y_2 = 4$$

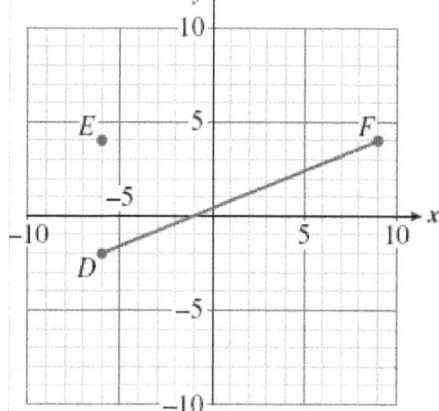

Now we can describe the formula for slope in an organized way. We compute Δy and Δx as directed distances. First, we observe that the coordinates of point E are $(-6, 4)$, or (x_1, y_2). From the figure, we can see that

$$\Delta y = y_2 - y_1 = 4 - (-2) = 6$$

and

$$\Delta x = x_2 - x_1 = 9 - (-6) = 15$$

Thus, the slope of the line segment joining D and F is

$$m = \frac{y_2 - y_1}{x_2 - x_1} = \frac{6}{15} = \frac{2}{5} \qquad ■$$

> **Caution!**   The subscript 1 on x_1, for instance, has nothing to do with the value of the coordinate; it merely identifies this coordinate as the x-coordinate of the first point. ■

> **RQ9.**   What do the subscripts in (x_1, y_1) mean?
> **RQ10.** Write a formula for computing Δy.

A New Formula for Slope

   The method described in Example 5 gives us a new formula for computing slope.

> **Two-Point Formula for Slope**
>   The slope of the line joining points $P_1(x_1, y_1)$ and $P_2(x_2, y_2)$ is
> $$m = \frac{y_2 - y_1}{x_2 - x_1} \qquad \text{if} \quad x_2 \neq x_1$$

We don't need a graph in order to use this formula, just the coordinates of two points on the line.

Example 6   Compute the slope of the line joining the points $(-6, 2)$ and $(3, -1)$.

Solution It doesn't matter which point is P_1 and which is P_2, so we choose P_1 to be $(-6, 2)$. Then $(x_1, y_1) = (-6, 2)$ and $(x_2, y_2) = (3, -1)$. Thus,

$$
\begin{aligned}
m &= \frac{y_2 - y_1}{x_2 - x_1} \\
&= \frac{-1 - 2}{3 - (-6)} = \frac{-3}{9} = \frac{-1}{3}
\end{aligned}
$$

Caution!   In Example 6, we can reverse the order of **both** subtractions to find

$$
\begin{aligned}
m &= \frac{y_1 - y_2}{x_1 - x_2} \\
&= \frac{2 - (-1)}{-6 - 3} = \frac{3}{-9} = \frac{-1}{3}
\end{aligned}
$$

the same answer as before. The order of the points does not matter, but we must be consistent and use the same order when computing Δy and Δx.

RQ11. What is wrong with this formula for slope: $m = \dfrac{y_2 - y_1}{x_1 - x_2}$

Skills Warm-Up

For Problems 1-8, find the negative reciprocal.

1. $\dfrac{3}{4}$ **2.** $\dfrac{-1}{3}$ **3.** -6 **4.** $\dfrac{11}{8}$

5. $1\dfrac{2}{3}$ **6.** $-2\dfrac{1}{2}$ **7.** -3.2 **8.** 0.625

For Problems 9-14, simplify.

9. $\dfrac{10 - 2}{2 - 8}$ **10.** $\dfrac{-5 - (-5)}{2 - 9}$ **11.** $\dfrac{3}{2}(4 - 7) + \dfrac{1}{2}$

12. $\dfrac{-6 - (-12)}{3 - (-5)}$ **13.** $-3(-4 - 2) - 6$ **14.** $\dfrac{5}{3}(5 - 8) + 3$

Answers: **1.** $\dfrac{-4}{3}$ **2.** 3 **3.** $\dfrac{1}{6}$ **4.** $\dfrac{-8}{11}$ **5.** $\dfrac{-3}{5}$ **6.** $\dfrac{2}{5}$ **7.** 0.3125 **8.** -1.6
9. $\dfrac{-4}{3}$ **10.** 0 **11.** -4 **12.** $\dfrac{3}{4}$ **13.** 12 **14.** -2

Homework 3.5

Skills Practice

■ In Problems 1-3, decide whether the lines are parallel perpendicular, or neither.

1. $3x - 4y = 2$
$8y - 6x = 6$

2. $5x + 3y = 1$
$3y - 5x = 5$

3. $2x = 4 - 5y$
$2y = 4x - 5$

4. The slopes of several lines are given below. Which of the lines are parallel to the graph of $y = 0.75x + 2$, and which are perpendicular to it?

a. $m = \dfrac{3}{4}$ **b.** $m = \dfrac{8}{6}$ **c.** $m = \dfrac{-20}{15}$ **d.** $m = \dfrac{-39}{52}$

e. $m = \dfrac{4}{3}$ **f.** $m = \dfrac{-16}{12}$ **g.** $m = \dfrac{36}{48}$ **h.** $m = \dfrac{9}{12}$

■ In Problems 5-7,
   a. Sketch a rough graph of each equation, and label its intercept.
   b. State the slope of each line.

5. $y = -3$ **6.** $2x = 8$ **7.** $x = 0$

■ In Problems 8-11, give the equation of the line described.

8. Horizontal, passes through $(6, -5)$
9. m is undefined, passes through $(2, 1)$
10. Parallel to $x = 4$, passes through $(-8, 3)$
11. The x-axis
12. Perpendicular to the y-axis, passes through $(4, -12)$

■ In Problems 13-16, state the y-intercept and the slope of the line.

13. $y = -6$ **14.** $y = -6x$ **15.** $x = -6y$ **16.** $x = 6 - y$

■ For Problems 17-20, compute the slope of the line joining the given points.

17. $(5, 2), (8, 7)$ **18.** $(3, -2), (0, 1)$ **19.** $(6, -2), (-3, -3)$ **20.** $(3, -5), (8, -5)$

■ For Problems 21-23,
   a. Find the x- and y-intercepts of each line.
   b. Use the intercepts to compute the slope of the line.

21. $\dfrac{x}{5} + \dfrac{y}{7} = 1$ **22.** $\dfrac{-x}{2.4} + \dfrac{y}{1.6} = 1$ **23.** $3x + \dfrac{2}{7}y = 1$

Applications

■ For Problems 24-25, you are given the slope of a line and one point on it. Use the grids to help you find the missing coordinates of the other points on the line.

24. $m = 2$, $(1, -1)$

    a. $(0, ?)$ **b.** $(?, -7)$

25. $m = \dfrac{-1}{2}$, $(2, -6)$

    a. $(6, ?)$ **b.** $(?, -3)$

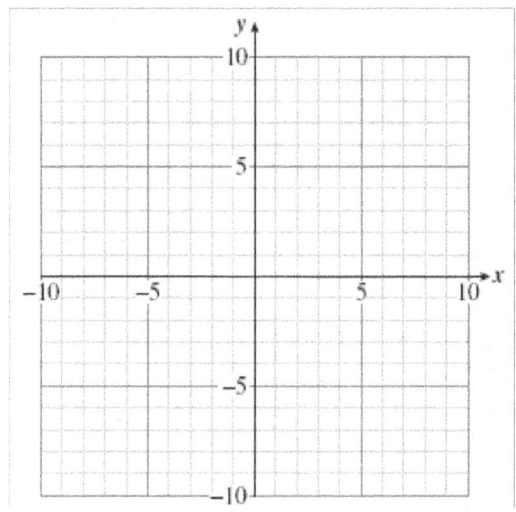
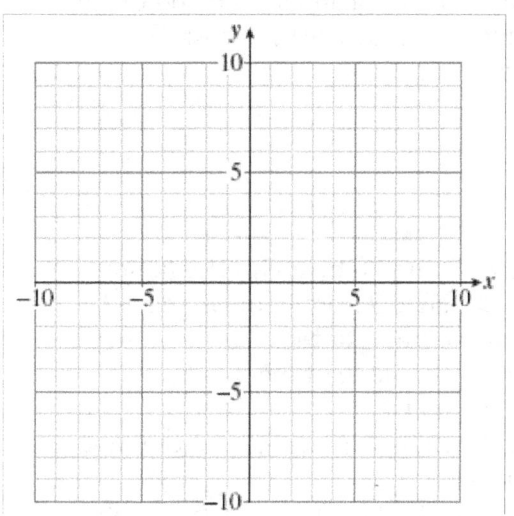

■ For Problems 26-29, calculate the slope of the line described, including units, and explain its meaning for the problem.

26. Rani and Larry start a college fund for their son, Colby. When Colby is five years old, the fund has $11,000. When he is 10 years old, the fund has grown to $17,000. The graph of the amount, A, in the fund when Colby is n years old is a straight line.

27. It cost Melanie $6.88 to copy and bind her 26-page paper at CopyQuik. It cost Nelson $8.80 to copy and bind his 50-page portfolio. The graph of the cost, C, of producing a document of length p pages at CopyQuik is a straight line.

28. Dean is cycling down Mt. Whitney. After five minutes, his elevation is 9200 feet, and after 21 minutes, his elevation is 1200 feet. The graph of Dean's elevation, h, after cycling for t minutes is approximately a straight line.

29. At 2 pm the temperature was 16°, and by midnight it had dropped to −14°. The graph of temperature, T, at h hours after noon is approximately a straight line.

30. The figure shows the wholesale cost, C, in
dollars, of p pounds of dry-roasted peanuts.
  a. What does the slope of this graph tell
          you about dry-roasted peanuts?
  b. Compute the slope, including units.
  c. If the Lone Star Barbecue increases its
          weekly order of peanuts from 100 to 120
          pounds, how much will the cost
          increase?

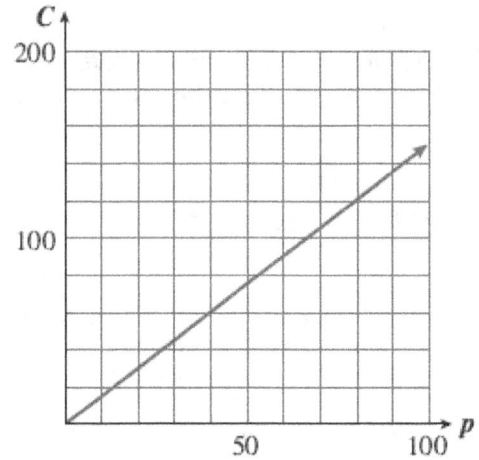

31. Roy is traveling home by train. The figure
shows his distance d from home in miles h
hours after the train started.
  a. Compute the slope of the graph,
          including units.
  b. What does the slope tell you about Roy's
          journey?
  c. Find the vertical intercept of the graph.
          What is its meaning for the problem?
  d. Compute the horizontal intercept of the
          graph. What is its meaning for the
          problem?

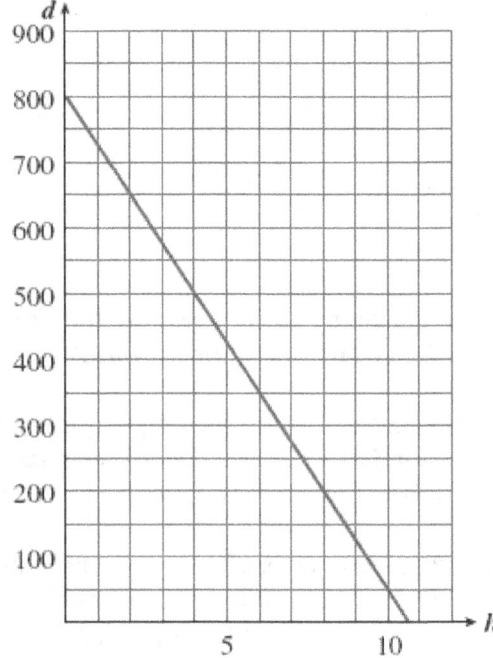

Chapter 3 Summary and Review

Section 3.1 Intercepts

- The **intercepts** of a line are the points where the graph crosses the axes.
- The intercepts of a graph often represent the starting or ending values for a particular variable.

> **Intercepts of a Graph**
> The points where a graph crosses the axes are called the **intercepts**.
> 1. To find the x-intercept, set $y = 0$ and solve for x.
> 2. To find the y-intercept, set $x = 0$ and solve for y.

> **To Graph a Linear Equation Using the Intercept Method:**
> 1. Find the x- and y-intercepts of the graph.
> 2. Draw the line through the two intercepts.
> 3. Find a third point on the graph as a check.
>   (Choose any convenient value for x and solve for y.)

Section 3.2 Ratio and Proportion

- A **ratio** is a type of quotient that allows us to compare two numerical quantities.

> **Ratio**
> The **ratio of** a **to** b is written $\dfrac{a}{b}$.

- A **rate** is a ratio that compares two quantities with different units.
- A **proportion** is a statement that two ratios are equal. In other words, a proportion is a type of equation in which both sides are ratios.
- We use cross-multiplying to solve proportions.

> **Property of Proportions**
> $$\text{If} \quad \frac{a}{b} = \frac{c}{d}, \quad \text{then} \quad ad = bc.$$

- Two variables are said to be **proportional** if their ratios are always the same.

Section 3.3 Slope

- The slope of a line measures the rate of change of y with respect to x.

> **Slope**
> The slope of a line is defined by the ratio
> $$\text{slope} = \frac{\text{change in } y\text{-coordinate}}{\text{change in } x\text{-coordinate}}$$
> as we move from one point to another on the line. In symbols,
> $$m = \frac{\Delta y}{\Delta x}$$

- Lines have constant slope. No matter what two points on a line we use to calculate its slope, we always get the same result.
- The slope of a horizontal line is zero, and the slope of a vertical line is undefined.

Section 3.4 Slope-Intercept Form

> **Slope-Intercept Form**
>   A linear equation written in the form
>
> $$y = mx + b$$
>
> is said to be in **slope-intercept** form. The coefficient m is the slope of the graph, and b is the y-intercept.

- The constant term in the slope-intercept form tells us the initial value of y, and the coefficient of x tells us the rate of change of y with respect to x.

> **To Graph a Line Using the Slope-Intercept Method**
> 1. Write the equation in the form $y = mx + b$.
> 2. Plot the y-intercept, $(0, b)$.
> 3. Write the slope as a fraction, $m = \dfrac{\Delta y}{\Delta x}$.
> 4. Use the slope to find a second point on the graph:
>   Starting at the y-intercept, move Δy units in the y-direction, then Δx units in the x-direction.
> 5. Find a third point by moving $-\Delta y$ units in the y-direction, then $-\Delta x$ units in the x-direction, starting from the y-intercept.
> 6. Draw a line through the three plotted points.

Section 3.5 Properties of Lines

> **Parallel and Perpendicular Lines**
> 1. Two lines are **parallel** if their slopes are equal, that is, if
>
> $$m_1 = m_2$$
>
>   or if both lines are vertical.
> 2. Two lines are **perpendicular** if the product of their slopes is -1, that is, if
>
> $$m_1 m_2 = -1$$
>
>   or if one of the lines is horizontal and one is vertical.

> **Horizontal and Vertical Lines**
> 1. The equation of the **horizontal line** passing through $(0, b)$ is
>
> $$y = b$$
>
> 2. The equation of the **vertical line** passing through $(a, 0)$ is
>
> $$x = a$$

> **Two-Point Formula for Slope**
> The slope of the line joining points $P_1(x_1, y_1)$ and $P_2(x_2, y_2)$ is
>
> $$m = \frac{y_2 - y_1}{x_2 - x_1} \quad \text{if} \quad x_2 \neq x_1$$

Review Questions

■ Use complete sentences to answer the questions in Problems 1-10.

1. Explain how to find the intercepts of a graph.
2. What is the difference between a ratio and a proportion?
3. State the fundamental property of proportions.
4. How can we tell if two variables are proportional?
5. Give a formula for slope using the Δ notation.
6. State the slope-intercept formula, and explain the meaning of the coefficients.
7. Explain the slope-intercept method for graphing a line.
8. What is the easiest way to find the slope of a line from its equation?
9. How can we tell if two lines are perpendicular?
10. What is the difference between a line whose slope is zero and a line whose slope is undefined?

Review Problems

■ Find the x-and y-intercept of each line, then graph the line.

1. $6x - 4y = 12$

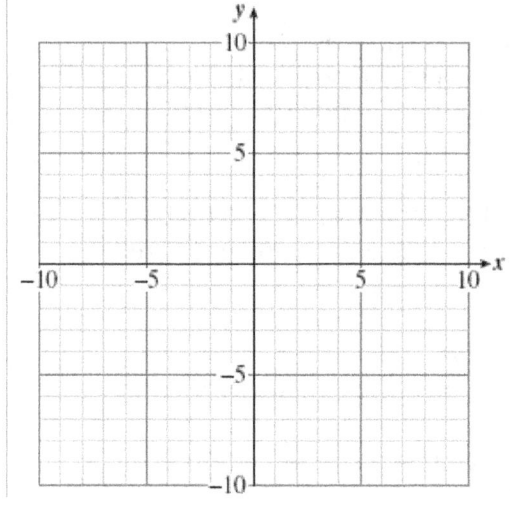

2. $y = -2x + 8$

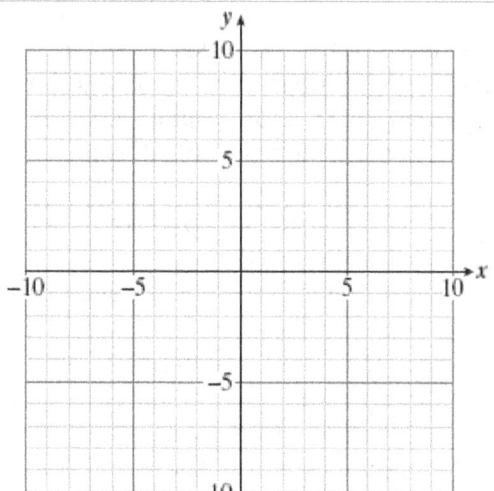

3. Monica has saved $7800 to live on while she attends college. She spends $600 a month.
   a. Write an equation for the amount, S, in Monica's savings account after m months.
   b. Write and solve an equation to answer the question: How long will it be before her savings are reduced to $2400?
   c. Explain what the intercepts of your equation mean in terms of the problem.

4. Ronen is on a rock-climbing expedition. He is climbing out of a deep gorge at a rate of 4 feet per minute, and right now his elevation is -156 feet.
   a. Write an equation for Ronen's elevation, h, after m minutes.
   b. Write and solve an equation to answer the question: When will his elevation be -20 feet?
   c. Explain what the intercepts of your equation mean in terms of the problem.

■ For Problems 5-6, write ratios or rates.

5. In an anthropology class of 35 students, 14 are men. What is the ratio of men to women in the class?

6. Zach's car went 210 miles on 8.4 gallons of gasoline, and Tasha's car went 204 miles on 8 gallons of gasoline. Which car had the higher rate of fuel consumption?

■ For Problems 7-8, solve the proportion.

7. $\dfrac{105}{y} = \dfrac{15}{17}$
8. $\dfrac{16}{q} = \dfrac{52}{86.125}$

■ For Problems 9-14, write and solve a proportion to answer the question.

9. At Van's Hardware, 36 metric frimbles cost $4.86. How many metric frimbles can you buy for $6.75?

10. Bob's weekly diet includes 70 grams of fat, 200 grams of protein, and 1800 grams of carbohydrates. Jenni's weekly diet includes 112 grams of fat. How many grams of protein and carbohydrates should she consume so that her diet has the same proportions as Bob's diet?

11. Simon joined a co-op that provides access to the Internet. His user bill, B, is proportional to the number of hours, h, that he is logged on to the net. Last month he was logged on for 28 hours, and his bill was $21. This month Simon's bill was $39. How many hours did he log on the Internet?

12. A small orchard that is 150 yards on each side contains 100 apple trees. The number of trees in an orchard is proportional to the area of the orchard. How many trees are in a larger orchard that is 600 yards on each side?

13. On a map of Arenac County, 3 centimeters represents 5 miles.
   a. What are the true dimensions of a rectangular township whose dimensions on the map are 6 centimeters by 9 centimeters?
   b. What is the perimeter of the township? What is the perimeter of the corresponding region on the map?
   c. What is the ratio of the perimeter of the actual township to the perimeter of the corresponding region on the map?
   d. What is the area of the township? What is the area of the corresponding region on the map?
   e. What is the ratio of the area of the actual township to the are of the corresponding region on the map?

14. On a map of Euclid County, $\frac{1}{3}$ inch represents 2 miles. Lake Pythagoras is represented on the map by a right triangle with sides 1 inch, $\frac{4}{3}$ inch, and $\frac{5}{3}$ inch.

   a. What are the true dimensions of Lake Pythagoras?

   b. What is the perimeter of Lake Pythagoras? What is the perimeter of the corresponding region on the map?

   c. What is the ratio of the perimeter of the actual lake to the perimeter of the corresponding region on the map?

   d. What is the area of Lake Pythagoras? What is the area of the corresponding region on the map?

   e. What is the ratio of the area of the actual lake to the are of the corresponding region on the map?

▨ For Problems 15-16, decide whether the two variables are proportional.

15. a.

D	0.5	1	2	4
V	0.25	2	16	64

b.

s	0.2	0.8	1.6	2.5
M	0.08	0.32	0.64	1

16. a.

Time	5	10	15	20
Cost	1.50	2.50	3.50	4.50

b.

Speed	30	40	50	60
Distance	42	56	70	84

▨ For Problems 17-18, use the graph to find the slope of each line, and illustrate Δx and Δy on the graph.

17. $3x + 2y = -7$

18. $3y = 5x$

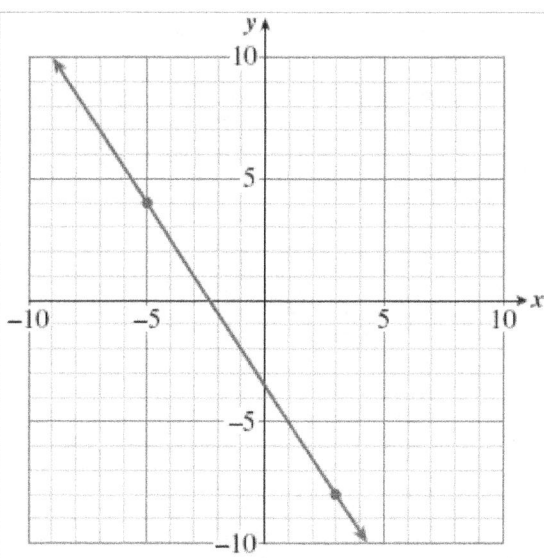
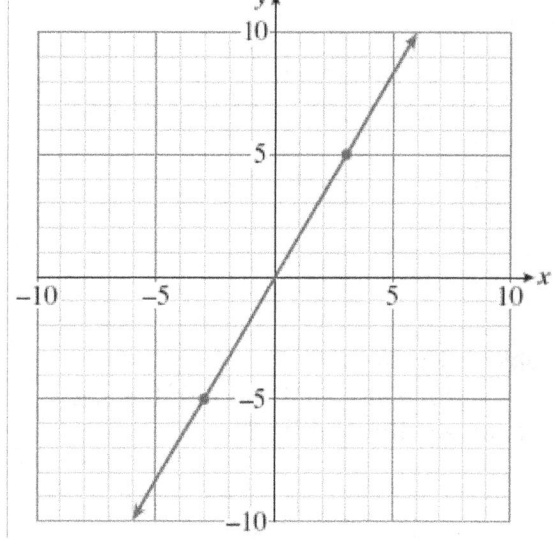

19. The Palm Springs aerial tramway ascends Mt. San Jacinto at a slope of approximately 0.516. It traverses a horizontal distance of about 11,380 feet. The elevation at the bottom of the tramway is 2643 feet. What is the elevation at the top?

20. The ruins of the Pyramid of Cholula in Guatemala have a square base 1132 feet on each side. The sides of the pyramid rise at a slope of about 0.32. How tall was the pyramid originally?

■ For Problems 21-24,

    a. Find the intercepts of each line.

    b. Use the intercepts to graph the line.

    c. Use the intercepts to find the slope of the line.

21. $6x - 3y = 18$

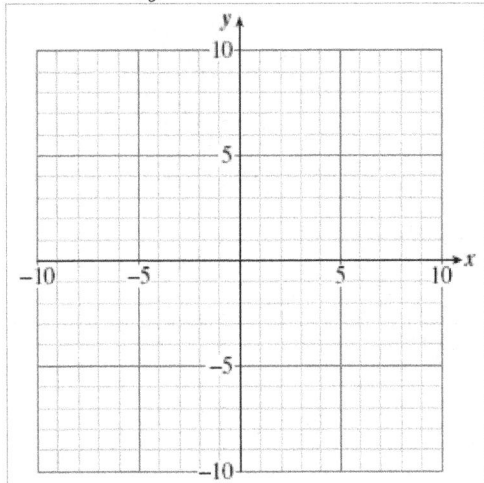

22. $4y + 9x = 36$

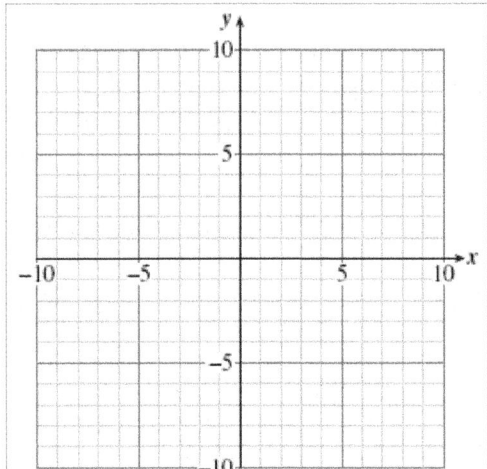

23. $\dfrac{x}{2} - \dfrac{y}{3} = 1$

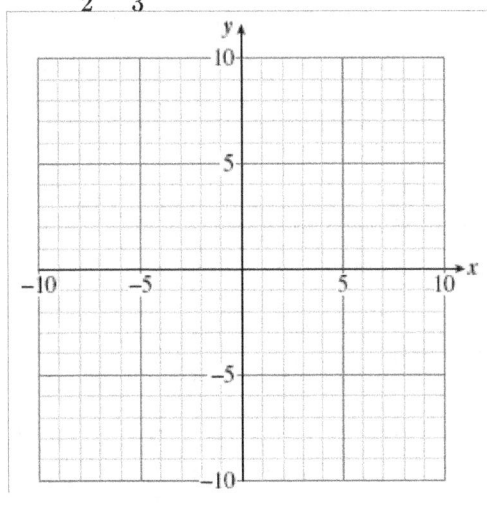

24. $y = 3x - 8$

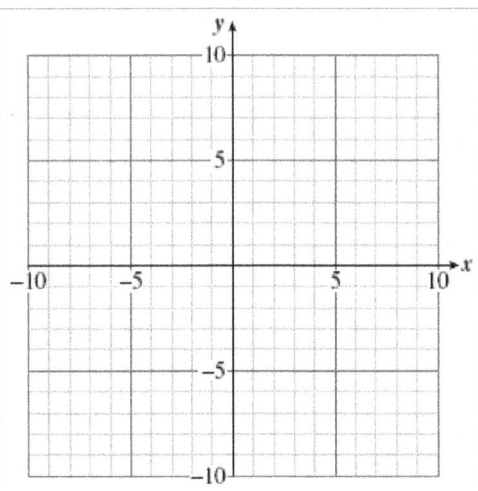

■ For Problems 25-26,

    a. Find the slope and y-intercept of the line.

    b. Write an equation for the line.

25.

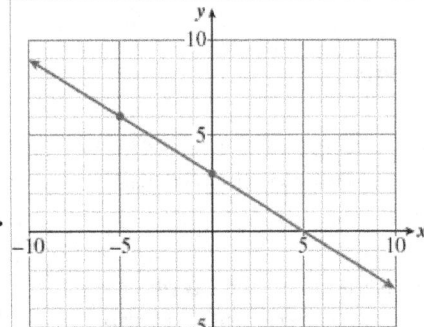

26.

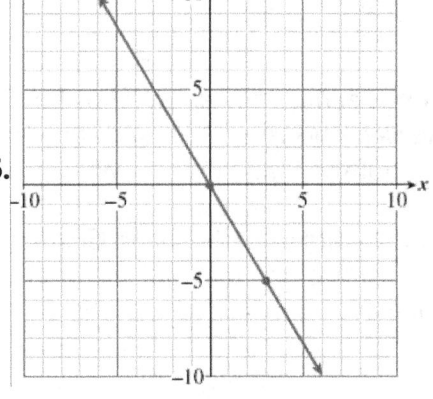

For Problems 27-30,
   a.   Find the slope and y-intercept of the line.
   b.   Graph the line using the slope-intercept method.

27. $2y + 5x = -10$

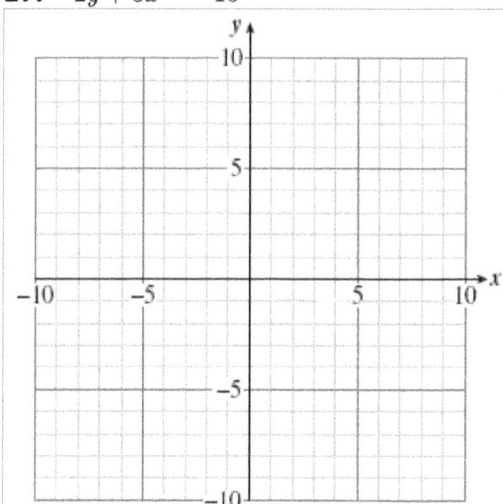

28. $y + 3x = 0$

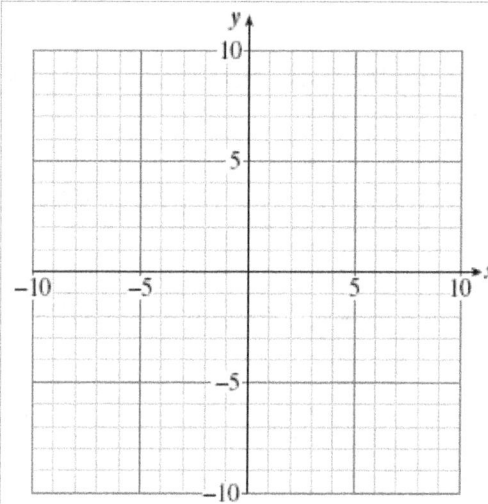

29. $3x = 4y - 8$

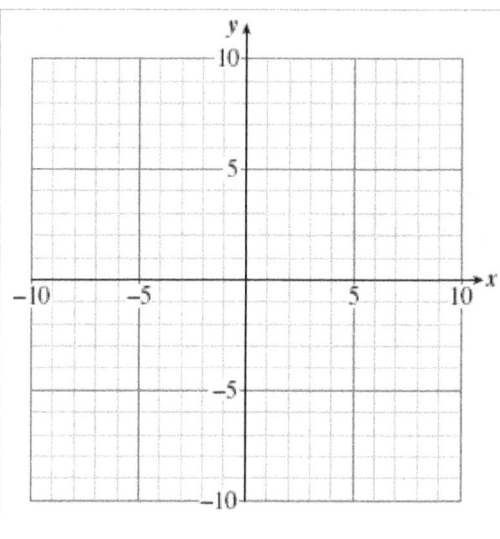

30. $\dfrac{x}{4} - \dfrac{y}{5} = 1$

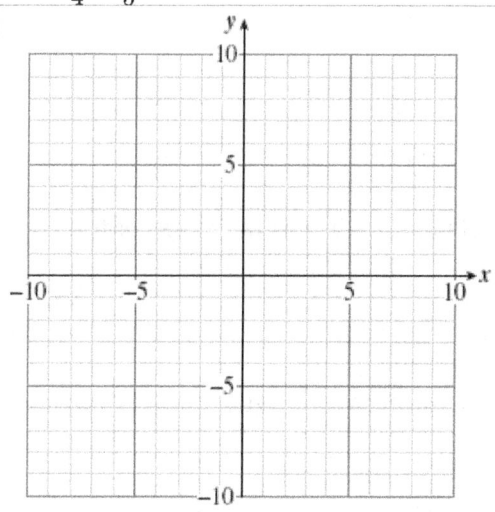

31. This year Francine bought a 3-foot tall blue spruce sapling for her front yard, and it is supposed to grow about 6 inches per year. Write an equation for the height, h, of the tree t years from now.

32. The price of a medium bowl of frozen yogurt is given by the equation $y = 1.35 + 0.85t$, where t is the number of toppings you select. Find the slope and the y-intercept of the graph, and explain what each means in terms of the problem.

33. Beryl is sailing in a hot air balloon. Her altitude at t minutes after 2 pm is given in feet by $h = 500 - 15t$. Find the slope and the y-intercept of the graph, and explain what each means in terms of the problem.

34. The amount of water in the municipal swimming pool is given in gallons by

$$y = 500,000 - 5000h$$

where h is the number of hours since they started draining the pool for the winter. State the slope and y-intercept of the equation, and explain their meaning in terms of the problem.

■ For Problems 35-36, graph the line with the given slope and passing through the given point.

35. $m = -\dfrac{3}{4}$, $(2, -1)$ **36.** $m = \dfrac{1}{3}$, $(0, -3)$

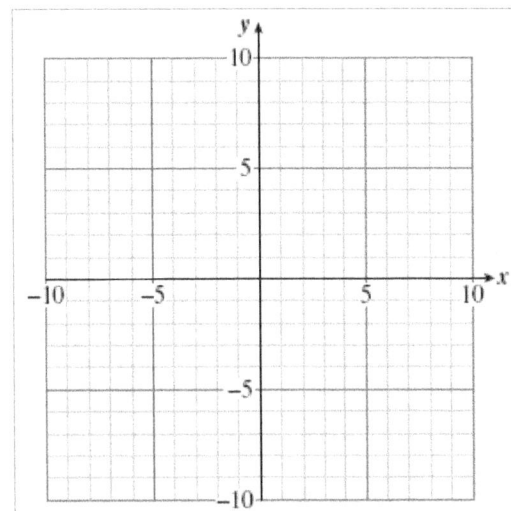   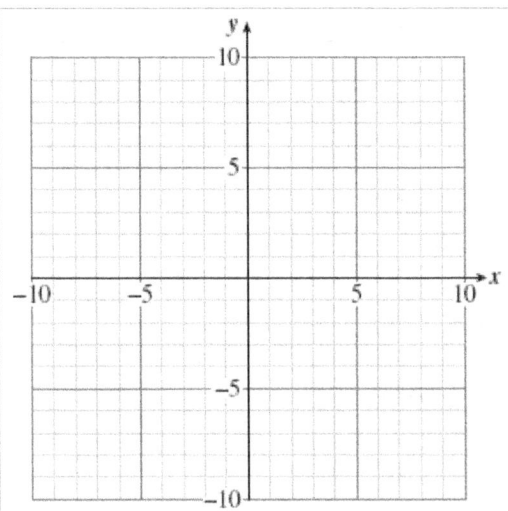

■ For Problems 37-40, find the slope of the line segment joining the points.

37. $(-1, 4)$, $(3, -2)$ **38.** $(5, 0)$, $(2, -6)$
39. $(6.2, 1.4)$, $(-2.1, 4.8)$ **40.** $(0, -6.4)$, $(-5.6, 3.2)$

■ For Problems 41-42, find the slope of the line described.

41. Vertical with x-intercept $(-5, 0)$
42. Perpendicular to the line $y = \frac{1}{3}x - 2$

43. It costs Delbert $50.40 to fill up the 12-gallon gas tank in his sports car, and $105 to fill the 25-gallon tank in his recreational vehicle.
   a. If you plot price, p, on the vertical axis, and gallons, g, on the horizontal axis, compute the slope of the line segment joining the two points.
   b. What does the slope tell you about the problem?

44. Find the slope and y-intercept of the line $y = 4$.

■ For Problems 45-48, find an equation for the line described. Write your answers in slope-intercept form if possible.

45. With x-intercept $(5, 0)$ and y-intercept $(0, -1)$.
46. Passing through the points $(0, -2)$ and $(3, -5)$.
47. Parallel to the y-axis and passing through the point $(-4, 2)$.
48. Horizontal and passing through the point $(-3, 8)$.

For Problems 49-50, decide whether the lines are parallel, perpendicular, or neither.

49. **a.** $y = \frac{1}{2}x + 3;\quad x - 2y = 8$ **b.** $4x - y = 6;\quad x + 4y = -2$

50. **a.** $5x + 3y = 1;\quad 5y - 3x = 6$ **b.** $4y - 5x = 1;\quad 2y - \frac{5}{2}x = 2$

For Problems 51-58, use the most convenient method to graph the equation.

51. $4x - 3y = 12$

52. $\frac{x}{6} - \frac{y}{12} = 1$

53. $50x = 40y - 20,000$

54. $1.4x + 2.1y = 8.4$

55. $3x - 4y = 0$

56. $x = -4y$

57. $4x = -12$

58. $2y - x = 0$

Chapter 4 Applications of Linear Equations

Lesson 4.1 The Distributive Law

- Apply the distributive law
- Write and simplify expressions using the distributive law
- Write an equation to model a situation
- Solve equations that involve the distributive law

Lesson 4.2 Systems of Linear Equations

- Decide whether an ordered pair is a solution of a system
- Solve a system of equations by graphing
- Identify a system as consistent, inconsistent, or dependent
- Write a system of equations to solve an applied problem

Lesson 4.3 Algebraic Solution of Systems

- Solve systems by the substitution method
- Solve systems by the elimination method
- Identify a system as consistent, inconsistent, or dependent
- Write a system of equations to solve an applied problem

Lesson 4.4 Applied Problems

- Use a system to solve problems about interest
- Use a system to solve problems about mixtures
- Use a system to solve problems about motion

Lesson 4.5 Point-Slope Form

- Use the point-slope form to graph a line
- Use the point-slope form to find the equation of a line
- Find the equation of a line through two points
- Use the point-slope form to find a linear model

4.1 The Distributive Law

Products

So far we have considered sums and differences of algebraic expressions. Now we take a look at simple products. To simplify a product such as $3(2a)$, we multiply the numerical factors to find

$$3(2a) = (3 \cdot 2)a = 6a$$

This calculation is an application of the associative law for multiplication. We can see why the product is reasonable by recalling that multiplication is repeated addition, so that

$$3(2a) = 2a + 2a + 2a = 6a$$
Three terms

■ **Example 1a.** $5(3x) = 15x$ **b.** $-2(-4h) = 8h$ ■

RQ1.   Explain the difference between the expressions $2b + 3b$ and $2(3b)$.

The Distributive Law

Another useful property for simplifying expressions is the distributive law. First consider a numerical example. Suppose we would like to find the area of the two rooms shown in the figure.
We can do this in two different ways:

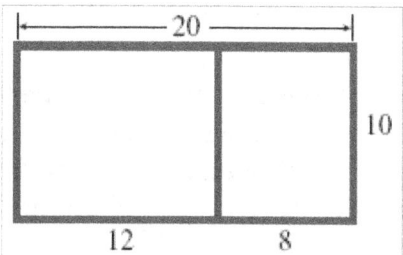

Method 1:   Think of the rooms as one large rectangle, with width 10 feet and length $12 + 8$ feet. The area is then given by

$$\text{Area} = \text{width} \times \text{length}$$
$$= 10\,(12 + 8) = 10\,(20) = 200 \text{ square feet}$$

Method 2:   Find the area of each room separately and add the two areas together. This gives us

$$10(12) + 10(8) = 120 + 80 = 200 \text{ square feet}$$

Of course, we get the same answer with either method.   However, looking at the first step in each calculation, we see that

$$10(12 + 8) = 10(12) + 10(8)$$

These are two ways to compute the expression $10(12 + 8)$. The first way follows the order of operations, adding $12 + 8$ first. The second way distributes the multiplication to each term inside parentheses; that is, we multiply each term inside the parentheses by 10.

$$10(12 + 8) = 10(12) + 10(8)$$

This equation is an example of the distributive law.

Distributive Law
   If a, b, and c are any numbers, then

$$a(b + c) = ab + ac$$

   If the terms inside parentheses are not like terms, we have no choice but to use the distributive law to simplify the expression.

■ Example 2   Simplify $-2(3x - 1)$

Solution We multiply each term inside parentheses by -2:
$$-2(3x - 1) = -2(3x) - 2(-1)$$
$$= -6x + 2$$
■

Caution!   Compare the three similar expressions:
$$-3x - 5x, \qquad -3(-5x), \qquad -3(-5 - x)$$

To simplify these expressions, we must first recognize the operations involved. The first is a sum of terms, the second is a product, and the third is an application of the distributive law. Thus,
$$-3x - 5x = -8x, \qquad -3(-5x) = 15x, \qquad -3(-5 - x) = 15 + 3x$$
■

RQ2.   When do we need the distributive law to simplify an expression?

Solving Equations and Inequalities

   If an equation or inequality contains parentheses, we use the distributive law to simplify each side before we begin to solve.

■ Example 3   Solve $25 - 6x = 3x - 2(4 - x)$

Solution We apply the distributive law to the right side of the equation.

$25 - 6x = 3x - 2(4 - x)$	**Apply the distributive law.**
$25 - 6x = \mathbf{3x} - 8 + \mathbf{2x}$	**Combine like terms.**
$25 - 6x = 5x - 8$	**Subtract $5x$ from both sides.**
$25 - 11x = -8$	**Subtract 25 from both sides.**
$-11x = -33$	**Divide both sides by -11.**
$x = 3$	**The solution is 3.**

You should verify each step of the solution, and check the answer.   ■

Caution!   The espression on the right side of the equation in Example 3 is made up of two terms:
$$3x \quad \text{and} \quad -2(4 - x)$$

Remember that we think of all terms as *added* together, so the minus sign in front of the 2 tells us that 2 is negative. When we apply the distributive law, we multiply each term inside parentheses by -2 to get $-8 + 2x$.   ■

   We know that a fraction bar often serves as a grouping device, like parentheses.

Example 4   Solve the proportion $\dfrac{7.6}{1.2} = \dfrac{x+3}{2x-1}$

Solution We apply the property of proportions and cross-multiply to get

$$7.6\,(2x - 1) = 1.2\,(x + 3)$$ **Apply the Distributive Law.**

$$15.2\,x - 7.6 = 1.2\,x + 3.6$$ **Subtract** $1.2x$ **from both sides and add 7.6 to both sides.**

$$14x = 11.2$$ **Divide both sides by 14.**

$$x = \dfrac{11.2}{14} = 0.8$$

We can check the solution by substituting $x = 0.8$ into the original proportion.

RQ3.   In Example 4, why did we put parentheses around $2x - 1$ and $x + 3$?

Algebraic Expressions

We can often simplify algebraic expressions by combining like terms.

Example 5   Delbert and Francine are collecting aluminum cans to recycle. They will be paid x dollars for every pound of cans they collect. At the end of three weeks, Delbert collected 23 pounds of aluminum cans, and Francine collected 47 pounds.
a. Write algebraic expressions for the amount of money Delbert made, and the amount Francine made.
b. Write and simplify an expression for the total amount of money Delbert and Francine made from aluminum cans.

Solutions a.   We multiply the number of pounds collected by the price per pound. Delbert made $23x$ dollars, and Francine made $47x$ dollars.
b. We add the amount of money Delbert made to the amount Francine made.

$$23x + 47x = 70x$$

RQ4.   In Example 5, what does x stand for? What does $23x$ stand for?

The distributive law is also helpful for writing algebraic expressions.

Example 6   The length of a rectangle is 3 feet less than twice its width, w.
a. Write an expression for the length of the rectangle in terms of w.
b. Write and simplify an expression for the perimeter of the rectangle in terms of w.
c. Suppose the perimeter of the rectangle is 36 feet. Write and solve an equation to find the dimensions of the rectangle.

Solutions a.   "Three feet less than twice the width" tells us to subtract 3 from twice the width. The length of the rectangle is $2w - 3$.

b. The perimeter of a rectangle is given by the formula $P = 2l + 2w$. We substitute our expression for the length to get

$$P = 2(2w - 3) + 2w \qquad \textbf{Apply the distributive law.}$$
$$= 4w - 6 + 2w \qquad \textbf{Combine like terms.}$$
$$= 6w - 6$$

c. We set our expression for the perimeter equal to 36.

$$6w - 6 = 36 \qquad \textbf{Add 6 to both sides.}$$
$$6w = 42 \qquad \textbf{Divide both sides by 7.}$$
$$w = 7$$

The width of the rectangle is 7 feet, and its length is $2(7) - 3 = 11$ feet.

RQ5. In Example 6, why did we multiply $2w - 3$ by 2?

Skills Warm-Up

Use a formula to write an equation. Then solve the equation.

1. 32.5% of the class are engineering majors. If there are 91 engineering majors, how many students are in the class?

2. A rectangular cookie sheet is 40 cm long and has a perimeter of 116 cm. How wide is the cookie sheet?

Write an equation you could solve to answer the question.

3. Erika bought a 50-pound bag of dog food and after 28 days she had 29 pounds left. How much dog food did she use per day?

4. A hamburger contains 60 calories less than two bags of fries. A hamburger and one bag of fries contains 780 calories. How many calories are in a bag of fries?

Answers:
1. $0.325x = 91$; 280
2. $2w + 2(40) = 116$; 18 cm
3. $50 - 28d = 29$; $\frac{3}{4}$ lb
4. $x + (2x - 60) = 780$; 280

Homework 4.1

Skills Practice

▊ In Problems 1-4, use the distributive law to remove parentheses. For some of these exercises you will use the distributive law in the form

$$(b+c)a = ba + ca$$

1. $5(2y - 3)$ **2.** $-2(4x + 8)$

3. $-(5b - 3)$ **4.** $(-6 + 2t)(-6)$

▊ In problems 5-6, simplify each product. Which product in each pair requires the distributive law?

5. a. $8(4c)$ **6. a.** $2(-8 - t)$

    b. $8(4 + c)$ **b.** $2(-8t)$

7. Which of the following is a correct application of the distributive law?
    a. $5(3a) = 15a$ **b.** $5(3 + a) = 15 + a$
    c. $5(3a) = 15(5a)$ **d.** $5(3 + a) = 15 + 5a$

8. Simplify each expression if possible. Then evaluate for $x = 3$, $y = 9$.
    a. $2(xy)$ **b.** $2(x + y)$
    c. $2 - xy$ **d.** $-2xy$

▊ In Problems 9-12, simplify and combine like terms.

9. $-6(x + 1) + 2x$ **10.** $5 - 2(4x - 9) + 9x$

11. $-4(3 + 2z) + 2z - 3(2z + 1)$ **12.** $3(-3a - 4) - 3a - 4(3a - 4)$

▊ In Problems 13-20, solve each equation or inequality.

13. $6(3y - 4) = -60$ **14.** $5w - 64 = -2(3w - 1)$

15. $-22c + 5(3c + 4) = 20 + 8c$ **16.** $6 - 4(2a + 3) \geq -6a + 2$

17. $4 - 3(2t - 4) > -2(4 - 3t)$ **18.** $0.25(x + 3) - 0.45(x - 3) = 0.30$

19. $\dfrac{a}{a + 2} = \dfrac{2}{3}$ **20.** $\dfrac{0.3}{0.5} = \dfrac{b + 2}{12 - b}$

21. Choose a value for the variable and show that the following pairs of expressions are **not** equivalent.
    a. $5(x + 3)$ and $5x + 3$ **b.** $-4(c - 3)$ and $-4c - 12$

22. a. Evaluate the expression $(2x + 7) - 4(4x - 2) - (-2x + 3)$ for $x = -3$.
    b. Simplify the expression in part (a) by combining like terms.
    c. Evaluate your answer to part (b) for $x = -3$. Check that you got the same answer for part (a).
    d. Are the expressions in parts (a) and (b) equivalent? Why or why not?

Applications

23. The length of a rectangle is 8 feet shorter than twice its width. Write expressions in terms of w, the width of the rectangle.
  a. The length of the rectangle
  b. The perimeter of the rectangle
  c. The area of the rectangle

24. Risa bought a skirt and a sweater that together cost $108. Then she bought a pair of shoes that cost twice as much as the sweater. Write and simplify expressions in terms of c, the cost of the skirt:
  a. The cost of the sweater
  b. The cost of the shoes
  c. The total amount Risa spent

25. Albert and Isaac left the same hotel at the same time, and each drove for 3 hours, but Albert drove 15 miles per hour faster than Isaac. Write and simplify expressions in terms of Isaac's speed, s.
  a. How far Isaac drove
  b. How far Albert drove
  c. If they drove in the same direction, how far apart they are now
  d. If they drove in opposite directions, how far apart they are now

26. A box of AlmondOats contains 15 ounces of cereal made of oat flakes and sliced almonds. Oat flakes cost 15 cents per ounce, and almonds cost 35 cents per ounce. Write and simplify expressions in terms of a, the number of ounces of almonds in the box.
  a. The number of ounces of oat flakes in the box
  b. The cost of the almonds
  c. The cost of the oat flakes
  d. The total cost of a box of AlmondOats

27. Trader Jim's Pomegranate Punch is 25% pomegranate juice, and Pomajoy is 60% pomegranate juice. Vera mixes some of each to make 8 quarts of juice drink. Write and simplify expressions in terms of x, the number of quarts of Pomajoy she uses.
  a. The number of quarts of Pomegranate Punch
  b. The number of quarts of pomegranate juice in the Pomajoy
  c. The number of quarts of pomegranate juice in the Pomegranate Punch
  d. The number of quarts of pomegranate juice in Vera's mixture

28. Premium ice cream is 12% butterfat, and chocolate syrup is 55% butterfat. A chocolate sundae, without whipped cream and a cherry, weighs 10 ounces. Write and simplify expressions in terms of x, the number of ounces of ice cream in the sundae.
  a. The number of ounces of chocolate syrup in the sundae
  b. The amount of fat in the ice cream
  c. The amount of fat in the chocolate syrup
  d. The total amount of fat in the sundae

■ In problems 29-30, find an algebraic expression for the area of the rectangle. Use the distributive law to write the expression in two ways.

29.

30.

31. The length of a rectangular vegetable garden is 6 yards more than twice its width.
   a. Write an expression for the perimeter of the garden in terms of its width.
   b. Ann bought 42 yards of fence to enclose the garden. What are its dimensions?

32. Revenue from the state lottery is divided between education and administrative costs in the ratio of 4 to 3. If the lottery revenue this year is $24,000,000, how much money will go to education?

33. An apple and a glass of milk together contain 260 calories.
   a. If an apple contains a calories, how many calories are in a glass of milk?
   b. Write an expression for the number of calories in two apples and three glasses of milk.
   c. If two apples and three glasses of milk contain 660 calories, find the number of calories in an apple.

34. Melody sold 47 tickets to a charity concert. Reserved seats cost $10 and open seating was $6 a ticket. Let x represent the number of reserved seats she sold. Write expressions in terms of x for:
   a. The number of open seating tickets Melody sold
   b. The amount of money Melody collected from reserved seating tickets
   c. The amount of money Melody collected from open seating tickets
   d. The total amount of money Melody collected

35. Melody from Problem 34 collected $330 from the sale of concert tickets. Write and solve an equation to answer the question: How many reserved seats did she sell?

4.2 Systems of Linear Equations
What is a System of Equations?

In many applications it is useful to write two or more linear equations.

Example 1 Delbert and Francine are buying appliances for their new home. They have a choice of two different refrigerators: a standard model that sells for $1000, or an energy-efficient model at a price of $1200. The standard model costs $6 per month to run, and the energy-efficient model costs $2 per month. Write linear equations for the total cost of each refrigerator after t months.

Solution Let S stand for the total cost of running the standard refrigerator for t months, so $S = mt + b$. The initial cost of the standard model is $1000, so $b = 1000$. The cost increases at a rate of $6 per month, so $m = 6$. Thus,

$$S = 6t + 1000$$

Let E stand for the total cost of running the energy-efficient refrigerator. For this model, the initial cost is $b = 1200$, and the total cost increases at a rate of $m = 2$ dollars per month. Thus,

$$E = 2t + 1200$$

Look Closer: When we consider two equations together, as in the example above, we often use the same variables for both equations, like this:

$$y = 6x + 1000$$
$$y = 2x + 1200$$

A pair of linear equations with the same variables is called a **system of linear equations**.

In Example 1, the standard refrigerator costs less initially, but costs more to run each month. Delbert and Francine want to know when the energy-efficient model will begin to pay for itself, or in other words, when the total cost of running the standard model will exceed the total cost of the energy-efficient model.

Example 2 When is the total cost of running the two refrigerators in Example 1 equal?

Solution We would like to know the value of t when $S = E$. We could use trial-and-error by evaluating both S and E for various values of t. The results of such a search are given in the table below.

t	10	20	30	40	50
S	1060	1120	1180	1240	1300
E	1220	1240	1260	1280	1300

We see that when $t = 50$, the total cost of both models is $1300. Thus, at the end of 50 months, or 4 years 2 months, the cost of the standard model equals the cost of the energy-efficient model.

In Example 2 we found an ordered pair, $(50, 1300)$, that makes both equations true.

> A **solution** to a system of equations is an ordered pair (x, y) that satisfies each equation in the system.

To check whether $(50, 1300)$ is really a solution of the system, we substitute the coordinates into each equation.

$$S = 6t + 1000 \qquad \text{Does } 1300 = 6(50) + 1000 ? \qquad \textbf{True.}$$
$$E = 2t + 1200 \qquad \text{Does } 1300 = 2(50) + 1200 ? \qquad \textbf{True.}$$

We have verified that the ordered pair $(50, 1300)$ is a solution to the system in Example 1.

> **RQ1.** What is a solution to a system of equations?
> **RQ2.** Decide whether the ordered pair $(2, 3)$ is a solution to the system
> $$3x - 4y = -6$$
> $$x + 2y = -4$$

Solving a System of Equations by Graphing

The solution of a system satisfies both equations in the system, so the point that represents the solution must lie on both graphs. It is the intersection point of the two lines described by the system. Thus, we can solve a system of equations by graphing the equations and looking for the point (or points) where the graphs intersect.

■ **Example 3**   Solve the system in Example 1 by graphing.

Solution The graph of the two equations is shown at right. The graph of S has a smaller initial value than E, but it increases more rapidly because its slope is greater. The two graphs intersect at the point labeled P. At this point the values of S and E are the same, approximately 1300. The t-coordinate of point P is 50, so the solution of the system is $(50, 1300)$. The total cost of each model is about $1300 over the first 50 months of operation.

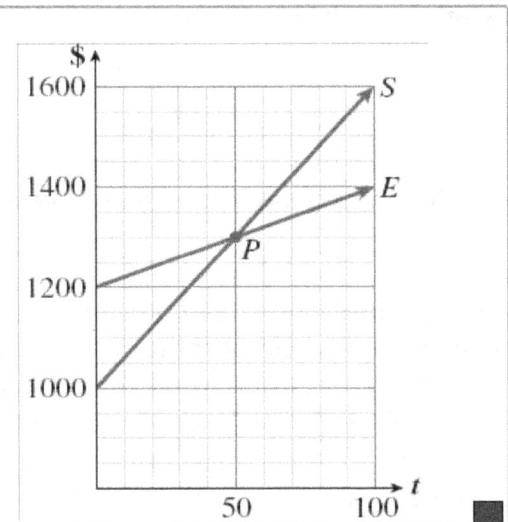

After 50 months, the total cost of the energy-efficient model will be less than the cost of the standard model.

> **RQ3.**   How do we find the solution to a system of equations by graphing?

Inconsistent and Dependent Systems

Not every system of equations has a solution.

■ **Example 4**   Delbert and Francine are also shopping for homeowners insurance. HomeLife offers a policy for semiannual premiums of $408, or they can join a co-op for

$50 and pay just $68 per month for comparable coverage. Which option is less expensive in the long run?

Solution   We'll write a system of equations and graph them, as we did in Example 3. If t stands for the number of months, then the total cost of the co-op insurance is

$$C = 50 + 68t$$

and the HomeLife policy costs

$$H = \frac{408}{6}t = 68t$$

(because a semiannual premium covers 6 months). The solution of the system represents the time when the costs of the two policies are equal. As you can see, the two lines are parallel, so the graphs do not intersect. This system has no solution. There is no point where the two policies cost the same amount; the HomeLife policy is always less expensive.

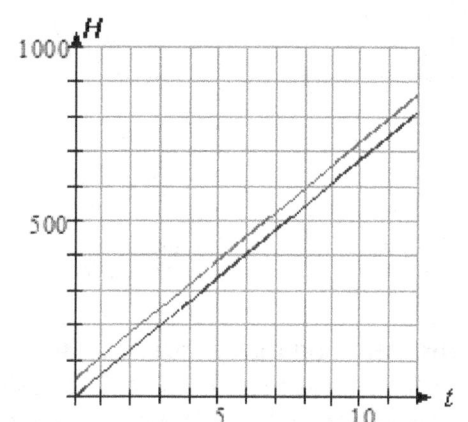

In Example 4, the two graphs never intersect, so the system has no solution. Such a system is called **inconsistent**. There is a third possibility: Both equations may have the same graph.

Example 5   Solve the system

$$y = 2x + 2$$
$$6x - 3y = -6$$

Solution   We graph the first equation by the slope-intercept method:

$$b = 2 \quad \text{and} \quad m = \frac{\Delta y}{\Delta x} = \frac{2}{1}$$

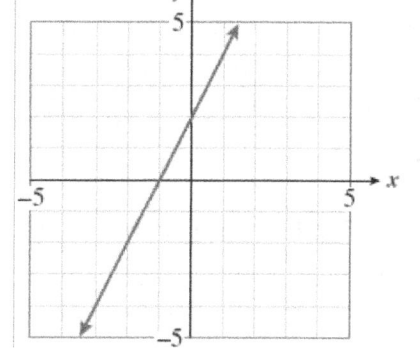

Its graph is shown at right. The second equation, which has the form $Ax + By = C$, is easier to graph by the intercept method. You can verify that its intercepts are $(0, 2)$ and $(-1, 0)$. The graph of the second line is identical to the first graph. Every solution to the first equation is also a solution to the second equation, so every point on the line is a solution of the system as well. The system has infinitely many solutions.

We can verify that the two equations in Example 5 are actually equivalent by solving the second equation for y in terms of x.

$$
\begin{array}{ll}
6x - 3y = -6 & \textbf{Add } 3y \textbf{ and 6 to both sides.} \\
6x + 6 = 3y & \textbf{Divide both sides by 3.} \\
2x + 2 = y &
\end{array}
$$

The second equation is the same as the first equation. A system in which the two equations are equivalent is called **dependent**. A dependent system has infinitely many solutions: Every point on the graph is a solution to the system.

We have now seen three types of systems. These are illustrated in the figure on the next page. Most of the systems we will study have exactly one solution. Such systems are called **consistent and independent**.

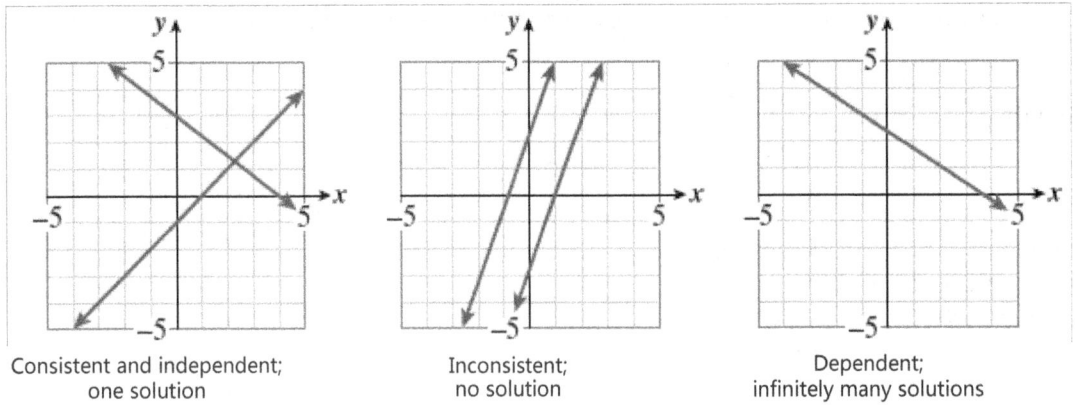

Consistent and independent;   Inconsistent;   Dependent;
     one solution   no solution   infinitely many solutions

1. **Consistent and independent system.**   The graphs of the two lines intersect in exactly one point. The system has exactly one solution.
2. **Inconsistent system**. The graphs of the equations are parallel lines and hence do not intersect. An inconsistent system has no solutions.
3. **Dependent system**. All the solutions of one equation are also solutions to the second equation, and hence are solutions of the system. The graphs of the two equations are the same line. A dependent system has infinitely many solutions.

RQ4.   What does the graph of an inconsistent system look like? How many solutions does it have?

RQ5.   What does the graph of a dependent system look like? How many solutions does it have?

Problem Solving with Systems

Many practical problems involve two or more unknown quantities. Often it is easier to solve these problems by using two variables and writing a system of equations.

Example 6   Allen has been asked to design a rectangular Plexiglas plate whose perimeter is 28 inches and whose length is three times its width. What should the dimensions of the plate be?

Solution Step 1   The dimensions of a rectangle are its length and width, so we let x represent the width of the plate and y represent its length.

Step 2   We must write two equations about the length and width of the plate. We know a formula for the perimeter of a rectangle, $P = 2l + 2w,$ so the first equation is

$$2x + 2y = 28$$

We also know that the length is three times the width, or $y = 3x.$ These two equations make a system.

$$2x + 2y = 28$$
$$y = 3x$$

Step 3 We use the intercept method to graph the first equation, $2x + 2y = 28$:

x	y
0	14
14	0

When $x = 0$, $y = 14$, and
when $y = 0$, $x = 14$.

We use the slope-intercept method to graph $y = 3x$:

$$b = 0 \qquad m = \frac{\Delta y}{\Delta x} = \frac{3}{1}$$

Their graphs are shown at right. The graphs intersect at approximately $(3.5, 10.5)$. You can verify that $x = 3.5$ and $y = 10.5$ are a solution to the system by checking that these values make *both* equations true.

Step 4 The width of the plate should be 3.5 inches and its length should be 10.5 inches.

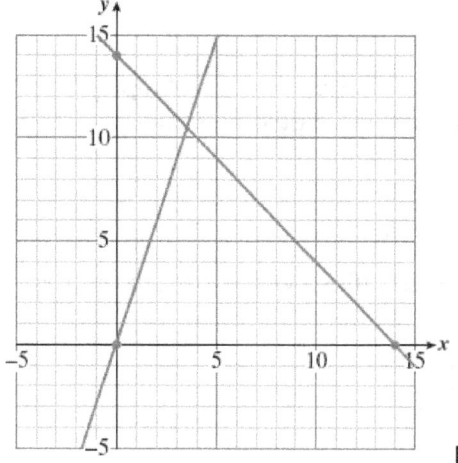

RQ6. When might you use a system of equations to solve an applied problem?

Skills Warm-Up

Write algebraic expressions to answer the questions.

1. How much interest is earned after 2 years on $d + 50$ dollars at 6% annual interest rate?
2. How much manganese is there in $8 - z$ grams of an alloy that is 35% manganese?
3. How far will a boat travel in 2 hours at a speed of $r + 3$ miles per hour?
4. How far will a train moving at 40 miles per hour travel in $3 + x$ hours?
5. A small plane has a top airspeed of v miles per hour. How far can the plane travel in 5 hours against a headwind of 15 miles per hour?
6. A fishing boat has a top speed in still water of 26 miles per hour. How far can the boat travel downstream in 3 hours if the speed of the current is w miles per hour?

Answers: **1.** $0.12(d + 50)$ **2.** $0.35(8 - z)$ **3.** $2(r + 3)$ **4.** $40(3 + x)$ **5.** $5(v - 15)$
6. $3(26 + w)$

Homework 4.2

Skills Practice

▮ In Problems 1-2, decide whether the given ordered pair is a solution of the system.

1. $x + 2y = -8$ $(4, -2)$
$2x - y = 4$

2. $3x + 2y = -5$ $(-3, -2)$
$x + 3y = -9$

▮ In Problems 3-5, decide which graphing technique, the intercept method or the slope-intercept method, would be easier to use. Explain why you made the choice you did.

3. $3x - 2y = 18$

4. $y = \dfrac{5x}{3} - 8$

5. $y = \dfrac{-5}{6}x$

▮ In Problems 6-10, solve the system by graphing.

6. $y = -3x$

$y = 3x - 6$

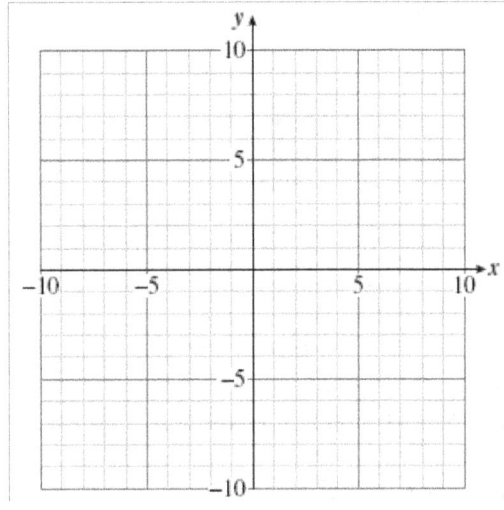

7. $y = \dfrac{3}{4}x - 2$

$2y - 5x = 10$

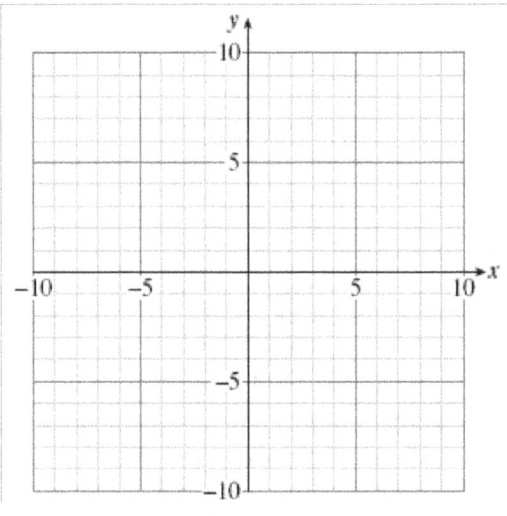

8. $x + y = 4$
$3x - 3y = 12$

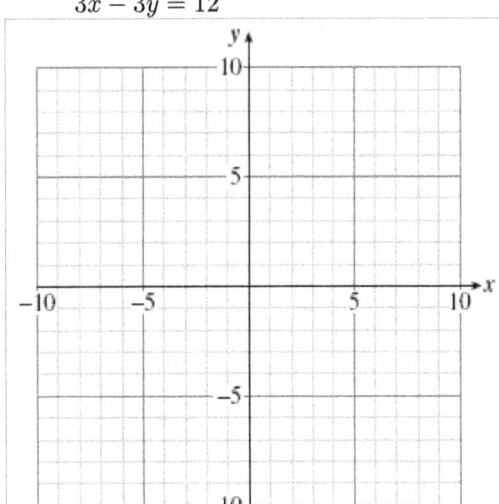

9. $2y = -3x$
$y = -2x - 1$

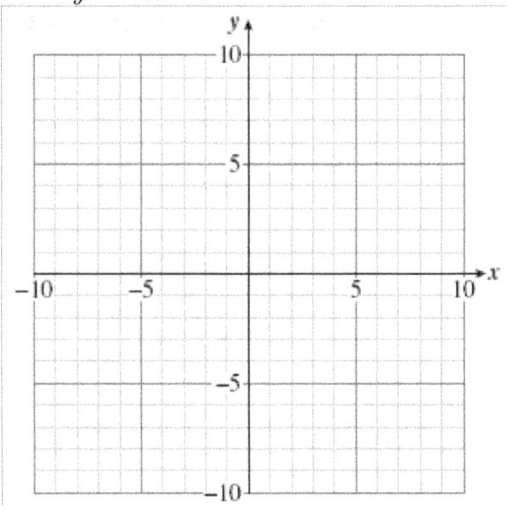

■ In Problems 10-11, decide whether the system is dependent or inconsistent.

10. $x + y = 4$
$2x + 2y = 12$

11. $y = -2x + 4$
$4x + 2y = 2$

12. Without graphing, explain how you can tell that the system has no solution.

$$3x = y + 3$$
$$6x - 2y = 12$$

Applications

13. Delbert has accepted a sales job and is offered a choice of two salary plans. Under Plan A he receives $20,000 a year plus a 3% commission on his sales. Plan B offers a $15,000 annual salary plus a 5% commission.

  a. Let x stand for the amount of Delbert's sales in one year. Write equations for his total annual earnings, E, under each plan.

  Plan A: $E =$

  Plan B: $E =$

  b. Fill in the tables, where x is given in thousands of dollars.

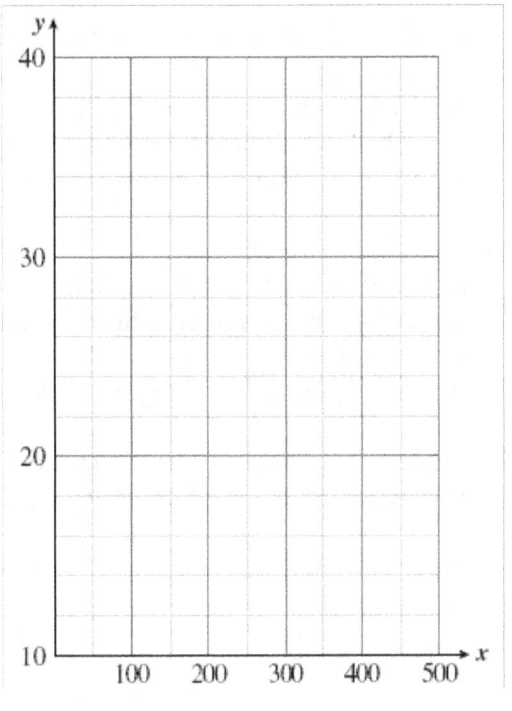

x	Earnings Under Plan A
0	
50	
100	
150	
200	
250	
300	
350	
400	

x	Earnings Under Plan B
0	
50	
100	
150	
200	
250	
300	
350	
400	

  c. Graph both of your equations on the grid. (Both axes are scaled in thousands of dollars.)

  d. For what sales amount do the two plans result in equal earnings for Delbert?

14. Orpheus Music plans to manufacture clarinets for schools. Their startup costs are $6000, and each clarinet costs $60 to make. They plan to sell the clarinets for $80 each.

  a. Let x stand for the number of clarinets Orpheus manufactures. Write equations for the total cost, C, of producing x clarinets, and the revenue, R, earned from selling x clarinets.

  Cost: $C =$

  Revenue: $R =$

b. Fill in the tables.

x	Cost
0	
50	
100	
150	
200	
250	
300	
350	
400	

x	Revenue
0	
50	
100	
150	
200	
250	
300	
350	
400	

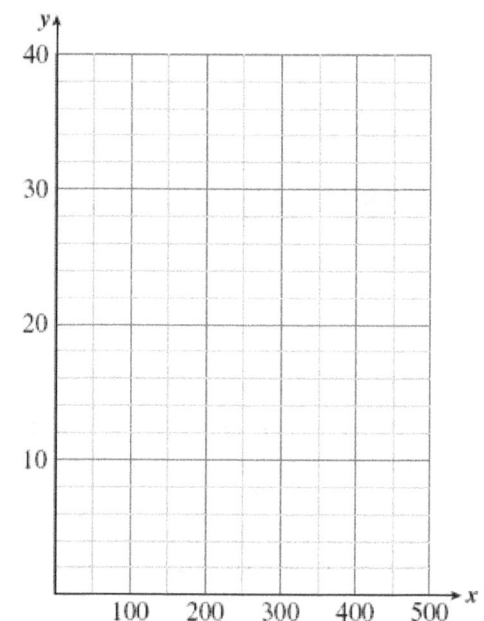

c. Graph both your equations on the grid. (The vertical axis is scaled in thousands of dollars.)
d. How many clarinets must Orpheus sell in order to break even?
e. What is their profit if they sell 500 clarinets? 200 clarinets?
f. Illustrate your answers to part (e) on your graph.

In Problems 15-16,
  a. Choose a variable for each of the unknown quantities.
  b. Write a system of equations in two variables to model the problem.
  c. Solve the system graphically.
  d. Verify your solution algebraically.

15. A bouquet of 4 roses and 8 carnations costs $14. A bouquet of 6 roses and 9 carnations costs $18. How much do one rose and one carnation cost?

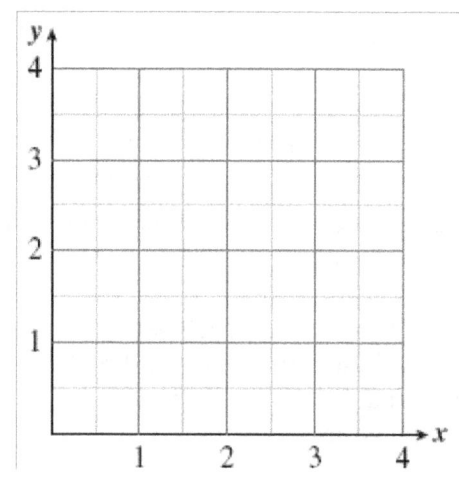

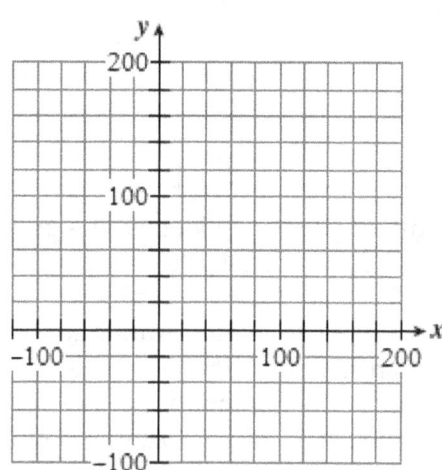

16. The vertex angle of an isosceles triangle is 15° less than each base angle. (In an isosceles triangle, the base angles are equal.) Find the measure of each angle of the triangle. (**Hint:** For your second equation, recall that the sum of the angles in any triangle is 180°.)

4.3 Algebraic Solution of Systems

In the previous Lesson, we compared the costs of operating two different refrigerators, a standard model and an energy-efficient model. We wrote a system of equations for this problem,

$$y = 6x + 1000$$
$$y = 2x + 1200$$

where x is the number of months the refrigerator has been running. We looked for a point on the graphs where the two y-values, which represent the costs, are equal. This point is the intersection point of the graphs, as shown in the figure. The x-coordinate of the intersection point shows that the costs are equal after 50 months.

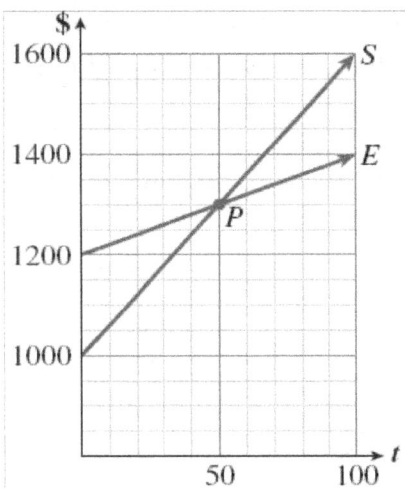

We can also solve the system using algebra.

■ **Example 1**   Solve the system algebraically.

$$y = 6x + 1000$$
$$y = 2x + 1200$$

Solution We are looking for the point where the two y-values are equal. Therefore, we set the two expressions for y equal, which gives us an equation in x to solve:

$6x + 1000 = 2x + 1200$	**Subtract $2x$ from both sides.**
$4x + 1000 = 1200$	**Subtract 1000 from both sides.**
$4x = 200$	**Divide both sides by 4.**
$x = 50$	

We find that $x = 50$, the same answer we got using graphing. Substituting $x = 50$ into either equation gives $y = 1300$. The solution is $(50, 1300)$. ■

Substitution Method

In Example 1, we solved the system by equating the two expressions for y. We can think of this procedure as substituting the expression for y from one equation into the other equation, like this:

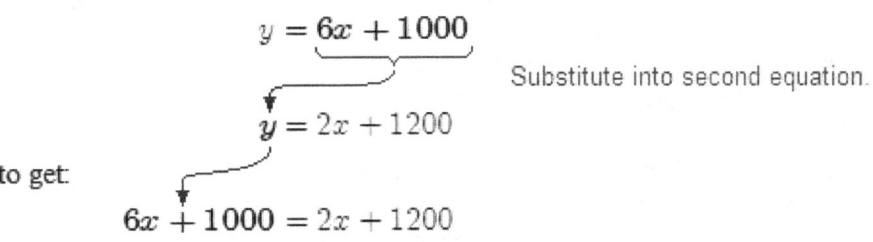

to get:

$$6x + 1000 = 2x + 1200$$

Look Closer: The method works well in this example because both equations in the system were already in the form $y = mx + b$. Sometimes we must solve one of the equations for y in terms of x before we can substitute.   ●

Example 2   Use substitution to solve the system
$$y + 3x = 1$$
$$2y + 5x = 5$$

Solution   We first solve one of the equations for y in terms of x. It is easier to solve the first equation for y, by subtracting $3x$ from both sides to get:

$$y = 1 - 3x$$

We now substitute the expression $1 - 3x$ for y in the *other* equation:

$$y = \underbrace{1 - 3x}$$

$$2y + 5x = 5$$

$$2(1 - 3x) + 5x = 5$$

This gives us an equation in one variable, which we can solve as usual. We begin by simplifying the left side.

$$2(1 - 3x) + 5x = 5 \qquad \text{**Apply the distributive law.**}$$
$$2 - 6x + 5x = 5 \qquad \text{**Combine like terms.**}$$
$$2 - x = 5 \qquad \text{**Subtract 2 from both sides.**}$$
$$-x = 3 \qquad \text{**Divide both sides by** } -1\text{.}$$
$$x = -3$$

This is the x-coordinate of the solution point. To find the y-coordinate, we substitute $x = -3$ into either of the original equations. It is easiest to use the equation we solved for y, namely $y = 1 - 3x$:

$$y = 1 - 3(-3) = 10 \qquad \text{**Substitute** } -3 \text{ **for** } x\text{.}$$

The solution to the system is $x = -3$, $y = 10$, or the point $(-3, 10)$. You can check that these values satisfy both of the original equations in the system. ■

RQ1.   What is the first step in the substitution method?
RQ2.   How can you check your answer for the solution of a system?

Here is a summary of the substitution method.

To Solve a System by Substitution:
1. Choose one of the variables in one of the equations. (It is best to choose a variable whose coefficient is 1 or −1.) Solve the equation for that variable.
2. Substitute the result of Step 1 into the *other* equation. This gives an equation in one variable.
3. Solve the equation obtained in Step 2. This gives the solution value for one of the variables.
4. Substitute this value into the result of Step 1 to find the solution value of the other variable.

Elimination Method

A second algebraic method for solving systems is called **elimination**. As with the substitution method, we try to obtain an equation in a single variable, but we do it by eliminating one of the variables in the system. We first put both equations into the general linear form, $Ax + By = C$.

■ **Example 3**   Solve the system

$$5x = 2y + 21$$
$$2y = 19 - 3x$$

Solution First, we rewrite each equation in the form $Ax + By = C$.

$5x = 2y + 21$ **Subtract** $2y$.	$2y = 19 - 3x$ **Add** $3x$.
$\underline{-2y = -2y}$	$\underline{+3x = \quad +3x}$
$5x - 2y = 21$	$3x + 2y = 19$

We add the equations together by adding the left side of the first equation to the left side of the second equation, and then adding the two right sides together, as follows:

$$
\begin{aligned}
5x - 2y &= 21 \\
\underline{3x + 2y} &= \underline{19} \\
8x \quad\;\; &= 40
\end{aligned}
$$

Note that the y-terms canceled, or were eliminated. We are left with an equation in x that is easy to solve:

$$8x = 40 \qquad \textbf{Divide both sides by 8.}$$
$$x = 5$$

We are not finished yet, because we must still find the value of y. We can substitute our value for x into either of the original equations, and solve for y. We'll use the second equation, $3x + 2y = 19$:

$$3(5) + 2y = 19 \qquad \textbf{Subtract 15 from both sides.}$$
$$2y = 4 \qquad\qquad \textbf{Divide by 2.}$$
$$y = 2$$

Thus, the solution is the point $(5, 2)$. ■

Look Ahead: You may have noticed that this method worked because the coefficients of y in the two equations were opposites, 2 and -2. This caused the y-terms to cancel out when we added the two equations together. What if the coefficients of neither x nor y are opposites? Then we must multiply one or both of the equations in the system by a suitable constant. ●

■ **Example 4**   Solve the system

$$4x + 3y = 7$$
$$3x + y = -1$$

Solution   We can choose to eliminate either the x-terms or the y-terms in a system. For this example, it will be faster to eliminate the y-terms. If we multiply each term of the second equation by -3, then the coefficients of y will be opposites:

$$-3(3x + y = -1) \quad \rightarrow \quad -9x - 3y = 3$$

Because we are applying the multiplication property of equality, we must multiply *each* term by -3, not just the y-term. We can then replace the second equation by its new version to obtain a new system, and we add the equations together:

$$\begin{array}{r} 4x + 3y = 7 \\ -9x - 3y = 3 \\ \hline -5x \quad\;\; = 10 \end{array}$$

The y-terms were eliminated, and we solve the resulting equation for x to get $x = -2$. Finally, we substitute $x = -2$ into either of the equations to find $y = 5$. The solution is $(-2, 5)$. ■

Caution! In Example 4, we multiplied both sides of the equation $3x + y = -1$ by -3. (This is an application of the Multiplication Property of Equality.) We must be careful to multiply *every term on both sides of the equation* by the same constant.   Otherwise, we won't have an equivalent equation -- its solutions will not be the same. ■

When we add a multiple of one equation to the other we are making a **linear combination** of the equations.

The method of elimination is also called the method of linear combinations. Sometimes it is necessary to multiply *both* equations by suitable constants in order to eliminate one of the variables.

■ **Example 5**   Use linear combinations to solve the system

$$\begin{array}{r} 5x - 2y = 22 \\ 2x - 5y = 13 \end{array}$$

Solution This time we choose to eliminate the x-terms. We must arrange things so that the coefficients of the x-terms are opposites, so we look for the smallest integer that both 2 and 5 divide into evenly. (This number is called the *lowest common multiple*, or LCM, of 2 and 5.) The LCM of 2 and 5 is 10. We want one of the coefficients of x to be 10, and the other to be -10. To achieve this, we multiply the first equation by 2 and the second equation by -5.

$$\begin{array}{rcl} \mathbf{2}\,(5x - 2y = 22) & \rightarrow & 10x - 4y = 44 \\ \mathbf{-5}\,(2x - 5y = 13) & \rightarrow & -10x + 25y = -65 \end{array}$$

Adding these new equations eliminates the x-term and yields an equation in y.

$$\begin{array}{r} 10x - \;\,4y = 44 \\ -10x + 25y = -65 \\ \hline 21y = -21 \end{array}$$

We solve for y to find $y = -1$. Finally, we substitute $y = -1$ into the first equation and solve for x.

$$\begin{array}{r} 5x - 2(-1) = 22 \\ 5x + 2 = 22 \\ x = 4 \end{array}$$

The solution to the system is $(4, -1)$. ■

> **RQ3.** What is a linear combination of expressions?
> **RQ4.** What is the first step in the elimination method?

Here are the steps for solving a system by elimination.

To Solve a System by Elimination:
1. Write each equation in the form $Ax + By = C$.
2. Decide which variable to eliminate. Multiply each equation by an appropriate constant so that the coefficients of that variable are opposites.
3. Add the equation from Step 2 and solve for the remaining variable.
4. Substitute the value found in Step 3 into one of the original equations and solve for the other variable.

Look Ahead: How do you know which method to use to solve a system, substitution or elimination? Both methods will work on any linear system. However, substitution will be easier if one of the variables in one of the equations has a coefficient of 1 or -1. Otherwise, the elimination method is usually more efficient.

> **RQ5.** How do we eliminate a variable from a system?
> **RQ6.** When is the substitution method easier than elimination?

Inconsistent and Dependent Systems

Recall that a system of two parallel lines has no solution and is called inconsistent. If the two equations in a system have the same graph, then every point on the graph is a solution and the system is called dependent. The elimination method will reveal whether the system falls into one of these two cases.

■ **Example 6**   Solve each system by elimination.

a. $3x - y = 2$
$\quad\;\; -6x + 2y = 3$

b. $x - 2y = 3$
$\quad\;\; 2x - 4y = 6$

Solutions a.   To eliminate the y-terms, we multiply the first equation by 2 and add:

$$\begin{array}{r} 6x - 2y = 4 \\ \underline{-6x + 2y = 3} \\ 0x + 0y = 7 \end{array}$$

Both variables are eliminated, and we are left with the false statement $0 = 7$. There are no values of x or y that will make this equation true, so the system has no solutions. The graph shows that the system is inconsistent.

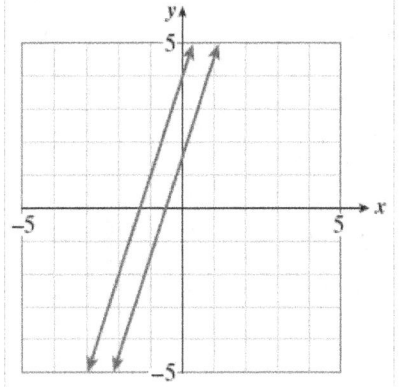

b. To eliminate the x-terms we multiply the first equation by -2 and add:

$$
\begin{array}{r}
-2x + 4y = -6 \\
2x - 4y = 6 \\
\hline
0x + 0y = 0
\end{array}
$$

We are left with the true but unhelpful equation $0 = 0$. The two equations are in fact equivalent (one is a constant multiple of the other), so the system is dependent. The graph of both equations is shown in the figure.

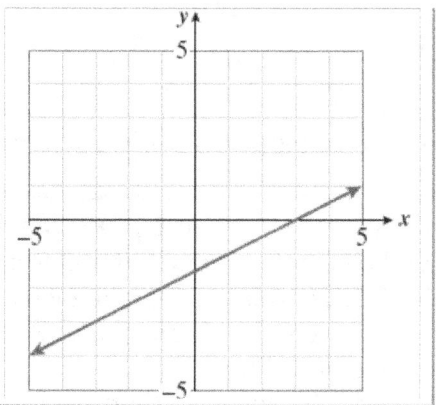

Example 5 illustrates a rule for identifying inconsistent and dependent systems.

When Using Elimination to Solve a System:

1. If combining the two equations results in an equation of the form

$$0x + 0y = k \quad (k \neq 0)$$

then the system is inconsistent.

2. If combining the two equations results in an equation of the form

$$0x + 0y = 0$$

then the system is dependent.

RQ7. How can you tell if a system is inconsistent?
RQ8. How can you tell if a system is dependent?

Skills Warm-Up

Simplify each expression. Write your answers in the form $ax + b$.

1. $2x + 3(4x - 2) + 1$ **2.** $7x - 5(2x + 3) - 2$ **3.** $\dfrac{3}{2}(4x - 6) - 3x$

4. $\dfrac{2}{3}(6x + 3) + x$ **5.** $\dfrac{2x + 5}{3} - 1$ **6.** $\dfrac{3x - 1}{4} + 2$

7. $x + 4 - \dfrac{3 - 5x}{2}$ **8.** $8 - x - \dfrac{4 + 2x}{3}$

Answers: **1.** $14x - 5$ **2.** $-3x - 17$ **3.** $3x - 9$ **4.** $5x + 2$ **5.** $\dfrac{2}{3}x + \dfrac{2}{3}$

6. $\dfrac{3}{4}x + \dfrac{7}{4}$ **7.** $\dfrac{7}{2}x + \dfrac{5}{2}$ **8.** $\dfrac{-5}{3}x - \dfrac{36}{3}$

Homework 4.3

Skills Practice

In Problems 1-6, solve the system by the substitution method.

1. $y = 2x$
$3x + y = 10$

2. $2a + 3b = 4$
$b = 2a + 4$

3. $2x - 3y = 4$
$x + 1 = 3y$

4. $2a + b = 16$
$4a - 4 = -3b$

5. $8r + s = 4$
$2r + 7s = -8$

6. $36a = 6 - 3b$
$2b = 3a - 5$

In Problems 7-12, solve the system by elimination.

7. $x + y = 5$
$x - y = 1$

8. $2a + b = 3$
$a + b = 2$

9. $3x + 2y = 7$
$x + y = 3$

10. $3a = 1 + 5b$
$2b = 14 + 6a$

11. $2x + 3y = -1$
$3x + 5y = -2$

12. $5z = 1 - 3w$
$5w = 2 - 7z$

In Problems 13-15, decide whether the system is inconsistent, dependent, or consistent.

13. $3x - 6y = 6$
$x - 2y = 3$

14. $8a = 6 + 12b$
$4 = 6a - 9b$

15. $3p = 1 + 7q$
$21q = 9p - 3$

In Problems 16-17, which method, elimination or substitution, would be easier to apply to each system?

16. $2u + 3v = -8$
$2u - 3v = -6$

17. $17a - b = 0$
$11a + 3b = 4$

Applications

For Problems 18-21,
   a. Choose variables for the two unknown quantities in the problem.
   b. Write a system of equations to model the problem.
   c. Solve the system and answer the question.

18. A hamburger and a chocolate shake together contain 1030 calories. Two shakes and three hamburgers contain 2710 calories. How many calories are there in one hamburger and in one chocolate shake?

19. Darryl is in charge of buying furniture for a new restaurant. Chairs cost $175 and tables cost $250. Darryl's budget allows $11,400 for tables and chairs, and he plans to buy four chairs for each table. How many of each can he buy?

20. The perimeter of a rectangle is 42 meters, and its width is 13 meters less than its length. Find the dimensions of the rectangle.

21. Three pounds of bacon and 2 pounds of coffee costs $17.80. Two pounds of bacon and 5 pounds of coffee costs $32.40. How much do 1 pound of bacon and 1 pound of coffee cost?

4.4 Problem Solving with Systems

In this section we explore some applications of systems of equations. We begin by reviewing some familiar formulas.

Interest

Recall the formula for calculating interest,

$$I = Prt$$

I is the interest you will earn after t years if you invest a principal of P dollars in an account that earns simple interest at an annual rate r. If we make two or more investments at the same time, then the total interest we earn is the sum of the interests earned on each investment separately.

Caution!   Notice the difference between interest, I, and interest rate, r. The **interest rate** is a *percentage*, such as 5%. The **interest** is the *amount* of money you earn, usually in dollars.   ■

Warm-Up 1   Harvey deposited $1200 in two accounts. He put $700 in a savings account that pays 6% annual interest rate, and the rest in his credit union, which pays 7% annual interest rate. How much will Harvey's investments earn in two years?

Step 1   Calculate the interest earned by each account.

Savings: $I = Prt =$
Credit union: $I = Prt =$

Step 2   Add the earnings from the two investments.

$$\begin{pmatrix} \text{Total} \\ \text{interest} \end{pmatrix} = \begin{pmatrix} \text{Interest from} \\ \text{stocks} \end{pmatrix} + \begin{pmatrix} \text{Interest from} \\ \text{bonds} \end{pmatrix}$$

Here is a similar example with variables.

Warm-Up 2   You have $5000 to invest for one year. You want to put part of the money into bonds that pay 7% interest rate and the rest of the money into stocks, which involve some risk but will pay 12% if the investment is successful.

a.   Fill in the table.

Amount invested in stocks	Amount invested in bonds	Interest from stocks	Interest from bonds	Total interest
$500				
$1000				
$3200				
$4000				

Now suppose you invest x dollars in the stocks and y dollars in the bonds.

b.   Sum of amounts invested: $x + y =$
c.   Interest earned on the stocks: $I = Prt =$
       Interest earned on the bonds: $I = Prt =$

d. Finally, suppose that you earned a total of $345 in interest from your two investments. Write an equation about this, using your expressions from part (c).

$$\left(\begin{array}{c}\text{Total}\\\text{interest}\end{array}\right) = \left(\begin{array}{c}\text{Interest from}\\\text{stocks}\end{array}\right) + \left(\begin{array}{c}\text{Interest from}\\\text{bonds}\end{array}\right)$$

RQ1. What formula do we need for interest problems?

RQ2. What is the difference between principal and interest?

■ **Example 1**   Mort invested money in two accounts, a savings plan that pays 8% interest and a mutual fund that pays 7% interest. He put twice as much money in the savings plan as in the mutual fund. At the end of the year Mort's total interest income was $345. How much did he invest in each account?

Solution The unknown quantities are the amounts Mort invested in his two accounts.

$$\begin{array}{ll}\text{Amount invested in savings:} & x\\\text{Amount invested in mutual fund:} & y\end{array}$$

Do not confuse the amount Mort invested in each account (the principal) with the amount he earned in interest! A table is a good way to keep these amounts straight.

	Principal	Interest Rate	Interest
Savings plan	x	0.08	$0.08x$
Mutual fund	y	0.07	$0.07y$

We must write two equations for the problem, one about the principal and one about the interest. From the statement "He put twice as much money in the savings plan as in the mutual fund," we write the equation

$$\left(\begin{array}{c}\text{Amount invested}\\\text{in savings plan}\end{array}\right) = 2 \cdot \left(\begin{array}{c}\text{Amount invested}\\\text{in mutual fund}\end{array}\right)$$

Equation about principal: $x \qquad\qquad = 2y$

From the statement "Mort's total interest income was $345," we write the equation

$$\left(\begin{array}{c}\text{Interest from}\\\text{savings plan}\end{array}\right) + \left(\begin{array}{c}\text{Interest from}\\\text{mutual fund}\end{array}\right) = \text{Total interest}$$

Equation about interest: $0.08x \quad + \quad 0.07y \quad = 345$

This gives us a system of equations.

$$x = 2y$$
$$0.08x + 0.07y = 345$$

Because the first equation is already solved for x in terms of y, we'll use the substitution method to solve the system. We substitute $2y$ for x in the second equation and solve for y.

$$\begin{array}{ll}0.08(2y) + 0.07y = 345 & \\0.16y + 0.07y = 345 & \textbf{Combine like terms.}\\0.23y = 345 & \textbf{Divide both sides by 0.23.}\\y = 1500 &\end{array}$$

Finally, we substitute $y = 1500$ into the first equation to find

$$x = 2y = 2(1500) = 3000$$

Thus, Mort invested $3000 in the savings plan and $1500 in the mutual fund.   ■

RQ3.   What were the two equation in Example 1 about?
RQ4.   Which method did we use to solve the system in Example 1?

Mixtures

A pharmacist has on hand 20 ounces of a certain drug at 40% strength, but she needs a small quantity of the drug at 75% strength for a prescription. She decides to add a pure form of the drug to the 40% solution. How much should she add to make a mixture of 75% strength?

To solve such problems, we review some properties of percent. We will need the percent formula,

$$P = rW$$

P stands for the part obtained when we take r percent of a whole amount, W.

Warm-Up 3   You have two jars of marbles. The first contains 40 marbles, of which 10 are red, and the second contains 60 marbles, of which 30 are red.

a. What percent of the marbles in the first jar are red? $r = \dfrac{P}{W} =$

What percent of the marbles in the second jar are red? $r = \dfrac{P}{W} =$

You pour both jars of marbles into a larger jar and mix them together.

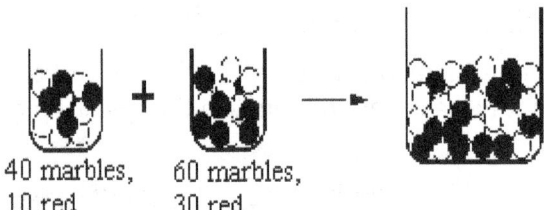

40 marbles,   60 marbles,
10 red   30 red

b. How many marbles total are in the larger jar?

How many red marbles are in the larger jar?

What percent of marbles in the larger jar are red? $r = \dfrac{P}{W} =$

c. Can we add the percents for the first two jars to get the percent red marbles in the mixture?

Caution! In Warm-Up 3, you should find that adding the percent of red marbles in the first two jars does **not** give the percent of red marbles in the mixture. That is,

$$25\% + 50\% \neq 40\%$$

In general, **we cannot add percents** unless they are percents of the same whole amount.

RQ5.   What formula do we need for mixture problems?
RQ6.   When can we add percents?

Warm-Up 4   In a local city council election, your favored candidate, Justine Honest, ran in a small district with two precincts. Ms. Honest won 30% of the 500 votes cast in Precinct 1 and 70% of the 300 votes cast in Precinct 2. Did Candidate Honest win a majority (more than 50%) of the votes in her district?

Step 1   Fill in the first two rows of the table, using information from the problem and the formula $P = rW$.

	Total votes (W)	Percent for Honest (r)	Votes for Honest (P)
Precinct 1			
Precinct 2			
Entire district			

Step 2   Add down to complete the first and third columns of the table.

Step 3   Fill in the last entry in the table to answer the question in the problem. Use the formula $P = rW$ again.

So far we have considered mixture problems involving discrete objects, such as marbles or votes. The same methods apply to mixtures of liquids.

Example 2   A chemist wants to produce 45 milliliters of a 40% solution of carbolic acid by mixing a 20% solution with a 50% solution. How many milliliters of each should he use?

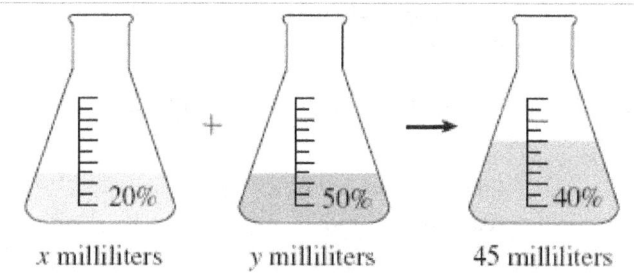

Solution We let x represent the number of milliliters of the 20% solution he needs and y the number of milliliters of the 50% solution. We use a table to organize the information. The first two columns contain the variables and information given in the problem: the number of milliliters of each solution and its strength as a percent.

	Number of Milliliters (W)	Percent Acid (r)	Amount of Acid (P)
20% Solution	x	0.20	
50% Solution	y	0.50	
Mixture	45	0.40	

We fill in the last column of the table by using the formula $P = rW$. The entries in this last column give the amount of the important ingredient (in this case, milliliters of acid) in each component solution and in the mixture.

	Number of Milliliters (W)	Percent Acid (r)	Amount of Acid (P)
20% Solution	x	0.20	$0.20x$
50% Solution	y	0.50	$0.50x$
Mixture	45	0.40	$0.40(45)$

Now we can write two equations about the mixture problem. The first equation is about the total number of milliliters mixed together. The chemist must mix x milliliters of one solution with y milliliters of the other solution and end up with 45 milliliters of the mixture, so

$$\text{Total amount of mixture:} \qquad x + y = 45$$

The second equation uses the fact that the acid in the mixture can only come from the acid in each of the two original solutions. We used the last column of the table to calculate how much acid was in each component, and we add these quantities to get the amount of acid in the mixture.

$$\text{Amount of acid:} \qquad 0.20x + 0.50y = 0.40(45)$$

These two equation make up a system:

$$x + y = 45$$
$$0.20x + 0.50y = 18$$

To simplify the system we first multiply the second equation by 100 to clear the decimals.

$$x + y = 45$$
$$20x + 50y = 1800$$

We solve the system by elimination. Multiply the first equation by $-20,$ and add the equations together.

$$-20x - 20y = -900$$
$$\underline{20x + 50y = 1800}$$
$$30y = 900$$

Solving for y, we find $y = 30$. Substitute $y = 30$ into the first equation to find

$$x + 30 = 45$$

or $x = 15$. The chemist needs 15 milliliters of the 20% solution and 30 milliliters of the 50% solution for the mixture.   ■

Look Closer: Notice that, in Example 2, once we have completed the table, it is easy to write a system of equations; we simply add down the first and third columns of the table. (Remember that we cannot add down the middle column, because percents don't add!)   ●

RQ7.   What were the two equation in Example 2 about?
RQ8.   Which method did we use to solve the system in Example 2?

Mixture problems are easy to solve if you complete a table first. The rows of the table represent the two components and the final mixture, and the columns are used to calculate the amount of the important ingredient in each. Here is a sample table that you can customize for the specifics of a particular problem.

	Total Amount (W)	Percent of Important Ingredient (r)	Amount of Important Ingredient (P)
First Component			
Second Component			
Mixture			

Motion

We can also use systems of equations to solve problems involving motion at a constant speed. We'll need the formula $D = RT$.

■ **Example 3**   Geologists can calculate the distance from their seismograph to the epicenter of an earthquake by timing the arrival of the P and S waves. They know that P waves travel at about 5.4 miles per second and S waves travel at 3.0 miles per second. If the P waves arrived 3 minutes before the S waves, how far away is the epicenter of the quake?

Solution Let x represent the distance from the seismograph to the epicenter. The time it took for the waves to arrive is also unknown, so we'll let y be the travel time for the P waves, in seconds. The travel time for the S waves is then $y + 180$ seconds.
We organize all this information in a table.

	Rate	Time	Distance
P waves	5.4	y	x
S waves	3.0	$y + 180$	x

We can now write two equations about the problem, one for the P waves and one for the S waves, using the formula $RT = D$.

$$5.4\,y = x$$
$$3.0\,(y + 180) = x$$

We solve the system by substitution. Substitute $5.4y$ for x in the second equation, and then solve for y.

$$
\begin{aligned}
3.0\,(y + 180) &= 5.4y & &\textbf{Apply the distributive law.}\\
3y + 540 &= 5.4y & &\textbf{Subtract } 3y \textbf{ from both sides.}\\
540 &= 2.4y & &\textbf{Divide both sides by 2.4.}\\
225 &= y
\end{aligned}
$$

Thus, it took the P waves 225 seconds to arrive at the seismograph. To solve for the distance x, substitute $y = 225$ into the first equation to find

$$x = 5.4\,(225) = 1215$$

The epicenter is located 1215 miles from the seismograph. ■

RQ9.   What formula do we need for motion problems?
RQ10. What did x and y represent in Example 3?

Warm-up Answers: **1.** $154   **2b.** $5000 **c.** $0.07x$, $0.12y$ **d.** $0.07x + 0.12y$
3a. 25%, 50%   **b.** 100, 40, 40%   **c.** No   **4.** No, 45%

Skills Warm-Up

For part (b) of each problem, write an algebraic expression in two variables.

1. **a.** How much interest will you earn in 1 year if you invest \$2400 in a T-bill that pays $7\frac{1}{2}\%$ interest and \$800 in a savings account that pays 4.8% interest?
   b. How much will you earn if you invest x dollars in the T-bill and y dollars in the savings account?
2. **a.** How far will you travel if you jog for 40 minutes at 9 miles per hour, and then jog for 30 minutes at 5 miles per hour?
   b. How far will you travel if you jog for 40 minutes at x miles per hour, and the jog for 30 minutes at y miles per hour?
3. **a.** How long will it take you to drive 180 miles on the highway at an average speed of 60 miles per hour, and then 30 miles on a gravel road at an average speed of 40 miles per hour?
   b. How long will it take you if you drive on the highway at x miles per hour and on the gravel road at y miles per hour?
4. **a.** How much nitrogen is in a mixture of 10 pounds of fertilizer that is 6% nitrogen and 4 pounds of fertilizer that is 60% nitrogen?
   b. How much nitrogen is in a mixture of x pounds of the first fertilizer and y pounds of the second fertilizer?

Answers: **1a.** \$281.40 **b.** $0.075x + 0.048y$ **2a.** 8.5 miles **b.** $\frac{2}{3}x + \frac{1}{2}y$
3a. 3.75 hours **b.** $\dfrac{180}{x} + \dfrac{30}{y}$ **4a.** 3 lbs. **b.** $0.06x + 0.60y$

Homework 4.4

Applications

■ Use systems to solve Problems 1-3 about interest.

1. Goodlife Insurance Company has $150,000 in fees from its clients to invest. They deposited part of the money into bonds that pay 6.5% annual interest and the rest into a mutual fund that pays 11.8% annual interest.
   a. Assign variables to the amount of money Goodlife deposited into each account, and write an equation about the sum of the deposits.
   b. Write expressions for the interest earned on each account after 1 year.
   c. Goodlife earned $12,930 interest in 1 year. Write an equation about that.
   d. Solve your system of equations to find out how much Goodlife invested in each account.

2. Mario borrowed $30,000 from two banks to open a print shop. The first loan charges 12% annual interest, and the second charges 15% interest. Mario's annual interest payment on both loans together is $3750. How much did he borrow at each rate?
   a. Assign variables to the unknown quantities and make a table like the one in Example 1.
   b. Write two equations for the problem, one about the principals and one about the interests.
   c. Solve your system and answer the question in the problem.

3. Stefan borrowed twice as much on his 7% car loan as on his 4% student loan. The annual interest on the car loan is $500 more than the interest on the student loan. How much did Stefan borrow on each loan?

■ Solve Problems 4-5 about mixtures. (No variables are needed!)

4. The chemistry department has 80 students, of whom 35% are women. The physics department has 60 students, of whom 15% are women.
   a. How many chemistry students are women? How many physics students are women?
   b. How many students are there in both departments? How many of them are women?
   c. What percent of the students in chemistry and physics are women?

	Number of Students (W)	Percent Women (r)	Number of Women (P)
Chemistry			
Physics			
Toral			

5. Pipette, a French chemistry student, has 30 milliliters of a 50% solution of acid. She wants to reduce the strength by adding 12 milliliters of a 15% solution of the same acid.
    a. How much acid is in the 30 milliliters of 50% solution? How much acid is in the 12 milliliters of 15% solution?
    b. How much acid is in the mixture? How many milliliters of the mixture are there?
    c. What percent of the mixture is acid?
    d. Fill in the table with your answers to parts (a)-(c).

	Number of Milliliters (W)	Percent Acid (r)	Amount of Acid (P)
50% Solution			
15% Solution			
Mixture			

■ Use a system of equations to solve Problems 6-7 about mixtures.

6. A pet store owner wants to mix a 12% saltwater solution and a 30% saltwater solution to obtain 45 liters of a 24% solution. How many liters of each ingredient does he need?
    a. Choose variables and make a table for the problem.
    b. Use your table to write two equations about the mixture.
    c. Solve your system and answer the question in the problem.

7. A newspaper poll of 400 people stated that 58% were in favor of a recycling program. It also said that 50% of the men and 70% of the women polled favored the program. How many women were polled?

■ Solve Problems 8-9 about motion.

8. Delbert and Francine leave Cedar Rapids at the same time and drive in opposite directions for 6 hours.
    a. Choose variables for Delbert and Francine's speeds and fill in the table.

	Rate	Time	Distance
Delbert			
Francine			

    b. Make a sketch showing Cedar Rapids, Delbert, and Francine. Label your sketch with the distance that each traveled.
    c. Francine drove 5 miles per hour slower than Delbert. After 6 hours, they are 570 miles apart. Write two equations about the problem.
    d. Solve your system to find Delbert's and Francine's speeds.

9. Bonnie left Dallas and drove north at 40 miles per hour. Three hours later Clyde
headed north form Dallas on the same road at 70 miles per hour until he caught up
with Bonnie.
a. Make a sketch showing Dallas, Bonnie, and Clyde.
b. Complete the table.

	Rate	Time	Distance
Bonnie		t	d
Clyde			

c. Use your table to write two equations about the problem.
d. Solve your system. How long did Bonnie drive before Clyde caught up? How far
had she driven?

■ Use a system of equations to solve Problems 10-11.

10. A yacht leaves San Diego and heads south, traveling at 25 miles per hour. Six hours
later a Coast Guard cutter leaves San Diego traveling at 40 miles per hour and pursues
the yacht. How long will it take the cutter to catch the yacht? How far will they have
traveled?

11. Byron and Ada conduct sight-seeing tours of New England by bicycle. Byron leads the
tour group at an average speed of 10 miles per hour, while Ada goes ahead at a speed
of 12 miles per hour to prepare lunch. Ada arrives at the lunch stop 40 minutes ($\frac{2}{3}$ of
an hour) before the tour group. How far did the tour bicycle that morning?

■ Solve by writing a single equation in two variables.

12. In Julio's history class, the final grade is
computed by adding 60% of the test
average to 40% of the term paper grade.
Julio's test average is 72.
a. Write an equation for Julio's final grade,
g, in terms of his term paper grade, t.
b. What will Julio's final grade be if he earns
a grade of 65 on the term paper? What
if he earns a grade of 80?
c. Use the graph at right to estimate what
grade Julio must make on the term
paper in order to earn a final grade of 80
in the class.
d. Use your equation from part (a) to verify
your answer algebraically.

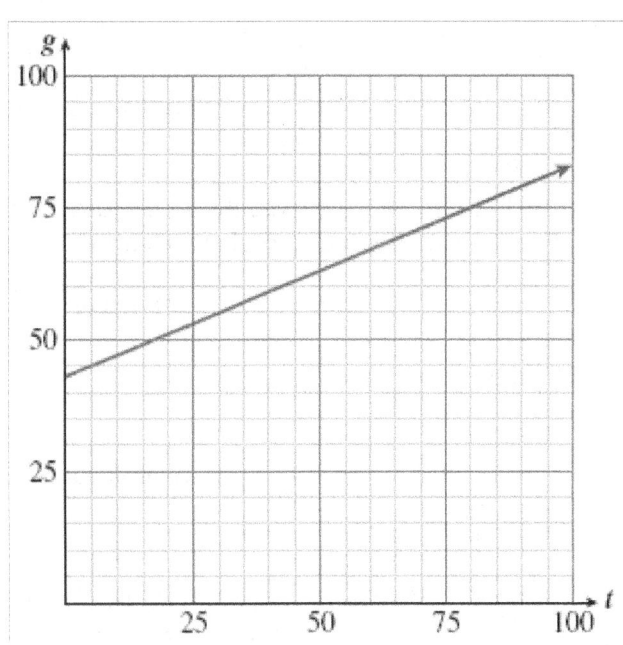

4.5 Point-Slope Form

Finding an Equation for a Line

The slope of a line is the same, no matter which points we use to compute it. Therefore, if we know the slope m of a line and any one point (x_1, y_1) on the line, then all other points (x, y) on the line must satisfy the slope formula:

$$\frac{y - y_1}{x - x_1} = m$$

This is, in fact, an equation for the line.

■ **Example 1a.**   Graph the line that passes through the point $(1, 3)$ and has slope -2.
b. Find an equation for the line that passes through the point $(1, 3)$ and has slope -2.

Solutions a.   We plot the point $(1, 3)$, then use the slope, $m = \dfrac{\Delta y}{\Delta x} = \dfrac{-2}{1}$, to find another point on the line. From the point $(1, 3)$, we move 2 units down and 1 unit to the right, arriving at $(-1, 4)$. We draw the line through these two points.

b. We use the formula

$$m = \frac{y - y_1}{x - x_1}$$

with $m = \dfrac{\Delta y}{\Delta x} = \dfrac{-2}{1}$ and $(x_1, y_1) = (1, 3)$ to get

$$\frac{-2}{1} = \frac{y - 3}{x - 1}$$

To simplify the equation, we cross-multiply.

$$1(y - 3) = -2(x - 1)$$
$$y - 3 = -2x + 2$$
$$y = -2x + 5$$

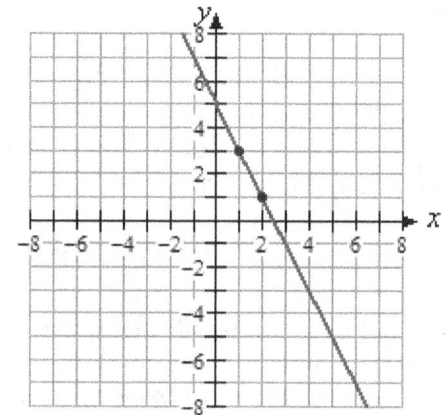

In Example 1 we used the slope formula in a new way: to find the equation of a line. We can simplify the formula by clearing the denominator:

$$(x - x_1)\,\frac{y - y_1}{x - x_1} = m\,(x - x_1)$$

to get

$$y - y_1 = m\,(x - x_1)$$

We call this the **point-slope formula** for linear equations.

Point-Slope Formula
　　To find an equation for the line of slope m passing through the point (x_1, y_1), use the **point-slope formula**

$$\frac{y - y_1}{x - x_1} = m$$

or

$$y - y_1 = m\,(x - x_1)$$

Look Closer: What is the difference between the slope formula and the point-slope formula?

the **slope formula** $m = \dfrac{y_2 - y_1}{x_2 - x_1}$

the **point-slope formula** $m = \dfrac{y - y_1}{x - x_1}$

They are really the same formula, but they are used for different purposes:

1. We use the slope formula to calculate the slope of a line when we know two points on the line. That is, we know $(x_1,\ y_1)$ and $(x_2,\ y_2)$, and we are looking for m.
2. We use the point-slope formula to find the equation of a line. That is, we know $(x_1,\ y_1)$ and m, and we are looking for $y = mx + b$.

RQ1.   Give two versions of the point-slope formula.
RQ2.   What is the point-slope formula used for?

◼ Example 2   Find an equation for the line that passes through the point $(1, 4)$ and is perpendicular to the line $4x - 2y = 6$.

Solution We first find the slope of the desired line, then use the point-slope formula to write its equation. The line we want is perpendicular to the given line, so its slope is the negative reciprocal of $m_1 = 2$, the slope of the given line. Thus

$$m_2 = \frac{-1}{m_1} = \frac{-1}{2}$$

Now use the point-slope formula with $m = \frac{-1}{2}$
and $(x_1, y_1) = (1, 4)$.

$y - y_1 = m\,(x - x_1)$

$y - 4 = \dfrac{-1}{2}\,(x - 1)$ Apply distributive law.

$y - 4 = \dfrac{-1}{2}\,x + \dfrac{1}{2}$ Add 4 to both sides.

$y = \dfrac{-1}{2}\,x + \dfrac{9}{2}$ $\dfrac{1}{2} + 4 = \dfrac{1}{2} + \dfrac{8}{2} = \dfrac{9}{2}$

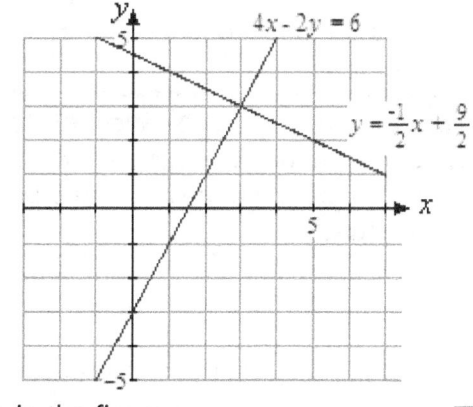

The given line and the perpendicular line are shown in the figure.   ◼

Caution! If we happen to know the y-intercept of a line, we can write its equation using the slope-intercept formula, $y = mx + b$. But if we don't know the y-intercept and instead know a different point, it is easier to use the point-slope formula. ◼

The Line Through Two Points

How many lines pass through two given points? There is only one. We can use the point-slope formula to can find its equation.

■ **Example 3**   Find an equation for the line that passes through $(2, -1)$ and $(-1, 3)$.

Solution We solve this problem in two steps:
First, we find the slope of the line, and then we use
the point-slope formula.
Step 1   Let $(x_1, y_1) = (2, -1)$ and
$(x_2, y_2) = (-1, 3)$. Using the slope formula, we find
that

$$m = \frac{y_2 - y_1}{x_2 - x_1}$$
$$= \frac{3 - (-1)}{-1 - 2} = \frac{4}{-3} = \frac{-4}{3}$$

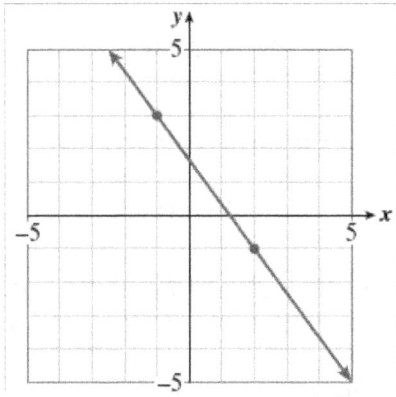

Step 2   We apply the point-slope formula with $m = \frac{-4}{3}$ and $(x_1, y_1) = (2, -1)$. (We can
use either point to find the equation of the line.) Then

$$\frac{y - y_1}{x - x_1} = m \qquad \text{becomes} \qquad \frac{y - (-1)}{x - 2} = \frac{-4}{3}$$

We cross-multiply to find

$$3(y + 1) = -4(x - 2) \qquad \textbf{Apply the distributive law.}$$
$$3y + 3 = -4x + 8 \qquad \textbf{Solve for } y.$$
$$3y = -4x + 5$$
$$y = \frac{-4}{3}x + \frac{5}{3}$$

The graph of the line is shown in the figure above.   ■

RQ3.   How many different lines pass through two given points?
RQ4.   What formula do we use to find the equation of a line through two points?

To Fit a Line through Two Points :
1. Compute the slope between the two points.
2. Substitute the slope and either point into the point-slope
    formula.

Applications

Variables that increase or decrease at a constant rate can be described by linear
equations.

■ **Example 4**   In 1993, Americans drank 188.6 million cases of wine. Wine
consumption increased at a constant rate over the next decade, and in 2003 we drank
258.3 million cases of wine.   (Source: LA Times, Adams Beverage Group)

a. Find a formula for wine consumption, W, in millions of cases, t years after 1990.
b. State the slope as a rate of change. What does the slope tell us about this problem?

Solutions a.   We have two data points of the form (t, W), namely $(t_1, W_1) = (3, 188.6)$
and $(t_2, W_2) = (13, 258.3)$. We use the point slope formula to fit a line through these two
points. First we compute the slope.

$$\frac{\Delta W}{\Delta t} = \frac{W_2 - W_1}{t_2 - t_1} = \frac{258.3 - 188.6}{13 - 3} = 6.97$$

Next, we substitute the slope $m = 6.97$ and either of the two data points into the point-slope formula.

$$W = W_1 + m(t - t_1)$$
$$W = 188.6 + 6.97(t - 3)$$
$$W = 167.69 + 6.97t$$

Thus, $W = 167.69 + 6.97t$.

b. The slope gives us the rate of change of W with respect to t. The units of the variables can help us interpret the slope in context.

$$\frac{\Delta W}{\Delta t} = \frac{258.3 - 188.6}{13 - 3} \frac{\text{millions of cases}}{\text{years}} = 6.97 \text{ millions of cases/year}$$

Over the ten years between 1993 and 2003, wine consumption in the US increased at a rate of 6.97 million cases per year.

RQ5. What are the two steps to fit a line through two points?

Skills Warm-Up

Use cross-multiplying to solve each proportion for y in terms of x.

1. $\dfrac{y}{x} = \dfrac{-5}{2}$

2. $\dfrac{y-3}{4} = \dfrac{x}{2}$

3. $\dfrac{x+1}{5} = \dfrac{y-1}{3}$

4. $-2 = \dfrac{y+6}{x}$

5. $\dfrac{y+2}{x-5} = \dfrac{3}{4}$

6. $\dfrac{-1}{3} = \dfrac{4-y}{1-x}$

Answers: **1.** $y = \dfrac{-5}{2}x$ **2.** $y = 2x + 3$ **3.** $y = \dfrac{3}{5}x + \dfrac{8}{5}$ **4.** $y = -2x - 6$

5. $y = \dfrac{3}{4}x - \dfrac{23}{4}$ **6.** $y = \dfrac{-1}{3}x + \dfrac{13}{3}$

Homework 4.5

Skills Practice

■ For Problems 1-6,
  a. Use the point-slope method to graph the line with the given slope and passing through the given point.
  b. Find an equation for the line. Write your equation in slope-intercept form.
  c. Find the x-intercept of the line.

1. $m = -2, \ (-3, 4)$

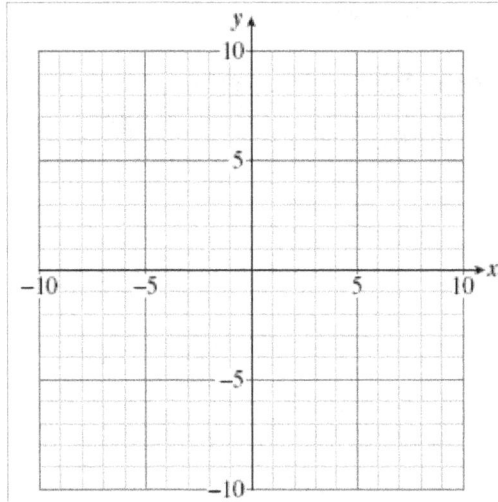

2. $m = \dfrac{1}{2}, \ (4, -3)$

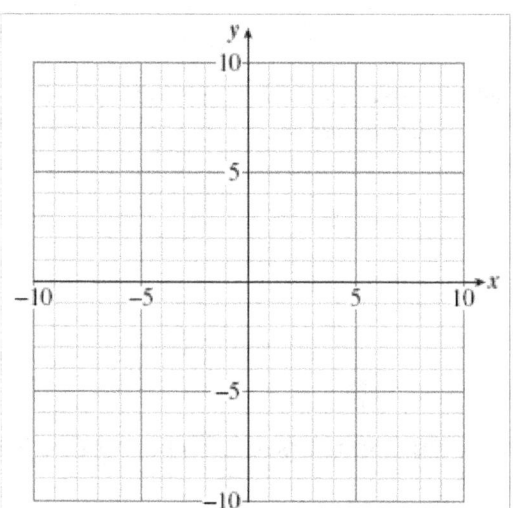

3. $m = \dfrac{-2}{3}, \ (-6, 2)$

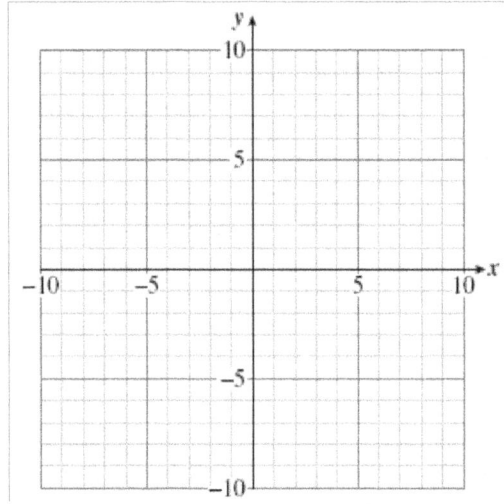

4. $m = 0, \ (-3, 5)$

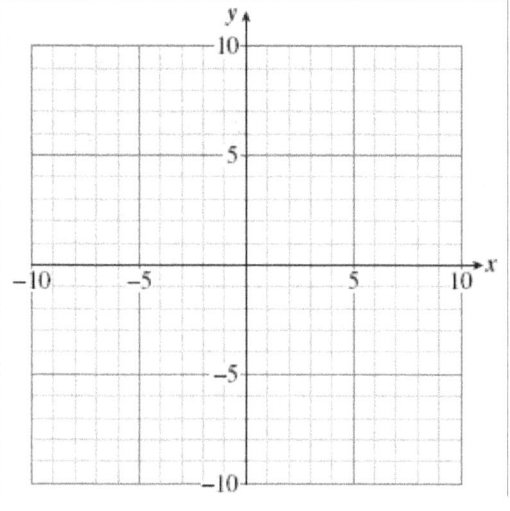

5. $m = \dfrac{-5}{2}$, $(6, -10)$

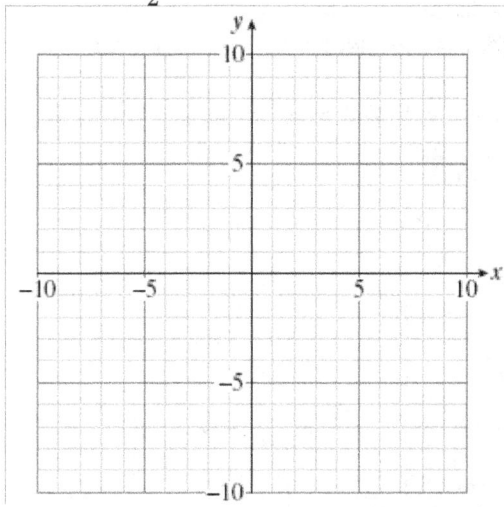

6. $m = 3$, $(1, -1)$

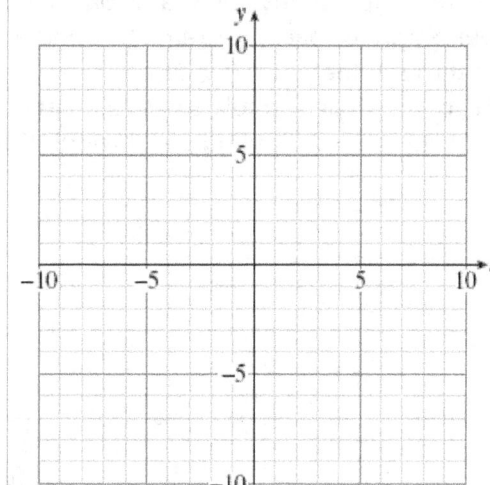

■ For Problems 7-9, without doing any calculations, give the slope of each line and the coordinates of one point on the line.

7. $y = \dfrac{3}{5}x - 7$ **8.** $y - 2 = 3(x + 5)$ **9.** $y = \dfrac{4}{5}x$

■ In Problems 10-12, find an equation for the line passing through the given points. Write your answer in slope-intercept form.

10. $(-2, 4), (1, 7)$ **11.** $(3, 5), (-3, -5)$ **12.** $(6, 4), (-2, 5)$

■ In Problems 13-16, find the equation of the line shown in the graph.

13.

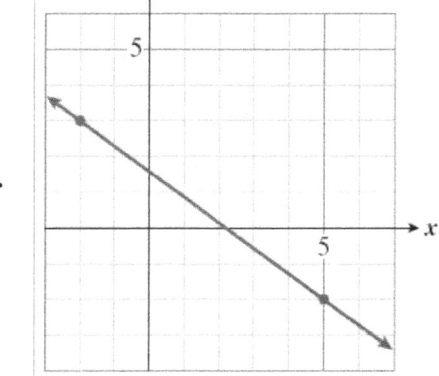

14.

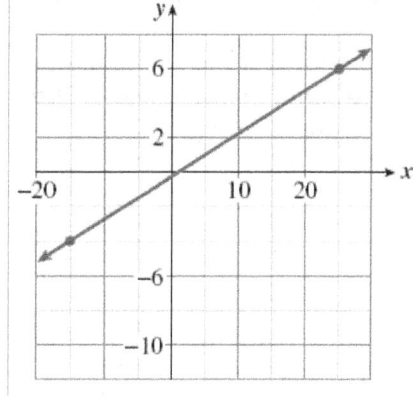

15.

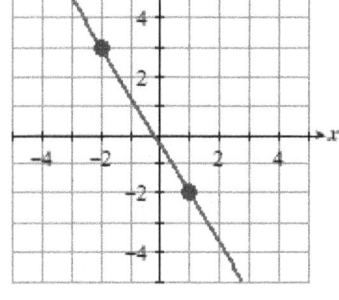

16.

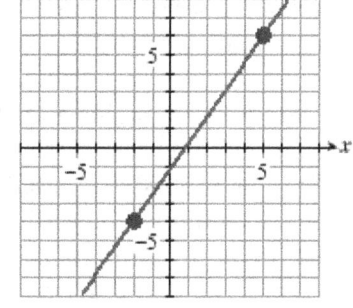

17. a. Put the equation $2y - 3x = 5$ into slope-intercept form, and graph the equation.
   b. What is the slope of any line that is parallel to $2y - 3x = 5$?
   c. On your graph for part (a), sketch by hand a line that is parallel to $2y - 3x = 5$ and passes through the point $(-3, 2)$.
   d. Use the point-slope formula to write an equation for the line that is parallel to the graph of $2y - 3x = 5$ and passes through the point $(-3, 2)$.

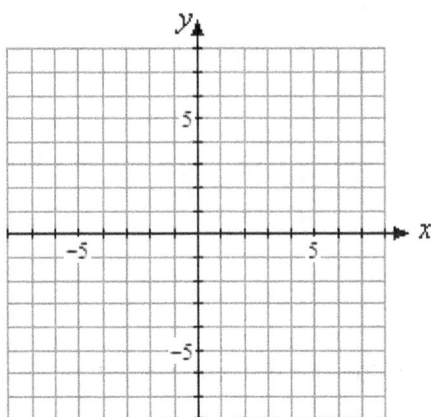   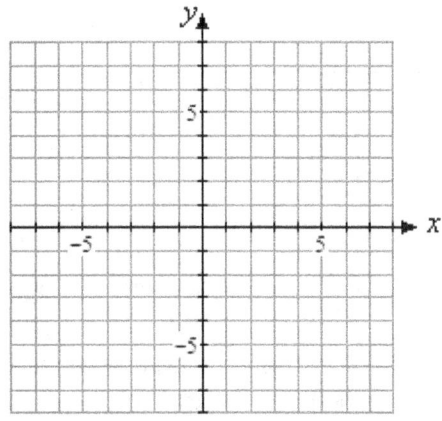

18. a. Put the equation $x - 2y = 5$ into slope-intercept form, and graph the equation.
   b. What is the slope of any line that is perpendicular to $x - 2y = 5$?
   c. On your graph for part (a), sketch by hand a line that is perpendicular to $x - 2y = 5$ and passes through the point $(4, -3)$.
   d. Use the point-slope formula to write an equation for the line that is perpendicular to the graph of $x - 2y = 5$ and passes through the point $(4, -3)$.

19. a. What is the slope of the line $y = 2x + 1$?
   b. What is the slope of a line parallel to $y = 2x + 1$?
   c. A line is parallel to $y = 2x + 1$ and passes through $(-2, 1)$. What is its equation?

20. a. What is the slope of the line $y = -3x - 2$?
   b. What is the slope of a line perpendicular to $y = -3x - 2$?
   c. A line is perpendicular to $y = -3x - 2$ and passes through $(1, 3)$. What is its equation?

Applications

In Problems 21-24, we'll find a linear model from two data points.
   Step 1   Make a table showing the coordinates of two data points for the model.
   Step 2   Find a linear equation in slope-intercept form relating the variables.
   Step 3   Use your equation to answer the questions.

21. Francine is driving into the mountains and stopping periodically to record the temperature, T, at various altitudes, h. The temperature at an altitude of 3200 feet is 77°, and the temperature at 8000 feet is 65°.
   a. What is the slope, including units? What does it tell you about the problem?
   b. What will the temperature be at 10,000 feet?
   c. What was the temperature at sea level?

22. Envirotech is marketing a new line of microwave clothes dryers. It cost them $45,000 to produce the first 100 dryers. When they had produced 180 dryers, their total cost was up to $61,000. They would like to know the total cost, C, of producing x dryers.
    a. What is the slope, including units? What does it tell you about the problem?
    b. Envirotech has budgeted $100,000 for microwave clothes dryers this year. How many can they produce?
    c. How much did Envirotech invest in development before they made the first dryer?

23. Flying lessons cost $645 for an 8-hour course and $1425 for a 20-hour course. Both prices include a fixed insurance fee. Express the cost, C, of flying lessons in terms of the length, h, of the course in hours.
    a. What is the slope, including units? What does it tell you about the problem?
    b. How much does a 10-hour course cost?
    c. How much is the fixed insurance fee?

24. On an international flight a passenger may check two bags each weighing 70 kilograms, or 154 pounds, and one carryon bag weighing 50 kilograms, or 110 pounds. Express the weight, p, of a bag in pounds in terms of its weight, k, in kilograms.
    a. What is the slope, including units? What does it tell you about the problem?
    b. What does a 50-pound bag weigh in kilograms?
    c. Why is the constant term in your equation equal to zero?

Chapter 4 Summary and Review

Section 4.1 The Distributive Law

> **Distributive Law**
>   If a, b, and c are any numbers, then
> $$a(b + c) = ab + ac$$

> **Steps for Solving Linear Equations**
> 1. Use the distributive law to remove any parentheses.
> 2. Combine like terms on each side of the equation.
> 3. By adding or subtracting the same quantity on both sides of the equation, get all the variable terms on one side and all the constant terms on the other.
> 4. Divide both sides by the coefficient of the variable to obtain an equation of the form $x = a$.

Section 4.2 Systems of Linear Equations

* A pair of linear equations in two variables

$$a_1 x + b_1 y = c_1$$
$$a_2 x + b_2 y = c_2$$

considered together is called a **system of linear equations**, or a **linear system**.

* A **solution** to a system of linear equations is an ordered pair (x, y) that satisfies each equation in the system. It is the intersection point of the two lines described by the system.

> **Linear Systems**
>   There are three types of linear systems:
> 1. **Consistent and independent system.**   The graphs of the two lines intersect in exactly one point. The system has exactly one solution.
> 2. **Inconsistent system.** The graphs of the equations are parallel lines and hence do not intersect. An inconsistent system has no solutions.
> 3. **Dependent system**. All the solutions of one equation are also solutions to the second equation, and hence are solutions of the system. The graphs of the two equations are the same line. A dependent system has infinitely many solutions.

Section 4.3 Algebraic Solution of Systems

> **To Solve a System by Substitution:**
> 1. Choose one of the variables in one of the equations. (It is best to choose a variable whose coefficient is 1 or -1.) Solve the equation for that variable.
> 2. Substitute the result of Step 1 into the *other* equation. This gives an equation in one variable.
> 3. Solve the equation obtained in Step 2. This gives the solution value for one of the variables.

> **4.** Substitute this value into the result of Step 1 to find the solution value of the other variable.

To Solve a System by Elimination:
1. Write each equation in the form $Ax + By = C$.
2. Decide which variable to eliminate. Multiply each equation by an appropriate constant so that the coefficients of that variable are opposites.
3. Add the equation from Step 2 and solve for the remaining variable.
4. Substitute the value found in Step 3 into one of the original equations and solve for the other variable.

When Using Elimination to Solve a System:
1. If combining the two equations results in an equation of the form

$$0x + 0y = k \quad (k \neq 0)$$

then the system is inconsistent.
2. If combining the two equations results in an equation of the form

$$0x + 0y = 0$$

then the system is dependent.

Section 4.4 Applications of Systems

- We can use tables to organize the information in applications involving interest, mixtures, or motion.
- The formula for calculating interest is

$$I = Prt$$

I is the interest you will earn after t years if you invest a principal of P dollars in an account that earns simple interest at an annual rate r.

- The formula for calculating percents is

$$P = rW$$

P stands for the part obtained when we take r percent of a whole amount, W.

- In general, **we cannot add percents** unless they are percents of the same whole amount.

- To solve problems involving motion at a constant speed we use the formula

$$D = RT$$

D stands for the distance you travel at speed R in time T.

Section 4.5 Point-Slope Form

Point-slope Formula
The equation for the line that passes through the point $(x_1,\, y_1)$ and has slope m is

$$y = y_1 + m(x - x_1)$$

or

$$y - y_1 = m(x - x_1) \quad \text{or} \quad \frac{y - y_1}{x - x_1} = m$$

To Fit a Line through Two Points :
1. Compute the slope between the two points.
2. Substitute the slope and either point into the point-slope formula.

Review Questions

▇ Use complete sentences to answer the questions in Problems 1-10.

1. State the two-point formula for slope.
2. How can you find the equation of a line when you know the slope of the line and one point on the line?
3. How can you find the solution to a linear system by graphing?
4. Name two algebraic methods for solving linear systems.
5. Suppose you are using the elimination method to solve a system. How can you tell if the system is dependent or inconsistent?
6. How would you label the columns when making a table for a problem about interest?
7. How would you label the columns when making a table for a problem about a mixture?
8. How would you label the columns when making a table for a problem about motion?
9. Under what conditions can we add percents?
10. Explain why the distributive law does not apply to the expression $-3(2ab)$.

Review Problems

▇ For Problems 1-2, simplify. Which product requires the distributive law?

1. **a.** $-5(-6m)$
   b. $-5(6 - m)$

2. **a.** $9(-3 - w)$
   b. $9(-3w)$

▇ For Problems 3-6, simplify.

3. $(4m + 2n) - (2m - 5n)$
4. $(-5c - 6) + (-11c + 15)$
5. $-7w - 2(4w - 13)$
6. $4(3z - 10) + 5(-z - 6)$

▇ For Problems 7-10, solve.

7. $5p + 10(17 - p) = 2p - 5$
8. $-3(k - 2) - 4(2k + 5) = 10 + 3k$
9. $4(3a - 7) < -18 + 2a$
10. $\dfrac{2x - 1}{x + 3} = \dfrac{3}{2}$

11. The length of a rectangle is 3 times its width.
    a. If the width of a rectangle is x, what is its length?
    b. Express the perimeter of the rectangle in terms of x.
    c. Suppose the perimeter of the rectangle is 48 centimeters. Find the dimensions of the rectangle.

12. Last Saturday, a total of 620 people attended the Gaslamp Theater at its two performances.
  a. If p people attended the matinee last Saturday, how many attended the evening performance?
  b. This Saturday, attendance at the matinee increased by 5%, and attendance at the evening performance increased by 8%. Write expressions in terms of p for the attendance at each performance.
  c. This Saturday, the total attendance was 663 people. How many people attended the matinee last Saturday?

13. Wheaton Elementary school plans to buy 30 computers. The computers with speakers cost $1200 each, and those without speakers cost $800 each. Let x represent the number of computers with speakers. Write expressions in terms of x for:
  a. The number of computers without speakers.
  b. The total cost of the computers with speakers.
  c. The total cost of the computers without speakers.
  d. The total cost of all 30 computers.
  e. If Wheaton Elementary has $28,800 to spend on the computers, how many of each kind can they buy?

14. The current in the Lazy River flows at 4 miles per hour.
  a. If your motorboat is traveling at speed v miles per hour, write expressions for your speed traveling upstream and your speed traveling downstream.
  b. You travel upstream for $\frac{3}{2}$ hours and stop for lunch at an island. Write an expression in terms of v for the distance you traveled upstream.
  c. You return to your starting point downstream in $\frac{1}{2}$ hour. Write an expression in terms of v for the distance you traveled downstream.
  d. Write an equation and solve it to find the speed, v, of your motorboat.

For Problems 15-16, decide whether the given point is a solution of the system.

15. $x + y = 8$ $(-2, 10)$
       $x - y = 2$

16. $8x + 3y = 21$ $(-3, 1)$
       $5x = y - 16$

For Problems 17-18, solve the system by graphing.

17. $x + y = 5$
       $2x - y = 4$

18. $x - y = 7$
       $y = \dfrac{-2}{3}x - 2$

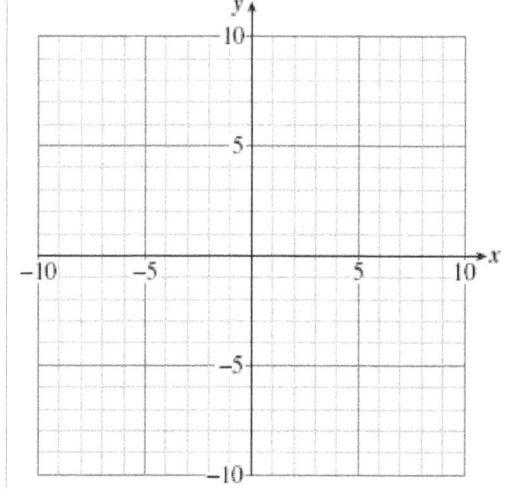

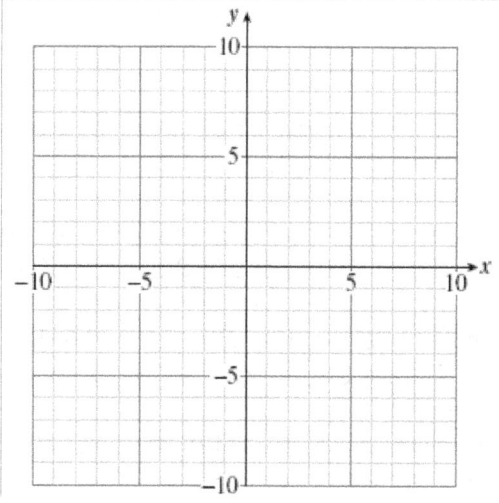

■ For Problems 19-20, solve by substitution.

19. $y = 2x + 1$
$2x + 3y = -21$

20. $x + 4y = 1$
$2x + 3y = -3$

■ For Problems 21-22, solve by elimination.

21. $2x + 7y = -19$
$5x - 3y = 14$

22. $4x + 3y = -19$
$5x + 15 = -2y$

■ For Problems 23-26, solve using substitution or elimination.

23. $x + 5y = 18$
$x - y = -3$

24. $x + 5y = 11$
$2x + 3y = 8$

25. $\dfrac{2}{3}x - 3y = 8$
$x + \dfrac{3}{4}y = 12$

26. $3x = 5y - 6$
$3y = 10 - 11x$

■ For Problems 27-30, decide whether each system is inconsistent, dependent, or consistent and independent.

27. $2x - 3y = 4$
$x + 2y = 7$

28. $2x - 3y = 4$
$6x - 9y = 4$

29. $2x - 3y = 4$
$6x - 9y = 12$

30. $x - y = 6$
$x + y = 6$

■ For Problems 30-42, solve by writing and solving a system of equations.

31. A health food store wants to produce 30 pounds of granola worth 80 cents per pound. They plan to mix cereal worth 65 cents per pound with dried fruit worth 90 cents per pound. How much of each should they use?

32. The perimeter of a rectangle is 50 yards and its length is 9 yards greater than its width. Find the dimensions of the rectangle.

33. Last year Veronica made $93 in interest from her savings accounts, one of which paid 6% interest, and the other paid 9%. This year her interest rates dropped to 4% and 8%, respectively, and she made $76 interest. How much does Veronica have invested in each account?

34. Marvin invested $300 more at 6% than he invested at 8%. His total annual income from his two investments is $242. How much did he invest at each rate?

35. How many pounds of an alloy containing 60% copper must be melted with an alloy containing 20% copper to obtain 8 pounds of an alloy containing 30% copper?

36. Jerry Glove came to bat only 20 times in the first half of the season and got hits 15% of the time. During the second half of the season, Jerry came to bat 140 times and got hits 35% of the time.
   a. How many hits did Jerry get in the first half of the season? How many hits did he get in the second half?

    b. How many times did Jerry come to bat all season? How many hits did he get?

    c. What percent of Jerry's at-bats resulted in hits?

    d. Who had a better batting average in the first half of the season, Joe Cleat from Problem 8 in Homework 4.3, or Jerry? Who had a better batting average in the second half? Who had the better batting average overall?

37. Alida and Steve are moving to San Diego. Alida is driving their car, and Steve is driving a rental truck. They start together, but Alida drives twice as fast as Steve. After 3 hours they are 93 miles apart. How fast is each traveling?

38. Jake rides for the Pony Express, covering his route in 6 hours and returning home, 8 miles per hour slower, in 9 hours. How far does Jake ride?

39. A math contest exam has 40 questions. A contestant scores 5 points for each correct answer, but loses 2 points for each wrong answer. Lupe answered all the questions and her score was 102. How many questions did she answer correctly?

40. A game show contestant wins $25 for each correct answer he gives but loses $10 for each incorrect response. Roger answered 24 questions and won $355. How many answers did he get right?

41. Barbara wants to earn $500 a year by investing $5000 in two accounts, a savings plan that pays 8% annual interest and a high-risk option that pays 13.5% interest. How much should she invest in each account?

42. An investment broker promises his client a 12% return on her funds. If the broker invests $3000 in bonds paying 8% interest, how much must he invest in stocks paying 15% interest to keep his promise?

43. a. Graph the line with slope $\dfrac{-5}{3}$ that passes through the point $(-2, 1)$.

    b. Find an equation in point-slope form for the line in part (a)

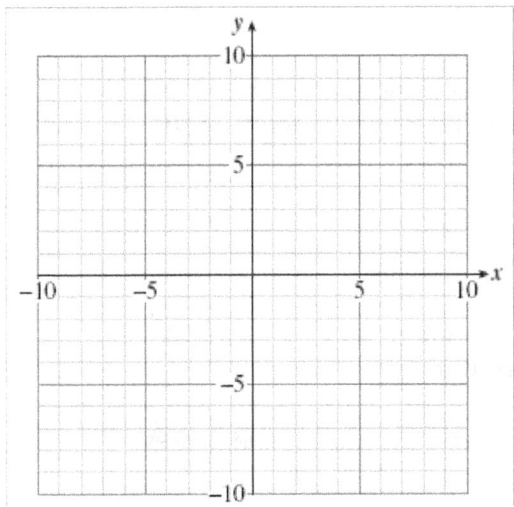 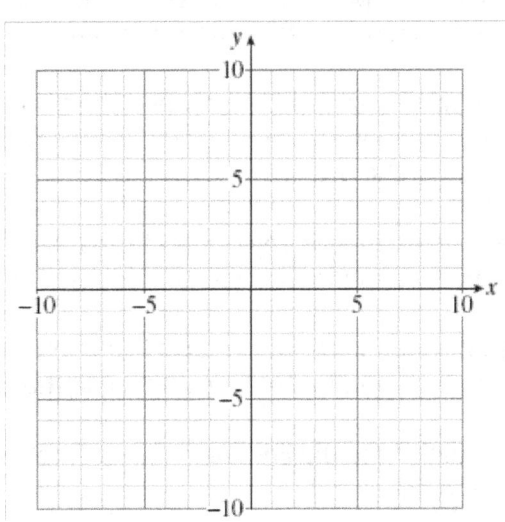

44. a. Graph the line of slope $\dfrac{6}{5}$ that passes through the point $(-3, -4)$.

    b. Find an equation in point-slope form for the line in part (a).

■ For Problems 45-46, find an equation for the line passing through the two points.

45. $(3, -5)$, $(-2, 4)$

46. $(0, 8)$, $(4, -2)$

47. An interior decorator bases her fee on the cost of a remodeling job. The table below shows her fee, F, for jobs of various costs, C, both given in dollars.

C	5000	10,000	20,000	50,000
F	1000	1500	2500	5500

  a. Write a linear equation for F in terms of C.
  b. Give the slope of the graph, and explain the meaning of the slope in terms of the decorator's fee.

48. Auto registration fees in Connie's home state depend on the value of the automobile. The table below shows the registration fee R for a car whose value is V, both given in dollars.

V	5000	10,000	15,000	20,000
R	135	235	335	435

  a. Write a linear equation for R in terms of V.
  b. Give the slope of the graph, and explain the meaning of the slope in terms of the registration fee.

■ For Problems 49-50,
  a. Make a table of values showing two data points.
  b. Find a linear equation relating the variables.
  c. State the slope of the line, including units, and explain its meaning in the context of the problem.

49. The population of Maple Rapids was 4800 in 1982 and had grown to 6780 by 1997. Assume that the population increases at a constant rate. Express the population P of Maple Rapids in terms of the number of years t since 1982.

50. Cicely's odometer read 112 miles when she filled up her 14-gallon gas tank and 308 when the gas gauge read half full. Express her odometer reading m in terms of the amount of gas g she used.

51. a. Write an equation for any line that is parallel to $2y = 5x - 3$.
  b. Write an equation for any line that is perpendicular to $2y = 5x - 3$.

52. a. Write an equation for the vertical line that passes through $(3, -6)$
  b. Write an equation for the horizontal line that passes through $(3, -6)$

53. Write an equation for the line that is parallel to the graph of $2x + 3y = 6$ and passes through the point $(1, 4)$.

54. Write an equation for the line that is perpendicular to the graph of $2x + 3y = 6$ and passes through the point $(1, 4)$.

Chapter 5 Exponents and Roots

Lesson 1 Exponents

- Compute powers
- Simplify expressions involving exponents
- Evaluate expressions involving exponents
- Combine like terms

Lesson 2 Square Roots and Cube Roots

- Compute square roots and cube roots
- Use radical notation
- Distinguish between rational and irrational numbers
- Distinguish between exact values and approximations
- Simplify expressions involving radicals

Lesson 3 Using Formulas

- Use formulas to compute volume and surface area
- Use the Pythagorean theorem
- Solve an equation in x^2
- Solve a formula for one variable in terms of the others

Lesson 4 Products of Binomials
- Apply the distributive law to simplify products
- Simplify products involving variables
- Compute the product of two binomials

5.1 Exponents

What is an Exponent?

The area of a rectangle is given by the formula $A = lw$. The variable l stands for the length of the rectangle, and w stands for for its width. The area tells us how many square tiles, one unit on a side, will fit inside the rectangle, as shown below.

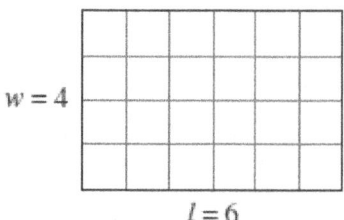

Area is the number of square tiles
that fit inside a rectangle

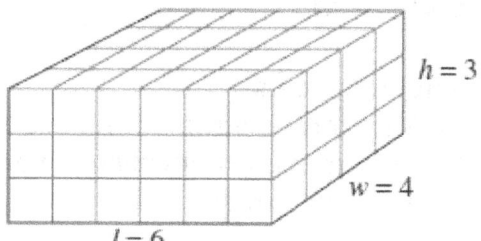

Volume is the number of cubes
that fit inside a box

Similarly, the volume of a box (measured in cubic units) is given by the formula $V = lwh$, where l, w, and h stand for the length, width, and height of the box. The volume tells us how many blocks, one unit on a side, will fit inside the box. The volume of the box shown above is

$$V = lwh = 6 \cdot 4 \cdot 3 = 72 \text{ cubic inches}$$

A **square** is a rectangle whose length and width are equal. If we use s, for side, to stand for both the length and width of the square, its area is given by

$$A = s \cdot s$$

A **cube** is a box whose length, width, and height are all equal, so its volume is given by

$$V = s \cdot s \cdot s$$

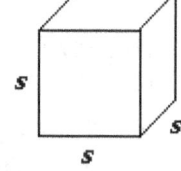

RQ1. What is the formula for the volume of a box?

Finding the area of a square or the volume of a cube involves repeated multiplication by the same number.   We indicate repeated multiplication with a symbol called an **exponent**.

> An **exponent** is a number that appears above and to the right of a particular factor. It tells us how many times that factor occurs in the expression. The factor to which the exponent applies is called the **base**, and the product is called a **power** of the base.

For example,

$$2^5 = 2 \cdot 2 \cdot 2 \cdot 2 \cdot 2 = 32$$

exponent, power, base, 5 factors of 2

We read the expression 2^5 as *the fifth power of 2,* or as *2 raised to the fifth power,* or simply as *2 to the fifth.*

Exponents
   An **exponent** indicates repeated multiplication.

$$a^n = a \cdot a \cdot a \cdot \cdots \cdot a \quad (n \text{ factors of } a)$$

where n is a positive integer.

RQ2.   What does an exponent tell us?

■ **Example 1**   Compute the following powers.

a. $6^2 = 6 \cdot 6 = 36$

b. $3^4 = 3 \cdot 3 \cdot 3 \cdot 3 = 81$

c. $\left(\dfrac{2}{3}\right)^3 = \left(\dfrac{2}{3}\right)\left(\dfrac{2}{3}\right)\left(\dfrac{2}{3}\right) = \dfrac{8}{27}$

d. $1.4^4 = 1.4(1.4)(1.4)(1.4) = 3.8416$ ■

Caution!   Note that 3^4 does not mean 3 times 4 or 12; we already have a way to write $3(4)$. Remember that an exponent or power indicates repeated multiplication of the base.   ■

Look Closer: Because the exponents 2 and 3 are used frequently, they have special names.

5^2 means $5 \cdot 5,$ and is read   **5 squared** or the square of 5

5^3 means $5 \cdot 5 \cdot 5,$ and is read   **5 cubed** or the cube of 5

These names come from the formulas for the area, A, of a square with side s and the volume, V, of a cube with edge s :

$$A = s^2 \quad \text{and} \quad V = s^3$$
                                                                      ●

Powers of Negative Numbers

   To show that a negative number is raised to a power, we enclose the negative number in parentheses.

For example, to indicate the square of −5, we write

$$(-5)^2 = (-5)(-5) = 25 \qquad \textbf{Exponent applies to } (-5).$$

If the negative number is *not* enclosed in parentheses, then the exponent applies only to the positive number, and the negative sign tells us that the power is negative. For example,

$$-5^2 = -5 \cdot 5 = -25 \qquad \textbf{Exponent applies only to 5}.$$

Look Closer: Note how the placement of parentheses changes the meaning of the expressions in Example 2.

Example 2   Compute each power.
a. -4^2 b. $(-4)^2$ c. $-(4)^2$ d. (-4^2)

Solutions a.   Only 4 is squared: $-4^2 = -4 \cdot 4 = -16$
b. The negative number is squared: $(-4)^2 = (-4)(-4) = 16$
c. Only 4 is squared: $-(4)^2 = -(4)(4) = -16$
d. Only 4 is squared, and the entire expression appears within parentheses:

$$(-4^2) = (-4 \cdot 4) = -16$$

RQ3.   How do we indicate that a negative number should be raised to a power?

Using a Calculator

Scientific calculators usually have a key labeled $\boxed{x^y}$ or $\boxed{y^x}$, called the power key, for computing powers. To compute 7^4 using the power key, we enter

$$7 \ \boxed{y^x} \ 4 \ \boxed{=}$$

Graphing calculators have a caret key, $\boxed{\wedge}$, for entering powers. On a graphing calculator, we enter 7^4 as

$$7 \ \boxed{\wedge} \ 4 \ \boxed{\textbf{ENTER}}$$

Also, many calculators have a key labeled $\boxed{x^2}$ for computing squares of numbers.

Example 3   Use a calculator to compute the powers.
a. $(1.2)^3$ b. $(-12)^4$

Solutions a.   We enter the following key strokes:

$$1.2 \ \boxed{y^x} \ 3 \ \boxed{=} \quad \text{or} \quad 1.2 \ \boxed{\wedge} \ 3 \ \boxed{\textbf{ENTER}}$$

to find that $(1.2)^3 = 1.728$.
b. If your calculator has a $\boxed{+/-}$ key, you can enter

$$12 \ \boxed{+/-} \ \boxed{y^x} \ 4 \ \boxed{=}$$

to find that $(-12)^4 = 20,736$. On a graphing calculator, we enter

$$\boxed{(} \ \boxed{-} \ 12 \ \boxed{)} \ \boxed{\wedge} \ 4$$

Caution!   When using a calculator to compute a power of a negative number, we must remember to enclose the number in parentheses. For example, which sequence will calculate the square of -3?

$$\boxed{-} \ 3 \ \boxed{x^2} = -9 \qquad \text{Exponent applies only to 3.}$$

$$\boxed{(} \ \boxed{-} \ 3 \ \boxed{)} \ \boxed{x^2} = 9 \qquad \text{Exponent applies to } (-3).$$

The first sequence tells the calculator to square 3, then make the result negative. The second sequence tells us to square square -3 (multiply -3 by itself).

Powers of Variables

We can also use exponents with variables. We must be careful to distinguish between a **product** of a variable and a **power** of a variable.

> An exponent on a variable indicates repeated multiplication, while a coefficient in front of a variable indicates repeated addition.

Example 4   Compare the expressions x^4 and $4x$.

Solution These two expressions are not the same!

$$x^4 = x \cdot x \cdot x \cdot x$$
$$\text{but} \quad 4x = x + x + x + x$$

The expressions x^4 and $4x$ are not equivalent; they are not equal for all values of x. (Can you think of one value of x for which they are equal?)

RQ4.   What is the difference between an exponent and a coefficient?

Order of Operations

How do exponents fit into the order of operations?

Example 5   Compare the expressions $2x^3$ and $(2x)^3$.

Solution In the first expression, only the x is cubed. Thus,

| $2x^3$ | means | $2xxx$ | **Exponent applies to the base, x, only.** |
| $(2x)^3$ | means | $(2x)(2x)(2x)$ | **Exponent applies to the base, $2x$.** |

The two expressions are not equivalent, as you can see by evaluating each for, say, $x = 5$:

$$2x^3 = 2(5^3) = 2(125) = 250$$
$$(2x)^3 = (2 \cdot 5)^3 = 10^3 = 1000$$

> An exponent applies only to its base, and not to any other factors in the product. If we want an exponent to apply to more than one factor, we must enclose those factors in parentheses.

Look Closer: Think about the operations in Example 5.
  - To evaluate $2x^3$, we compute the power x^3 first, and then the product, $2 \cdot x^3$.
  - To evaluate $(2x)^3$, we compute the product $2 \cdot x$ inside parentheses first, and then compute the power.

When simplifying an expression, we perform powers before multiplications, but after operations within parentheses. Thus, we include exponents in the order of operations as follows.

> **Order of Operations**
> 1. Perform any operations inside parentheses, or above or below a fraction bar.
> 2. Compute all indicated powers.
> 3. Perform all multiplications and divisions in the order in which they occur from left to right.
> 4. Perform additions and subtractions in order from left to right.

■ **Example 6**   Simplify $-2(8 - 3 \cdot 4)^2$

Solution   We simplify the expression within parentheses first.

$$-2(8 - 3 \cdot 4)^2 = -2(8 - 12)^2 \qquad \textbf{Multiply first, then subtract.}$$
$$= -2(-4)^2 \qquad \textbf{Compute the power, then multiply.}$$
$$= -2(16) = -32 \qquad\qquad\qquad ■$$

RQ5.   In the order of operations, when do we evaluate powers?

It is especially important to follow the order of operations when evaluating an expression.

> When we substitute a negative number for the variable, we enclose the negative number in parentheses.

■ **Example 7**   Evaluate each expression for $x = -6$.
a. x^2 **b.** $-x^2$ **c.** $2 - x^2$ **d.** $(2x)^2$

Solutions a.   We replace x by (-6), and then square. This means that -6 gets squared, not just 6. Thus,

$$x^2 = (-6)^2 = (-6)(-6) = 36$$

The parentheses are essential for this calculation. It would be incorrect to write $x^2 = -6^2 = -36$.
b. In this expression, only x is squared. The negative sign is applied to the result.

$$-x^2 = -(-6)^2 = -36$$

c. We replace x by (-6) to get

$$2 - x^2 = 2 - (-6)^2 \qquad \textbf{Compute the power first.}$$
$$= 2 - 36 = -34$$

d. When we replace x by (-6), for clarity we also change the existing parentheses to brackets.

$$(2x)^2 = [2(-6)]^2 \qquad \textbf{Multiply inside brackets first.}$$
$$= [-12]^2 = 144 \qquad\qquad ■$$

Like Terms

In Chapter 2 we learned how to add or subtract like terms. For example,

$$8x - 3x = 5x$$

but $8x - 3y$ cannot be simplified

We can also combine like powers of the same variable. For instance,

$$8x^2 - 3x^2 = 5x^2$$

Look Closer: Notice that when we add like terms, we do not alter the exponent; only the coefficient of the power changes. Can we add different powers of the same variable? The answer to this question is No. For example,

$$8x^2 - 3x \quad \text{cannot be simplified}$$

For most values of x, the numbers x and x^2 are different. Thus, $8x^2$ and $3x$ are not like terms, and they cannot be combined.   ●

■ Example 8   Combine like terms where possible.

a. $-6a^3 + 10a^3$ b. $5w^2 + 3w^3$

Solutions a.   The exponents on the two terms are the same, so they can be combined. We add the coefficients, $-6 + 10 = 4$, and leave the powers unchanged:

$$-6a^3 + 10a^3 = 4a^3$$

b. The exponents on the terms are different, so they are not like terms. They cannot be combined.   ■

Caution!   When adding like terms, we do not add the exponents. For example,

$$4x^2 + 3x^2 = 7x^4 \quad \text{is incorrect.}$$ ■

RQ6.   How do we combine like terms?

In the next Example we compare adding expressions and multiplying expressions.

■ Example 9   Simplify the expressions.

a. Add: $3a + 5a$ b. Multiply: $(3a)(5a)$

Solutions a.   These are like terms, so they can be combined. We add the coefficients, $3 + 5 = 8$, and leave the variable unchanged:

$$3a + 5a = 8a$$

b. This product can be written as

$$3 \cdot a \cdot 5 \cdot a$$

We use the commutative law to rearrange the factors and multiply to find

$$3 \cdot 5 \cdot a \cdot a = 15a^2$$ ■

Skills Warm-Up

■ Follow the order of operations to simplify each expression.

1. $-2[-3(-5) - 8(4)]$

2. $[-8 + 6(-4)(-3)][5 - (-2)]$

3. $\dfrac{4}{3}(-6)(6 - 9)(6 - 9)$

4. $\dfrac{3}{8}(-7 - 5)(-4 - 4)$

5. $-2.4(-3) + (8 - 4.5)(-7.2)$

6. $-9.6 - 3.2(-8 - 2.4)(-3)$

Answers: 1. 34   **2.** 448   **3.** −72   **4.** −36   **5.** −18   **6.** −109.44

Homework 5.1

Skills Practice

■ Compute the powers in Problems 1-2.

1a. 4^3 **b.** 5^3 **c.** 5^4 **2a.** $\left(\dfrac{2}{3}\right)^4$ **b.** $\left(\dfrac{4}{5}\right)^3$ **c.** $\left(\dfrac{11}{9}\right)^2$

3. Use a calculator to compute the powers. Round your answers to the nearest hundredth.

   a. $(3.1)^3$ **b.** $(2.6)^4$ **c.** $(0.8)^4$

■ For Problems 4-5, simplify.

4a. -5^2 **b.** -5^3 **c.** $(-5)^2$ **d.** $(-5)^3$
5a. $-(-2)^2$ **b.** $-(-2)^3$ **c.** $-2^3 - 2^2$ **d.** $-(2^3 - 2)^2$

6. Evaluate for $x = -2$.
   a. $5x^3$ **b.** $5x^2$ **c.** $5 - x^2$ **d.** $5 - x^3$

7. Evaluate for $a = -3$, $b = -4$.
   a. ab^3 **b.** $a - b^3$ **c.** $(a - b^2)^2$ **d.** $ab(a^2 - b^2)$

■ Simplify each pair of expressions in Problems 8-11.

8. a. $x + x + x$ **9. a.** $5a \cdot 5a$
   b. $x \cdot x \cdot x$ **b.** $5a + 5a$

10. a. $-q - q - q$ **11. a.** $-3m - 3m$
   b. $-q(-q)(-q)$ **b.** $(-3m)(-3m)$

■ In Problems 12-15, one of the two statements is true for all values of x, and the other is not. By trying some values of x, decide which statement is true.

12. a. $x + x = 2x$ **13. a.** $x \cdot x = 2x$
   b. $x + x = x^2$ **b.** $x \cdot x = x^2$

x	$x + x$	$2x$	x^2
3			
5			
−4			
−1			

x	$x \cdot x$	$2x$	x^2
4			
6			
−3			
−1			

14. a. $x^2 + x^2 = x^4$
  b. $x^2 + x^2 = 2x^2$

x	$x^2 + x^2$	x^4	$2x^2$
2			
3			
−2			
−1			

15. a. $x + x^2 = x^3$
  b. $x \cdot x^2 = x^3$

x	$x + x^2$	$x \cdot x^2$	x^3
1			
4			
−3			
−1			

■ Using what you learned in Problems 12-15, simplify the expressions in Problems 16-21 if possible, or state "cannot be simplified."

16. $5a^2 - 7a^2$ **17.** $3t - 2t^2$ **18.** $-m^2 - m^2$
19. $3k(4k)$ **20.** $3k + 4k^2$ **21.** $3k^2 + 4k^2$

■ In Problems 22-24, simplify by combining like terms.

22. $6b^3 - 2b^3 - (-8b^3)$ **23.** $(2y^3 - 4y^2 - y) + (6y^2 + 2y + 1)$
24. $(5x^3 + 3x^2 - 4x + 8) - (2x^3 - 4x - 3)$

■ In Problems 25-28, explain why the calculation is **incorrect**, and give the correct answer.

25. $6w^3 + 8w^3 \rightarrow 14w^6$ **26.** $6 + 3x^2 \rightarrow 9x^2$
27. $4t^2 + 7 - (3t^2 - 5) \rightarrow t^2 + 2$ **28.** $5b^2 - 3b \rightarrow 2b$

Applications

■ For Problems 29-31, translate into an algebraic expression, then simplify. Round to two decimal places if necessary.

29. The square of the sum of 3 and 4
30. 5 more than x to the third power
31. 25% of the cube of h

32. Myra sells mugs at the sidewalk fair every week. Her revenue from selling x mugs is $12x - 0.3x^2$ dollars, and the cost of producing x mugs is $50 + 3x$ dollars.
  a. Write an expression for the profit Myra earns from selling x mugs.
  b. Find Myra's profit from the sale of 10 mugs, from 15 mugs, and from 20 mugs.

33. The owner of the Koffee Shop pours the remainder of her old house blend, which is 30% Colombian beans, into a 50-pound bin, and fills it up with her new house blend, which is 25% Colombian beans. Let h stand for the number of pounds of the old house blend. Write algebraic expressions to answer the questions below.
  a. How many pounds of Colombian beans are in the old house blend?
  b. How many pounds of new blend does she pour into the bin?
  c. How many pounds of Colombian beans are in this amount of new house blend?
  d. How many pounds of Colombian beans are in the 50-pound bin?

■ For Problems 34-35, recall that two algebraic expressions are called **equivalent** if they are equal for every value of their variables.

34. a. Explain why the expressions $3z^2$ and $(3z)^2$ are not equivalent.
b. Explain why the expressions $(3z)^2$ and $9z^2$ are equivalent.

35. a. Find two values of x for which $2x = x^2$.
b. Find four values of x for which $2x \neq x^2$.
c. Is $2x$ equivalent to x^2?

■ For Problems 36-39, simplify mentally, without using pencil, paper, or calculator!

36. a. Multiply 3.5 by 100.
b. Multiply 3.5 by 1000.
c. Multiply 3.5 by 10,000.

37. a. 0.074×10^2 **b.** 0.074×10^3 **c.** 0.074×10^4

38. a. 24×10^2 **b.** 8.91×10^5 **c.** 0.003×10^4

39. a. $3 \cdot 10 + 9$ **b.** $2 \cdot 10^2 + 3 \cdot 10 + 4$
c. $3 \cdot 10^3 + 4 \cdot 10^2 + 5 \cdot 10 + 6$

40. Find the perimeter of the triangle.

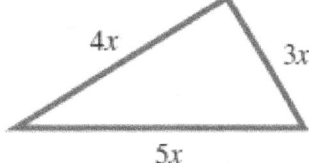

41. Find the area of the triangle.

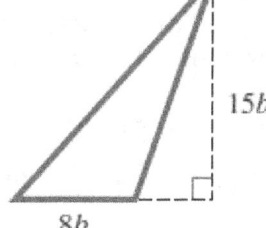

42. Find the area and perimeter of the rectangle.

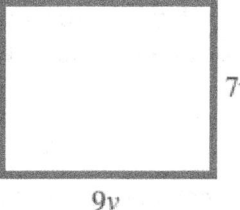

5.2 Square Roots and Cube Roots

Suppose we would like to draw a square whose area is 25 square inches. How long should each side of the square be? Because the formula for the area of a square is $A = s^2$, the side s should satisfy the equation $s^2 = 25$. We want a number whose square is 25. As you can probably guess, the length of the square should be 5 inches, because $5^2 = 25$. We say that 5 is a **square root** of 25.

> **Square Root**
>   The number s is called a **square root** of a number b if $s^2 = b$.

> Finding a square root of a number is the opposite of squaring a number.

■ **Example 1**   Find a square root of 25, and a square root of 144.

Solution   5   is a square root of   25 because $5^2 = 25$
               12 is a square root of 144 because $12^2 = 144$ ■

Look Closer: 5 is not the only square root of 25, because $(-5)^2 = 25$ as well. Thus, 25 has two square roots, 5 and −5.

> Every positive number has two square roots, one positive and one negative.

> **RQ1.**   What is the square root of a number n?

Radicals

The positive square root of a number is called the **principal square root**. The symbol $\sqrt{}$ denotes the positive or principal square root. Thus, we may write

$$\sqrt{25} = 5 \quad \text{and} \quad \sqrt{144} = 12$$

The symbol $\sqrt{}$ is called a **radical** sign, and the number inside is called the **radicand**. Square roots are often called **radicals**.

What about the other square root, the negative one? If we want to indicate the negative square root of a number, we place a negative sign outside the radical sign, like this:

$$-\sqrt{16} = -4 \quad \text{and} \quad -\sqrt{49} = -7$$

If we want to refer to both square roots, we use the symbol $\pm$, read plus or minus. For example,

$$\pm\sqrt{36} = \pm 6, \quad \text{which means} \quad 6 \text{ or } -6$$

Note that zero has only one square root: $\sqrt{0} = 0$.

■ **Example 2**   Find each square root.

a. $-\sqrt{81} = -9$ **b.** $\pm\sqrt{\dfrac{64}{121}} = \pm\dfrac{8}{11}$ ■

RQ2.	What is the positive square root of a number called?
RQ3.	What is a radicand?

Every positive number has two square roots, and zero has exactly one square root. What about the square root of a negative number? For example, can we find $\sqrt{-4}$? The answer is No, because the square of any number is positive (or zero). Try this yourself: The only reasonable candidates for $\sqrt{-4}$ are 2 and -2, but

$$2^2 = \underline{\hspace{3cm}}$$
$$(-2)^2 = \underline{\hspace{3cm}}$$

We cannot find the square root of a negative number. We say that the square root of a negative number is **undefined**.

RQ4.	How do we find the square root of a negative number?

Rational and Irrational Numbers

A **rational number** is one that can be expressed as a quotient (or ratio) of two integers, where the denominator is not zero.

The term "rational" has nothing to do with being reasonable or logical; it comes from the word ratio. Thus, any fraction such as

$$\frac{2}{3}, \quad \frac{-4}{7}, \quad \text{or} \quad \frac{15}{8}$$

is a rational number. Integers are also rational numbers, because any integer can be written as a fraction with a denominator of 1. (For example, $6 = \frac{6}{1}$.) All of the numbers we have encountered before this chapter are rational numbers.

RQ5.	What is a rational number?

Every fraction can be written in decimal form.

Decimal Form of a Rational Number
The decimal representation of a rational number has one of two forms.
  1. The decimal representation terminates, or ends.
  2. The decimal representation repeats a pattern.

Example 3   Write the decimal form for each rational number.

a. $\dfrac{3}{4}$

b. $\dfrac{4}{11}$

Solutions a.   We can use a calculator or long division to divide the numerator by the denominator to find $\frac{3}{4} = 0.75$.
b. We divide 4 by 11 to find $\frac{4}{11} = 0.363636... = 0.\overline{36}$. The line over the digits 36 is called a repeater bar, and it indicates that those digits are repeated forever.

> **RQ6.** How can you recognize a rational number in its decimal form?

What about the decimal form for $\sqrt{5}$? If you use a calculator with an eight-digit display, you will find

$$\sqrt{5} \approx 2.236\ 068$$

However, this number is only an approximation, and not the exact value of $\sqrt{5}$. (Try squaring $2.236\ 068$ and you will see that

$$2.236\ 068^2 = 5.000\ 000\ 100\ 624$$

which is not exactly 5, although it is close.) In fact, no matter how many digits your calculator or computer can display, you can never find an exact decimal equivalent for $\sqrt{5}$. $\sqrt{5}$ is an example of an irrational number.

> An **irrational number** is one that cannot be expressed as a quotient of two integers.

Look Closer: There is no terminating decimal fraction that gives the exact value of $\sqrt{5}$. The decimal representation of an irrational number never ends, and does not repeat any pattern! The best we can do is round off the decimal form and give an approximate value. Nonetheless, an irrational number still has a precise location on the number line, just as a rational number does. The figure below shows the locations of several rational and irrational numbers on a number line.

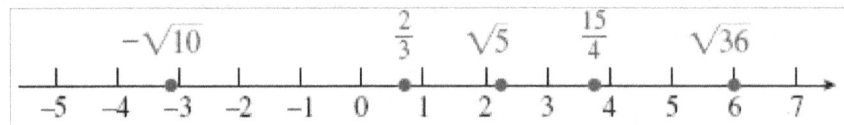

> Each point on a number line corresponds either to a rational number or an irrational number, and these numbers fill up the number line completely. The rational and irrational numbers together make up the **real numbers**, and the number line is sometimes called the **real line**.

> **RQ7.** What are the rational and irrational numbers together called?

It is important that you understand the distinction between an exact value and an approximation.

■ **Example 4a.** We cannot write down an exact decimal equivalent for an irrational number.

$$\sqrt{5} \qquad \text{indicates the exact value of the square root of } 5$$
$$2.236068 \qquad \text{is an approximation to the square root of } 5$$

b. We often use a decimal approximation for a rational number.

$$\frac{2}{3} \qquad \text{indicates the exact value of 2 divided by 3}$$
$$0.666667 \qquad \text{is an approximation for } \frac{2}{3} \qquad ■$$

Of course, even though many radicals are irrational numbers, some radicals, such as $\sqrt{16} = 4$ and $\sqrt{\frac{9}{25}} = \frac{3}{5}$, represent integers or fractions. Integers such as 9 and 25, whose square roots are whole numbers, are called **perfect squares**.

Order of Operations

When we evaluate algebraic expressions that involve radicals, we must follow the order of operations as usual. Square roots occupy the same position as exponents in the order of operations: They are computed after parentheses but before multiplication.

■ **Example 5**   Find a decimal approximation to three decimal places for $8 - 2\sqrt{7}$

Solution You may be able to enter this expression into your calculator just as it is written. If not, you must enter the operations in the proper order. The expression has two terms, 8 and $-2\sqrt{7}$, and the second term is the product of $\sqrt{7}$ with -2. We should not begin by subtracting 2 from 8, because multiplication precedes subtraction. First, we find an approximation for $\sqrt{7}$:

$$\sqrt{7} \approx 2.6457513$$

Do not round off your approximations at any intermediate steps in the problem or you will lose accuracy at each step! You should be able to work directly with the value on your calculator's display.

Next, we multiply our approximation by -2 to find

$$-2\sqrt{7} \approx -5.2905026$$

Finally, we add 8 to get

$$8 - 2\sqrt{7} \approx 2.7084974$$

Rounding to three decimal places gives 2.708. ■

RQ8.   When do we evaluate roots in the order of operations?

Caution!   In Example 5, it is not true that $8 - 2\sqrt{7}$ is equal to $6\sqrt{7}$. The order of operations tells us that we must perform the multiplication $2\sqrt{7}$ first, then subtract the result from 8. You can verify that $6\sqrt{7} \approx 15.874$, which is not the same answer we got in Example 5. ■

We can now update the order of operations by modifying Steps 1 and 2 to include radicals.

Order of Operations
1. Perform any operations inside parentheses, under a radical, or above or below a fraction bar.
2. Compute all indicated powers and roots.
3. Perform all multiplications and divisions in the order in which they occur from left to right.
4. Perform additions and subtractions in order from left to right.

Cube Roots

Imagine a cube whose volume is 64 cubic inches. What is the length, c, of one side of this cube? Because the volume of a cube is given by the formula $V = c^3$, we must find a number that satisfies

$$c^3 = 64$$

We are looking for a number c whose cube is 64. With a little trial and error we can soon discover that $c = 4$. The number c is called the **cube root** of 64, and is denoted by $\sqrt[3]{64}$.

Cube Root
   The number c is called a **cube root** of a number b if $c^3 = b$.

■ **Example 6a.** $\sqrt[3]{-64} = -4$ because $(-4)^3 = -64$
b. $\sqrt[3]{9}$ is an irrational number approximately equal to 2.08, because $2.08^3 = 8.998912$ ■

Recall that every positive number has two square roots, and that negative numbers do not have square roots. The situation is different with cube roots.

   Every number has exactly one cube root. The cube root of a positive number is positive, and the cube root of a negative number is negative.

Just as with square roots, some cube roots are irrational numbers and some are not. Cube roots are treated the same as square roots in the order of operations.

RQ9. What is the cube root of a number n?
RQ10. How many cube roots does a negative number have?

Skills Warm-Up

■ Solve each equation for x.

1. $3x + 5 = 17$
2. $3x + k = 17$
3. $\dfrac{x}{4} - 9 = -4$
4. $\dfrac{x}{4} - m = -4$
5. $2x - 3 = 4x + 7$
6. $2x - c = 4x + 7$
7. $6x + 1 = 3(2 - x)$
8. $bx + 1 = 3(2 - x)$

Answers: 1. 4   **2.** $\dfrac{17 - k}{3}$ **3.** 20   **4.** $4(-4 + m)$ **5.** -5 **6.** $\dfrac{-c - 7}{2}$
7. $\dfrac{5}{9}$ **8.** $\dfrac{5}{b + 3}$

Homework 5.2

Skills Practice

■ For Problems 1-3, simplify. Do not use a calculator!

1a. $4 - 2\sqrt{64}$

b. $\dfrac{4 - \sqrt{64}}{2}$

2a. $\sqrt{9 - 4(-18)}$

b. $\sqrt{\dfrac{4(50) - 56}{16}}$

3a. $5\sqrt[3]{8} - \dfrac{\sqrt[3]{64}}{8}$

b. $\dfrac{3 + \sqrt[3]{-729}}{6 - \sqrt[3]{-27}}$

■ For Problems 4-9, give a decimal approximation rounded to thousandths.

4. $5\sqrt{3}$

5. $\dfrac{-2}{3}\sqrt{21}$

6. $-3 + 2\sqrt{6}$

7. $2 + 6\sqrt[3]{-25}$

8. $\dfrac{8 - 2\sqrt{2}}{4}$

9. $3\sqrt{3} - 3\sqrt{5}$

■ In Problems 10-12, choose the best approximation. Do not use a calculator!

10. $\sqrt{72}$ **a.** 7   **b.** 8   **c.** 36   **d.** 64

11. $\sqrt{13}$ **a.** 6   **b.** 6.5   **c.** 3.5   **d.** 4

12. $\sqrt{134}$ **a.** 11   **b.** 11.5   **c.** 12   **d.** 15

13. Each number below is approximately the square root of a whole number. Find the whole number.

    a. 9.220   **b.** 12.961   **c.** 63.891

14. a. Evaluate $\sqrt{x^2}$ for $x = 3, 5, 8,$ and 12.
    b. Simplify $\sqrt{a^2}$ if $a \geq 0$.

■ For Problems 15-18, use the definitions of square root and cube root to simplify each expression. Do not use a calculator!

15. a. $(\sqrt{16})(\sqrt{16})$

b. $\sqrt{29}(\sqrt{29})$

c. $\sqrt{x}(\sqrt{x})$

16. a. $(\sqrt{7})^2$

b. $(\sqrt[3]{20})^3$

c. $(\sqrt[3]{4})(\sqrt[3]{4})(\sqrt[3]{4})$

17. a. $\dfrac{6}{\sqrt{6}}$

b. $\dfrac{-15}{\sqrt{15}}$

c. $\dfrac{2m}{\sqrt{m}}$

18. a. $(\sqrt{2b})(\sqrt{2b})$

b. $(3\sqrt{a})(3\sqrt{a})$

c. $(2\sqrt[3]{b})^3$

■ Put each set of numbers in order from smallest to largest. Try not to use a calculator.

19. $\dfrac{5}{4},\ 2,\ \sqrt{8},\ 2.3$

20. $2\sqrt{3},\ 3,\ \dfrac{23}{6},\ \sqrt{6}$

In Problems 21-23, evaluate the expression for the given values of x. Round your answers to three decimal places if necessary.

21. $\sqrt{x^2 - 4}$ $x = 3, \sqrt{5}, -2$ **22.** $\sqrt{x} - \sqrt{x + 3}$ $x = 1, 4, 100$

23. $2x^2 - 4x$ $x = -3, \sqrt{2}, \frac{1}{4}$

Applications

24. The distance m in miles that you can see on a clear day from a height of h miles is given by the formula $m = 89.4\sqrt{h}$. How far can you see from an airplane flying at an altitude of 4.7 miles?

25. The period of a pendulum (the time it takes to complete one full swing) is given in seconds by the formula $T = 6.28\sqrt{\dfrac{L}{32}}$, where L is the length of the pendulum in feet. What is the period of the Foucault pendulum in the United Nations Headquarters in New York, whose length is 75 feet?

26. What is a rational number? An irrational number? Give examples of each.

27. Give three examples of square roots that represent rational numbers, and three that are irrational.

28. Identify each number as rational or irrational.

  a. $\sqrt{6}$ **b.** $\dfrac{-5}{3}$ **c.** $\sqrt{16}$ **d.** $\sqrt{\dfrac{5}{9}}$ **e.** 6.008   **f.** $3.\overline{23}$

In Problems 29-32, decide whether the two expressions are equivalent by completing the table.

29.   Does $(a + b)^2 = a^2 + b^2$?

a	b	$a + b$	$(a + b)^2$	a^2	b^2	$a^2 + b^2$
2	3					
3	4					
1	5					
−2	6					

30.   Does $\sqrt{a^2 + b^2} = a + b$?

a	b	$a + b$	a^2	b^2	$a^2 + b^2$	$\sqrt{a^2 + b^2}$
3	4					
2	5					
1	6					
−2	−3					

31. Does $\sqrt{a+b} = \sqrt{a} + \sqrt{b}$?

a	b	$a+b$	$\sqrt{a+b}$	$\sqrt{a}$	$\sqrt{b}$	$\sqrt{a}+\sqrt{b}$
2	7					
4	9					
1	5					
9	16					

32. Does $(\sqrt{a} + \sqrt{b})^2 = a + b$?

a	b	$a+b$	$\sqrt{a}$	$\sqrt{b}$	$\sqrt{a} + \sqrt{b}$	$(\sqrt{a} + \sqrt{b})^2$
4	9					
1	4					
3	5					
6	10					

5.3 Using Formulas

Volume and Surface Area

In Section 5.1 we used exponents to calculate the area of a square and the volume of a cube.

> • **Volume** is the amount of space contained within a three-dimensional object. It is measured in cubic units, such as cubic feet or cubic centimeters.
> • **Surface area** is the sum of the areas of all the faces or surfaces that contain a solid. It is measured in square units.

We saw that the area of a square whose side has length s units is given by $A = s^2$ square units, and the volume of a cube of side s units is gven by $V = s^3$ cubic units. Many other useful formulas involve exponents. The figure below shows several solid objects, along with formulas for their volumes and surface areas.

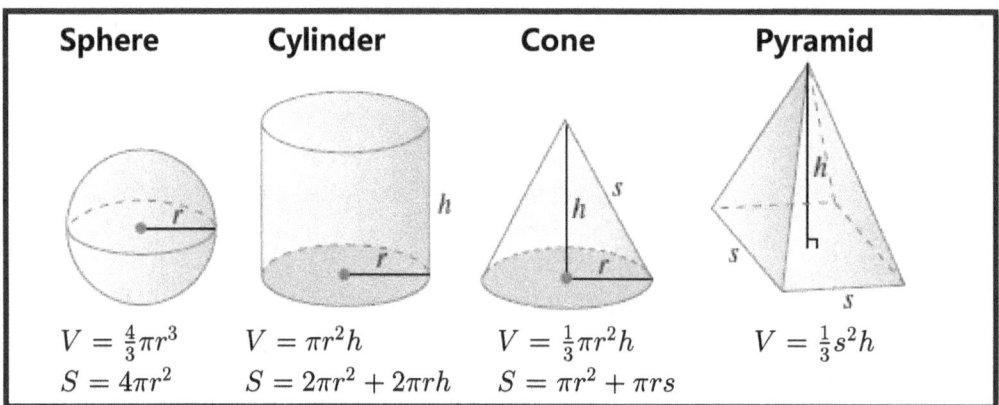

Sphere	Cylinder	Cone	Pyramid
$V = \frac{4}{3}\pi r^3$	$V = \pi r^2 h$	$V = \frac{1}{3}\pi r^2 h$	$V = \frac{1}{3}s^2 h$
$S = 4\pi r^2$	$S = 2\pi r^2 + 2\pi rh$	$S = \pi r^2 + \pi rs$	

> ■ **Example 1**   At the Red Deer Pub and Microbrewery there is a spherical copper tank in which beer is brewed. If the tank is 4 feet in diameter, how much beer does it hold?
>
> **Solution** The formula for the volume of a sphere is $V = \frac{4}{3}\pi r^3$, where r is the radius of the sphere. The tank has diameter 4 feet, so its radius is 2 feet. We substitute $r = 2$ into the formula, and simplify, following the order of operations. (If your calculator does not have a key for π, you can use the approximation $\pi \approx 3.14$.)
>
> $$V = \frac{4}{3}\pi(2)^3 \qquad \textbf{Compute the power first.}$$
> $$= \frac{4\pi(8)}{3} \qquad \textbf{Multiply by } 4\pi, \textbf{ divide by 3.}$$
> $$= 33.51...$$
>
> The tank holds approximately 33.5 cubic feet of beer, or about 251 gallons.   ■

> **RQ1.**   What are the units of volume?
> **RQ2.**   What are the units of surface area?

Look Closer: What does it mean for the volume of a round tank to be measured in "cubic" units? If we pour all the beer in the tank into a rectangular box, that box will hold 33 and a half cubes that measure 1 foot on each side.   ●

Solving Equations with x^2

Taking a square root is the opposite of squaring a number. Thus, we can undo the squaring operation by taking square roots. For example, to solve the equation

$$x^2 = 64$$

we take the square root of each side. Saying that x^2 equals 64 is the same as saying that x is a square root of 64. Remember that every positive number has two square roots, so we write

$$x^2 = 64 \qquad \textsf{Take square roots of both sides.}$$
$$x = \pm\sqrt{64}$$
$$x = \pm 8$$

The equation has two solutions, 8 and −8.

■ **Example 2**   The volume of a can of soup is 582 cubic centimeters, and its height is 10.5 centimeters. What is its radius?

Solution   The volume of a cylinder is given by the formula $V = \pi r^2 h$. We substitute 582 for V and 10.5 for h, then solve for r.

$$582 = \pi r^2 (10.5) \qquad \textsf{Divide both sides by } \pi.$$
$$185.256 = r^2 (10.5) \qquad \textsf{Divide both sides by 10.5.}$$
$$17.643 = r^2 \qquad \textsf{Take square roots.}$$
$$\pm 4.2 = r$$

Because the radius of a soup can cannot be a negative number, we discard the negative solution in this application. The radius of the can is 4.2 centimeters. ■

The equations above, which involve the square of the variable (such as x^2 or r^2), are called **quadratic equations**, and we shall see more about them later.

RQ3.   What operation is the opposite of squaring a number?

Pythagorean Theorem

A triangle in which one of the angles is a right angle, or 90°, is called a **right triangle**. The side opposite the right angle is the longest side of the triangle, and is called the **hypotenuse**. The other two sides of the triangle are called the **legs**.

If we know the lengths of any two sides of a right triangle, we can find the third side using a formula called the Pythagorean Theorem. We use the variable c for the length of the hypotenuse, and the lengths of the legs are denoted by a and b (it doesn't matter which is which).

Pythagorean Theorem

If c stands for the length of the hypotenuse of a right triangle, and the lengths of the two legs are represented by a and b, then
$$a^2 + b^2 = c^2$$

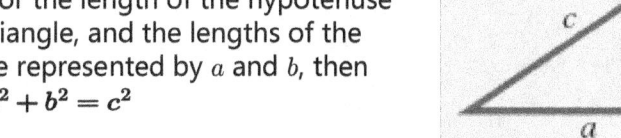

The Pythagorean theorem says that the square of the hypotenuse of a right triangle is equal to the sum of the squares of the two legs.

■ **Example 3**   The length of a rectangle is 17 centimeters and its diagonal is 20 centimeters long. What is the width of the rectangle?

Solution The diagonal of the rectangle is the hypotenuse of a right triangle, as shown in the figure. We are looking for the width of the rectangle, which forms one of the legs of the right triangle. We use the Pythagorean Theorem.

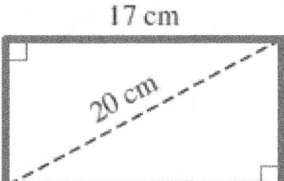

$$a^2 + b^2 = c^2 \qquad \textbf{Substitute 20 for } c \textbf{ and 17 for } b.$$
$$a^2 + (17)^2 = (20)^2$$

We compute the powers, then solve for a.

$$a^2 + 289 = 400 \qquad \textbf{Subtract 289 from both sides.}$$
$$a^2 = 111 \qquad \textbf{Take the square root.}$$
$$a = \pm\sqrt{111}$$

We use a calculator to evaluate the square root and find that the width of the rectangle is approximately 10.536 centimeters.   ■

RQ4.   What is a right triangle, and what is the hypotenuse?
RQ5.   What do we use the Pythagorean theorem for?

Solving for One Variable

In the formula in Example 4, the variable P occurs more than once. We substitute the value for P each time it occurs.

■ **Example 4**   If a principal of P dollars is invested at interest rate r, the formula $A = P + Prt$ gives the amount of money in an account after t years. (The "amount," A, is the original principal plus the interest earned.) How long will it take an investment of $5000 to grow to $7000 in an account that pays $6\frac{1}{2}\%$ interest?

Solution We substitute 7000 for A, 5000 for P, and 0.065 for r in the formula, and then solve the resulting equation for t.

$$7000 = 5000 + 5000\,(0.065)\,t \qquad \textbf{Simplify the right side.}$$
$$2000 = 5000 + 325\,t \qquad \textbf{Subtract 5000 from both sides.}$$
$$2000 = 325t \qquad \textbf{Divide both sides by 325.}$$
$$6.15 \approx t$$

However, the answer to the question in the problem is not 6.15 years! Because interest is paid only once a year, the investor must wait until the end of the seventh year before the account contains at least $7000.   ■

Caution!   To simplify the right side of the equation

$$7000 = 5000 + 5000(0.65)t$$

it is *not* correct to add the two 5000's and write $7000 = 10,000(0.65)t$. Remember that multiplication comes before addition in the order of operations.

If we plan to use a formula more than once with different values for the variables, it may be faster in the long run to solve for one of the variables in terms of the others. We have already done something like this when put a linear equation into slope-intercept form: in that situation we solved for y in terms of x.

Example 5   Solve the formula $A = P + Prt$ for t.

Solution For this problem, we treat t as the unknown and treat all the other variables as if they were constants. We begin by isolating the term containing t on one side of the equation.

$$A = P + Prt \qquad \text{Subtract } P \text{ from both sides.}$$
$$A - P = P + Prt - P$$
$$A - P = Prt \qquad \text{Divide both sides by } Pr.$$
$$\frac{A - P}{Pr} = \frac{Prt}{Pr}$$
$$\frac{A - P}{Pr} = t$$

We now have a new formula for t in terms of the other variables.

Caution!   In the expression $P + Prt$ in Example 5, the terms P and Prt are not like terms. They cannot be combined.

Skills Warm-Up

Follow the order of operations to simplify.

1. $3(-5) - 2^3$

2. $6(7 - 4)^2$

3. $-6 - 2 \cdot 4^2$

4. $(5 - 3)^4(3 - 6)^3$

5. $4(4 - 4^2)$

6. $-3(-2)^2 - 5$

7. $-(3 \cdot 2)^2 - 5$

8. $3 - 4(-2)^3(-3)$

9. $\dfrac{-5^2 + 1}{4} + \dfrac{2(-3)^3}{-6}$

10. $\dfrac{2^2(-3^2)}{1 - 3^2} - \dfrac{7^2 - 6^2}{(1 - 3)^2}$

11. $\dfrac{3^3 - 3}{(3 - 5)^3} - \dfrac{2(-2)^3 - 8}{-2^2(8 - 2^2)}$

12. $3 \cdot \dfrac{5^3 - (-10)^2}{3^2 - 3(5)} \cdot \dfrac{2^5 + 4}{3^2 - 2^2}$

Answers: **1.** -23 **2.** 54 **3.** -38 **4.** -432 **5.** -48 **6.** -17 **7.** -41
8. -93 **9.** 3 **10.** $\frac{5}{4}$ **11.** $\frac{-9}{2}$ **12.** -90

Homework 5.3

Skills Practice

■ Write an expression for the shaded area in each combination of squares and circles.

1.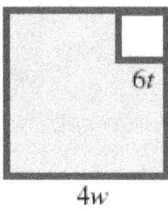

2.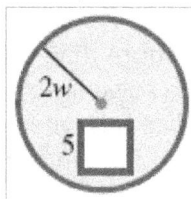

■ Write an algebraic expression for the volume of each figure.

3.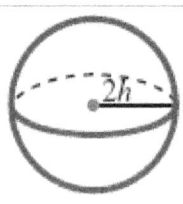

4.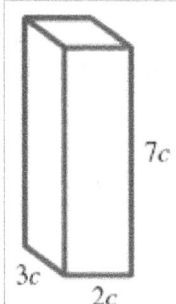

■ Find an expression for the surface area of each box. (**Hint:** Each box has six faces: top and bottom, front and back, left and right. Find the areas of all six faces, then add them.)

5.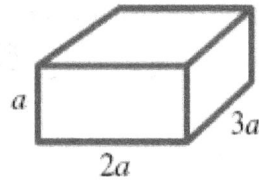

6.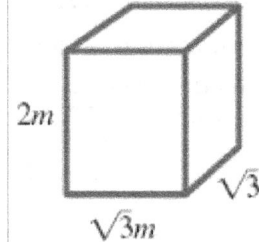

■ Find the unknown side or sides for the right triangles in Problems 7-11.

7.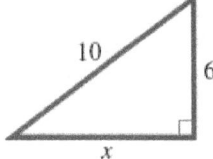

8.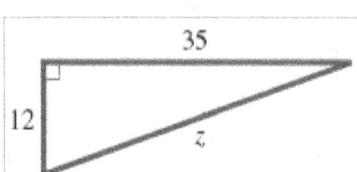

9.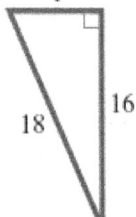

10.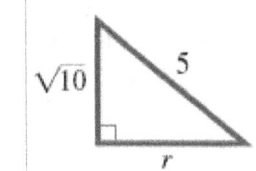

11.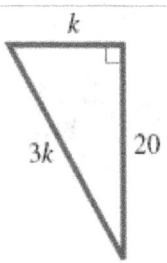

■ Decide whether a triangle with the given sides is a right triangle.

12. 9 in, 16 in, 25 in   **13.** 5 m, 12 m, 13 m   **14.** 5^2 ft, 8^2 ft, 13^2 ft

■ For Problems 15-18, explain whythe equation is an incorrect application of the Pythagorean theorem for the figure.

15.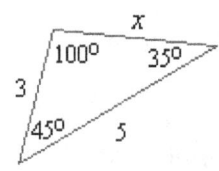

$$3^2 + x^2 = 5^2$$

16.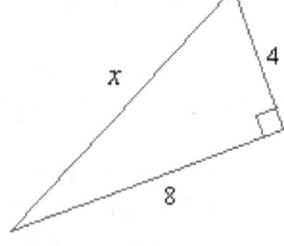

$$8 + 4 = x$$

17.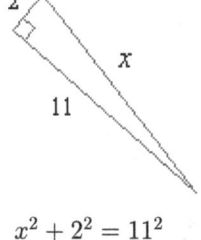

$$x^2 + 2^2 = 11^2$$

18.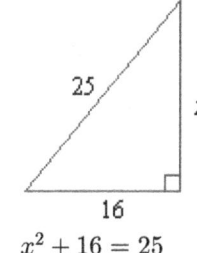

$$x^2 + 16 = 25$$

■ For Problems 19-22, solve the formula for the variable indicated.

19. $v = lwh$ for w **20.** $E = \dfrac{mv^2}{2}$ for m

21. $A = \dfrac{h}{2}(b + c)$ for h **22.** $F = \dfrac{9}{5}C + 32$ for C

23. $A = \pi rh + 2\pi r^2$ for h **24.** $\dfrac{x}{a} + \dfrac{y}{b} = 1$ for x

Applications
■ For Problems 25-28,
  a. Assign a variable and and write an equation.
  b. Solve your equation and answer the question.

25. Juliet's balcony is 24 feet above the ground, and there is a 10-foot moat at the base of the wall. How long a ladder will Romeo need to reach the balcony?

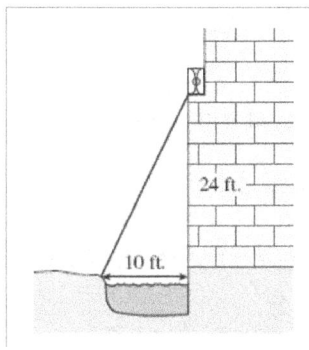

Juliet's balcony

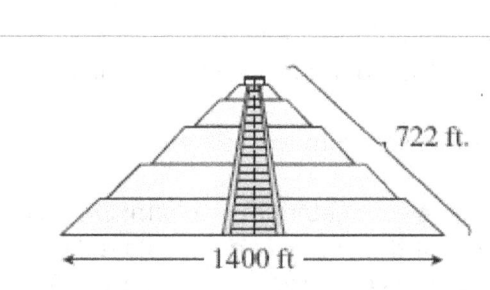

Quetzalcoatl pyramid

26. Marlene visited the Quetzalcoatl pyramid near Mexico City last summer. She measured the base and found it is about 1400 feet on each side. She unrolled a ball of string as she climbed the face of the pyramid, and it was about 722 feet to the top. How tall is the pyramid?

27. A baseball diamond is a square whose sides are 90 feet long. What is the straight-line distance from home plate to second base?

28. The light house is 5 miles east of Gravelly Point. The marina is 7 miles south of Gravelly Point. How far is it from the light house to the marina?

29. A surveyor would like to know the distance across a lake. She picks a spot P on a line perpendicular to the width of the lake, and measures the two distances shown in the figure below left. How wide is the lake?

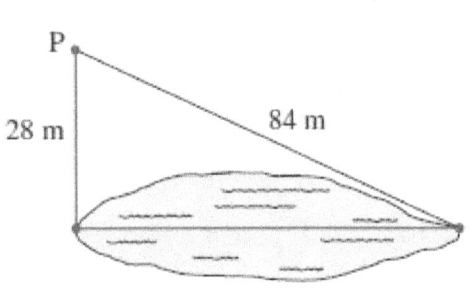

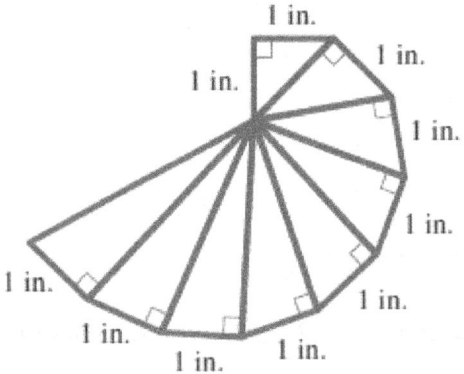

30. Find the length of each hypotenuse in the spiral figure shown above right.

▇ Problems 31-34 involve an area, a circumference, or a volume. Decide which measure is appropriate, and then solve the problem.

31. A circular pizza pan has a diameter of 14 inches. How much pizza dough is needed to cover it?

32. To find the diameter of a large tree in his yard, Delbert wraps a string around the trunk and then measures its length. If the string is 72.25 inches long, what is the diameter of the tree?

33. The first solo transatlantic balloon crossing was completed in 1984 in a helium-filled balloon called Rosie O'Grady. The diameter of the balloon was 58.7 feet. Assuming that the balloon was approximately spherical, calculate its volume.

34. To calculate the area of a large circular fish pond, Graham measures the distance around the edge of the pond as 75.4 feet. Calculate the area of the pond.

▇ In Problems 35-36, use the formulas for volume and surface area.

35. The dome on the new planetarium is a hemisphere (half a sphere) of radius 40 feet.
    a. Draw a sketch of the dome, and label its radius.
    b. How much space is enclosed within the dome?
    c. What is the surface area of the dome?

36. a. How much sheet steel is needed to make a cylindrical can with radius 5 centimeters and height 16 centimeters? (**Hint:** Should you calculate the volume or the surface area of the can?)

   b. How long is a cylindrical section of concrete pipe if its diameter is 3 feet and its surface area is approximately 848.25 square feet?

   c. Draw a sketch of the pipe in part (b), and label its dimensions.

5.4 Products of Binomials

In Section 5.1, we learned to simplify the sum of two algebraic expressions by combining like terms. In this section we see how to multiply algebraic expressions.

Products of Variables

Recall that we can simplify a product such as $3(2x)$ because

$$3(2x) = 2x + 2x + 2x = 6x$$
Three terms

This calculation is actually an application of the associative property:

$$3(2x) = (3 \cdot 2)x = 6x$$

In a similar way, we can simplify the product $(3b)(4b)$ by applying the commutative and associative properties:

$$(3b)(4b) = 3 \cdot b \cdot 4 \cdot b = 3 \cdot 4 \cdot b \cdot b = (3 \cdot 4) \cdot (b \cdot b) = 12b^2$$

Look Closer: You can convince yourself that $(3b)(4b)$ is equivalent to $12b^2$ by substituting some values for b; for example, if $b = 2$, then

$$(3 \cdot 2)(4 \cdot 2) = (6)(8) = 48, \quad \text{and} \quad 12(2)^2 = 12 \cdot 4 = 48 \qquad \bullet$$

> The commutative and associative properties tell us that we can multiply the factors of a product in any order.

■ **Example 1**   Simplify the product or power.

a. $(5a)(-3a)$ **b.** $(2x)^3$ **c.** $(xy^2)(4x^2)$

Solutions a. We apply the commutative property:

$$(5a)(-3a) = 5(-3) \cdot a \cdot a = -15a^2$$

b. To cube an expression means to multiply three copies of the expression together:

$$(2x)^3 = (2x)(2x)(2x) = 2 \cdot 2 \cdot 2 \cdot x \cdot x \cdot x = 8x^3$$

c. We rearrange the factors to group each variable together:

$$(xy^2)(4x^2) = x \cdot y \cdot y \cdot 4 \cdot x \cdot x = 4 \cdot x \cdot x \cdot x \cdot y \cdot y = 4x^3y^2 \qquad ■$$

Caution!   When we add like terms, we do not change the variable in the terms; we combine the coefficients. For example,

$$3a + 2a = 5a$$

When we multiply expressions, we multiply the coefficients and we multiply the variables:

$$3a(2a) = 3(2) \cdot a \cdot a = 6a^2 \qquad ■$$

RQ1.   Explain the difference between $5x - 2x$ and $5x(-2x)$.

Using the Distributive Law

We can use the areas of rectangles to investigate products of algebraic expressions. Recall that we find the area of a rectangle by multiplying its length times its width, $A = lw$. We have already used rectangles to visualize the distributive law. Here are some examples.

■ **Example 2** Calculate the area of the rectangle by adding the areas of each piece. Then use the distributive law to find the product of the algebraic expressions.

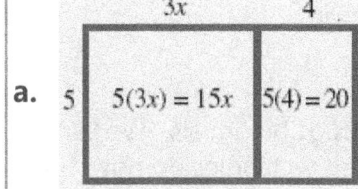

a.
| | 3x | 4 |
| 5 | $5(3x) = 15x$ | $5(4) = 20$ |

Area $= 15x + 20$

a. $5(3x + 4) = 5(3x) + 5(4)$

$ = 15x + 20$

b.

Area $= 2x^2 + 18x$

b. $2x(x + 9) = 2x(x) + 2x(9)$

$ = 2x^2 + 18x$

c.

Area $= 12b^2 + 21b$

c. $3b(4b + 7) = 3b(4b) + 3b(7)$

$ = 12b^2 + 21b$ ■

RQ2. State the distributive law, and explain what it means.

At this stage it will be helpful to introduce some terminology.

- An algebraic expression with only one term, such as $2x^3$, is called a **monomial**.
- An expression with two terms, such as $x^2 - 16$, is called a **binomial.**
- An expression with three terms is a **trinomial**.
- The expression $ax^2 + bx + c$ is thus called a **quadratic trinomial**, because it involves the square of the variable.

Caution!   Notice the difference between $(3a)(2a)$ and $3a(2 + a)$:

- $(3a)(2a)$ is the product of two monomials, and we use the commutative property to simplify it:

$$(3a)(2a) = 3 \cdot a \cdot 2 \cdot a = 3 \cdot 2 \cdot a \cdot a = 6a^2$$

- $3a(2 + a)$ is the product of a monomial and a binomial, and we use the distributive law to simplify it:

$$3a(2 + a) = 3a(2) + 3a(a) = 6a + 3a^2$$ ■

RQ3. Explain the terms monomial, binomial, and trinomial.

Multiplying Binomials

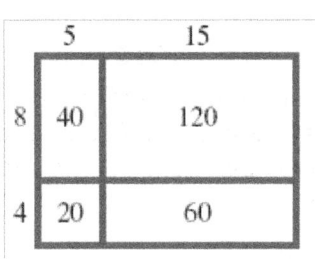

Consider the rectangle shown at right. As you can see, it is divided into four smaller rectangles. You can verify that we get the same answer when we compute its area in two different ways: We can add up the areas of the four smaller rectangles, or we can find the length and width of the entire large rectangle and then find their product:

Sum of four sub-rectangles:
$$\textbf{Area} = 8(5) + 8(15) + 4(5) + 4(15)$$
$$= 40 + 120 + 20 + 60$$
$$= 240$$

One large rectangle:
$$\textbf{Area} = (8+4)(5+15)$$
$$= (12)(20)$$
$$= 240$$

Look Ahead: Our goal in this Lesson is to understand products of binomials. We can use rectangles to illustrate, or model, the product of two binomials. The rectangles do not have to be drawn exactly to scale; they are merely tools for visualizing products. With a small stretch of the imagination, we can use rectangles to represent negative numbers as well.   ●

■ **Example 3a.**   Use a rectangle to represent the product $(x-4)(x+6)$.
b. Write the product as a quadratic trinomial.

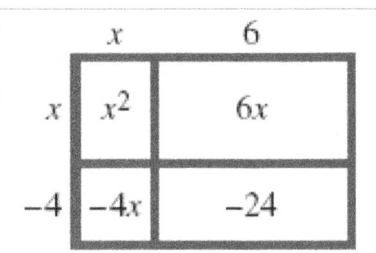

Solutions a.   We let the first factor, $(x-4)$, represent the width of the rectangle, and the second factor, $(x+6)$, represent its length.
b. We find the area of each sub-rectangle, as shown in the figure. Then we add the areas together.
$$\text{Area} = x^2 + 6x - 4x - 24$$
$$= x^2 + 2x - 24$$

We say that $(x+6)(x-4)$ is the **factored form** of the product, and $x^2 + 2x - 24$ is the **expanded form**.

> **RQ4.**   When we use a rectangle to model the product of two binomials, what do the two binomials reresent? What does their product represent?

The Four Terms in a Binomial Product

Using a rectangle to multiply binomials illustrates how the distributive law works. Analyzing the rectangle method will help us in the next Lesson, when we reverse the process to **factor** a quadratic trinomial. Let us take a closer look at the example above.

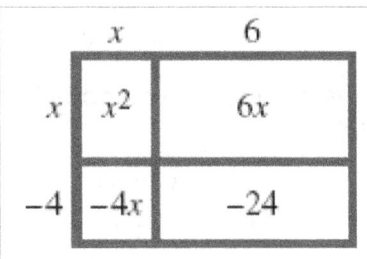

Look Closer: In Example 3 we computed the product $(x-4)(x+6)$. The top row of the rectangle corresponds to

$$x(x+6) = x^2 + 6x$$

and the bottom row corresponds to

$$-4(x+6) = -4x - 24$$

Thus, we multiply each term of the first binomial by each term of the second binomial, resulting in four multiplications in all:

$$(x-4)(x+6) = x^2 + 6x - 4x - 24$$

Each term of the product corresponds to the area of one of the four sub-rectangles, as shown below.

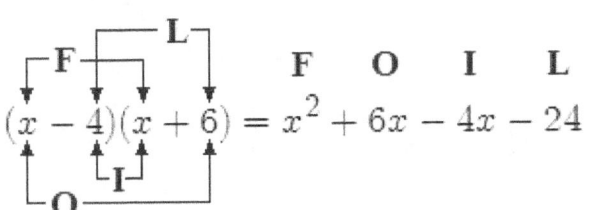   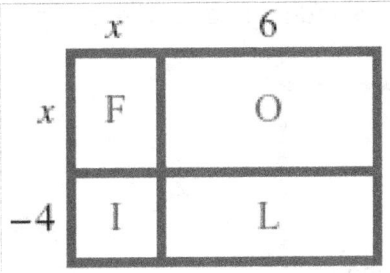

The letters **F, O, I, L** indicate the four steps in computing the product:

  1. **F** stands for the product of the **First** terms in each binomial.
  2. **O** stands for the product of the **Outer** terms.
  3. **I** stands for the product of the **Inner** terms.
  4. **L** stands for the product of the **Last** terms.

Note how each term of the trinomial product arises from the binomial factors:

$$(x-4)(x+6) = \quad x^2 \quad + 2x \quad - 24$$
$$\qquad\qquad\qquad \textbf{F} \quad \textbf{O+I} \quad \textbf{L}$$

• The **quadratic** term in the product comes from the **First** terms.   • The **linear** or x-term of the product is the sum of **Outer** and **Inner**.   • The **constant** term of the product comes from the **Last** terms.

We say that $(x+6)(x-4)$ is the **factored form** of the product, and $x^2 + 2x - 24$ is the **expanded form**.

RQ5.   What does the acronym FOIL stand for?

Products involving variables often arise in working with binomials.

■ **Example 4**   Compute the product $(3x-5)(4x+2)$, and write your answer as a quadratic trinomial.   **Solution**   We multiply each term of the first factor by each term of the second factor, using the "FOIL" template to keep track of the products. $(3x-5)(4x+2) = 3x(4x) + 3x(2) - 5(4x) - 5(2)$ $\qquad\qquad\qquad\quad \textbf{F} \qquad \textbf{O} \qquad \textbf{I} \qquad \textbf{L}$ $\qquad\qquad = 12x^2 + 6x - 20x - 10 \qquad$ **Combine like terms.** $\qquad\qquad = 12x^2 - 14x - 10$ ■

RQ6.   In the "FOIL" representation of a binomial product, which terms are like terms?

Skills Warm-Up

Find the area and perimeter of each figure.

1.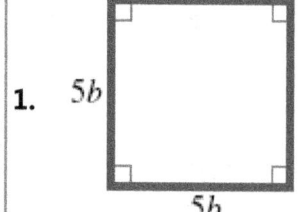

2.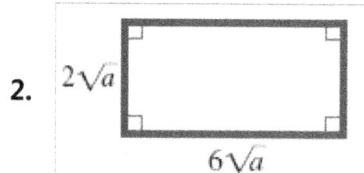

3.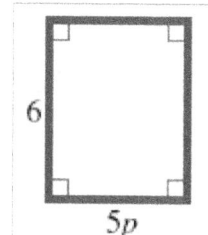

4.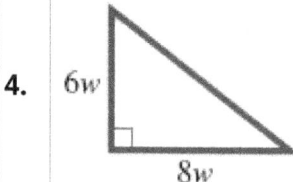

5.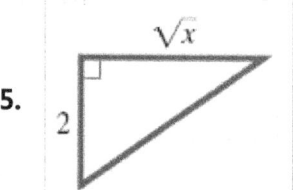

6.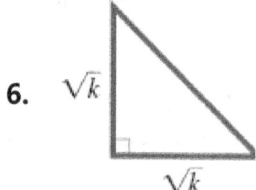

Answers: 1. $25b^2$, $20b$ **2.** $12a$, $16\sqrt{a}$ **3.** $30p$, $12 + 10p$ **4.** $48w^2$, $24w$
5. $\sqrt{x}$, $2 + \sqrt{x} + \sqrt{x+4}$ **6.** $\frac{k}{2}$, $2\sqrt{k} + \sqrt{2k}$

Homework 5.4

Skills Practice

■ In Problems 1-4, simplify each product or power.

1a. $3(4n)$ **b.** $3n(4n)$ **c.** $(4n)^3$
2a. $6x(-5x^2)$ **b.** $-6x^2(-5x)$ **c.** $(-5x)^2$
3a. $(4p)(-4p)$ **b.** $-(4p)^2$ **c.** $-4(-p)^2$

■ In Problems 4-6, simplify each expression.

4a. $-8x(5t)$ **b.** $-8x(5+t)$ **c.** $-8x(-5-t)$
5a. $3n(-4n)$ **b.** $3n - 4n$ **c.** $(3n - 4)n$
6a. $2x(-5x^2)$ **b.** $2(x - 5x^2)$ **c.** $2x - (5x)^2$

■ In Problems 7-8,
   a. Write a product (length × width) for the area of each rectangle.
   b. Use the distributive law to compute the product.

7. 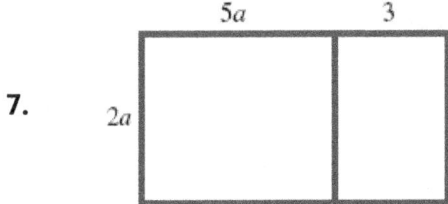   **8.**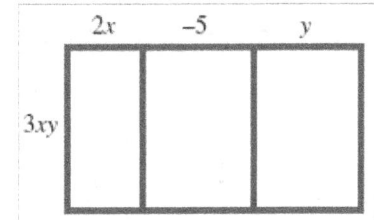

■ In Problems 9-14, compute the product.

9. $-2b(6b - 2)$ **10.** $(6a - 5)(3a)$ **11.** $3v(5v - 2v^2)$
12. $-4x^2(2x + 3y)$ **13.** $(y^3 + 3y - 2)(2y)$ **14.** $-xy(2x^2 - xy + 3y^2)$

■ In Problems 15-18, simplify.

15. $2a(x + 3) - 3a(x - 3)$ **16.** $2x(3 - x) + 2(x^2 + 1) - 2x$
17. $ax(x^2 + 2x - 3) - a(x^3 + 2x^2)$ **18.** $3ab^2(2 + 3a) - 2ab(3ab + 2b)$

■ In Problems 19-20, write two different expressions for the area of each rectangle:
   a. as the sum of four small areas,
   b. as one large rectangle, using the formula Area = length × width.

19. 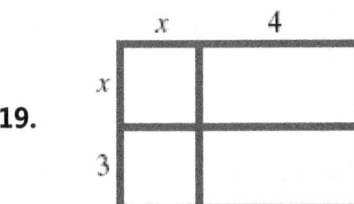   **20.**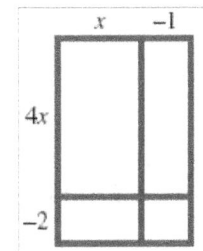

■ In Problems 21-23,
    a. Use a rectangle to represent each product.
    b. Write the product as a quadratic trinomial.

21. $(a - 5)(a - 3)$ **22.** $(y + 1)(3y - 2)$ **23.** $(5x - 2)(4x + 3)$

■ In Problems 24-26, use rectangles to help you multiply these binomials in two variables.

24. $(x + 2y)(x - y)$ **25.** $(3s + t)(2s + 3t)$ **26.** $(2x - a)(x - 3a)$

■ In Problems 27-30, compute the product: Multiply the binomials together first, then multiply the result by the numerical coefficient.

27. $2(3x - 1)(x - 3)$ **28.** $-3(x + 4)(x - 1)$
29. $-(4x + 3)(x - 2)$ **30.** $5(2x + 1)(2x - 1)$

Applications

■ In Problems 31-32, find the product without a calculator by using rectangles.

31. 36×42 **32.** 82×16

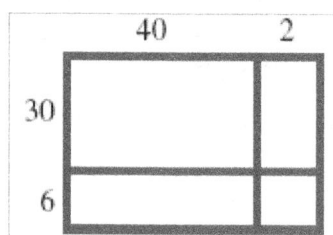

                                          (Make your own drawing)

■ In Problems 33-34,
    a. Find the linear term in the product.
    b. Shade the sub-rectangles that correspond to the linear term.

33. $(x + 6)(x - 9)$ **34.** $(2x - 5)(x + 4)$

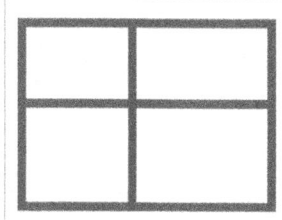 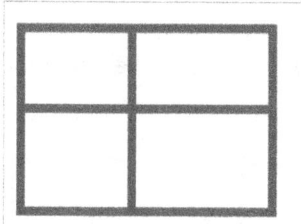

■ In Problems 35-37, compute the product. What do you notice? Explain why this happens.

35. $(x + 3)(x - 3)$ **36.** $(x - 2a)(x + 2a)$ **37.** $(3x + 1)(3x - 1)$

In Problems 38-40, compute the product.

38. $(w+4)(w+4)$ **39.** $(z-6)(z-6)$ **40.** $(3a-2c)(3a-2c)$

41. a. Complete the table below.
   b. Decide whether the two expressions, $(a-b)^2$ and a^2-b^2, are equivalent.

a	b	$a-b$	$(a-b)^2$	a^2	b^2	a^2-b^2
5	3					
2	6					
−4	−3					

42. Explain why $(x-y)^2$ cannot be simplified to x^2-y^2.

In Problems 43-44, write the area of the square in two different ways:
  a. as the sum of four smaller areas,
  b. as one large square, using the formula Area = (length)^2.

43.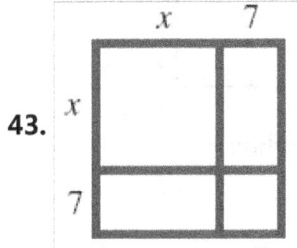

44.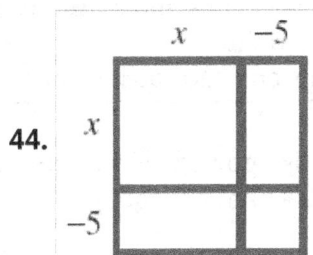

45. Is $(x+4)^2$ equivalent to x^2+4^2? Explain why or why not, and give a numerical example to justify your answer.

In Problems 46-48, compute the product.

46. $(x-2)^2$ **47.** $(2x+1)^2$ **48.** $(3x-4y)^2$

In Problems 49-50, use the Pythagorean theorem to write an equation about the sides of the right triangle.

49.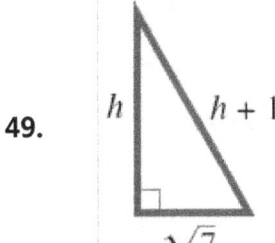

50.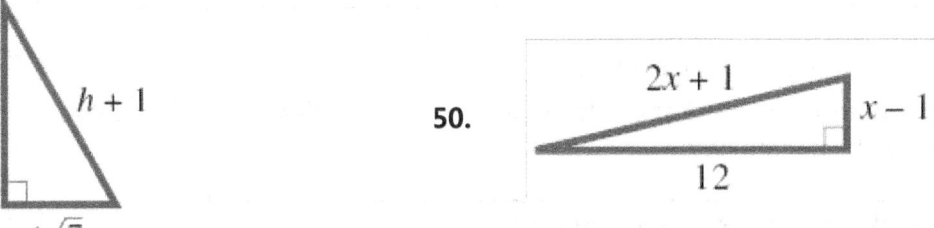

Chapter 5 Summary and Review

Section 5.1 Exponents

- An **exponent** tells us how many times its **base** occurs as a factor in an expression.
- To indicate that a negative number is raised to a power, we must enclose the negative number in parentheses.
- An exponent on a variable indicates repeated multiplication, while a coefficient in front of a variable indicates repeated addition.
- An exponent applies only to its base, and not to any other factors in the product. If we want an exponent to apply to more than one factor, we must enclose those factors in parentheses.

Order of Operations
1. Perform any operations inside parentheses, or above or below a fraction bar.
2. Compute all indicated powers.
3. Perform all multiplications and divisions in the order in which they occur from left to right.
4. Perform additions and subtractions in order from left to right.

- We can add or subtract like powers of the same variable by combining their coefficients; the exponent does not change.

Section 5.2 Square Roots and Cube Roots

Square Root
  s is called a **square root** of a number b if $s^2 = b$

- Every positive number has two square roots, one positive and one negative. The positive square root of a number is called the **principal square root**.
- We cannot take the square root of a negative number.

Cube Root
  c is called a **cube root** of a number b if $c^3 = b$

- Every number has exactly one cube root.
- A **rational number** is one that can be expressed as a quotient (or ratio) of two integers, where the denominator is not zero.

Decimal Form of a Rational Number
  The decimal representation of a rational number has one of two forms.
  1. The decimal representation terminates, or ends.
  2. The decimal representation repeats a pattern.

- The rational and irrational numbers together make up the **real numbers**, and the number line is sometimes called the **real line**.

Order of Operations
1. Perform any operations inside parentheses, under a radical, or above or below a fraction bar.
2. Compute all indicated powers and roots.
3. Perform all multiplications and divisions in the order in which they occur from left to right.
4. Perform additions and subtractions in order from left to right.

Section 5.3 Using Formulas

• **Volume** is the amount of space contained within a three-dimensional object, and is measured in cubic units, such as cubic feet or cubic centimeters. **Surface area** is the sum of the areas of all the faces or surfaces that contain a solid, and is measured in square units.

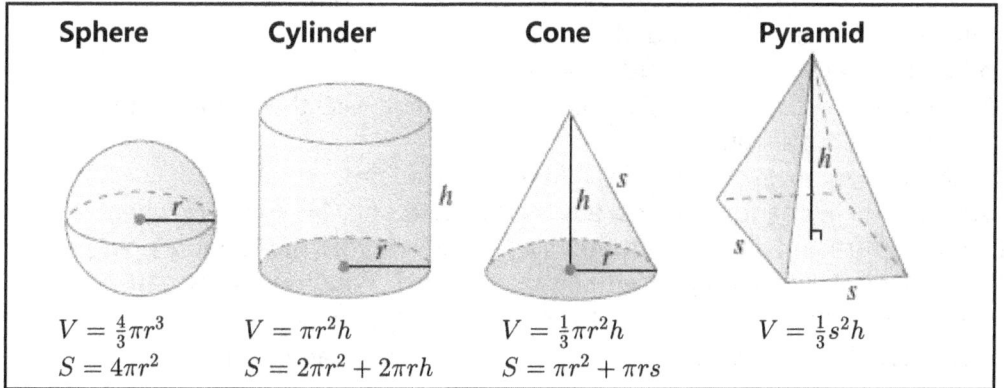

Sphere	Cylinder	Cone	Pyramid
$V = \frac{4}{3}\pi r^3$	$V = \pi r^2 h$	$V = \frac{1}{3}\pi r^2 h$	$V = \frac{1}{3}s^2 h$
$S = 4\pi r^2$	$S = 2\pi r^2 + 2\pi r h$	$S = \pi r^2 + \pi r s$	

• Taking a square root is the opposite of squaring a number. We can undo the squaring operation by taking square roots.

Pythagorean Theorem

If c stands for the length of the hypotenuse of a right triangle, and the lengths of the two legs are represented by a and b, then
$$a^2 + b^2 = c^2$$

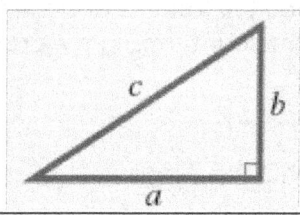

• If we plan to use a formula more than once with different values for the variables, it may be faster in the long run to solve for one of the variables in terms of the others.

Section 5.4 Products of Binomials

• When we add like terms, we do not change the variable in the terms; we combine the coefficients. When we multiply expressions, we multiply the coefficients and we multiply the variables.

• An algebraic expression with only one term is called a **monomial**. An expression with two terms is called a **binomial**, and an expression with three terms is a **trinomial**. The expression $ax^2 + bx + c$ is thus called a **quadratic trinomial**.

- We use the distributive law to expand the product of two binomials. The letters
  F, O, I, L indicate how each term of the product arises from the binomial factors:
    1. **F** stands for the product of the **First** terms in each binomial.
    2. **O** stands for the product of the **Outer** terms.
    3. **I** stands for the product of the **Inner** terms.
    4. **L** stands for the product of the **Last** terms.

Review Questions

▓ Use complete sentences to answer the questions in Problems 1-15.

1. Explain the difference between -3^2 and $(-3)^2$.
2. If $s^2 = 17$, then s is called a _____ of 17.
3. If $s^3 = 17$, then s is called a _____ of 17.
4. What is the difference between the square of 4 and the square root of 4?
5. What is the difference between $x^2 = 9$ and $x = \sqrt{9}$?
6. What is the difference between $-\sqrt{25}$ and $\sqrt{-25}$?
7. Explain the difference between a coefficient and an exponent. Use the expressions $2x^3$ and $3x^2$ to illustrate your explanation.
8. Is the statement $\sqrt{81} = \sqrt{9} = 3$ correct? Explain why or why not.
9. When asked to state the Pythagorean theorem, a classmate says, "$a^2 + b^2 = c^2$ when the sides of a triangle are a, b, and c." Another student claims that the formula does not work for all triangles. Do you agree with either student? What necessary part or parts of the theorem are not mentioned by either of the two classmates?
10. A classmate says that $\sqrt[3]{2}$ must be a rational number because it can be written as $\frac{\sqrt[3]{2}}{1}$. Do you agree or disagree? Explain.
11. Give an example of a negative irrational number.
12. What do you get if you square $\sqrt[3]{5}$ and then multiply the result by $\sqrt[3]{5}$? Explain how to get the correct answer without using a calculator.
13. What do volume and surface area measure? What are the units of each?
14. How can you tell from its decimal form whether a number is rational or irrational?
15. What are the four terms in the product $(a + b)(c + d)$?

Review Problems

▓ For Problems 1-6, simplify.

1. **a.** $3(-7)^2$ **b.** $3^2(-7^2)$

   c. $3^2 - 7^2$ **d.** $(3 - 7)^2$

2. **a.** $\sqrt{13^2 - 5^2}$ **b.** $\sqrt{13^2} - \sqrt{5^2}$

   c. $\sqrt{13^2}\,\sqrt{5^2}$ **d.** $\sqrt{(13 - 5)^2}$

3. **a.** $\sqrt{\left(\frac{4}{5}\right)^2 + \left(\frac{3}{5}\right)^2}$ **b.** $\sqrt{\left(\frac{4}{5}\right)^2} + \sqrt{\left(\frac{3}{5}\right)^2}$

   c. $\sqrt{\left(\frac{4}{5}\right)^2}\,\sqrt{\left(\frac{3}{5}\right)^2}$ **d.** $\sqrt{\left(\frac{4}{5} + \frac{3}{5}\right)^2}$

4. a. $2 - \sqrt[3]{64}$ **b.** $2\sqrt[3]{-64}$

    c. $\left(\sqrt[3]{101}\right)^3$ **d.** $\sqrt[3]{7} \cdot \sqrt[3]{7} \cdot \sqrt[3]{7}$

5. a. $\left(\sqrt[3]{8}\right)^2$ **b.** $\sqrt[3]{8^2}$

    c. $\left(\sqrt{9}\right)^3$ **d.** $\sqrt{9^3}$

6. a. $(\sqrt{3x})(\sqrt{3x})$ **b.** $3\sqrt{x}\,(3\sqrt{x})$

    c. $(5\sqrt{a})^2$ **d.** $(\sqrt[3]{B})^2(\sqrt[3]{B})$

■ For Problems 7-8, identify each number as rational, irrational or undefined.

7a. $\sqrt[3]{-5}$ **b.** $\sqrt{-5^2}$ **c.** $\sqrt{(-5)^2}$ **d.** $\left(\sqrt[3]{-5}\right)^2$

8a. $-3.1\overline{6}$ **b.** -3.16 **c.** $-\sqrt{10}$ **d.** $-\pi$

■ For Problems 9-10, decide whether each number is rational or irrational. Then find a decimal form for each number. If it is irrational, round to hundredths.

9. a. $\sqrt{300}$ **b.** $5 + \sqrt[3]{15}$ **10. a.** $\sqrt[3]{512}$ **b.** $\dfrac{7}{\sqrt{81}}$

■ For Problems 11-14, simplify.

11. $4 - 2 \cdot 3^2$ **12.** $-2 - 3(-3)^3 - 2$

13. $\dfrac{-6 - \sqrt{6^2 - 4(2)(4)}}{2(2)}$ **14.** $18 - 2\sqrt[3]{\dfrac{4}{3}(48)}$

■ For Problems 15-16, give a decimal approximation rounded to thousandths.

15. $-8 - 5\sqrt{6}$ **16.** $\dfrac{3 + \sqrt[3]{3}}{3}$

■ For Problems 17-26, evaluate the expression for the given values of the variables. Round to three decimal places any answers which are irrational numbers.

17. $a = 3$:   **a.** $5 - a^2$ **b.** $(5 - a)^2$

18. $a = -3$:   **a.** $5 - a^2$ **b.** $(5 - a)^2$

19. $w = -2$:   **a.** $-w^2$ **b.** $(-w)^2$

20. $w = -2$:   **a.** $\sqrt{-w}$ **b.** $\sqrt[3]{-w}$

21. $x = -4$:   **a.** $3 - x^2$ **b.** $3(-x)^2$

22. $a = 6,\ b = -2$:   **a.** $\dfrac{3b^3}{6} - \dfrac{4 - a^2}{8}$ **b.** $\dfrac{(a - b)^2}{ab^2}$

23. $x = 13,\ y = -5$:   **a.** $\sqrt{x^2 - y^2}$ **b.** $\left(\sqrt{x}\right)^2 + \left(\sqrt{-y}\right)^2$

24. $a = 2,\ b = -3,\ c = 1$:   **a.** $\sqrt{b^2 - 4ac}$ **b.** $\dfrac{-b + \sqrt{b^2 - 4ac}}{2a}$

25. $x = 3,\ y = 4,\ z = 5$:   **a.** $\sqrt[3]{x^3 + y^3 + z^3}$ **b.** $\sqrt[3]{x^3} + \sqrt[3]{y^3} + \sqrt[3]{z^3}$

26. $m = 3,\ n = 2$:   **a.** $m^3 + n^3$ **b.** $(m + n)(m^2 - mn + n^2)$

27. What is the smallest positive integer (other than 1) that is both a perfect square and a perfect cube?

28. Find the cube root of 27^2 without using a calculator.

■ For Problems 29-32,
  a. Find the unknown side or sides for each right triangle.
  b. Find the perimeter of the triangle.
  c. Find the area of the triangle.

29.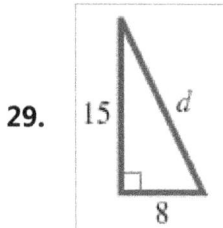

30.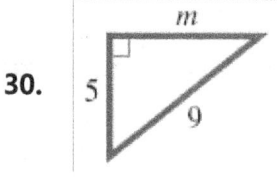

31.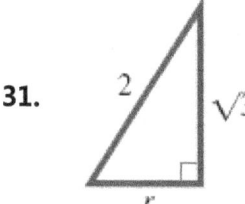

32.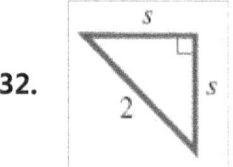

■ For Problems 33-34, find the volume and surface area of the box.

33.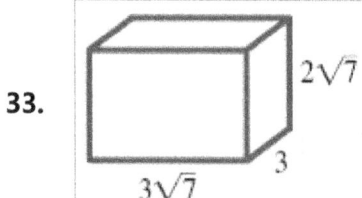

34.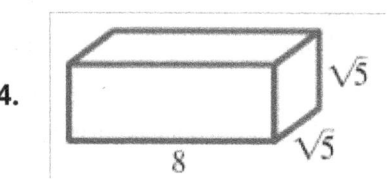

■ For Problems 35-36, write an algebraic expression for the volume of the figure.

35.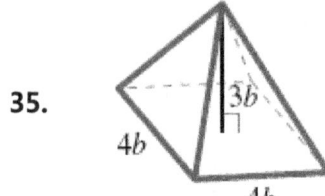

36.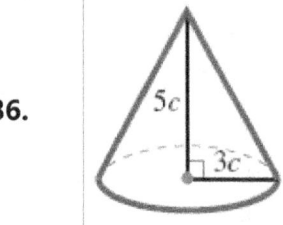

■ For Problems 37-38, write expressions for the volume and surface area of the figure.

37.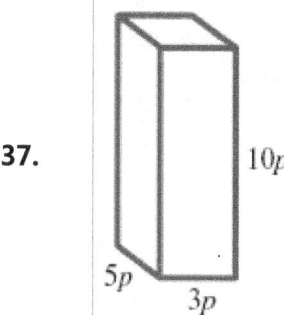

$10p$

$5p$

$3p$

38.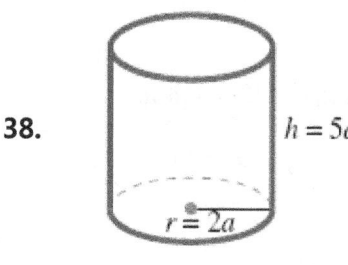

$h = 5a$

$r = 2a$

■ For Problems 39-42, solve the formula for the indicated variable.

39. $s = vt + \dfrac{1}{2}at^2$ for v

40. $A = \dfrac{h}{2}(b + c)$ for c

41. $S = \dfrac{n}{2}(a + f)$ for a

42. $s = vt + \dfrac{1}{2}at^2$ for a

■ For Problems 43-50, draw a sketch, write an equation, and solve. Round your answer to the nearest tenth.

43. Francine walks along the edge of a rectangular yard from the northwest corner to the southeast corner. Delbert walks diagonally across the same yard. If the yard is 30 yards long and 20 yards wide, how much farther does Francine walk than Delbert?

44. Rani kayaks due west for 2500 meters, then due south for another 1000 meters. If she now kayaks directly back to her starting point, how long is her total trip?

45. a. The volume of a spherical communications satellite is 65.45 cubic feet. What is the radius of the satellite?
   b. What is the surface area of the satellite in part (a)?

46. a. Francine's coffee cup is shaped like a cylinder 10 centimeters tall with a radius of 3.36 centimeters. What is its volume in cubic centimeters?
   b. One fluid ounce is about 29.56 cubic centimeters. Find the volume of Francine's coffee cup in ounces, rounded to the nearest ounce.
   c. Delbert's coffee cup is also 10 centimeters tall, but its volume is 24 fluid ounces. What is its radius?

47. A rectangle is 7 meters wide and 9 meters long. What is the length of its diagonal?

48. The size of a TV screen is the length of its diagonal. What is the size of a TV whose screen measures 12 inches by 16 inches?

49. A 26-meter guy wire is attached to a radio antenna for support. One end of the wire is attached to the antenna 24 meters above the ground, and the other is attached to an iron ring in a cement slab on the ground. How far is the ring from the base of the antenna?

50. Find the height of an equilateral triangle whose sides are each 8 centimeters long.

■ For Problems 51-54, compute each product.

51. $-5x(4 - 3x)$

52. $(3xy - 4x^2 + 2)(-2xy)$

53. $3xy(2x - 4 - y)$

54. $-2a(-a^2 - 2a + 4)$

■ For Problems 55-56, simplify if possible.

55a. $6t^2 - 8t^2$

b. $6t - 8t^2$

c. $6t(-8t^2)$

56a. $w^2 + w^2$

b. $-w - w$

c. $w^2 - w$

■ For Problems 57-60, simplify.

57. $6a(2a - 1) - (3a^2 - 3a)$

58. $4b - 2(3 - b^2) - b(b - 3)$

59. $5a(2a - 3) - 4(3a^2 - 2) + 6a$

60. $2x(x - 3y) - 3y(x - 3y)$

■ For Problems 61-62, fill in the missing algebraic expressions.

61. $-2n(3n + ?) = -6n^2 - 2n$

62. $5b(6 + ?) = 30b - 35b^2$

■ For Problems 63-64, write the area of the rectangle in two different ways:
a. as the sum of four smaller areas,
b. as one large rectangle, using the formula Area = length × width.

63.

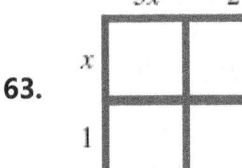

64.

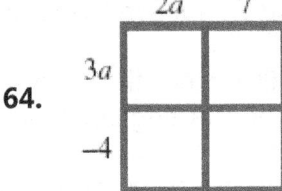

■ For Problems 65-76, compute the product.

65. $(x - 2)(x + 4)$

66. $(2y + 1)(3y - 2)$

67. $(5a + 1)(5a - 1)$

68. $(n - 7)(n - 7)$

69. $(2q + 5)(2q + 5)$

70. $(3c - 8)(3c + 8)$

71. $(u - 5)(u - 2)$

72. $(3r + 2)(r - 4)$

73. $(a + 6)^2$

74. $(6y + 5)(6y - 5)$

75. $(2a - 5c)(3a + 2c)$

76. $-3(x - 4)(2x + 5)$

77. Use the given areas to find the area of the large square shown below.

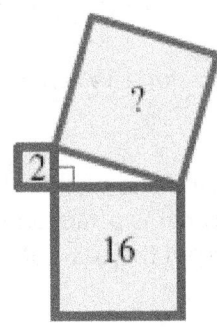

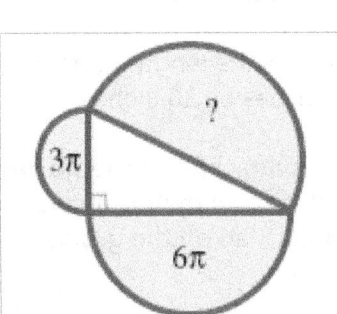

78. Use the given areas to find the area of the large semicircle shown above.

Chapter 6 Quadratic Equations

Lesson 1 Extracting Roots

- Identify a quadratic equation
- Graph a parabola by plotting points
- Use a graph to solve a quadratic equation
- Solve a quadratic equation by extracting roots
- Solve a quadratic formula for one variable

Lesson 2 Some Quadratic Models

- Use the zero-factor principle
- Factor out a common factor
- Solve quadratic equations
- Graph a parabola by plotting points
- Interpret quadratic models

Lesson 3 Solving Quadratic Equations by Factoring

- Use the zero-factor principle
- Graph a parabola by plotting points
- Factor quadratic trinomials
- Solve a quadratic equation by factoring
- Solve applied problems involving quadratic equations

Lesson 4 Graphing Quadratic Equations

- Find the x-intercepts of a parabola
- Find the vertex of a parabola
- Find the y-intercept of a parabola
- Sketch the graph of a quadratic equation

Lesson 5 Quadratic Formula

- Use the quadratic formula to solve equations
- Solve applied problems using the quadratic formula
- Use the quadratic formula to find the x-intercepts of a parabola
- Determine how many x-intercepts the graph has

6.1 Extracting Roots

So far you have learned how to solve linear equations. In linear equations, the variable cannot have any exponent other than 1, and for this reason such equations are often called **first-degree**. In this chapter we'll consider second-degree equations, or **quadratic equations**.

> **Quadratic Equations**
> A **quadratic equation** can be written in the standard form
> $$ax^2 + bx + c = 0$$
> where a, b, and c are constants and a is not zero.

■ Example 1   Some examples of quadratic equations are
$$2x^2 + 5x - 3 = 0, \qquad 7t - t^2 = 0, \qquad \text{and} \qquad 3w^2 = 16$$

The first equation is already written in standard form, with $a = 2$, $b = 5$, and $c = 3$. In standard form, the other two equations are

$$-t^2 + 7t + 0 = 0, \qquad \text{so } a = -1, b = 7, \text{ and } c = 0$$
$$3w^2 + 0w - 16 = 0, \qquad \text{so } a = 3, b = 0, \text{ and } c = -16 \qquad ■$$

The numbers a, b, and c are called **parameters**. They are the coefficients of, respectively, the quadratic term (the x^2-term), the linear term (the x-term), and the constant term.

> **Caution!**   The x^2-term and the x-term in a quadratic expression are not like terms, so they cannot be combined.   ■

> **RQ1.**   What is a quadratic equation? Give an example.

Graphs of Quadratic Equations

Quadratic equations appear in a variety of applications. A quadratic equation in two variables has the form

$$y = ax^2 + bx + c, \ a \neq 0$$

Its graph is not a straight line, but a curve called a **parabola**. The simplest, or basic, parabola is the graph of $y = x^2$, shown in the figure.

x	-3	-2	-1	0	1	2	3
y	9	4	1	0	1	4	9

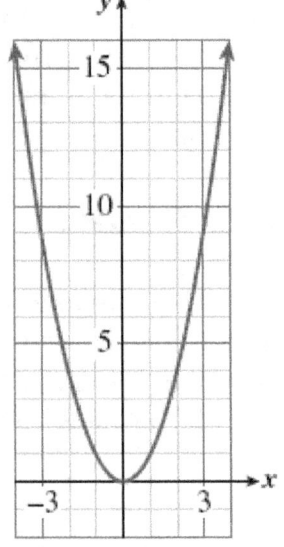

Look Closer: In the equation for the basic parabola, $y = x^2$, we have $a = 1$, $b = 0$, and $c = 0$. If the parameters have different values, the appearance or location of the graph will be altered. Example 2 illustrates a parabola that is shifted downward by 4 units.   ●

Example 2   Graph $y = x^2 - 4$

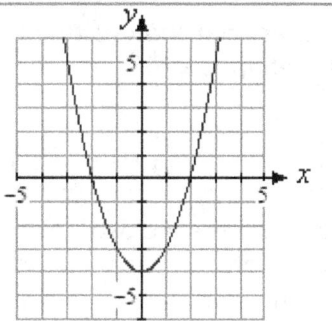

Solution We make a table of values and plot the points. The graph is shown at right.

x	-3	-2	-1	0	1	2	3
y	5	0	-3	-4	-3	0	5

Note that in this equation, $c = -4$.

RQ2.   The basic parabola is the graph of what equation?

Solving Quadratic Equations

Consider a simple quadratic equation,

$$x^2 = 16$$

This equation tells us that x is a number whose square is 16. Thus, x must be a square root of 16. Now, 16 has two square roots, 4 and -4, so these are the solutions of the equation.

We can think of the solution process in the following way. Because x is squared in the equation, we perform the opposite operation, or take square roots, in order to solve for x.

$x^2 = 16$ **Take square roots of both sides.**

$x = \pm\sqrt{16}$ **Simplify.**

$x = \pm 4$

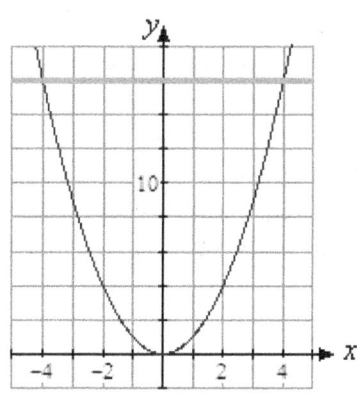

Look Closer: Remember that every positive number has two square roots, so the equation has two solutions, $x = 4$ and $x = -4$. We can also see the two solutions on a graph. The figure at right shows the graph of $y = x^2$ and a horizontal line at $y = 16$. There are two points on the graph with y-coordinate 16, and their x-coordinates are 4 and -4.

Example 3   Solve each equation without using a calculator.

a. $t^2 = 10,000$ **b.** $b^2 = \dfrac{4}{9}$ **c.** $h^2 = 0.04$ **d.** $m^2 = (-17)^2$

Solutions   For each equation, we find the square root of the right side. (That is, find a number whose square gives the right side.) Remember that every positive number has two square roots.

a. Because $100^2 = 10,000$, $t = \pm 100$
b. Because $\left(\frac{2}{3}\right)^2 = \frac{4}{9}$, $b = \pm\frac{2}{3}$
c. Because $0.2^2 = 0.04$, $h = \pm 0.2$
d. Because $(-17)^2 = (17)^2$, $m = \pm 17$

RQ3.   Why do the equations in Example 3 have two solutions?

Extraction of Roots

The figure at right shows a graph of the quadratic equation $y = 2x^2 - 5$.

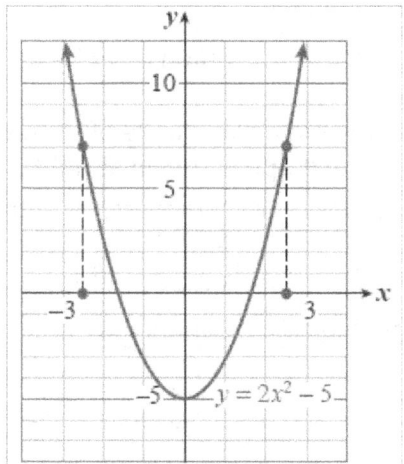

x	−3	−2	−1	0	1	2	3
y	13	3	−3	−5	−3	3	13

You can see that there are two points on the graph for each y-value greater than −5. For example, the two points with y-coordinate 7 are shown. To solve the quadratic equation

$$2x^2 - 5 = 7$$

we need only find the x-coordinates of these points. From the graph, the solutions appear to be about 2.5 and −2.5.

How can we solve this equation algebraically? We use the fact that taking square roots is the opposite of squaring.

■ **Example 4**   Solve the equation $2x^2 - 5 = 7$ algebraically.

Solution   We first solve for x^2 as follows.

$$2x^2 - 5 = 7 \qquad \textbf{Add 5 to both sides.}$$
$$2x^2 = 12 \qquad \textbf{Divide both sides by 2.}$$
$$x^2 = 6$$

Once we have isolated x^2, we take the square root of each side to find

$$x = \pm\sqrt{6}$$

The exact solutions are thus $\sqrt{6}$ and $-\sqrt{6}$. (Don't forget that every positive number has two square roots.) We can also find decimal approximations for the solutions using a calculator. Rounded to two decimal places, the solutions are 2.45 and −2.45.   ■

RQ4a. What are the exact solutions to the equation $2x^2 - 5 = 7$?
  b. What are the approximate values of those solutions, rounded to hundredths?

This method for solving quadratic equations is called **extraction of roots**. The method applies to quadratic equation of the form $ax^2 + c = 0$, where the linear term bx is missing.

Extraction of Roots
    To solve a quadratic equation of the form

$$ax^2 + c = 0$$

1. Isolate x^2 on one side of the equation.
2. Take the square root of each side.

RQ5.   Explain how to solve the equation $ax^2 + c = 0$.
RQ6.   Why does the equation $ax^2 + c = 0$ have two solutions?

Look Closer: In the next Example we compare *evaluating* a quadratic expression and *solving* a quadratic equation.

Example 5   A cat falls off a tree branch 20 feet above the ground. Its height t seconds later is given by $h = 20 - 16t^2$.
a. What is the height of the cat 0.5 second later?
b. How long does the cat have to get in position to land on its feet before it reaches the ground?

Solutions a.   In this question, we are given the value of t and asked to find the corresponding value of h. To do this we evaluate the formula for $t = 0.5$. We substitute 0.5 for t into the formula, and simplify.

$$h = 20 - 16(0.5)^2 \qquad \textbf{Compute the power.}$$
$$= 20 - 16(0.25) \qquad \textbf{Multiply, then subtract.}$$
$$= 20 - 4 = 16$$

The cat is 16 feet above the ground after 0.5 second. You can also use your calculator to simplify the expression for h by entering

$$20 \boxed{-} \; 16 \boxed{\times} \; 0.5 \boxed{x^2} \; \boxed{\textbf{ENTER}}$$

b. We would like to find the value of t when the height, h, is known. We substitute $h = 0$ into the equation to obtain

$$0 = 20 - 16t^2$$

To solve this equation we use extraction of roots. We first isolate t^2 on one side of the equation.

$$16t^2 = 20 \qquad \textbf{Divide by 16.}$$
$$t^2 = \frac{20}{16} = 1.25$$

Next, we take the square root of both sides of the equation to find

$$t = \pm\sqrt{1.25} \approx \pm 1.118$$

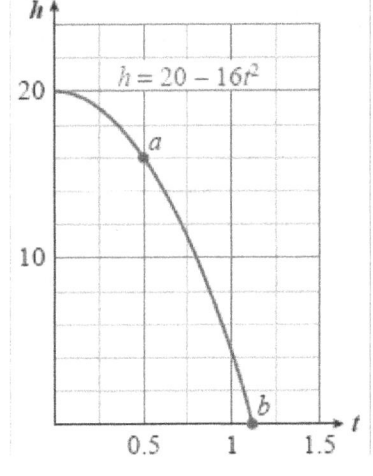

Only the positive solution makes sense here, so the cat has approximately 1.12 seconds to be in position for landing. A graph of the cat's height after t seconds is shown at right. The points corresponding to parts (a) and (b) are labeled.

RQ7.   In Example 5a, did we solve an equation or evaluate an expression?

Solving Formulas

Extraction of roots can be used to solve some formulas with a quadratic term.

Example 6   The formula $V = \frac{1}{3}\pi r^2 h$ gives the volume of a cone in terms of its height and radius. Solve the formula for r in terms of V and h.

Solution   Because the variable we want is squared, we use extraction of roots. First, we multiply both sides by 3 to clear the fraction.

$$3V = \pi r^2 h \qquad \textbf{Divide both sides by } \pi h.$$

$$\frac{3V}{\pi h} = r^2 \qquad \textbf{Take square roots.}$$

$$\pm \sqrt{\frac{3V}{\pi h}} = r$$

Because the radius of a cone must be a positive number, we use only the positive square root: $r = \sqrt{\dfrac{3V}{\pi h}}$.

Skills Warm-Up

Evaluate each expression if possible.

1a. $\sqrt{16}$ **b.** $-\sqrt{16}$ **c.** $\sqrt{-16}$

2a. $\sqrt{16^2}$ **b.** $\sqrt{(-16)^2}$ **c.** $\sqrt{-16^2}$

3a. $\left(\sqrt{16}\right)^2$ **b.** $\left(-\sqrt{16}\right)^2$ **c.** $\left(\sqrt{-16}\right)^2$

4a. 16^2 **b.** -16^2 **c.** $(-16)^2$

Answers: 1a. 4   **b.** −4   **c.** not a real number   **2a.** 16   **b.** 16
c. not a real number   **3a.** 16   **b.** 16   **c.** not a real number   **4a.** 256
b. −256   **c.** 256

Homework 6.1

Skills Practice

1. Which of the following are quadratic equations?

   a. $1 - 4x^2 = 2x$ **b.** $6a - 2a^3 = a^2$ **c.** $2b - 3 = 4b - b^2$

▪ For Problems 2-5, complete the table of values and graph the quadratic equation.

2. $y = x^2 + 4$

x	-3	-2	-1	0	1	2	3
y							

3. $y = x^2 - 4$

x	-3	-2	-1	0	1	2	3
y							

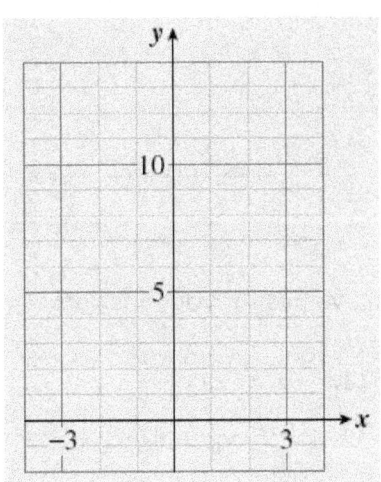

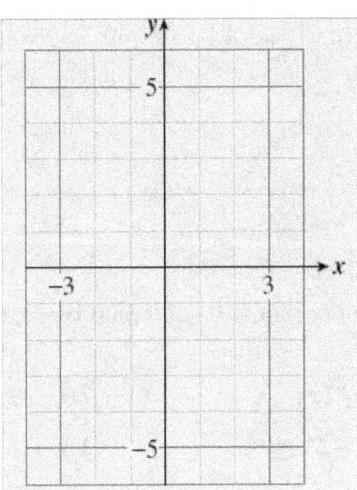

4. $y = 4 - x^2$

x	-3	-2	-1	0	1	2	3
y							

5. $y = \dfrac{x^2}{4}$

x	-4	-2	-1	0	1	2	4
y							

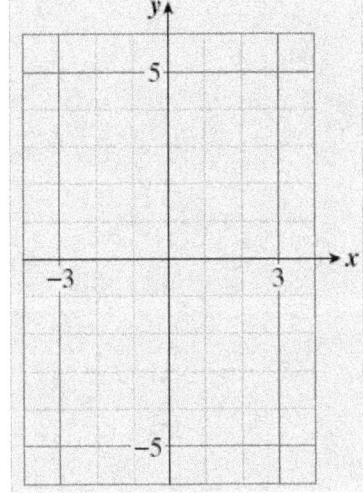

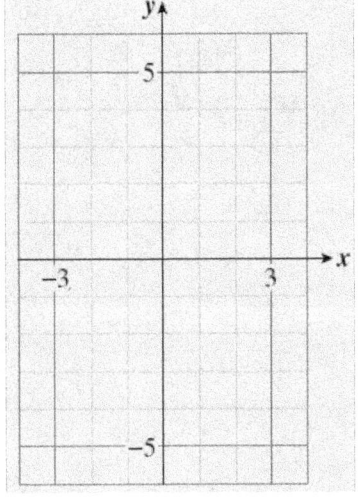

■ For Problems 6-8,
　a. Graph the parabola in part (a).
　b. Use the graph to solve the equation in part (b).

6. a. $y = \frac{1}{2}x^2$ **7.**
　　b. $\frac{1}{2}x^2 = 8$

　　a. $y = x^2 + 2$
　　b. $x^2 + 2 = 11$

8. a. $y = 6 - 2x^2$
　　b. $6 - 2x^2 = 4$

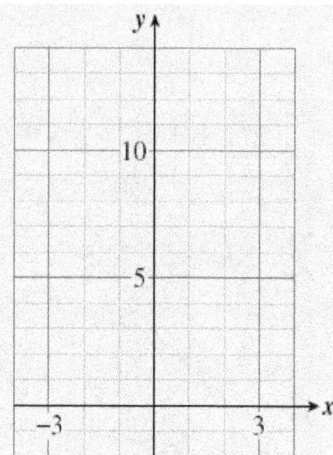
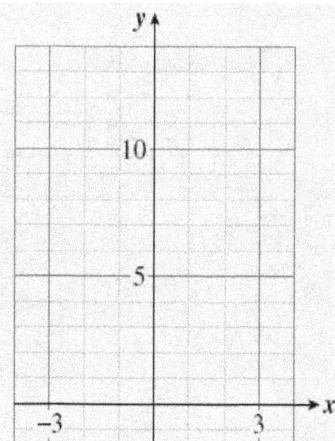
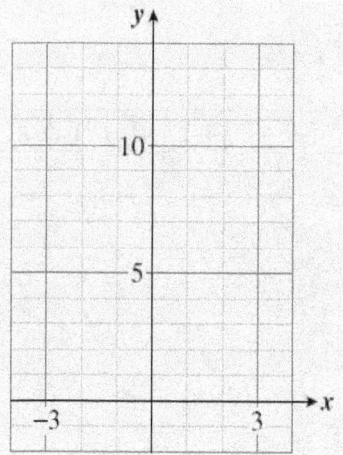

■ For Problems 9-14, solve by extracting roots. Give exact values for your answers

9. $x^2 = 121$

10. $98 = 2a^2$

11. $9x^2 = 25$

12. $3g^2 - 54 = 0$

13. $\dfrac{2x^2}{3} = 4$

14. $400 + \dfrac{k^2}{4} = 625$

■ For Problems 15-20, solve by extracting roots. Round your answers to two decimal places.

15. $2.4m^2 = 126$

16. $55 - 3z^2 = 7$

17. $2x^2 - 200 = x^2 + 25$

18. $3t^2 - 16 = 16t^2$

19. $1.5x^2 = 0.7x^2 + 26.2$

20. $5x^2 - 97 = 3.2x^2 - 38$

■ For Problems 21-24, solve the formula for the indicated variable.

21. $A = 4\pi r^2$ 　for r

22. $F = \dfrac{1}{2}mv^2$ 　for v

23. $d = 6 + kv^2$ 　for v

24. $h = 100 - \frac{1}{2}gt^2$ 　for t

Applications

25. At the Custom Pizza shop you can buy their special smoked chicken pizza in any size you like. The cost, C, of the pizza in dollars is given by the equation

$$C = \frac{1}{4}r^2$$

where r is the radius of the pizza.

a. Complete the table of values and graph the equation on the grid at right.

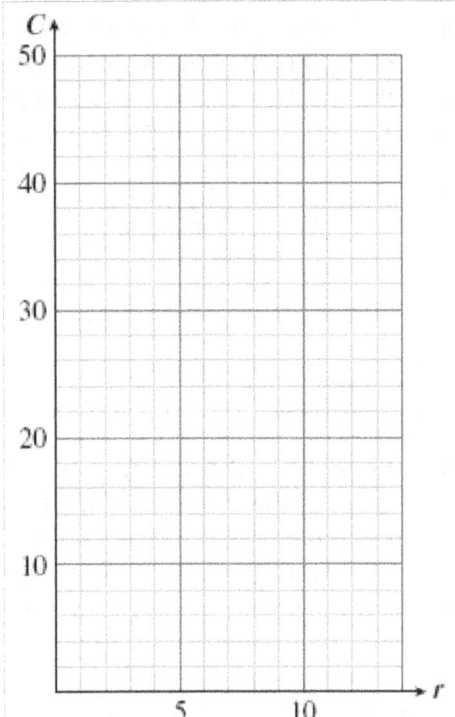

r	C
0	
1	
2	
4	
6	
9	
10	
14	

b. How much does a pizza of radius 3 inches cost? Locate this point on your graph.

c. Use your graph to find out how big a pizza you can buy for $16.

d. What does the point $(6, 9)$ tell you about pizzas?

26. The faster a car moves, the more difficult it is to stop. The distance, d, in meters, required to stop a car traveling at velocity v, in kilometers per hour, is given by $d = 0.005v^2$.

a. Complete the table of values and graph the equation on the grid above right.

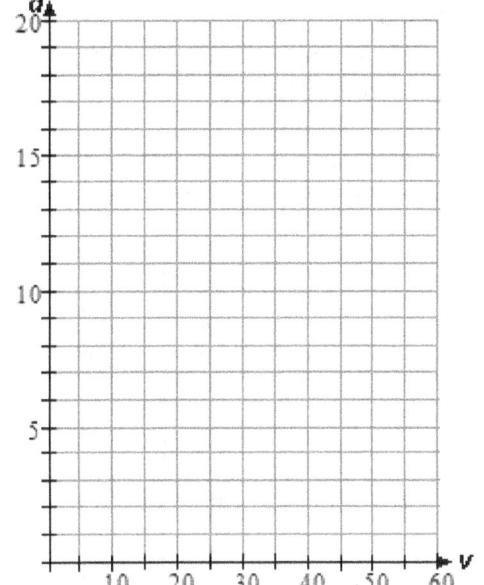

Velocity (kph)	5	10	15	20	40	60
Distance (meters)						

b. What distance does a car moving at 40 kilometers per hour require to stop? Locate `this point on your graph.

c. Use your graph to estimate the speed of a car that stopped in 12 meters.

d. What does the point $(20, 2)$ tell you about stopping distances?

■ For Problems 27-29, use the Pythagorean theorem. Give exact answers, and then approximate values rounded to thousandths.

27. a. Find the height of an equilateral triangle of side 6 feet, as shown at below left. (**Hint:** The altitude divides the base into two segments of equal length.)
   b. Find the area of the triangle in part (a).

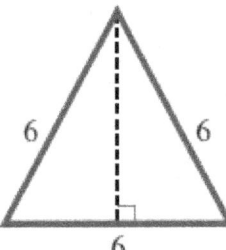

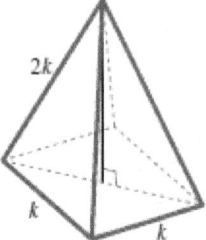

28. a. The pyramid shown above right has a square base. If $k = 2$ centimeters, find the length of the diagonal of the base.
   b. Find the height of the pyramid. (**Hint:** The altitude of the pyramid divides the diagonal of the base into two segments of equal length. Use your answer to part (a).)
   c. Find the volume of the pyramid.

29. Akemi wants to know the distance across a small lake, from point A to point B. She locates the point C on one side of the lake so that the line from A to C is perpendicular to the line from A to B. She then measures the distance from A to C as 2 miles. If the distance from B to C is x miles, write an expression for the distance from A to B across the lake.

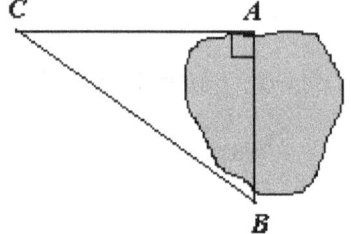

30. If you drop a stone from a bridge 100 feet above the water, the height h of the stone t seconds after you drop it is given in feet by $h = 100 - 16t^2$.
   a. Complete the table and sketch a graph of the equation.

t	h
0	
0.5	
1	
1.25	
1.5	
1.75	
2	
2.25	
2.5	

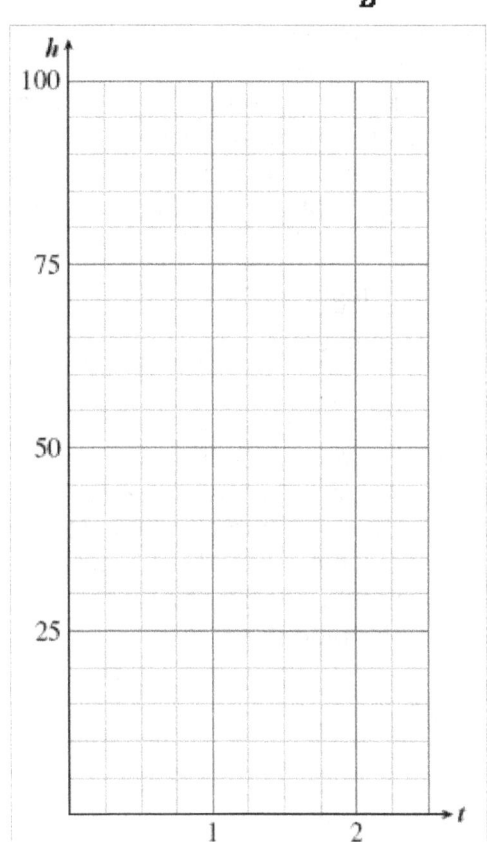

b. What is the height of the stone after 2 seconds? Locate this point on your graph.

c. Use the graph to find out when the height of the stone is 75 feet.

d. Write an equation you could solve to answer the question in part (c).

6.2 Some Quadratic Models

In this section we consider some applications that lead to quadratic equations. The first application involves revenue.

Revenue is the amount of money a company takes in from selling a product. To find the total revenue from the sale of a product, we multiply the price of one item by the number of items sold.

> ### Revenue = (price per item) · (number of items sold)

For example, if a snack bar sells 30 sandwiches at $4.25 each, their revenue from sandwiches is

$$\text{Revenue} = (\$4.25 \text{ per sandwich})(30 \text{ sandwiches}) = \$127.50$$

Look Ahead: Usually, the number of items that consumers will buy depends on the price of the item: the higher a company sets the price, the fewer items it is likely to sell. How does a company's revenue depend upon the price it charges? A good way to study such a problem is to consider the graph of the equation.

■ **Example 1**   Rick works as a personal trainer at his gym. By experimenting with his fee, he discovers that if he charges p dollars per hour, he attracts $40 - p$ clients.
a. Write a quadratic equation for Rick's revenue, R, in terms of his fee, p.
b. Graph your equation for R.
c. What should Rick charge if he wants to make $300 revenue? How many clients will he attract?
d. Use your graph to find the fee Rick should charge in order to earn the largest possible revenue. How many clients will he attract at that fee?

Solutions a.   Rick's revenue is given by

$$\text{Revenue} = (\text{hourly fee}) \cdot (\text{number of clients})$$
$$= p(40 - p) = 40p - p^2$$

b. We make a table of values and plot the points to obtain the parabola shown in the figure.

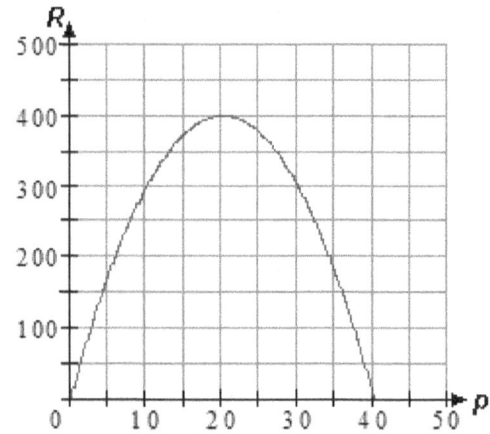

p	R	
10	300	$40(10) - 10^2$
20	400	$40(20) - 20^2$
30	300	$40(30) - 30^2$
40	0	$40(40) - 40^2$

c. There are two points on the graph with R-coordinate 300. Those points have p-coordinates 10 and 30. If he charges $10 per hour, he will attract $40 - 10 = 30$ clients, and if he charges \$30 per hour, he will attract $40 - 30 = 10$ clients. In either case, his revenue will be \$300.

d. The largest value of R that appears on the graph occurs at the point $(20, 400)$, so when $p = 20$, $R = 400$. If Rick charges \$20 per hour, he will make \$400 in revenue. He will attract $40 - p = 40 - 20 = 20$ clients at that price.

RQ1.   State a formula for calculating revenue.

Look Closer: In Example 1 we found a quadratic equation for the revenue Rick earns as a personal trainer. The equation was

$$R = p(40 - p) = 40p - p^2$$

The graph of this equation is shown at right.

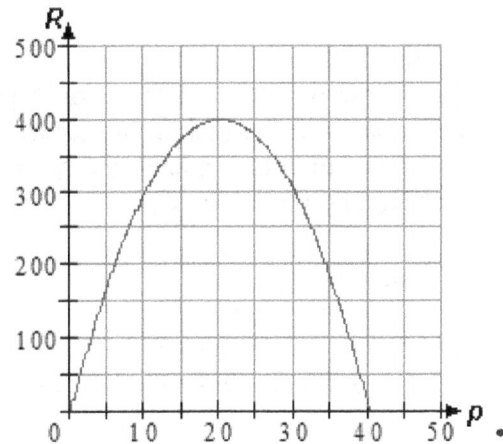

If Rick charges $p = 0$ dollars for each umbrella, he will not earn any revenue, so $R = 0$. You can see this fact in the graph, which passes through the point $p = 0$, $R = 0$. But there is another point on the graph where $R = 0$; it has p-coordinate 40. If Rick charges 40 dollars an hour, he will not earn any revenue. (Why does this happen?)

The Zero Factor Principle

Look again at the equation for revenue,

$$R = p(40 - p)$$

The expression for revenue has two factors. The first factor is p, and when $p = 0$ we have

$$R = 0(40 - 0) = 0$$

so the revenue is zero as well. The second factor, $40 - p$, is zero when $p = 40$. If Rick charges \$40 an hour, his revenue is

$$R = 40(40 - 40) = 0$$

(Recall that $40 - p$ represents the number of clients that Rick will attract if he charges p dollars an hour. Apparently, no one will pay \$40 an hour for a personal trainer.)

Both of these values, $p = 0$ and $p = 40$, are solutions of the quadratic equation

$$p(40 - p) = 0$$

A quadratic equation has a solution when either of its factors is equal to zero. This is a consequence of the following fact about numbers.

Zero-Factor Principle
       If the product of two numbers is zero, then one (or both) of the numbers must be zero. Using symbols,

$$\textbf{If } \boldsymbol{AB = 0,} \textbf{ then either } \boldsymbol{A = 0} \textbf{ or } \boldsymbol{B = 0.}$$

The zero-factor principle also applies to algebraic expressions: If the product of two factors is zero, then one of the factors must be zero.

■ **Example 2**   For what values of x is the expression $3x(x-5)$ equal to zero?

Solution   The expression is equal to zero if either of its factors equals zero. The factor $3x$ equals zero if $x = 0$, and $x - 5$ equals zero if $x = 5$. So $3x(x-5)$ equals zero if $x = 0$ or $x = 5$. ■

> **RQ2.**   If you multiply two numbers together and the product is zero, what can you say about the original numbers?

Factoring

In order to use the zero-factor principle to solve quadratic equations, we must be able to write quadratic expressions in **factored form**. This process is called **factoring**, and it is the reverse of multiplying factors together. Here are some examples comparing multiplying and factoring.

Multiplying	**Factoring**
1. $2 \cdot 3 = 6$	**1.** $6 = 2 \cdot 3$
2. $3(2x-5) = 6x - 15$	**2.** $6x - 15 = 3(2x-5)$
3. $(x-5)(x+2) = x^2 - 3x - 10$	**3.** $x^2 - 3x - 10 = (x-5)(x+2)$

The first type of factoring we consider is the reverse of the distributive law. Consider the expression $6x - 15$. The expression has two terms, $6x$ and -15, and each term is divisible by 3:

$$6x = 3 \cdot 2x \qquad \text{and} \qquad -15 = 3 \cdot -5$$

We say that 3 is a **common factor** for the expression $6x - 15$ because it is a factor of each term. Using the distributive law, we can write

$$6x - 15 = 3 \cdot 2x - 3 \cdot 15 = 3(2x - 15)$$

We have factored out a common factor, namely 3, from $6x - 15$.

> **RQ3.**   What does it mean to factor an expression?
> **RQ4.**   Factoring is the reverse process for which operation?

■ **Example 3**   Factor: $36t^2 - 63t$

Solution   We look for the largest common factor for the two terms. Both coefficients are divisible by 9, and t divides evenly into both terms. Thus, we factor out $9t$ and write the expression as

$$36t^2 - 63t = 9t(\,?-?\,)$$

You may be able to see the missing factors at a glance, but if not, you can divide each term of the original expression by the common factor, as follows:

$$\frac{36t^2}{9t} = 4t \qquad \text{and} \qquad \frac{63t}{9t} = 7$$

Fill in these factors for the question marks to obtain the factored form,

$$36t^2 - 63t = 9t(4t - 7)$$ ■

You can always check your factorization by computing the product; it should be the same as the original expression.

> **RQ5.**   Explain the meaning of "common factor."

Solving Quadratic Equations by Factoring

Now let's see how to solve a quadratic equation using factoring.

> First, we must arrange the terms so that **one side of the equation is zero**; the zero-factor principle applies only to zero products.

■ **Example 4**   Solve $2x^2 = 4x$

Solution First, we write the equation in standard form by subtracting $4x$ from both sides.

$$2x^2 - 4x = 0$$

Next, we factor the left side of the equation. We can take out a common factor of $2x$.

$$2x(x - 2) = 0$$

Now, we apply the zero-factor principle: In order for the product to be zero, one of the two factors must be zero. We set each factor equal to zero, and solve for x.

$$2x = 0 \quad \text{or} \quad x - 2 = 0$$
$$x = 0 \quad \text{or} \quad x = 2$$

Thus, the two solutions are $x = 0$ and $x = 2$. You should check that both of these values satisfy the original equation.   ■

> **RQ6.**   What is the first step in solving a quadratic equation by factoring?

Caution!   We cannot solve the equation in Example 4 by dividing both sides by $2x$. If we do that, we get the equation $x = 2$. This is one of the solutions, but we have lost the second solution, $x = 0$. We should **never divide both sides of an equation by the variable**, because we risk losing one of the solutions. Instead, we follow the steps below to solve the equation by factoring.   ■

> **RQ7.**   Why is it wrong to divide both sides of the equation $2x(x - 2) = 0$ by $2x$?

Skills Warm-Up

■ Mental Exercise: Evaluate each expression for the given values of the variable. Try not to use pencil, paper, or calculator.

1. $2x(x - 3), \quad x = 0, 1, 2, 3$

2. $(x + 1)(x - 6), \quad x = -1, 3, 6, 9$

3. $2(x + 2)(x + 4), \quad x = -4, -2, 0, 2$

4. $3n(2n - 1), \quad n = -1, 0, \frac{1}{2}, 1$

Answers: 1. 0, −4, −4, 0 **2.** 0, −12, 0, 30 **3.** 0, 0, 16, 48 **4.** 9, 0, 0, 3

Homework 6.2

Skills Practice

■ For Problems 1-3, apply the zero-factor principle to find the solutions.

1. $(x+1)(x-4) = 0$ **2.** $0 = p(p+7)$ **3.** $(2v+3)(4v-1) = 0$

■ For Problems 4-5,
    a. Verify that the factored form for each expression is correct.
    b. Use the factored form to find the solutions of the original equation.
    c. Check algebraically that your answers to part (b) actually are solutions of the
        original equation.

4. $x^2 - 6x - 27 = 0$
    factored form:
    $(x-9)(x+3) = 0$

5. $x^2 - 81 = 0$
    factored form:
    $(x-9)(x+9) = 0$

■ For Problems 6-8, find the largest common factor for the monomials.

6. $8x^2,\ 12x$ **7.** $6a^2b,\ 9ab^2$ **8.** $10w^2,\ 15w$

■ In Problems 9-12, fill in the missing algebraic expressions.

9. $4x(2x + \mathbf{?}) = 8x^2 - 12x$
11. $6h(\mathbf{?} + \mathbf{?}) = 24h - 18h^2$

10. $\mathbf{?}(a - 3) = 2a^2 - 6a$
12. $\mathbf{?}(4m - 9) = -12m^2 + 27m$

■ For Problems 13-15, factor out the largest common factor.

13. $6a^2 - 8a$ **14.** $-18v^2 - 6v$ **15.** $4h - 9h^2$

■ For Problems 16-18, factor the right side of the formula.

16. $d = k - kat$ **17.** $A = 2rh - \pi r^2$ **18.** $V = \pi r^2 h - \frac{1}{3}s^2 h$

■ For Problems 19-24, solve the equation.

19. $10x^2 - 15x = 0$
22. $4x - 6x^2 = 0$

20. $20x^2 = x$
23. $x^2 + x = 0$

21. $0 = 144x + 3x^2$
24. $\frac{1}{3}x^2 = \frac{2}{3}x$

25. a. Complete the table of values for the equation $y = -x^2 + 6x$, then graph the equation on the grid.

x	−1	0	1	2	3	4	5	6	7
y									

b. Use your graph to find the solutions of the equation $-x^2 + 6x = 5$.

c. Verify algebraically that your answers to part (a) are really solutions.

d. Use your graph to find the solutions of the equation $-x^2 + 6x = 0$.

e. Verify algebraically that your answers to part (a) are really solutions.

f. Compute the product $-x(x - 6)$. What do you get?

g. What is the value of the expression $-x(x - 6)$ when $x = 0$? When $x = 6$?

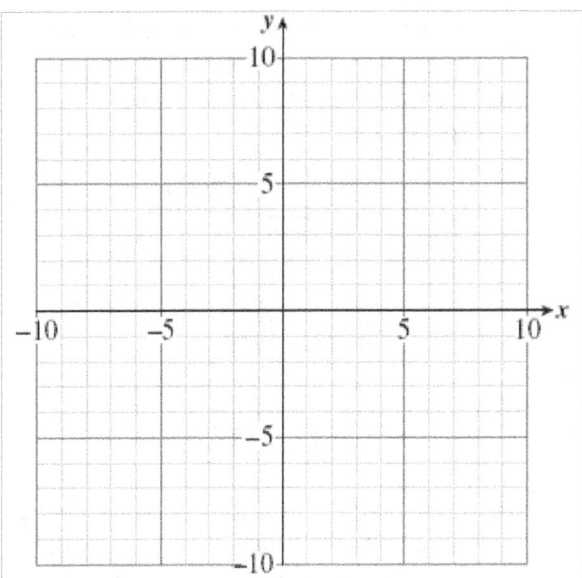

26. a. Complete the table of values for the equation $y = x^2 + 2x - 8$, then graph the equation on the grid.

x	−5	−4	−3	−2	−1	0	1	2	3
y									

b. Use your graph to find the solutions of the equation $x^2 + 2x - 8 = 7$.

c. Verify algebraically that your answers to part (a) are really solutions.

d. Use your graph to find the solutions of the equation $x^2 + 2x - 8 = 0$.

e. Verify algebraically that your answers to part (a) are really solutions.

f. Compute the product $(x + 4)(x - 2)$. What do you get?

g. What is the value of the expression $(x + 4)(x - 2)$ when $x = -4$? When $x = 2$?

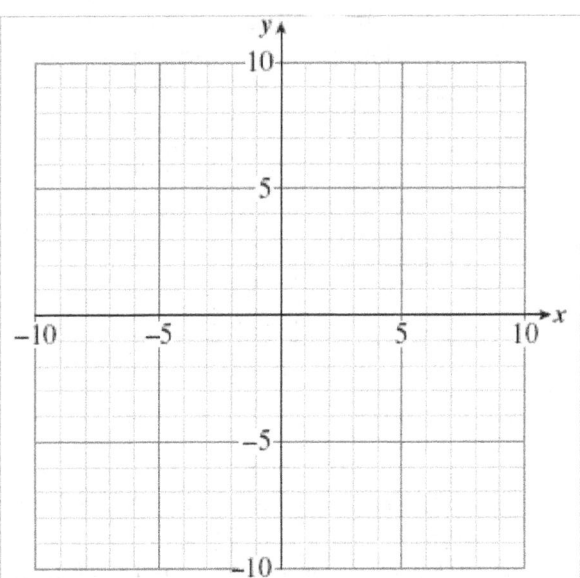

Applications

27. The revenue you can earn by making and selling n jade bracelets is given in dollars by

$$R = -2n^2 + 80n$$

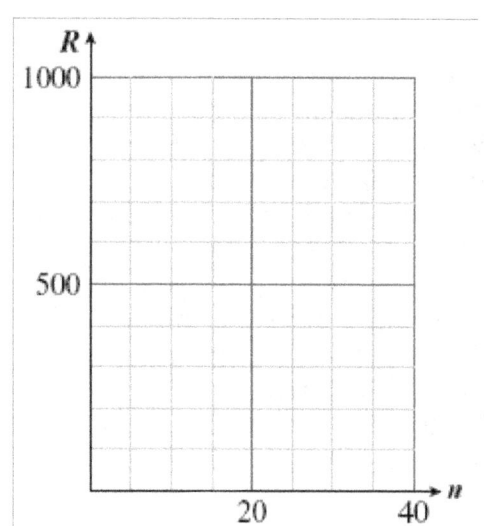

    a. Factor n from the expression for R. How much should you charge per bracelet in order to sell n bracelets?

    b. Complete the table and sketch a graph.

n	Price per Bracelet	R
0		
5		
10		
20		
30		
40		

    c. What is the maximum revenue that you can earn? How many bracelets should you make in order to earn that revenue?

    d. How much revenue will you earn if you charge $40 for a bracelet?

28. The height in feet of a football t seconds after being kicked from the ground is given by

$$h = -16t^2 + 80t$$

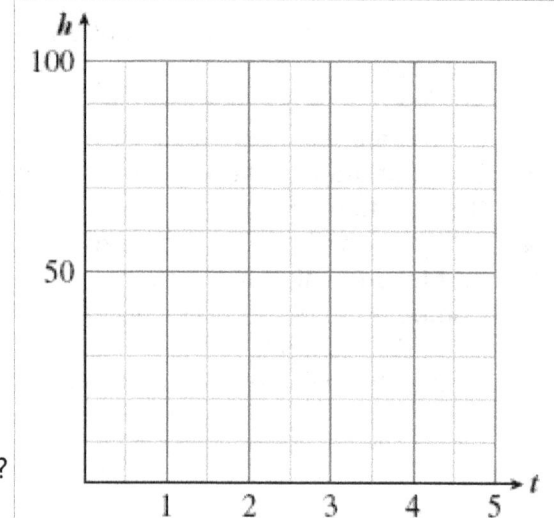

    a. Factor the expression for h.

    b. Complete the table and sketch the graph.

t	h
0	
1	
2	
2.5	
3	
4	
5	

    c. What is the maximum height of the football? When does it reach this height?

    d. When does the football fall back to the ground?

29. Sportsworld sells $180 - 3p$ pairs of their name-brand running shoes per week when they charge p dollars per pair.
   a. Write an equation for Sportsworld's revenue, R, in terms of p.
   b. Fill in the table, and graph the equation.

p	R
0	
10	
20	
30	
40	
50	
60	

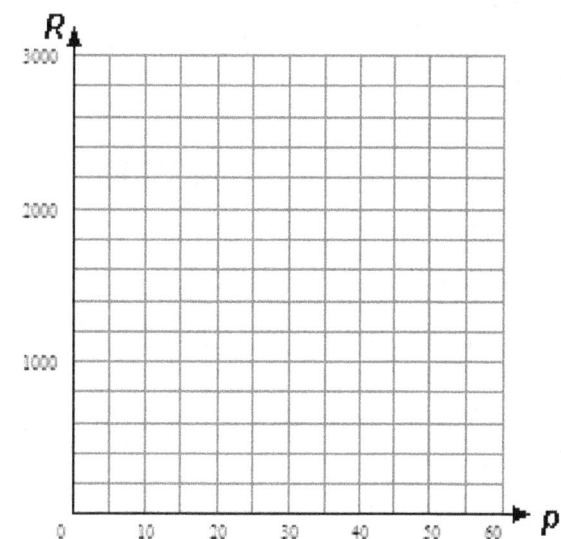

   c. At what price(s) will Sportsworld's revenue be zero?
   d. At what price will Sportsworld's revenue be maximum? What is their maximum revenue?

For Problems 30-32, find the x-intercepts of the graph of the equation.

30. $y = (3x - 4)(x + 2)$ **31.** $y = 3x^2 + 12x$ **32.** $y = -7x - 4x^2$

For Problems 33-34, graph all three equations on the same grid. What do you observe?

33. $y = x$
$y = 6 - x$
$y = 6x - x^2$

34. $y = x + 3$
$y = 3 - x$
$y = x^2 - 9$

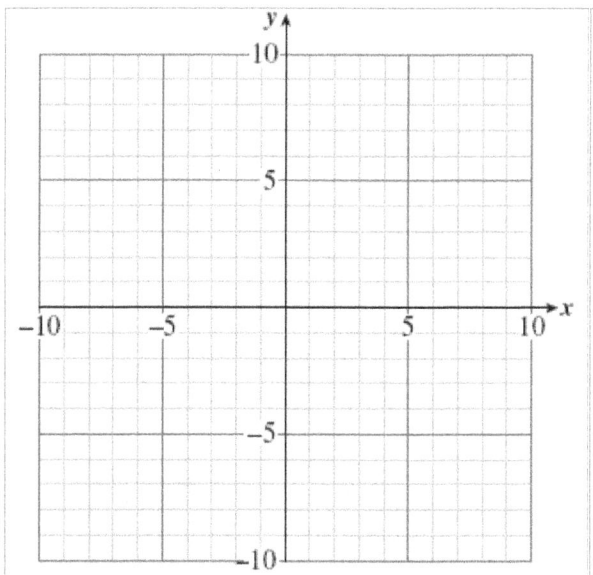

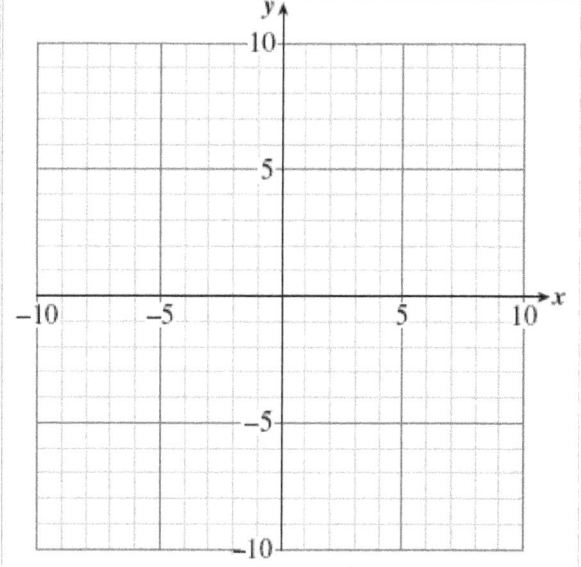

■ For Problems 35-36, find the mistake in the steps for solution, then write a correct solution.

35. $4x^2 = 12x$

$$\frac{4x^2}{4x} = \frac{12x}{4x}$$

$$x = 3$$

36. $9x^2 - 4 = 16$

$$3x - 2 = 4$$

$$3x = 6$$

$$x = 2$$

6.3 Solving Quadratic Equations by Factoring

In Section 6.2 we applied the Zero-Factor Principle to solve quadratic equations of the form $ax^2 + bx = 0$. For reference, here is the statement of the Zero-Factor Principle.

> **Zero-Factor Principle**
>   If the product of two numbers is zero, then one (or both) of the numbers must be zero. Using symbols,
>
> **If $AB = 0$, then either $A = 0$ or $B = 0$.**

In order to solve more general quadratic equations, of the form $ax^2 + bx + c = 0$, we will need to factor quadratic trinomials. Here is an example of how we can use the principle to solve a quadratic equation.

■ **Example 1a.**   Compute the product $(x + 5)(x - 4)$.
b. Solve the equation $x^2 + x - 20 = 0$.

Solutions a.   We use the FOIL method to apply the distributive law to the product.

$$(x + 5)(x - 4) = x^2 - 4x + 5x - 20$$
$$= x^2 + x - 20$$

b. From part (a) we see that $x^2 + x - 20$ can also be written in factored form as $(x + 5)(x - 4)$. Thus, the equation $x^2 + x - 20 = 0$ is equivalent to

$$(x + 5)(x - 4) = 0$$

We can apply the zero-factor principle to this equation: If the product is zero, then one of the factors must be zero. Thus, either

$$x + 5 = 0 \quad \text{or} \quad x - 4 = 0$$

This means that either $x = -5$ or $x = 4$. These are the solutions of the equation, as we verify below.
Check: Substitute each value into the original equation.

$x = -5 :$ $(-5)^2 + (-5) - 20 = 25 - 5 - 20 = 0$
$x = 4 :$ $(4)^2 + (4) - 20 = 16 + 4 - 20 = 0$ ■

> **RQ1.**   Why can't we solve the equation $x^2 + x - 20 = 0$ in Example 1 by extraction of roots?

Factoring Quadratic Trinomials

In Section 6.2 we learned to factor quadratic expressions of the form $ax^2 + bx$. Now we consider quadratic trinomials of the form $x^2 + bx + c$. Many of these can be factored into the product of two binomials. For example,

$$x^2 + 7x + 12 = (x + 3)(x + 4)$$

How can we come up with the factors of the quadratic trinomial? Let's recall how the product is obtained. We apply the distributive law and multiply each term of the first binomial by each term of the second binomial. We use the word "FOIL" to label the four terms of the product: First, Outside, Inside, and Last.

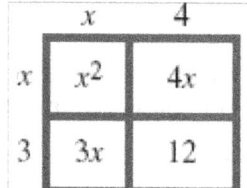

$$(x + 3)(x + 4) = x^2 + 4x + 3x + 12$$
$$\text{F} \qquad \text{O}+\text{I} \qquad \text{L}$$

RQ2.   Where does the constant term of the trinomial appear in the product box?
RQ3.   Where does the linear term appear?

We make several observations about this product.

---

1. The quadratic term, x^2, comes from the product of the **First** terms in each binomial.
2. The constant term, 12, is the product of the **Last** terms.
3. The middle term, $7x$, is the sum of two terms, the product of the **Inside** terms and the product of the **Outside** terms.

We can see that the first term of each factor must be x, so we only need to fill in the blanks below with the correct numbers:

$$x^2 + 7x + 12 = (x + \underline{\quad})(x + \underline{\quad})$$

If the two numbers are p and q, then

$$(x + p)(x + q) = x^2 + qx + px + pq$$
$$= x^2 + (p + q)x + pq$$

We see that the constant term of the trinomial is the product pq, and the linear term of the trinomial has as coefficient the sum $p + q$.

Thus, to factor $x^2 + 7x + 12$, we look for two integers p and q whose sum is 7 and whose product is 12. With a little trial and error, we discover that the two integers are 3 and 4. So the factored form is

$$x^2 + 7x + 12 = (x + 3)(x + 4)$$

To Factor a Quadratic Trinomial
   To factor $x^2 + bx + c$, we look for two numbers p and q so that

$$pq = c \text{ and } p + q = b$$

■ **Example 2**   Factor: $a^2 + 13a + 40$

Solution The factored form looks like

$$a^2 + 13a + 40 = (a + p)(a + q)$$

where $pq = 40$ and $p + q = 13$. To find the numbers p and q, it may help to list all possible pairs of numbers whose product is 40. Then we can check each pair of numbers to see which pair has a sum of 13.

$$
\begin{array}{ccc}
p & q & p+q \\
1 & 40 & 41 \\
2 & 20 & 22 \\
4 & 10 & 14 \\
5 & 8 & 13
\end{array}
$$

We see that $5 \cdot 8 = 40$ and $5 + 8 - 13$, so the correct factorization is

$$a^2 + 13a + 40 = (a + 5)(a + 8)$$

We can also write the factored form as $(a + 8)(a + 5)$; the order of the factors does not matter. Aside from rearranging the factors, there is only one correct factorization for these quadratic trinomials.

RQ4.   To factor $x^2 + 126x + 3393$, we look for two numbers whose sum is _____ and whose product is _____ .

Look Ahead: In the Activities we'll see how to factor trinomials with negative coefficients. Here are the results:

> **Sign Patterns for Factoring Quadratic Trinomials**
>
> Assume that b, c, p, and q are positive integers.
>
> **1.** $x^2 + bx + c = (x + p)(x + q)$
>
> If all the coefficients of the trinomial are positive, then both p and q are positive.
>
> **2.** $x^2 - bx + c = (x - p)(x - q)$
>
> If the linear term of the trinomial is negative and the other two terms positive, then p and q are both negative.
>
> **3.** $x^2 \pm bx - c = (x + p)(x - q)$
>
> If the constant term of the trinomial is negative, then p and q have opposite signs.

Example 3   Factor $x^2 + 3x - 18$

Solution   We look for two integers p and q for which $pq = -18$ and $p + q = 3$. Because pq is negative, p and q must have opposite signs. Because $p + q$ is positive, the choices are

$$-1 \text{ and } 18, \quad -2 \text{ and } 9, \quad \text{or} \quad -3 \text{ and } 6$$

Because $-3 + 6 = 3$, the correct factorization is

$$x^2 + 3x - 18 = (x - 3)(x + 6)$$

You can check the factorization by multiplying the two binomials to see that their product is in fact $x^2 + 3x - 18$.

Solving a Quadratic Equation

Now we can use factoring to solve a quadratic equation.

■ **Example 4**   Solve the equation $x^2 + 3x = 18$.

Solution   We must have the right side equal to zero if we want to apply the zero-factor principle, so we begin by subtracting 18 from both sides to get

$$x^2 + 3x - 18 = 0$$ **Factor the left side.**

$$(x + 6)(x - 3) = 0$$ **Apply the zero-factor principle: set each factor equal to zero.**

$$x + 6 = 0 \qquad x - 3 = 0$$ **Solve each equation.**

$$x = -6 \qquad x = 3$$

The solutions are -6 and 3. You can check that each of these solutions satisfies the original equation.   ■

Caution!   Before factoring and applying the Zero-Factor principle, we must write the equation in standard form, so that **one side of the equation is zero**. Thus, it would be incorrect to write

$$x(x + 3) = 18$$

We cannot apply the Zero-Factor principle to this equation.   ■

RQ5.   Why can't we solve the equation $x(x + 3) = 18$ by setting each factor equal to 18?

Here is our method for solving a quadratic equation by factoring.

To Solve a Quadratic Equation by Factoring:
1. Write the equation in standard form,

$$ax^2 + bx + c = 0 \qquad (a \neq 0)$$

2. Factor the left side of the equation.
3. Apply the zero-factor principle; that is, set each factor equal to zero.
4. Solve each equation to obtain two solutions.

■ **Example 5**   Solve the equation $2x^2 + 6x = 36$.

Solution   You may notice that each term of this equation is twice the corresponding term of the equation in Example 3. We can factor out a common factor of 2 from the left side of the standard form of this equation.

$$2x^2 + 6x - 36 = 0$$ **Factor out 2 from the left side.**

$$2(x^2 + 3x - 18) = 0$$

Then we can divide both sides of the equation by 2.

$$\frac{2(x^2 + 3x - 18)}{2} = \frac{0}{2}$$
$$x^2 + 3x - 18 = 0$$

This new equation is the same as the equation in Example 3. Thus, the solutions are the same as the solutions in Example 3, namely, −6 and 3. ◼

Look Closer: Multiplying the equation in Example 3 by a factor of 2 does not affect the solutions of the equation. We can divide both sides of a quadratic (or any) equation by a nonzero constant factor while solving. •

Application

Delbert is standing at the edge of a 240-foot cliff. He throws his algebra book upwards off the cliff with a velocity of 32 feet per second. The height of his book above the ground (at the base of the cliff) after t seconds is given by the formula

$$h = -16t^2 + 32t + 240$$

where h is in feet. The figure shows a graph of the equation. From the graph, we see that it takes the book 6 seconds to reach the ground. Because $h = 0$ when the book reaches the ground, 5 is one of the solutions of the quadratic equation

$$0 = -16t^2 + 32t + 240$$

Can we solve this equation algebraically, without using a graph?

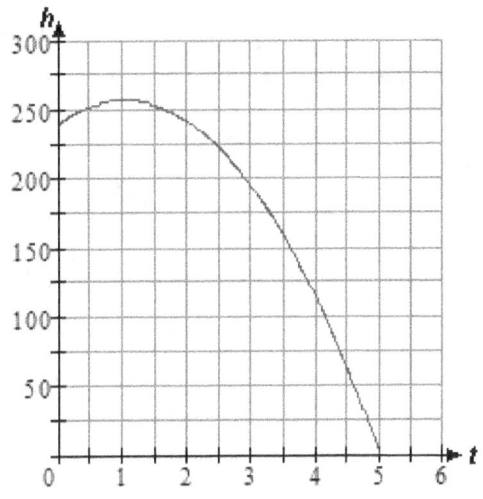

◼ **Example 6** The equation

$$h = -16t^2 + 32t + 240$$

gives the height of a book above the ground t seconds after being thrown off a cliff. How long will it take the book to reach the ground?

Solution We set $h = 0$ to obtain the equation

$$-16t^2 + 32t + 240 = 0$$

It is easier to factor if the coefficient of t^2 is positive, so we factor out −16.

$$-16(t^2 - 2t - 15) = 0 \qquad \textbf{Divide both sides by } -16.$$
$$t^2 - 2t - 15 = 0$$

Now we are ready to factor the left side of the equation. You can verify that the factorization is

$$(t + 3)(t - 5) = 0$$

Applying the zero-factor principle yields

$$t + 3 = 0 \qquad t - 5 = 0 \qquad \textbf{Solve each equation.}$$
$$t = -3 \qquad\quad t = 5$$

The solutions are -3 and 5. Because a negative time does not make sense for this problem, we discard that solution. The book takes 5 seconds to reach the ground. ■

RQ6. In Example 6, why did we set $h = 0$?

Skills Warm-Up

 **a.** Find the linear term of each product mentally.
  b. Find the constant term.

1. $(x + 1)(x + 3)$ **2.** $(a - 6)(a - 3)$ **3.** $(p + 7)(p - 4)$
4. $(q - 8)(q + 2)$ **5.** $(3w + 4)(w + 2)$ **6.** $(2z - 5)(3z + 2)$

Answers: 1a. $4x$ **b.** 3   **2a.** $-9a$ **b.** 18   **3a.** $3p$ **b.** -28 **4a.** $-6q$
b. -16 **5a.** $10w$ **b.** 8   **6a.** $-11z$ **b.** -10

Homework 6.3

Skills Practice

■ In Problems 1-3, the area of each rectangle represents a product.
   a. Fill in the missing expressions.
   b. Write the area as a product of binomials, then compute the product.

1.

	x	?
x	x^2	6x
?	5x	30

2.

	x	-3
x	x^2	-3x
?	?	27

3.

	x	?
?	x^2	2x
?	?	16

■ For Problems 4-6, list all the ways to write the number as a product of two whole numbers.

4.   24

5.   60

6.   40

■ For Problems 7-12,
   a. Factor the quadratic trinomial.
   b. Solve the quadratic equation.

7a. $n^2 + 10n + 16$
  b. $n^2 + 10n + 16 = 0$

8a. $h^2 + 26h + 48$
  b. $h^2 + 26h + 48 = 0$

9a. $a^2 - 8a + 12$
  b. $a^2 - 8a + 12 = 0$

10a. $t^2 - 15t + 36$
  b. $t^2 - 15t + 36 = 0$

11a. $x^2 - 3x - 10$
  b. $x^2 - 3x - 10 = 0$

12a. $a^2 + 8a - 20$
  b. $a^2 + 8a - 20 = 0$

■ For Problems 13-21, factor if possible. If the trinomial cannot be factored, say so.

13. $x^2 - 17x + 30$
16. $t^2 - 9t - 20$
19. $a^2 + 48 - 2a$

14. $x^2 + 4x + 2$
17. $q^2 - 5q - 6$
20. $32 - 12b + b^2$

15. $y^2 - 44y - 45$
18. $n^2 + 6 - 5n$
21. $4c - 60 + c^2$

■ For Problems 22-24, factor completely.

22. $3b^2 - 33b + 72$

23. $4x^2 - 20x - 144$

24. $42 - 8m - 2m^2$

■ In Problems 25-32, solve the quadratic equation.

25. $x^2 + 3x - 10 = 0$
28. $0 = n^2 - 14n + 49$
31. $(x - 2)(x + 1) = 4$

26. $t^2 + t = 42$
29. $5q^2 = 10q$
32. $(x - 8)^2 = 12 + 4(10 - 6x)$

27. $2x^2 - 10x = 12$
30. $x(x - 4) = 21$

Applications

33. An architect is designing rectangular offices to be 3 yards longer than they are wide, to allow for storage space. Let w represent the width of one office.

    a. Draw a sketch of the floor of an office and label its dimensions.

    b. Write an equation for the area, A, of one office in terms of its width.

    c. Complete the table and graph your equation for A.

w	A
-6	
-5	
-3	
-2	
-1	
0	
1	
2	

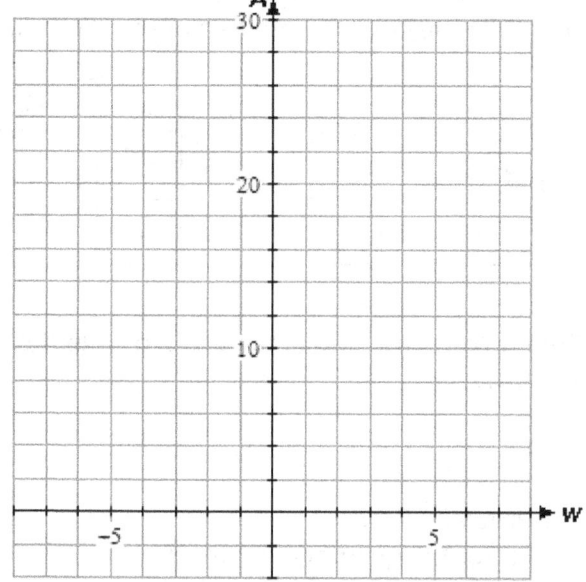

    c. The main office should have an area of 28 square yards. Write an equation for this requirement.

    d. Solve your equation. You should get two solutions. Which one makes sense for this application?

34. Icon Industries produces all kinds of electronic equipment. The cost, C, of producing a piece of equipment depends on the number of hours, t, it takes to build it, where

$$C = 8t^2 - 32t - 16$$

How many hours does it take to build a transformer if it costs \$80 to produce?

For Problems 35-36,

    a. Draw and label a sketch to illustrate the problem,

    b. Write an equation for the problem,

    c. Solve your equation and complete the problem.

35. The area of a computer circuit board must be 60 square centimeters. The length of the circuit board should be 2 centimeters shorter than twice its width. Find the dimensions of the circuit board.

36. A paper airplane in the shape of a triangle is 40 square inches in area. Its base is 11 inches longer than its altitude. Find the base and altitude of the triangle.

37. What is wrong with the following solution to the quadratic equation?

$$x^2 - 6x + 8 = 0$$
$$x^2 = 6x - 8$$
$$x = \frac{6x - 8}{x}$$

6.4 Graphing Quadratic Equations

So far we have graphed quadratic equations by plotting points. But we can use some algebraic techniques to make the process easier.

x-Intercepts of a Parabola

In Lesson 6.2 we saw that the solutions of the equation

$$40p - p^2 = 0$$

namely, $p = 0$ and $p = 40$, are also the horizontal intercepts of the graph of

$$R = 400p - p^2$$

This is the same connection we saw between the x-intercepts of a line and the solutions of its equation. Recall that the x-intercept of the line $y = mx + b$ is the point where the graph crosses the x-axis. To find the x-intercepts of a line, we set $y = 0$ and solve for x. The same strategy applies to quadratic equations.

---

x-Intercepts of a Parabola

To find the x-**intercepts** of the graph of

$$y = ax^2 + bx + c$$

we set $y = 0$ and solve the equation

$$ax^2 + bx + c = 0$$

Example 1a. Find the x intercepts of the graph of $y = 4x - x^2$.
b. Sketch the graph.

Solutions a. We set $y = 0$ and solve the equation

$$
\begin{aligned}
4x - x^2 &= 0 & &\text{Factor the left side.} \\
x(4 - x) &= 0 & &\text{Set each factor equal to 0.} \\
x = 0 \quad 4 - x &= 0 & &\text{Solve each equation.} \\
x = 0 \quad\quad x &= 4 &
\end{aligned}
$$

The x-intercepts are the points $(0, 0)$ and $(4, 0)$.
b. We make a table of values that includes the x-intercepts.

x	−1	0	1	2	3	4	5
y	−5	0	3	4	3	0	−5

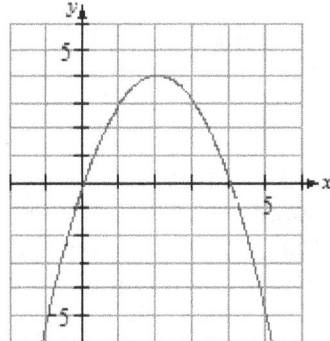

We plot the points and connect them with a parabola, as shown in the figure.

RQ1. How do we find the x-intercepts of the graph of $y = ax^2 + bx + c$?

The Vertex

The graph of a quadratic equation $y = ax^2 + bx + c$ is a smooth curve, called a parabola, that bends upwards or downwards.

> The high or low point of a parabola is called its **vertex**.

For example, the vertex of the basic parabola $y = x^2$ is the point $(0, 0)$.

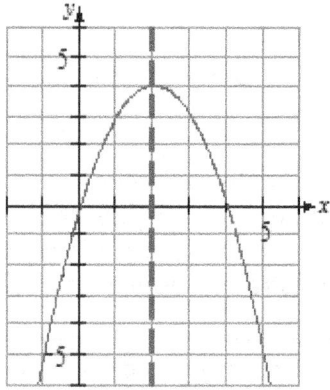

Look again at the graph of $y = 4x - x^2$ from Example 1, shown at right. Notice that the graph is symmetric about a vertical line (called the **axis of symmetry**) that passes through the vertex. All the parabolas we'll study have a vertical axis of symmetry.

Look Closer: Because of this symmetry, the x-intercepts are located at equal distances on either side of the axis of symmetry. Or we can say that the x-coordinate of the vertex is exactly halfway between the two x-intercepts.

> **RQ2.**   What is the vertex of a parabola?

> We can locate the vertex of a parabola by taking the average of its x-intercepts.

■ **Example 2**   Find the vertex of the graph of $y = 4x - x^2$.

Solution   In Example 1 we found that the x-intercepts of the graph are $x = 0$ and $x = 4$. The x-coordinate of the vertex is the average of these two numbers.

$$x = \frac{0 + 4}{2} = 2$$

The x-coordinate of the vertex is $x = 2$. To find the y-coordinate of the vertex, we substitute $x = 2$ into the equation of the parabola.

$$y = 4(2) - 2^2 = 8 - 4 = 4$$

The coordinates of the vertex are $(2, 4)$, as you can see in the graph above.   ■

> **The Vertex of a Parabola**
> 1. The x-coordinate of the vertex is the average of the x-intercepts.
> 2. To find the y-coordinate of the vertex, substitute its x-coordinate into the equation of the parabola.

> **RQ3.**   How can we find the x-coordinate of the vertex of a parabola?
> **RQ4.**   How can we find the y-coordinate of the vertex of a parabola?

Graphing Parabolas

By locating the x-intercepts and the vertex of the graph, we can make a quick sketch of a parabola.

Example 3   Sketch a graph of $y = x^2 - 8x + 7$.

Solution   First we find the x-intercepts: we substitute $y = 0$ into the equation, and solve for x.

$$x^2 - 8x + 7 = 0 \qquad \textsf{Factor the left side.}$$
$$(x - 7)(x - 1) = 0 \qquad \textsf{Apply the zero-factor principle.}$$
$$x - 7 = 0 \qquad x - 1 = 0 \qquad \textsf{Solve each equation.}$$
$$x = 7 \qquad\quad x = 1$$

The x-intercepts are the points $(7, 0)$ and $(1, 0)$. Next, we locate the vertex of the graph. The x-coordinate of the vertex is the average of the x-intercepts, so

$$x = \frac{1 + 7}{2} = 4$$

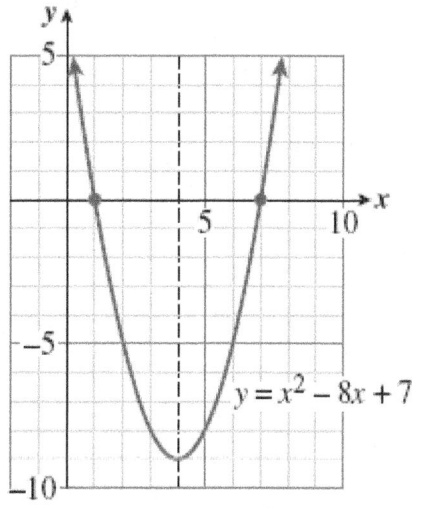

You can check that $x = 4$ is halfway between the two x-intercepts of the graph shown above. To find the y-coordinate of the vertex, we evaluate the formula for the parabola at $x = 4$.

$$y = 4^2 - 8(4) + 7$$
$$= 16 - 32 + 7 = -9$$

The vertex of the parabola is the point $(4, -9)$. We can also find the y-intercept of the graph by substituting $x = 0$ into the equation.

$$y = 0^2 - 8(0) + 7 = 7$$

The y-intercept is the point $(0, 7)$. We plot the vertex and the intercepts and draw a smooth curve through them. The completed graph is shown above.

RQ5.   How can we find the y-intercept of the graph of $y = ax^2 + bx + c$?

By combining the techniques we studied in this Lesson, we write the following guidelines for sketching an accurate graph of a quadratic equation.

To Graph a Quadratic Equation, $y = ax^2 + bx + c$
1. Find the x-intercepts: set $y = 0$ and solve for x.
2. Find the vertex: the x-coordinate is the average of the x-intercepts. Find the y-coordinate by substituting the x-coordinate into the equation of the parabola.
3. Find the y-intercept: set $x = 0$ and solve for y.
4. Draw a parabola through the points. The graph is symmetric about a vertical line through the vertex.

Skills Warm-Up

a. Solve the equation.
b. Write the equation in the form $ax + b = 0$.
c. Graph the equation $y = ax + b$.
d. Find the x-intercept of your graph. Compare with your answer to part (a).

1. $2x + 5 = 11$

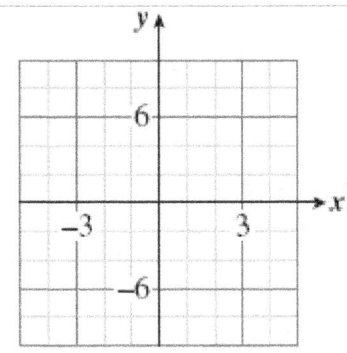

2. $2x - 3 = 5x + 9$

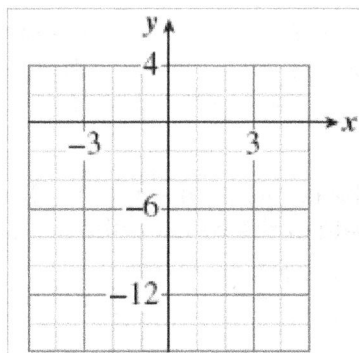

3. $0.7x + 0.2(100 - x) = 0.3(100)$

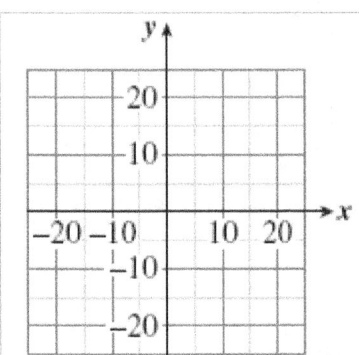

4. $4(7 - 4x) = -2(6x - 5) - 6$

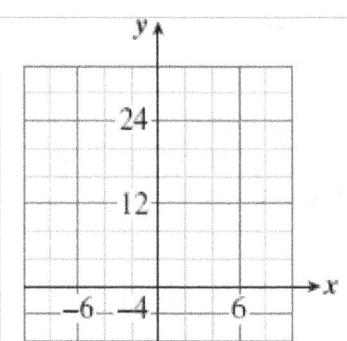

Answers: **1.** $x = 3$ **2.** $x = -4$ **3.** $x = 20$ **4.** $x = 6$

Homework 6.4

Skills Practice

■ Without graphing, find the x-intercepts of the graph of each equation.

1. $y = (x - 3)(2x + 5)$ **2.** $y = 2x^2 - 6x$ **3.** $y = 8x - 3x^2$

■ For Problems 4-5, find the intercepts and graph the two parabolas on the same grid.

4. **a.** $y = x^2 + 1$ **5.** **a.** $y = -x^2 - 3$
    b. $y = x^2 - 4$ **b.** $y = 1 - x^2$

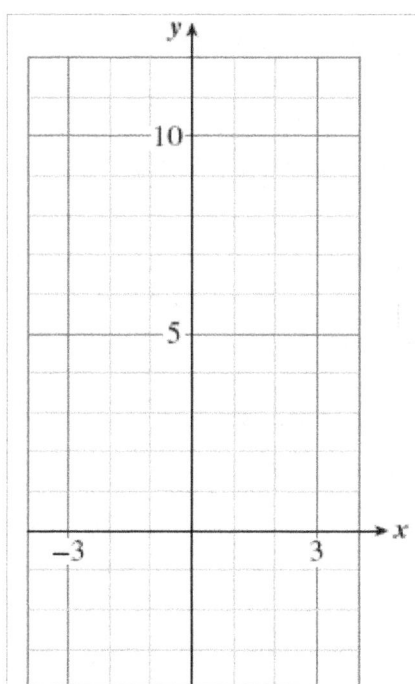
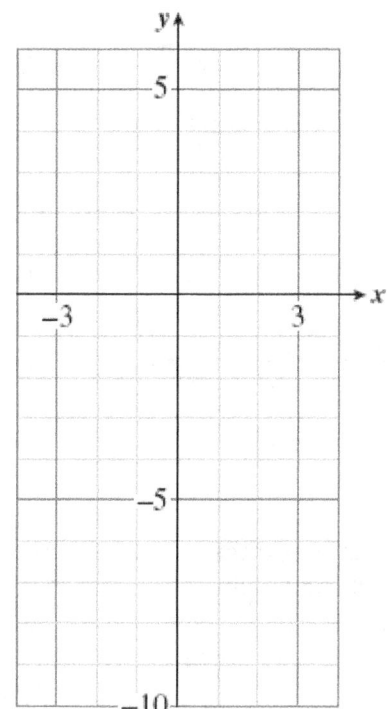

6. **a.** For each graph in Problems 4 and 5, give the coordinates of the vertex.
        Then use your answers to part (a) to help you answer the following questions. In
        these questions, k is a positive constant.
    b. What is the vertex of the graph of $y = ax^2 + k$?
    c. What is the vertex of the graph of $y = ax^2 - k$?

■ For Problems 7-8,
    a. Make a table of values and sketch the graph.
    b. What is the vertex of the graph?

7. $y = (x + 2)^2$

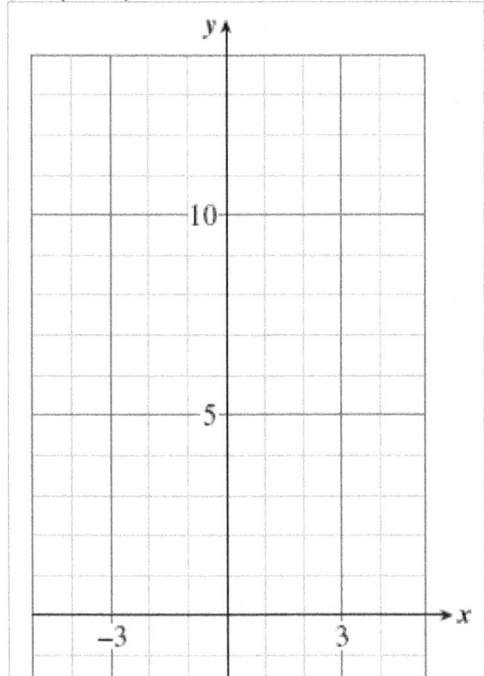

8. $y = (x - 3)^2$

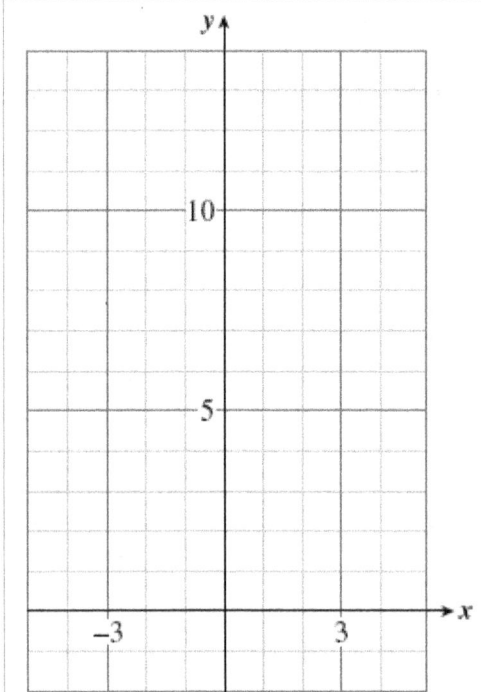

■ For Problems 9-12,
    a. Find the x-intercepts and the y-intercept of the graph.
    b. Find the vertex of the graph.
    c. Sketch the graph.

9. $y = x^2 + 2x$

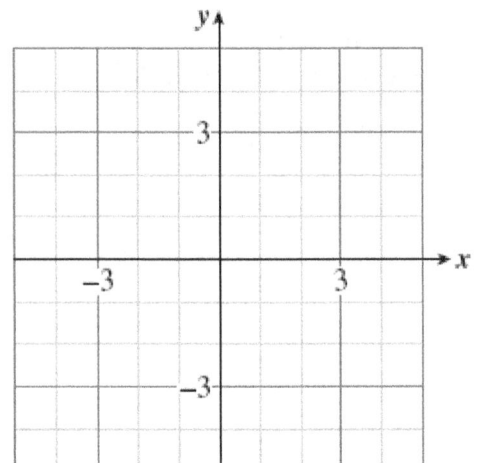

10. $y = x^2 - 2x - 3$

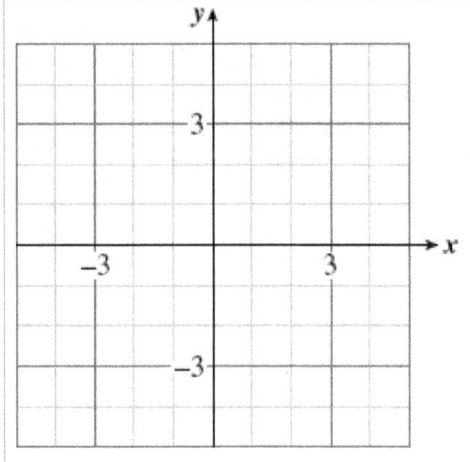

11. $y = 9 - x^2$

12. $y = -2x^2 + 12x - 10$

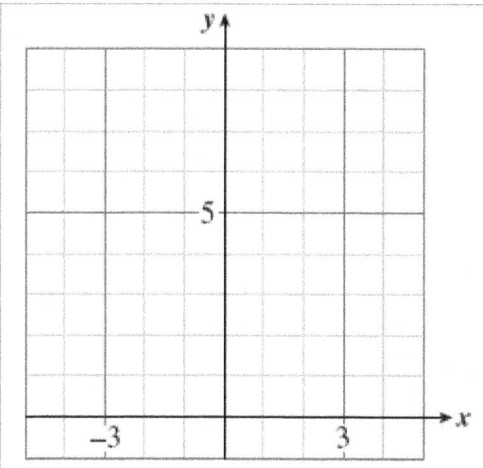

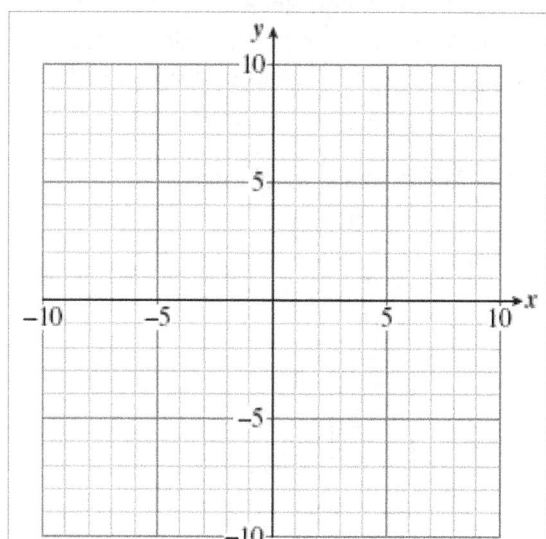

For Problems 13-14,
   a. Find the x- and y-intercepts of each parabola.
   b. Find the vertex of each parabola.
   c. Graph the pair of parabolas on the same grid and compare.

13. a. $y = x^2 - x - 2$
   b. $y = -x^2 + x + 2$

14. a. $y = x^2 - 2x - 15$
   b. $y = x^2 + 2x - 15$

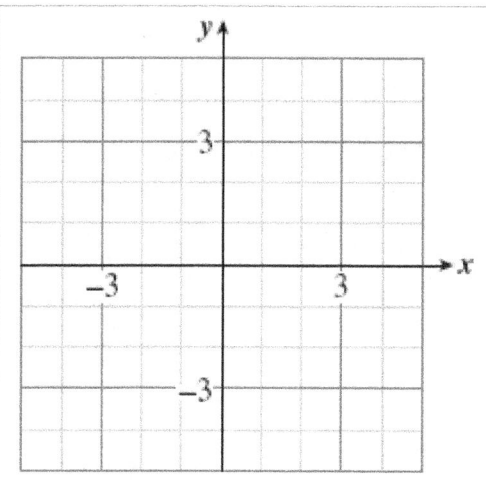

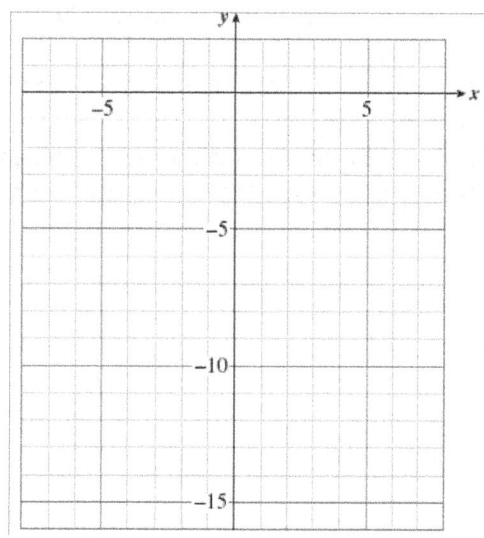

Applications

The bridge over the Rushing River at Marionville is 48 feet high. Francine stands on the bridge and tosses a rock into the air off the edge of the bridge. The height of the rock above the water t seconds later is given in feet by

$$h = 48 + 32t - 16t^2$$

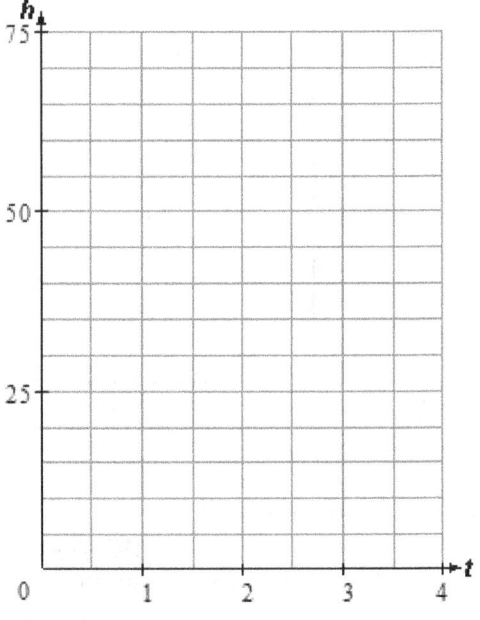

15. a. Complete the table of values.

t	h
0	
0.5	
1	
1.5	
2	
2.5	
3	

    b. Sketch a graph of the equation on the grid.

    c. Use the graph to estimate the height of the rock after 1.75 seconds. Verify your answer algebraically by substituting $t = 1.75$ into the equation for h.

16. a. Use your graph to answer the question: When is the rock about 40 feet above the water?

    b. Write an equation that you could solve to answer the question in part (a).

17. How long is the rock more than 60 feet high?

18. a. What is the highest point the rock reaches?

    b. After reaching its highest point, how long is the rock falling before it hits the water?

6.5 The Quadratic Formula

The graph of the equation

$$y = x^2 + 2x - 5$$

is shown at right. By reading the graph, we estimate the x-intercepts at approximately 1.5 and -3.5. To find their exact values, we must solve the equation

$$x^2 + 2x - 5 = 0$$

But, as you can check, the trinomial $x^2 + 2x - 5$ cannot be factored.

So far we know two methods for solving quadratic equations: extracting roots and factoring. Neither of these methods applies to every quadratic equation. In this Lesson we consider a technique that works on all quadratic equations.

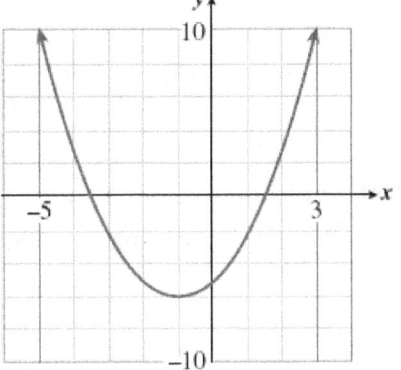

Using the Quadratic Formula

The graph of the quadratic equation

$$y = ax^2 + bx + c$$

depends upon its coefficients, $a, b,$ and c. In particular, the x-intercepts of the graph are determined by the values of these coefficients. In fact, there is a formula that uses the coefficients to calculate the solutions of the equation $ax^2 + bx + c = 0$. It is called the **quadratic formula**.

The Quadratic Formula

The solutions of the equation

$$ax^2 + bx + c = 0, \quad a \neq 0$$

are given by the formula

$$x = \frac{-b \pm \sqrt{b^2 - 4ac}}{2a}$$

Look Closer: The symbol $\pm$ is used in this context to combine two similar formulas into one. It means that the quadratic equation has two solutions, namely

$$\frac{-b + \sqrt{b^2 - 4ac}}{2a} \quad \text{and} \quad \frac{-b - \sqrt{b^2 - 4ac}}{2a} \qquad \bullet$$

Example 1   Use the quadratic formula to solve the equation

$$x^2 + 2x - 5 = 0$$

Solution We first identify the coefficients $a, b,$ and c:

$$a = 1, \quad b = 2, \quad c = -5$$

We substitute the values of $a, b,$ and c into the quadratic formula, and simplify according to the order of operations. We start with the expression under the radical.

$$x = \frac{-b \pm \sqrt{b^2 - 4ac}}{2a}$$

$$= \frac{-2 \pm \sqrt{2^2 - 4(1)(-5)}}{2(1)} \qquad \textbf{Simplify under the radical.}$$

$$= \frac{-2 \pm \sqrt{4 + 20}}{2} = \frac{-2 \pm \sqrt{24}}{2}$$

The solutions of the equation are $\dfrac{-2 + \sqrt{24}}{2}$ and $\dfrac{-2 - \sqrt{24}}{2}$. These are the exact values of the solutions. We can use a calculator to approximate each solution to hundredths.

$$\frac{-2 + \sqrt{24}}{2} \approx \frac{-2 + 4.90}{2} = 1.50$$

$$\frac{-2 - \sqrt{24}}{2} \approx \frac{-2 - 4.90}{2} = -3.50$$

These values are very close to our estimates from the graph. ■

RQ1. What should you do if a quadratic equation cannot be solved by factoring?

Caution! Before we use the quadratic formula, we must write the equation in standard form so that we can identify the coefficients a, b, and c. ■

RQ2. How do we find the values of a, b, and c in the quadratic formula?

If some of the coefficients are fractions, it helps to clear the fractions before applying the quadratic formula. The fastest way to clear all the fractions is to multiply each term of the equation by the lowest common denominator, or LCD, of all the fractions involved.

■ **Example 2a.** Solve $x = \dfrac{2}{3} - \dfrac{x^2}{6}$

b. Find decimal approximations to two decimal places for the solutions.

Solutions a. We multiply each term of the equation by the LCD, 6, to get

$$6(x) = 6\left(\frac{2}{3} - \frac{x^2}{6}\right) \qquad \textbf{Apply the distributive law.}$$

$$6x = 4 - x^2$$

Next, we write the equation in standard form and identify the coefficients.

$$x^2 + 6x - 4 = 0$$
$$a = 1, \ b = 6, \ c = -4$$

We substitute the coefficients into the quadratic formula, and simplify.

$$x = \frac{-b \pm \sqrt{b^2 - 4ac}}{2a} = \frac{-6 \pm \sqrt{6^2 - 4(1)(-4)}}{2(1)} \qquad \textbf{Simplify under the radical.}$$

$$= \frac{-6 \pm \sqrt{52}}{2}$$

The exact values of the solutions are $x = \dfrac{-6 + \sqrt{52}}{2}$ and $x = \dfrac{-6 - \sqrt{52}}{2}$.

b. We use a calculator to approximate each solution.

$$\frac{-6 + \sqrt{52}}{2} \approx \frac{-6 + 7.21}{2} = 0.605$$

$$\frac{-6 + \sqrt{52}}{2} \approx \frac{-6 - 7.21}{2} = -6.605$$

To two decimal places the solutions are 0.61 and −6.61.

RQ3. If the coefficients in a quadratic equation are fractions, what should you do?

Of course, if a quadratic equation can be solved by factoring, then the quadratic formula will give the same solutions. You can use whichever technique seems faster for a particular equation.

Non-Real Solutions

The graph of

$$y = x^2 - 4x + 5$$

has no x-intercepts, as you can see in the figure. Nonetheless, we can use the quadratic formula to solve the equation

$$x^2 - 4x + 5 = 0$$

Applying the formula with $a = 1$, $b = -4$, and $c = 5$, we find

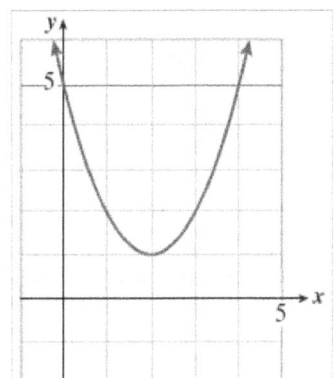

$$x = \frac{-b \pm \sqrt{b^2 - 4ac}}{2a} = \frac{-(-4) \pm \sqrt{(-4)^2 - 4(1)(5)}}{2(1)}$$

$$= \frac{4 \pm \sqrt{16 - 20}}{2} = \frac{4 \pm \sqrt{-4}}{2}$$

We cannot continue with the calculation because $\sqrt{-4}$ is undefined; it is not equal to any real number. This quadratic equation does not have any real-valued solutions, and that is why the graph of $y = x^2 - 4x + 5$ does not have any x-intercepts.

The solutions to the equation above are a type of number called **complex numbers**. If a quadratic equation does not have real-valued solutions, then its graph will not have x-intercepts.

> Every quadratic equation has two solutions. The nature of those solutions determines the nature of the x-intercepts of the graph. There are three possibilities.

> **x-Intercepts of $y = ax^2 + bx + c$**
>   The x-intercepts of the graph of $y = ax^2 + bx + c$ are the solutions of $ax^2 + bx + c = 0$.
> **1.** Both solutions are real numbers, and unequal. The graph has two x-intercepts.
> **2.** The solutions are real and equal. The graph has one x-intercept, which is also its vertex.
> **3.** Both solutions are non-real complex numbers. The graph has no x-intercepts.

> **RQ4.** If a quadratic equation has no real solutions, what does that tell you about the graph?

■ **Example 3**   How many x-intercepts does the parabola have?

a. $y = x^2 + 5$ **b.** $y = x^2 - 6x + 9$

Solutions a.   To find the x-intercepts, we solve the equation

$$x^2 + 5 = 0 \qquad \textbf{Solve by extraction of roots.}$$
$$x^2 = -5$$
$$x = \pm\sqrt{-5}$$

Because the square root of a negative number is undefined, this parabola has no x-intercepts.

b. We solve the equation

$$x^2 - 6x + 9 = 0$$
$$(x - 3)(x - 3) = 0$$
$$x - 3 = 0 \qquad x - 3 = 0$$
$$x = 3 \qquad\quad x = 3$$

This parabola has one x-intercept, $(3, 0)$. ■

> **RQ5.** If the solutions of a quadratic equation are equal, what does that tell you about the graph?

Skills Warm-Up

■ Solve each equation by the easiest method.

1. $x^2 = 36$ **2.** $x^2 - 9x = 36$ **3.** $x^2 - 9x = 0$
4. $9x^2 - 36 = 0$ **5.** $3x^2 - 24x - 36 = 0$ **6.** $3x^2 = 31x - 36$

Answers: **1.** $6, -6$ **2.** $12, -3$ **3.** $0, 9$ **4.** $2, -2$ **5.** $2, 6$ **6.** $\frac{4}{3}, 9$

Homework 6.5

Skills Practice

■ For Problems 1-4,
   a. Solve by using the quadratic formula.
   b. Give approximate values for your solutions, rounded to hundredths.

1. $x^2 - 8x + 4 = 0$

2. $3s^2 + 2s = 2$

3. $n^2 = n + 1$

4. $-4z^2 + 2z + 1 = 0$

■ For Problems 5-7, solve the equation two ways:
   a. Use the quadratic formula.
   b. Use either factoring or extracting roots.

5. $3t^2 - 5 = 0$

6. $z = 3z^2$

7. $2w^2 + 6 = 7w$

■ For Problems 8-11, solve the equation.

8. $\dfrac{x^2}{6} + x = \dfrac{2}{3}$

9. $v^2 + 3v - 2 = 9v^2 - 12v + 5$

10. $m^2 - 3m = \dfrac{1}{3}(m^2 - 1)$

11. $-0.2x^2 + 3.6x - 9 = 0$

Applications

12. Delbert throws a penny from the top of the Texas Building in Fort Worth. After t seconds, the height of the penny is given in feet by

$$h = -16t^2 + 8t + 380$$

   a. When does the penny pass a window 300 feet above the ground?
   b. How long does it take the penny to reach the ground?

13. The volume of a cedar chest is 12,000 cubic inches. The chest is 20 inches high, and its length is 5 inches less than three times its width. Find the dimensions of the chest.

14. The perimeter of a rectangle is 42 inches and its diagonal is 15 inches. Find the dimensions of the rectangle.

■ In Problems 15-16, the figure is a right triangle. Find the unknown sides.

15.

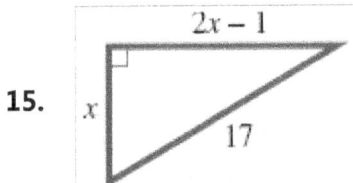

16.

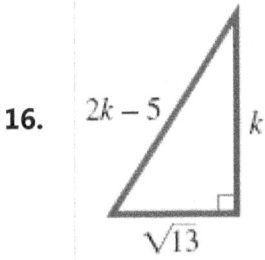

■ For Problems 17-20,
   a. Find the x-intercepts of the parabola. (Round your answers to hundredths.)
   b. Find the vertex of the parabola.
   c. Sketch the graph.

17. $y = x^2 - 2x - 2$

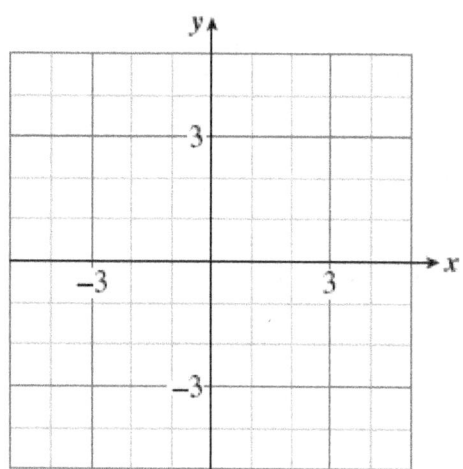

18. $y = -3x^2 + 2x - 1$

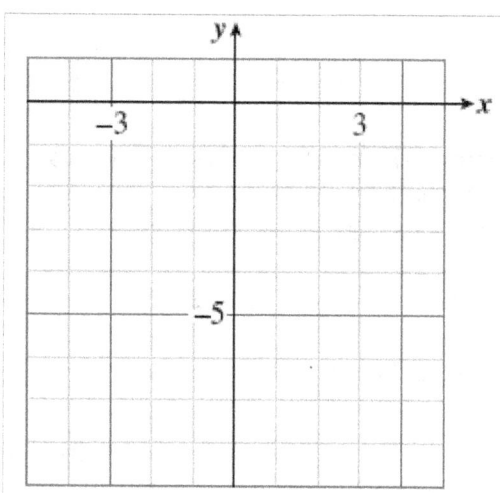

19. $y = x^2 - 4x - 1$

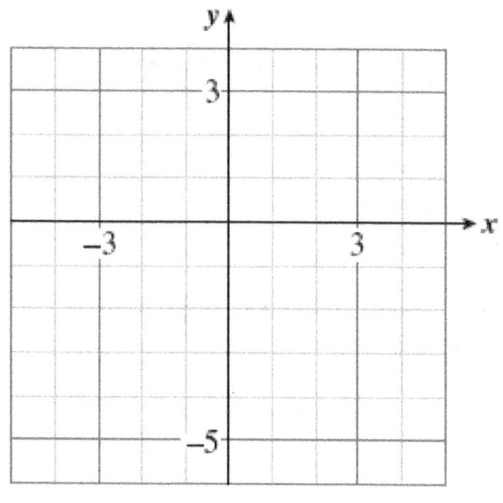

20. $y = x^2 - 0.6x - 7.2$

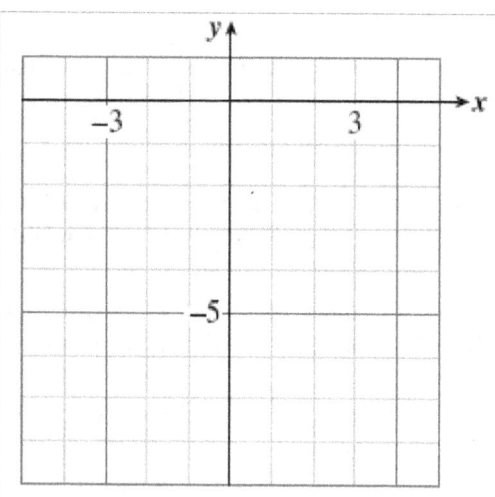

Chapter 6 Summary and Review

Section 6.1 Extracting Roots

> **Quadratic Equations**
> A **quadratic equation** can be written in the standard form
> $$ax^2 + bx + c = 0$$
> where a, b, and c are constants and a is not zero.

- The numbers a, b, and c are called **parameters**. They are the coefficients of, respectively, the quadratic term (the x^2-term), the linear term (the x-term), and the constant term.
- The graph of $y = ax^2 + bx + c$ is a curve called a **parabola**. The simplest, or basic, parabola is the graph of $y = x^2$.

> **Extraction of Roots**
> To solve a quadratic equation of the form
> $$ax^2 + c = 0$$
> 1. Isolate x^2 on one side of the equation.
> 2. Take the square root of each side.

Section 6.2 Some Quadratic Models

- Revenue = (price per item) · (number of items sold)

> **Zero-Factor Principle**
> If the product of two numbers is zero, then one (or both) of the numbers must be zero. Using symbols,
> **If $AB = 0$, then either $A = 0$ or $B = 0$.**

- In order to use the zero-factor principle to solve quadratic equations, we must be able to write quadratic expressions in **factored form**. This process is called **factoring**, and it is the reverse of multiplying factors together.
- To solve a quadratic equation using factoring, we must arrange the terms so that **one side of the equation is zero**. Then we set each factor equal to zero, and solve for x.

Section 6.3 Solving Quadratic Equations by Factoring

> **To Factor a Quadratic Trinomial**
> To factor $x^2 + bx + c$, we look for two numbers p and q so that
> $$pq = c \text{ and } p + q = b$$

> **To Solve a Quadratic Equation by Factoring:**
> 1. Write the equation in standard form,
>
> $$ax^2 + bx + c = 0, \qquad (a \neq 0)$$
>
> 2. Factor the left side of the equation.
> 3. Apply the zero-factor principle; that is, set each factor equal to zero.
> 4. Solve each equation to obtain two solutions.

Section 6.4 Graphing Quadratic Equations

- To find the x-intercepts of the graph of $y = ax^2 + bx + c$, we set $y = 0$ and solve the equation $ax^2 + bx + c = 0$.

- The high or low point of a parabola is called the **vertex** of the graph. The parabola has **axis of symmetry** that runs through the vertex.

> **The Vertex of a Parabola**
> 1. The x-coordinate of the vertex is the average of the x-intercepts.
> 2. To find the y-coordinate of the vertex, substitute its x-coordinate into the equation of the parabola.

> **To Graph a Quadratic Equation, $y = ax^2 + bx + c$**
> 1. Find the x-intercepts: set $y = 0$ and solve for x.
> 2. Find the vertex: the x-coordinate is the average of the x-intercepts. Find the y-coordinate by substituting the x-coordinate into the equation of the parabola.
> 3. Find the y-intercept: set $x = 0$ and solve for y.
> 4. Draw a parabola through the points. The graph is symmetric about a vertical line through the vertex.

Section 6.5 The Quadratic Formula

> **The Quadratic Formula**
> The solutions of the equation
>
> $$ax^2 + bx + c = 0, \quad a \neq 0$$
>
> are given by the formula
>
> $$x = \frac{-b \pm \sqrt{b^2 - 4ac}}{2a}$$

- If some or all of the coefficients are fractions, it helps to clear the fractions before applying the quadratic formula. The fastest way to clear all the fractions is to multiply each term of the equation by the lowest common denominator, or LCD, of all the fractions involved.

- Every quadratic equation has two solutions. The nature of those solutions determines the nature of the x-intercepts of the graph. There are three possibilities.

---

x-Intercepts of $y = ax^2 + bx + c$

    The x-intercepts of the graph of $y = ax^2 + bx + c$ are the solutions of $ax^2 + bx + c = 0$.

1. Both solutions are real numbers, and unequal. The graph has two x-intercepts.

2. The solutions are real and equal. The graph has one x-intercept, which is also its vertex.

3. Both solutions are non-real complex numbers. The graph has no x-intercepts.

Review Questions

▰ Use complete sentences to answer the questions in Problems 1-15.

1. What is the standard form for a quadratic equation?
2. What are parameters? What is a parabola?
3. Write down the steps for solving a quadratic equation by extraction of roots.
4. State the zero factor principle. What is it used for?
5. Explain how to factor out a common factor.
6. Francine factored $3x^2 - 2xy + x$ and got $x(3x - 2y)$. Why is this incorrect?
7. State the steps for solving a quadratic equation by factoring.
8. Explain how to factor a quadratic trinomial.
9. How do we find the x-intercepts of a parabola?
10. If you know the x-intercepts of a parabola, how can you find the coordinates of its vertex?
11. If you cannot solve a quadratic equation by facoring, what should you do?
12. How many solutions does a quadratic equaiton have?
13. State the quadratic formula.
14. What is the difference between a quadratic equation and the quadratic formula?
15. Does every parabola have two x-interepts? Why or why not?

Review Problems

▰ For Problems 1-6, solve by extraction of roots. Round to thousandths any answers that are irrational numbers.

1.	$9x^2 - 4 = 0$	**2.**	$3 + 2t^2 = 9$	**3.**	$5y^2 - 12 = 2y^2$
4.	$q^2 - 3 = 5 - q^2$	**5.**	$9k^2 + 21 = 25$	**6.**	$6a^2 + 3 = 4a^2 + 19$

7. The distance, d, that a penny will fall in t seconds is given in feet by

$$d = 16t^2$$

    a. If you drop a penny from a height of 144 feet, how long will it take to reach the ground?

    b. How far will the penny fall in $1\frac{1}{2}$ seconds?

8. **a.** How much water would you need to fill a spherical tank of radius 3 feet?

    b. A biosphere must contain 7000 cubic inches of space. What is the radius of the biosphere?

■ For Problems 9-20, factor.

9. $24x^2 - 18x$

10. $-32y^2 + 24y$

11. $a^2 - 18a + 45$

12. $x^2 - 14x - 51$

13. $2t^2 - t - 10$

14. $3b^2 - 14b + 8$

15. $14w - 5 + 3w^2$

16. $-3 + 2p^2 - 5p$

17. $z^2 - 121$

18. $81 - 4t^2$

19. $6x^2 + 21x + 9$

20. $8y^2 - 6y - 2$

■ For Problems 21-26, solve.

21. $0 = m^2 + 10m + 25$

22. $b^2 - 25 = 0$

23. $4p^2 = 16p$

24. $11t = 6t^2 + 3$

25. $(x - 5)(x + 1) = -8$

26. $2q(3q - 1) = 4$

■ For Problems 27-30, solve by using the quadratic formula.

27. $2t^2 + 6t + 3 = 0$

28. $\dfrac{x^2}{4} + 1 = \dfrac{13}{12}x$

29. $0.5x^2 - 0.3x - 0.25$

30. $\dfrac{2}{3}v^2 + \dfrac{5}{6}v = \dfrac{1}{6}$

■ For Problems 31-34, solve for the specified variable.

31. $V = \dfrac{s^2 h}{3}$ for s

32. $A = \dfrac{\pi d^2}{4}$ for d

33. $C = bh^2 r$ for h

34. $G = \dfrac{np}{r^2}$ for r

35. Audrey launches her experimental hydraulic rocket from the top of her apartment building. The height of the rocket after t seconds is given in feet by

$$h = -16t^2 + 40t + 80$$

How long is the test flight before the rocket hits the ground?

36. The formula $S = \dfrac{1}{2}n^2 + \dfrac{1}{2}n$ gives the sum of the first n counting numbers. How many counting numbers must you add to get a sum of 325?

37. The Corner Market sells $160 - 2p$ pounds of bananas per week if they charge p cents per pound.
   a. Write an expression for the market's weekly revenue, R, from bananas.
   b. It costs the market $80 + 24p$ cents to buy and display the bananas. Write an expression for the market's profit, M, from selling the of bananas.
   c. If the market made a profit of \$22 (or 2200 cents) on bananas last week, how much did they charge per pound?

38. The city park used 136 meters of fence to enclose its rectangular rock garden. The diagonal path across the middle of the garden is 52 meters long. What are the dimensions of the garden? (**Hint:** make a sketch.)

39. a. Sketch three rectangles of different sizes but such that the length of the rectangle is always twice its width.

   b. If the width of a particular such rectangle is w inches, write an equation for its area, A, in terms of w.

   c. Make a table of values for A in terms of w, and graph your equation.

   d. If the area of one of these rectangles is 48 square inches, find its width. Locate the point corresponding to this rectangle on your graph.

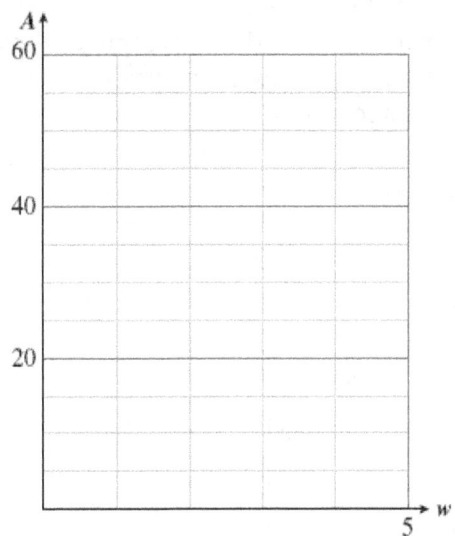

40. a. Sketch three equilateral triangles of different sizes.

   b. The area, A, of an equilateral triangle is given by

$$A = \frac{\sqrt{3}}{4}s^2$$

where s is the length of one side of the triangle. Rewrite the formula for the area using an approximation to three decimal places.

   c. Make a table of values for A in terms of s, and graph your equation.

   d. If the area of a particular equilateral triangle is 15.6 square inches, find the length of its side. Locate the point corresponding to this triangle on your graph.

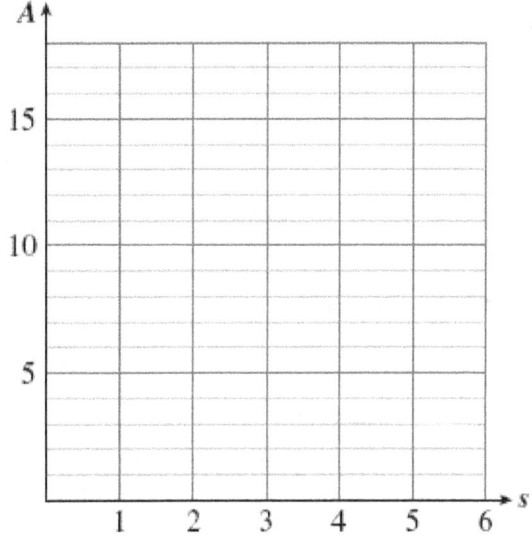

For Problems 41-46,
a. Find the x- and y-intercepts of the graph.
b. Find the vertex of the graph.
c. Sketch the graph.

41. $y = \dfrac{1}{2}x^2$

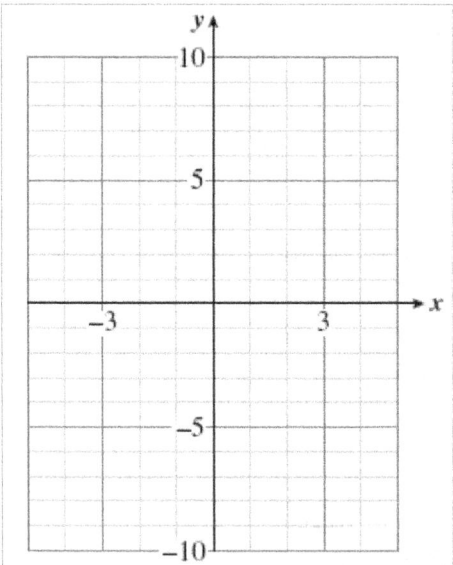

42. $y = x^2 - 4$

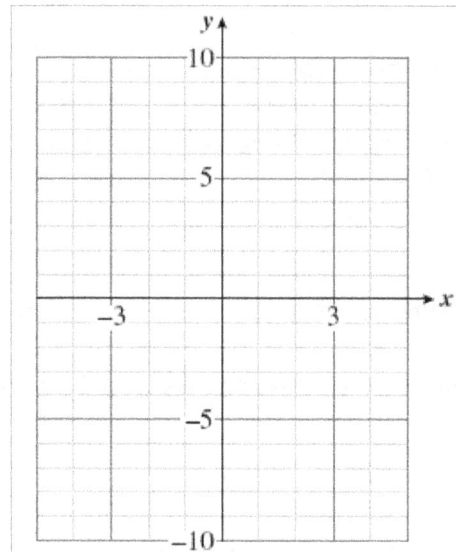

43. $y = x^2 - 9x$

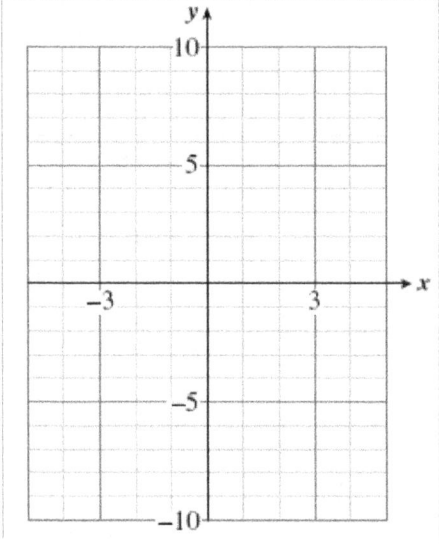

44. $y = -2x^2 - 4x$

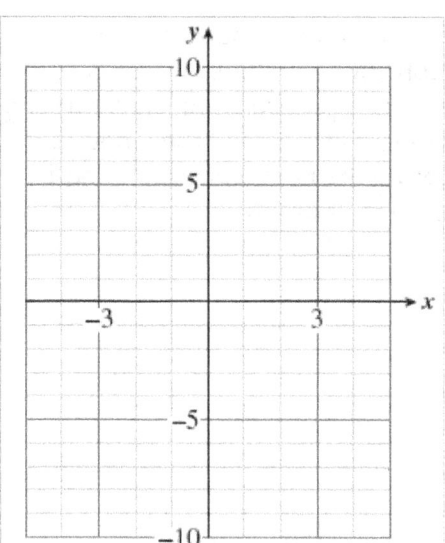

45. $y = x^2 + 6x$

46. $y = x^2 + 3x - 4$

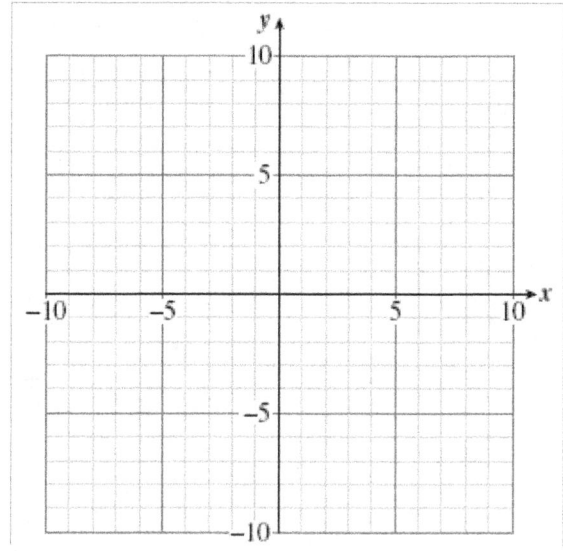

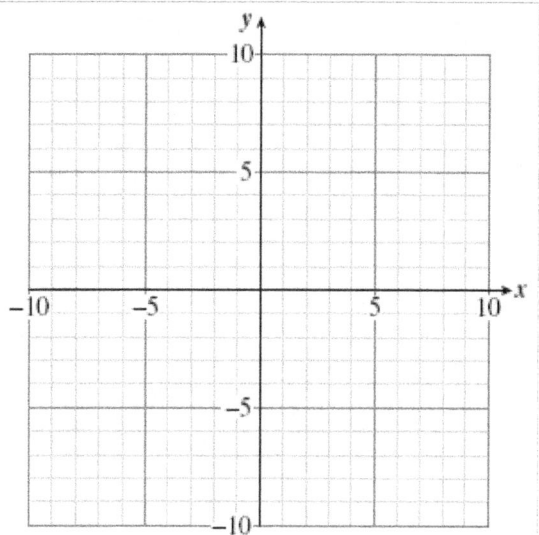

 For Problems 47-48, find the unknown sides of the right triangle.

47.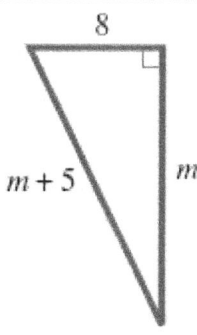

48.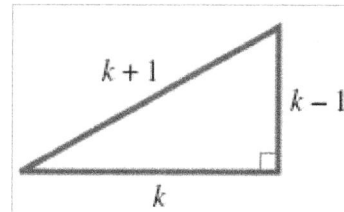

Chapter 7 Polynomials

Lesson 1 Polynomials

- Identify a polynomial
- Evaluate a polynomial
- Add and subtract polynomials

Lesson 2 Products of Polynomials

- Apply the first law of exponents
- Multiply monomials
- Multiply a polynomial by a monomial
- Multiply polynomials

Lesson 3 More about Factoring

- Apply the second law of exponents
- Divide monomials
- Factor a quadratic trinomial $ax^2 + bx + c$
- Factor a quadratic trinomial in two variables
- Factor out a binomial common factor
- Combine techniques to factor completely

Lesson 4 Special Products and Factors

- Identify squares of monomials
- Use the fromulas for squares of binomials and difference of squares
- Factor squares of binomials and difference of squares
- Know that the sum of squares cannot be factored

7.1 Polynomials

> A **polynomial** is an algebraic expression with several terms. Each term is a power of a variable (or a product of powers) with a constant coefficient.

Look Closer: The exponent must be a whole number, which means that a polynomial has no radicals containing variables, and no variables in the denominators of fractions.   •

> ■ **Example 1**   The following expressions are polynomials.
>
> $$4x^3 + 2x^2 - 7x - 5 \qquad \frac{1}{2}at^2 + vt$$
> $$2a^2 - 6ab + 3b^2 \qquad \pi r^2 h$$ ■

> Recall that: An algebraic expression with only one term is called a **monomial**. An expression with two terms is called a **binomial**, and an expression with three terms is a **trinomial**.

> ■ **Example 2**   The expression $2x^3$ is a monomial (it has only one term).
>   The expression $x^2 - 16$ is a binomial (it has two terms).
>   The expression $ax^2 + bx + c$ is called a **quadratic trinomial.**   ■

> **RQ1.**   Explain the terms monomial, binomial, trinomial, and polynomial.

Degree

In a term containing only one variable, the exponent of the variable is called the **degree** of the term. (The degree of a constant term is zero, for reasons that will become clear in Chapter 9.) For example,

$3x^2$	is of second degree
$8y^3$	is of third degree
$4x$	is of first degree (the exponent on x is 1)
5	is of zero degree

The degree of a polynomial in one variable is the largest exponent that appears in any term.

> ■ **Example 3**   The degree of a polynomial does not depend on the number of terms in the polynomial.
>
> | | |
> |---|---|
> | $2x + 1$ | is of first degree |
> | $3y^2 - 2y + 2$ | is of second degree |
> | $m - 2m^2 + m^5$ | is of fifth degree | ■

> **RQ2.**   What is the difference between a polynomial of degree four and a polynomial with four terms?

Polynomials in one variable are usually written in **descending powers** of the variable. The term with the largest exponent comes first, then the term with the next highest exponent, and so on down to the constant term, if there is one.

■ **Example 4**   Write the polynomial $x + 3x^4 - 5 - 2x^2$ in descending powers.

Solution   We start with the highest power. We must be careful to keep the correct sign of each term.

$$3x^4 - 2x^2 + x - 5$$

Exponents decrease from left to right.

Constant term is written last.

RQ3.   What does it mean to write a polynomial in descending powers?

Evaluating Polynomials

We evaluate polynomials the same way we evaluate any other algebraic expression: by substituting the given values for the variables and then following the order of operations to simplify.

■ **Example 5**   Evaluate $16t^3 - 6t + 20$ for $t = \frac{3}{2}$.

Solution   Substitute $\frac{3}{2}$ for t, and follow the order of operations to simplify.

$$16\left(\frac{3}{2}\right)^3 - 6\left(\frac{3}{2}\right) + 20 \qquad \text{Compute the power.}$$

$$= 16\left(\frac{27}{8}\right) - 6\left(\frac{3}{2}\right) + 20 \qquad \text{Perform all multiplications.}$$

$$= 54 - 9 + 20 \qquad \text{Add.}$$

$$= 65$$

Like Terms

Like terms are any terms that are exactly alike in their variable factors. The exponents on the variable factors must also match. For example,

$$x^3 \quad \text{and} \quad 5x^2 \qquad \textbf{Not like terms}$$
$$2x^2y \quad \text{and} \quad 2xy^2 \qquad \textbf{Not like terms}$$

are not like terms because their variable factors are different. However,

$$\frac{1}{2}x^2y \quad \text{and} \quad -3yx^2 \qquad \textbf{Like terms}$$

are like terms, because x^2y and yx^2 are equivalent expressions.

Recall that: To add or subtract like terms, we add or subtract their numerical coefficients. The variable factors of the terms remain unchanged.

RQ4.   What is the numerical coefficient of a term?

Look Closer: Which of the following two expressions can be simplified?

$$3x^2 + 5x^3 \qquad \text{and} \qquad 3x^2 + 5x^2$$

In the first expression, the two terms have different exponents, even though the base, x, is the same. Powers of the same variable with different exponents are not like terms. The first expression, $3x^2 + 5x^3$, cannot be simplified.

In the second expression, $3x^2$ and $5x^2$ are like terms. Thus,

$$3x^2 + 5x^2 = 8x^2$$

Note that we do *not* change the exponent on x in the sum; it is still 2.

Example 6   Simplify by combining any like terms:

$$x^3 + 2x^2 - 2x^3 - (-4x^2) + 4x$$

Solution The terms x^3 and $-2x^3$ are like terms; so are $2x^2$ and $-4x^2$. We group the like terms together and combine them.

$$\mathbf{x^3 - 2x^3} + 2x^2 - (-4x^2) + 4x = \mathbf{-x^3} + 6x^2 + 4x$$

The last term, $4x$, is not combined with any other terms.

Adding Polynomials

To add two polynomials we remove parentheses and combine like terms.

Example 7 $(4a^3 - 2a^2 - 3a + 1) + (2a^3 + 4a - 5)$ **Remove parentheses.**
$$= 4a^3 - 2a^2 - 3a + 1 + 2a^3 + 4a - 5 \qquad \textbf{Combine like terms.}$$
$$= 6a^3 - 2a^2 + a - 4$$

We can also use a vertical format to add polynomials.

Example 8   Add $4x^2 + 2 - 7x$ and $5x - 5 + 2x^2$

Solution We first write each polynomial in descending powers of x. Then we write the second polynomial beneath the first, aligning like terms.

$$
\begin{array}{r}
4x^2 - 7x + 2 \qquad \textbf{Combine like terms vertically.}\\
+\ \underline{2x^2 + 5x - 5} \\
6x^2 - 2x - 3
\end{array}
$$

Subtracting Polynomials

If an expression in parentheses is preceded by a minus sign, we must change the sign of **each** term within parentheses when we remove the parentheses. This rule applies when we subtract polynomials.

Example 9   Subtract the polynomials: $(4x^2 + 2x - 5) - (2x^2 - 3x - 2)$

Solution We first remove the parentheses, remembering to change the sign of **each** term of the subtracted polynomial.

$$(4x^2 + 2x - 5) - (2x^2 - 3x - 2) \qquad \text{Each sign is changed.}$$
$$= 4x^2 + 2x - 5 - 2x^2 + 3x + 2 \quad \text{Combine like terms.}$$
$$= 2x^2 + 5x - 3$$

Caution!   In Example 9, the following second step is **incorrect**

$$4x^2 + 2x - 5 - 2x^2 - 3x - 2$$

because we have not changed the sign of **each** term inside parentheses.

To use a vertical format for subtraction, we change the sign of *each term* in the bottom, or subtracted, polynomial.

Example 10   Subtract $2n^2 - 3n - 2$ from $4n^2 + 2n - 5$

Solution

$$\begin{array}{r} 4n^2 + 2n - 5 \\ - \underline{(2n^2 - 3n - 2)} \end{array} \quad \rightarrow \quad \begin{array}{c} \textbf{Change signs} \\ \textbf{of each term.} \end{array} \quad \rightarrow \quad \begin{array}{r} 4n^2 + 2n - 5 \\ + \underline{-2n^2 + 3n + 2} \\ 2n^2 + 5n - 3 \end{array}$$

RQ5.   When subtracting two polynomials, what must we do before combining like terms?

Application

Applied problems may involve addition or subtraction of polynomials. In particular, to calculate the profit it earns by selling its product, a company must subtract the cost of producing the goods from the revenue it earns by selling them. We can state this as a formula.

$$\boxed{\textbf{Profit} = \textbf{Revenue} - \textbf{Cost}}$$

Example 11   It costs The Cookie Company $200 + 2x$ dollars to produce x bags of cookies per week, and they earn $8x - 0.01x^2$ dollars from the sale of x bags of cookies.
a. Write a polynomial for the profit earned by The Cookie Company on x bags of cookies.
b. Find the company's profit on 300 bags of cookies, and on 600 bags.

Solutions a. Applying the profit formula to this situation, we subtract polynomials to find

$$\text{Profit} = (8x - 0.01x^2) - (200 + 2x) \qquad \textit{Change signs of second polynomial.}$$
$$= 8x - 0.01x^2 - 200 - 2x \qquad \textit{Combine like terms.}$$
$$= -0.01x^2 + 6x - 200$$

b. Evaluate the profit polynomial for $x = 300$ and $x = 600$. For 300 bags of cookies, the profit is

$$-0.01(300)^2 + 6(300) - 200 = -900 + 1800 - 200 = 700$$

or $700. For 600 bags of cookies, the profit is

$$-0.01(600)^2 + 6(600) - 200 = -3600 + 3600 - 200 = -200$$

The company loses $200 if they produce 600 bags of cookies.

Skills Warm-Up

■ Replace the comma in each pair by the proper symbol: $>$, $<$, or $=$.

1. $(-2)^3$, -2^3

2. -5^4, $(-5)^4$

3. $7 - 3^2$, 4^2

4. $2 + 4^3$, 6^3

5. $4 \cdot 2^3$, 8^3

6. $2 \cdot 5^2$, 10^2

7. $\dfrac{1}{2}$, $\left(\dfrac{1}{2}\right)^2$

8. $\dfrac{3}{4}$, $\left(\dfrac{3}{4}\right)^2$

Answers: **1.** $=$ **2.** $<$ **3.** $<$ **4.** $<$ **5.** $<$ **6.** $<$ **7.** $>$ **8.** $>$

Homework 7.1

Skills Practice

1. Which of the following are not allowed in a polynomial?
   a. More than three terms
   b. Coefficients that are fractions
   c. Division by a variable
   d. A term without variables

2. Give an example of each type of polynomial.
   a. A monomial of fourth degree.
   b. A binomial of first degree
   c. A trinomial of degree 2
   d. A monomial of degree 0

▦ Which of the expressions in Problems 3 and 4 are polynomials? If it is not a polynomial, explain why not.

3. a. $5x^4 - 3x^2$
   b. $3x + 1 + \dfrac{2}{x^2}$

   c. $\dfrac{1}{2a^2 + 5a - 6}$
   d. $\dfrac{2}{3}t^2 + \dfrac{1}{4}t^3 + \dfrac{5}{8}$

4. a. $27y^8$
   b. $2a^2 - 6ab + 3b^2$

   c. $\dfrac{3}{x^4} - \dfrac{7}{x^3} + \dfrac{5}{3}$
   d. $9z^9 - \dfrac{1}{2}z^2 + 8z^6$

5. Give the degree of each polynomial.
   a. $x^2 + 4x - \dfrac{1}{4}$
   b. $y - 2.8y^7$
   c. $\dfrac{z}{4} - 3z^4 + 4z^3$

▦ For Problems 6-7, write the polynomial in descending powers of x.

6. $x - 1.9x^3 + 6.4$

7. $6xy - 2x^2 + 2y^3$

▦ For Problems 8-11, evaluate the polynomial for the given values of the variables.

8. $2 - z^2 - 2z^3$ for $z = -2$

9. $2a^4 + 3a^2 - 3a$ for $a = 1.6$

10. $-abc^2$ for $a = -3, b = 2, c = 2$

11. $x^2 - 3x + 2$ for $x = \sqrt{3}$

▦ For Problems 12-13, simplify by combining like terms.

12. $6b^3 - 2b^3 - (-8b^3)$

13. $6x - 3y + 5xy - 6y + xy$

▦ For Problems 14-16, explain why the calculation is **incorrect**, and give the correct answer.

14. $6w^3 + 8w^3 \rightarrow 14w^6$

15. $6 + 3x^2 \rightarrow 9x^2$

16. $4t^2 + 7 - (3t^2 - 5) \rightarrow t^2 + 2$

▦ For Problems 17-18, add or subtract the polynomials.

17. $(2y^3 - 4y^2 - y) + (6y^2 + 2y + 1)$

18. $(5x^3 + 3x^2 - 4x + 8) - (2x^3 - 4x - 3)$

▦ For Problems 19-20, use a vertical format to add or subtract the polynomials.

19. Add $8x^2 - 3x + 4$ to $-2x^2 + 5x - 7$

20. Subtract $4x^2 - 3x - 1$ from $-3x^2 + 4x - 2$

Applications

21. Brenda is flying at an altitude of 4000 feet when she starts her descent for landing. After traveling x miles horizontally, her altitude is given in feet by

$$h = 125x^3 - 750x^2 + 4000$$

    a. What is Brenda's altitude after traveling 2 miles horizontally?

    b. Evaluate the polynomial for $x = 4$. What does this mean?

22. The height of a box is 4 inches less than its width, w, and its length is 8 inches greater than its width. Write a polynomial for the volume of the box.

23. Suppose you want to choose three items from a list of n possible items. The number of different ways you can make your choice is given by the polynomial

$$\frac{1}{6}n^3 - \frac{1}{2}n^2 + \frac{1}{3}n$$

    a. How many ways can you pick 3 elective courses from a list of 8 approved courses?

    b. How many ways can you pick 3 cards from a deck of 52?

    c. How many different 3-person committees can be chosen from a club with 20 members?

24. After you apply the brakes, a small car traveling at s miles per hour can stop in approximately $0.04s^2 + 0.6s$ feet. Will a car traveling at 50 miles per hour on the freeway avoid hitting a stalled car in the same lane 100 feet ahead?

25. Evaluate each polynomial for $n = 10$. Try to do the calculations mentally. What do you notice?

    a. $5n^2 + 6n + 7$ **b.** $5n^3 + n^2 + 3n + 3$

    c. $n^3 + 1$ **d.** $8n^4 + 8n$

26. Expand each expression by removing parentheses. What do you notice?

    a. $x[x(x + 3) + 4] + 1$ **b.** $x(x[x(x - 7) - 5] + 8) - 3$

27. a. Evaluate the expression in Problem 26a for $x = 2$. Can you do this mentally? Is it easier to evaluate the expression before or after expanding it?

    b. Use a calculator to evaluate the expression for $x = 0.8$.

28. GreatOutdoors sells specialty tents for high-altitude camping. Their cost for producing x tents is given by

$$C = x^3 - 12x^2 + 80x + 180$$

and their revenue from selling x tents is

$$R = 2800x - 2x^2$$

    a. Write a polynomial for the profit GreatOutdoors makes from selling x tents. (**Hint:** Recall that Profit = Revenue − Cost.)

    b. Find GreatOutdoors' profit from selling 10 tents, 20 tents, and 50 tents.

29. The Flying Linguine Brothers are working on a new act for the circus. Mario swings from a trapeze and catches Alfredo, who has somersaulted off a trampoline, in mid-air. Mario's height at time t seconds is given approximately by $12t^2 - 24t + 34$ feet, and Alfredo's height at time t is $-16t^2 + 32t + 6$.
   a. Write a polynomial for the difference in height between Mario and Alfredo at any time t.
   b. Find the difference in height between Mario and Alfredo when they start (at time $t = 0$), and after $\frac{1}{2}$ second.
   c. When will Mario and Alfredo be at the same height?

30. Ralph and Waldo start in towns that are 20 miles apart and travel in opposite directions for 2 hours. Ralph travels 30 miles per hour faster than Waldo. Let w stand for Waldo's speed and write algebraic expressions to answer the following questions.
   a. How far did Waldo travel?
   b. What was Ralph's speed?
   c. How far did Ralph travel?
   d. What is the distance between Ralph and Waldo after the 2 hours?

31. Let T_m stand for the sum of the squares of the first m integers. For example,

$$T_1 = 1^2 = 1$$
$$T_2 = 1^2 + 2^2 = 5$$
$$T_3 = 1^2 + 2^2 + 3^2 = 14$$

and so on.
   a. Fill in the table showing the first 10 values of T_m.
   b. Evaluate the polynomial $\frac{1}{3}m^3 + \frac{1}{2}m^2 + \frac{1}{6}m$ for integer values of m from 1 to 10, and fill in the table.
   c. Compare your answers to parts (a) and (b).

m	T_m	$\frac{1}{3}m^3 + \frac{1}{2}m^2 + \frac{1}{6}m$
1		
2		
3		
4		
5		
6		
7		
8		
9		
10		

32. Evaluate $(2x + y - z) + (3x - 4y + 6z) - (5x - 9y - 3z)$ for $x = 2.8$, $y = -3.6$, $z = 1.8$
   (**Hint:** There is a hard way and an easy way.)

7.2 Products of Polynomials

Products of Powers

Suppose we would like to multiply two powers together. For instance, consider the product $(x^3)(x^4)$, which can be written as

$$(x^3)(x^4) = xxx \cdot xxxx = x^7$$

because x occurs as a factor 7 times. We see that the number of x's in the product is the *sum* of the number of x's in each factor.

On the other hand, if we'd like to multiply x^3 times y^4, we cannot simplify the product because the two powers do not have the same base:

$$(x^3)(y^4) = xxx \cdot yyyy = x^3 y^4$$

These observations illustrate the following rule.

> **First Law of Exponents**
> To multiply two powers with the same base, we add the exponents and leave the base unchanged. In symbols,
>
> $$a^m \cdot a^m = a^{m+n}$$

> ■ **Example 1**   In each product below, we keep the same base and add the exponents.
> **a.** $x^2 \cdot x^6 = x^8$ **b.** $5^2 \cdot 5^6 = 5^8$ ■

> **Caution!**   Each product below is **incorrect**.
>
> $3^4 \cdot 3^3 \to 9^7$ The **base** does not change. The correct product is 3^7.
> $t^3 \cdot t^5 \to t^{15}$ We **add** the exponents. The correct product is t^8.   ■

> **RQ1.**   How do we simplify the product of two powers with the same base?
> How do we simplify the product of two powers with different bases?

Products of Monomials

We can use the first law of exponents to multiply two monomials together.

> ■ **Example 2**   Multiply $(2x^2 y)(5x^4 y^3)$
>
> **Solution** Rearrange the factors to group together the numerical coefficients and the powers of each base.
>
> $$(2x^2 y)(5x^4 y^3) = (2)(5)\ x^2 x^4\ yy^3$$
>
> Multiply the coefficients together, and use the first law of exponents to find the products of the variable factors.
>
> $$(2)(5)x^2 x^4 yy^3 = 10x^6 y^4$$ ■

| RQ2. | Which law of algebra allows us to rearrange the factors in a product? |

Multiplying by a Monomial

To multiply a polynomial by a monomial, we use the distributive law.

Example 3 Multiply $-3xy^2(4x^2 - 2xy + 2)$

Solution We apply the distributive law to multiply each term of the polynomial by the monomial $-3xy^2$.

$$-3xy^2(4x^2 - 2xy + 2) = -3xy^2(4x^2) - 3xy^2(-2xy) - 3xy^2(2)$$
$$= -3 \cdot 4 \cdot x \cdot x^2 \cdot y^2 - 3(-2) \cdot x \cdot x \cdot y^2 \cdot y - 3 \cdot 2 \cdot xy^2$$
$$= -12x^3y^2 + 6x^2y^3 - 6xy^2$$

To simplify each term, we group together the coefficients and powers with the same base.

| RQ3. | Which algebraic law do we use when we multiply a polynomial by a monomial? |

Products of Polynomials

In Chapter 5 we found the product of two binomials using the "FOIL" method, a special case of the distributive law. We can also use the distributive law to help us compute products of two or more polynomials.

Example 4 Multiply $(2x - 1)(3x^2 - x + 2)$

Solution We multiply each term of the first polynomial by each term of the second polynomial. This involves six multiplications: We first multiply each term of the trinomial by $2x$, then multiply each term by -1.

$$(2x - 1)(3x^2 - x + 2) = 2x(3x^2) + 2x(-x) + 2x(2) - 1(3x^2) - 1(-x) - 1(2)$$
$$= 6x^3 - 2x^2 + 4x - 3x^2 + x - 2 \qquad \text{Combine like terms.}$$
$$= 6x^3 - 5x^2 + 5x - 2$$

| RQ4. | How many terms are there in the product of two trinomials? |

Look Closer: If a product contains both polynomial and monomial factors, it is a good idea to multiply the polynomial factors together first, and save the monomial factor for last.

Example 5 Multiply $2x(x + 2)(3x - 5)$

Solution We begin by multiplying the binomial factors, $(x + 2)(3x - 5)$.

$$2x[(x+2)(3x-5)] = 2x[3x^2 - 5x + 6x - 10]$$
$$= 2x(3x^2 + x - 10)$$

Next we use the distributive law to multiply the result by $2x$.

$$2x(3x^2 + x - 10) = 6x^3 + 2x^2 - 20x$$

RQ5. To compute the product $3a(a+6)(4a-1)$, which factors should we multiply first?

In a product of three or more polynomials, we start by multiplying together any two of the three factors.

Example 6   Multiply $(3a-1)(a+2)(2a-3)$

Solution   We begin by multiplying together the last two binomials, $(a+2)(2a-3)$.

$$(3a-1)[(a+2)(2a-3)] = (3a-1)[2a^2 - 3a + 4a - 6]$$
$$= (3a-1)(2a^2 + a - 6)$$

Now we use the distributive law to multiply each term of the trinomial by each term of the binomial, as shown in Example 4.

$$(3a-1)(2a^2 + a - 6) = 6a^3 + 3a^2 - 18a - 2a^2 - a + 6 \qquad \textbf{Combine like terms.}$$
$$= 6a^3 + a^2 - 19a + 6$$

Skills Warm-Up

Simplify.

1. $3(b-4) - 2b(3-2b)$
2. $5x - 2x(1-2x) - 3(x-2)$
3. $6 + 3[x - 2x(x-4)]$
4. $a - 2[a - 2(a-2)]$
5. $4[-2(t - \frac{1}{2})(t-1)]$
6. $5[-6(w + \frac{2}{3})(w - \frac{3}{2})]$

Answers: 1. $4b^2 - 3b - 12$ **2.** $4x^2 + 6$ **3.** $-2x^2 + 27x + 6$ **4.** $3a - 8$

Homework 7.2

Skills Practice

1. Apply the first law of exponents to find the product.
  a. $x^3 \cdot x^6$ **b.** $5^6 \cdot 5^8$ **c.** $b^3(b)(b^5)$

2. Find a value of n that makes the expressions equivalent.
  a. $y^3 \cdot y^n = y^8$ **b.** $a^n \cdot a^4 = a^8$ **c.** $3 \cdot 3^n = 3^3$

3. Simplify if possible.
  a. $2x^4(-3x^4)$ **b.** $-x^4(-2x^2)$ **c.** $-x^4 \cdot y^3$ **d.** $-3x^5(3y^5)$

4. Simplify if possible.
  a. $2x^4 - 3x^4$ **b.** $-x^4 - 2x^2$ **c.** $-x^4 + y^3$ **d.** $-3x^5 + 3y^5$

For the pair of expressions in Problems 5 and 6, find
  a. their product,   **b.** their sum.

5. $-3z^4$, $-7z^4$ **6.** $-9cd^3$, $-cd^3$

For the rectangles in Problems 7 and 8, find
  a. the perimeter,   **b.** the area.

7.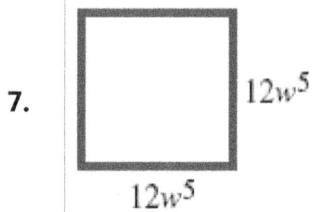

$12w^5$ / $12w^5$

8.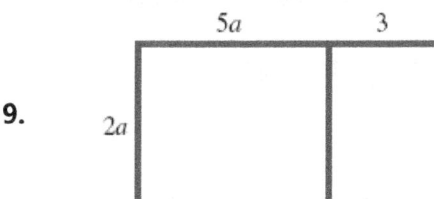

$4t^6$ / $13t^6$

For Problems 9-10,
  a. Write a product (length × width) for the area of each rectangle.
  b. Use the distributive law to compute the product.

9.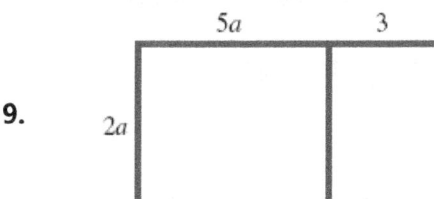

10.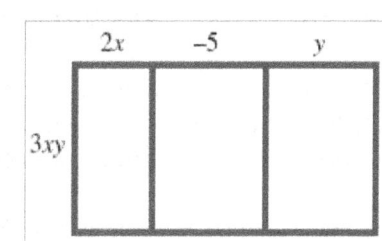

For Problems 11-14, simplify each expression if possible. If it cannot be simplified, say so.

11a. $x^2 + x^2$ **b.** $x^2(x^2)$ **c.** $x^2 - x^2$ **d.** $x^2(-x^2)$
12a. $-x - x$ **b.** $-x(-x)$ **c.** $-x^2 - x^2$ **d.** $-x^2(-x^2)$
13a. $x + x^2$ **b.** $x(x^2)$ **c.** $x^2 - x$ **d.** $x^2(-x)$
14a. $-x^3 \cdot x^3$ **b.** $x^3(-x^2)$ **c.** $x^3 - x^2$ **d.** $(-x)^3(-x)^2$

For Problems 15-16, multiply.

15. $-xy(x^2 + xy + y^2)$ **16.** $(6 - st + 3s^2t^2)(-3s^2t^2)$

▊ For Problems 17-18, simplify.

17. $ax(x^2 + 2x - 3) - a(x^3 + 2x^2)$ **18.** $3ab^2(2 + 3a) - 2ab(3ab + 2b)$

▊ For Problems 19-21, use rectangles to help you multiply the binomials in two variables.

19. $(x + 2y)(x - y)$ **20.** $(3s + t)(2s + 3t)$ **21.** $(2x - a)(x - 3a)$

▊ For Problems 22-25, compute the product: Multiply the binomials together first, then multiply the result by the numerical coefficient.

22. $2(3x - 1)(x - 3)$ **23.** $-3(x + 4)(x - 1)$
24. $-(4x + 3)(x - 2)$ **25.** $5(2x + 1)(2x - 1)$

▊ For Problems 26-29, multiply.

26. $4a(a - 1)(a + 5)$ **27.** $s^2t^2(2s + t)(3s - t)$
28. $(x - 2)(x^2 - 3x + 2)$ **29.** $(3x - 1)(9x^2 - 3x + 1)$

30. a. Multiply $(x + 1)(x + 2)(x + 3)$.
  b. Evaluate the product for $x = -1$, $x = -2$, and $x = -3$.

31. a. Multiply $(x - 2)(x + 4)(x + 5)$.
  b. Find three values of x for which the product is equal to zero.

32. Simplify mentally, without using paper, pencil, or calculator.
  a. $10^3(8 \cdot 10^4)$ **b.** $(3 \cdot 10^2)(2 \cdot 10^2)$ **c.** $(3.3 \cdot 10^2)(2 \cdot 10^2)$

Applications

33. The sum of two numbers is 16.
  a. If one of the numbers is n, write an expression for the other number.
  b. Write a polynomial for the product of the two numbers.

34. A large wooden box is 3 feet longer than it is wide, and its height is 2 feet shorter than its width.
  a. If the width of the box is w, write expressions for its length and its height.
  b. Write a polynomial for the volume of the box.
  c. Write a polynomial for the surface area of the box.

35. If you count by even numbers, such as 6, 8, 10, et cetera, you are listing consecutive even integers.
  a. If your first even integer is n, what is the next even integer?
  b. What is the next even integer after that?
  c. Write a polynomial for the product of the three even integers in parts (a) and (b).

▊ In Lesson 7.3 we'll use rectangles to factor quadratic trinomials. Problems 36-38 review representing a product as the area of a rectangle.
  a. Compute each product.
  b. Illustrate each product as the area of a rectangle.

36. $(3y + 4)(y + 2)$ **37.** $(2w - 6)(4w + 3)$ **38.** $(3t + 5)(2t + 3)$

7.3 More About Factoring

Quotients of Powers

We can reduce a fraction by dividing numerator and denominator by any common factors. For example,

$$\frac{10}{15} = \frac{2 \cdot \cancel{5}}{3 \cdot \cancel{5}} = \frac{2}{3}$$

We first factored the numerator and denominator of the fraction and then canceled the 5's by dividing. We can apply the same technique to quotients of powers.

■ **Example 1a.**   Simplify $\dfrac{a^5}{a^3}$ **b.** Simplify $\dfrac{a^4}{a^8}$

Solutions a.   We first write the numerator and denominator in factored form. Then we divide any common factors from the numerator and denominator.

$$\frac{a^5}{a^3} = \frac{\cancel{a} \cdot \cancel{a} \cdot \cancel{a} \cdot a \cdot a}{\cancel{a} \cdot \cancel{a} \cdot \cancel{a}} = \frac{a^2}{1} = a^2$$

You may observe that the exponent of the quotient can be obtained by subtracting the exponent of the denominator from the exponent of the numerator. In other words,

$$\frac{a^5}{a^3} = a^{5-3} = a^2$$

b. In this quotient, the larger power occurs in the denominator.

$$\frac{a^4}{a^8} = \frac{\cancel{a} \cdot \cancel{a} \cdot \cancel{a} \cdot \cancel{a}}{\cancel{a} \cdot \cancel{a} \cdot \cancel{a} \cdot \cancel{a} \cdot a \cdot a \cdot a \cdot a} = \frac{1}{a^4}$$

We see that we can subtract the exponent of the numerator from the exponent of the denominator. That is,

$$\frac{a^4}{a^8} = \frac{1}{a^{8-4}} = \frac{1}{a^4}$$ ■

These examples suggest the following property.

Second Law of Exponents

To divide two powers with the same base, we subtract the smaller exponent from the larger one, and keep the same base.
1. If the larger exponent occurs in the numerator, put the power in the numerator.
2. If the larger exponent occurs in the denominator, put the power in the denominator.

In symbols,

1. $\dfrac{a^m}{a^n} = a^{m-n}$ if $n < m$

2. $\dfrac{a^m}{a^n} = \dfrac{1}{a^{n-m}}$ if $n > m$

Quotients of Monomials

To divide one monomial by another, we apply the second law of exponents to the powers of each variable.

Example 2   Divide $\dfrac{3x^2y^4}{6x^3y}$

Solution We consider the numerical coefficients and the powers of each base separately. We use the second law of exponents to simplify each quotient of powers.

$$\frac{3x^2y^4}{6x^3y} = \frac{3}{6} \cdot \frac{x^2}{x^3} \cdot \frac{y^4}{y}$$

Subtract exponents on each base.

$$= \frac{1}{2} \cdot \frac{1}{x^{3-2}} \cdot y^{4-1}$$

$$= \frac{1}{2} \cdot \frac{1}{x} \cdot \frac{y^3}{1} = \frac{y^3}{2x}$$

Multiply factors.

Greatest Common Factors

Now we consider several techniques for factoring polynomials. The first of these is factoring out the **greatest common factor** (GCF).

The **greatest common factor** (GCF) is the largest factor that divides evenly into each term of the polynomial: the largest numerical factor and the highest power of each variable.

Example 3   Find the greatest common factor for $4a^3b^2 + 6ab^3 - 18a^2b^4$

Solution The largest integer that divides evenly into all three coefficients is 2. The highest power of a is a^1, and the highest power of b is b^2. Thus, the GCF is $2ab^2$. Note that the exponent on each variable of $2ab^2$ is the *smallest* exponent that appears on that variable among the terms of the polynomial.

Once we have found the greatest common factor for the polynomial, we can write each term as a product of the GCF and another factor. For example, the GCF of $8x^2 - 6x$ is $2x$, and we can write

$$8x^2 - 6x = \mathbf{2x} \cdot 4x - \mathbf{2x} \cdot 3$$

We can then use the distributive law to write the expression on the right side as a product:

$$\mathbf{2x} \cdot 4x - \mathbf{2x} \cdot 3 = \mathbf{2x}(4x - 3)$$

We say that we have factored out the greatest common factor from the polynomial.

For more complicated polynomials, we can divide the GCF into each term to find the remaining factors.

Example 4   Factor $4a^3b^2 + 6ab^3 - 18a^2b^4$

Solution The GCF for this polynomial is $2ab^2$, as we saw in Example 3. We factor out the GCF from each term and write the polynomial as a product,

$$2ab^2(\qquad\qquad)$$

To find the factor inside parentheses, we divide each term of the polynomial by the GCF.

$$\frac{4a^3b^2}{2ab^2} = \boldsymbol{2a^2}, \quad \frac{6ab^3}{2ab^2} = \boldsymbol{3b}, \quad \frac{-18a^2b^4}{2ab^2} = \boldsymbol{-9ab^2}$$

We apply the distributive law to factor $2ab^2$ from each term.

$$4a^3b^2 + 6ab^3 - 18a^2b^4 = 2ab^2 \cdot \boldsymbol{2a^2} + 2ab^2 \cdot \boldsymbol{3b} - 2ab^2 \cdot \boldsymbol{9ab^2}$$
$$= 2ab^2(2a^2 + 3b - 9ab^2)$$

■

Look Closer: Sometimes the greatest common factor for a polynomial is not a monomial, but may instead have two or more terms.

Example 5   Factor $x^2(2x + 1) - 3(2x + 1)$

Solution The given expression has two terms, and $(2x + 1)$ is a factor of each. We factor out the entire binomial $(2x + 1)$.

$$x^2(2x + 1) - 3(2x + 1) = (2x + 1)(\qquad\qquad)$$

To find the factor inside the parentheses, divide $(2x + 1)$ into each term of the given expression.

$$\frac{x^2(2x + 1)}{(2x + 1)} = x^2, \quad \frac{-3(2x + 1)}{(2x + 1)} = -3$$

Thus,

$$x^2(2x + 1) - 3(2x + 1) = (2x + 1)(x^2 - 3)$$

■

Factoring Quadratic Trinomials by Guess-and-Check

So far, we can factor quadratic trinomials of the form $x^2 + bx + c = 0$, where $a = 1$. What if the coefficient of x^2 is not 1? Sometimes, if the coefficients are small, we can factor a quadratic expression by the **guess-and-check** method.

Example 6   Factor $2x^2 - 7x + 3$ into a product of two binomials.

Solution We begin by factoring the quadratic term, $2x^2$, which can only be factored as x times $2x$, so we can fill in the First terms in each binomial.

$$2x^2 - 7x + 3 = (2x \underline{\qquad})(x \underline{\qquad})$$

Next, we factor the constant term, 3, which can only be factored as 3 times 1. Because the linear term is negative, we make both factors negative: $3 = -3(-1)$. Finally, we have to decide which number appears as the Last term in each binomial. We check **O + I** for each possibility.

$$(2x - 3)(x - 1) \quad O + I = -2x - 3x = -5x$$
$$(2x - 1)(x - 3) \quad \boldsymbol{O + I = -6x - x = -7x}$$

The second choice gives the correct middle term, so the factorization is

$$2x^2 - 7x + 3 = (2x - 1)(x - 3)$$

■

> **RQ3.**   When we factor a quadratic trinomial by the guess-and-check method, how do we check for the correct middle term?

Remember that you can always check your factorization by multiplying the factors together.

Quadratic Trinomials in Two Variables

So far, we have factored quadratic trinomials in one variable, that is, polynomials of the form

$$ax^2 + bx + c$$

The method we learned can also be used to factor trinomials in two variables of the form

$$ax^2 + bxy + cy^2$$

In this expression, the first and last terms are quadratic terms, while the middle term is a cross-term consisting of the product of the two variables.

■ **Example 7**   Factor $x^2 + 5xy + 6y^2$

Solution   As usual, we begin by factoring the first term into x times x.

$$x^2 + 5xy + 6y^2 = (x + \underline{\quad})(x + \underline{\quad})$$

Next we look for factors of the last term, $6y^2$. In order to obtain the xy-term in the middle, we need a y in each factor. Thus the possibilities are

$$y \text{ and } 6y \quad \text{ or } \quad 2y \text{ snd } 3y$$

We'll check the sum $O + I$ for each possibility.

$$(x + y)(x + 6y) \quad O + I = 6xy + xy = 7xy$$
$$(x + 2y)(x + 3y) \quad O + I = 3xy + 2xy = \boldsymbol{5xy}$$

The second possibility gives the correct middle term, so the factorization is

$$x^2 + 5xy + 6y^2 = (x + 2y)(x + 3y)$$

■

Combining Factoring Techniques

We should always begin factoring by checking to see if there is a common factor that can be factored out.

■ **Example 8**   Factor completely: $2b^3 + 8b^2 + 6b$

Solution   We begin by factoring out the greatest common factor, $2b$.

$$2b^3 + 8b^2 + 6b = 2b(b^2 + 4b + 3)$$

The remaining factor, $b^2 + 4b + 3$, is a quadratic trinomial that can be factored. We look for two numbers p and q so that $pq = 3$ and $p + q = 4$. You can check that $p = 3$ and $q = 1$ will work. Thus, $b^2 + 4b + 3 = (b + 3)(b + 1)$, and

$$2b^3 + 8b^2 + 6b = 2b(b + 3)(b + 1)$$

RQ4. What should always be the first step in factoring a polynomial?

Optional Extension: A Property of Binomial Products

If the coefficients in a quadratic trinomial $ax^2 + bx + c$ are not prime numbers, the guess-and-check method may be time-consuming. In that case, we can use another technique that depends upon the following property of binomial products.

Example 9a.   Compute the product $(3t + 2)(t + 3)$ using the area of a rectangle.
b. Verify that the products of the diagonal entries are equal.

Solutions a.   We construct a rectangle with sides $3t + 2$ and $t + 3$, as shown below. We see that the product of the two binomials is

$$3t^2 + 9t + 2t + 6 = 3t^2 + 11t + 6$$

b. Now let's compute the product of the expressions along each diagonal of the rectangle:

$$3t^2 \cdot 6 = 18t^2 \quad \text{and} \quad 9t \cdot 2t = 18t^2$$

	t	3
3t	$3t^2$	9t
2	2t	6

The two products are equal. This is not surprising when you think about it, because each diagonal product is the product of *all four* terms of the binomials, namely $3t$, 2, t, and 3, just multiplied in a different order. You can see where the diagonal entries came from in our example:

$$18t^2 = 3t^2 \cdot 6 = 3t \cdot t \cdot 2 \cdot 3$$
$$18t^2 = 9t \cdot 2t = 3t \cdot 3 \cdot 2 \cdot t$$

Product of Binomials
   When we represent the product of two binomials by the area of a rectangle, **the products of the entries on the two diagonals are equal.**

RQ5.   In Example 9, why are the products on the two diagonals equal?

Look Ahead: We can use rectangles to help us factor quadratic trinomials. Recall that factoring is the opposite or reverse of multiplying, so we must first understand how multiplication works.

Look Closer: Look carefully at the rectangle for the product

$$(3x+4)(x+2) = 3x^2 + 10x + 8$$

shown at right.

- The quadratic term of the product, $3x^2$, appears in the upper left sub-rectangle.
- The constant term of the product, 8, appears in the lower right sub-rectangle.
- The linear term, $10x$, is the sum of the other two sub-rectangles.

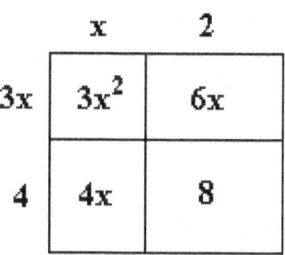

Factoring Quadratic Trinomials by the Box Method

Now we'll factor the trinomial

$$3x^2 + 10x + 8$$

We'll try to reverse the steps for multiplication. Instead of starting with the factors on the outside of the rectangle, we begin by filling in the areas of the sub-rectangles.

Step 1   The quadratic term, $3x^2$, goes in the upper left, and the constant term, 8, goes in the lower right, as shown in the figure.

What about the other two entries? We know that their sum must be $10x$, but we don't know what expressions go in each! This is where we use our observation that the products on the two diagonals are equal.

Step 2   We compute the product of the entries on the first diagonal:

$$D = 3x^2 \cdot 8 = 24x^2$$

The product of the entries on the other diagonal must also be $24x^2$. We now know two things about those entries:

1. Their product is $24x^2$ and
2. Their sum is $10x$

Step 3   To find the two unknown entries, we list all the ways to factor $D = 24x^2$, then choose the factors whose sum is $10x$.

Factors of $D = 24x^2$		Sum of Factors
x	$24x$	$x + 24x = 25x$
$2x$	$12x$	$2x + 12x = 14x$
$3x$	$8x$	$3x + 8x = 11x$
$4x$	$6x$	$4x + 6x = 10x$

Step 4   We see that the last pair of factors, $4x$ and $6x$, has a sum of $10x$. We enter these factors in the remaining sub-rectangles. (It doesn't matter which one goes in which spot.) We now have all the sub-rectangles filled in, as shown at right.

Finally, we work backwards to discover what length and width produce the areas of the four subrectangles.

Step 5   We factor each row of the rectangle, and write the factors on the outside. Start with the top row, factoring out x and writing the result, $3x + 4$, at the top, as shown at right. We get the same result when we factor 2 from the bottom row. The final rectangle is shown at right, and the factors of $3x^2 + 10x + 8$ appear as the length and width of the rectangle.
Our factorization is thus

$$3x^2 + 10x + 8 = (x + 2)(3x + 4)$$

	$3x$	4
x	$3x^2$	$4x$
2	$6x$	8

RQ6.   Which terms of the quadratic trinomial go into the upper right and lower left sub-rectangles of the box?

RQ7.   Why do we list the possible factors of D?

■ **Example 10**   Factor $2x^2 - 11x + 15$.

Solution Step 1   Enter $2x^2$ and 15 on the diagonal of the rectangle, as shown in the figure.

Step 2   Compute the diagonal product:

$$D = 2x^2 \cdot 15 = 30x^2$$

$2x^2$	
	15

Step 3   List all possible factors of D, and compute the sum of each pair of factors. (Note that both factors must be negative.)

Factors of $D = 30x^2$		Sum of Factors
$-x$	$-30x$	$-x - 30x = -31x$
$-2x$	$-15x$	$-2x - 15x = -17x$
$-3x$	$-10x$	$-3x - 10x = -13x$
$-5x$	$-6x$	$-5x - 6x = -11x$

The correct factors are $-5x$ and $-6x$.

Step 4   Enter the factors $-5x$ and $-6x$ into the rectangle.

Step 5   Factor $2x$ from the top row of the rectangle, and write the result, $x - 3$, at the top, as shown below.

	$2x^2$	$-6x$
	$-5x$	15

	x	-3
$2x$	$2x^2$	$-6x$
	$-5x$	15

	x	-3
$2x$	$2x^2$	$-6x$
-5	$-5x$	15

Finally, factor $x - 3$ from the bottom row, and write the result, -5, on the left. The correct factorization is

$$2x^2 - 11x + 15 = (2x - 5)(x - 3)$$
■

RQ8.   What do we do after we have filled in all the sub-rectangles of the box?

RQ9.   Where do the factors of the quadratic trinomial appear?

Here is a summary of our factoring method.

To Factor $ax^2 + bx + c$ **Using the Box Method :**
1. Write the quadratic term ax^2 in the upper left sub-rectangle, and the constant term c in the lower right.
2. Multiply these two terms to find the diagonal product, D.
3. List all possible factors px and qx of D, and choose the pair whose sum is the linear term, bx, of the quadratic trinomial.
4. Write the factors px and qx in the remaining sub-rectangles.
5. Factor each row of the rectangle, writing the factors on the outside. These are the factors of the quadratic trinomial.

Skills Warm-Up

Use the given areas to find the length and width of each rectangle.

1.

$6x^2$	$9x$
$10x$	15

2.

$8t^2$	$-14t$
$-12t$	21

3.

$12m^2$	$-10m$
$30m$	-25

4.

$9a^2$	$21a$
$-21a$	-49

Answers: **1.** $3x + 5,\ 2x + 3$ **2.** $2t - 3,\ 4t - 7$ **3.** $2m + 5,\ 6m - 5$ **4.** $3a - 7,\ 3a + 7$

Homework 7.3

Skills Practice

1. Find each quotient by using the second law of exponents.

   a. $\dfrac{a^6}{a^3}$
   b. $\dfrac{3^9}{3^4}$
   c. $\dfrac{z^6}{z^9}$

2. Choose a value for the variable and evaluate to show that the following pairs of expressions are **not** equivalent.

   a. $t^2 \cdot t^3, \ t^6$
   b. $\dfrac{v^8}{v^2}, \ v^4$
   c. $\dfrac{n^3}{n^5}, \ n^2$

■ For Problems 3-5, divide.

3. $\dfrac{2x^3y}{8x^4y^5}$
4. $\dfrac{-12bx^4}{8bx^2}$
5. $\dfrac{-15x^3y^2}{-3x^3y^4}$

■ For Problems 6-7, factor out a negative monomial.

6. $-b^2 - bc - ab$
7. $-4k^4 + 4k^2 - 2k$

■ For Problems 8-11, factor out the greatest common factor.

8. $2x^4 - 4x^2 + 8$
9. $16a^3b^3 - 12a^2b + 8ab^2$
10. $9x^2 - 12x^5 + 3x^3$
11. $14x^3y - 35x^2y^2 + 21xy^3$

■ For Problems 12-13, factor out the common factor.

12. $2x(x+6) - 3(x+6)$
13. $3x^2(2x+3) - (2x+3)$

■ For Problems 14-16, factor by the guess-and-check method.

14. $2x^2 + 11x + 5$
15. $5t^2 + 7t + 2$
16. $3x^2 - 8x + 5$

■ For Problems 17-28, factor completely.

17. $2x^2 + 10x + 12$
18. $4a^2b + 12ab - 7b$
19. $4z^3 + 10z^2 + 6z$
20. $18a^2b - 9ab - 27b$
21. $x^2 - 5xy + 6y^2$
22. $x^2 + 4xy + 4y^2$
23. $x^2 + 4ax - 77a^2$
24. $4x^3 + 12x^2y + 8xy^2$
25. $9a^3b + 9a^2b^2 - 18ab^3$
26. $2t^2 - 5st - 3s^2$
27. $4b^2y^2 + 5by + 1$
28. $12ab^2 + 15a^2b + 3a^3$

■ Use the box method to factor the quadratic trinomials in Problems 29-34.

29. $2x^2 - 13x + 18$
30. $5x^2 + 16x - 16$
31. $6h^2 + 7h + 2$
32. $9n^2 - 8n - 1$
33. $6t^2 - 5t - 25$
34. $5x^2 - 14x - 24$

Applications

■   For Problems 35-40, solve the equation by factoring.

35. $3n^2 - n = 4$ **36.** $11t = 6t^2 + 3$ **37.** $1 = 4y - 4y^2$
38. $12z^2 + 26z = 10$ **39.** $y(3y + 4) = 4$ **40.** $(2x - 1)(x - 2) = -1$

41. The cost C of producing a wool rug depends on the number of hours t it takes to weave it, where

$$C = 3t^2 - 4t + 100$$

How many hours did it take to weave a rug that costs $120?

42. Steve's boat locker is 2 feet longer than twice its width. Find the dimensions of the locker if the 13-foot mast of Steve's boat will just fit diagonally across the floor of the locker.

■   Mental exercise: Find the other factor of each quadratic trinomial without using pencil, paper, or calculator.

43. $b^2 + 8b - 240 = (b - 12)(\_\_\_\_\_)$ **44.** $n^2 - 97n - 300 = (n - 100)(\_\_\_\_\_)$
45. $3u^2 - 17u - 6 = (u - 6)(\_\_\_\_\_)$ **46.** $2t^2 - 21t + 54 = (t - 6)(\_\_\_\_\_)$

7.4 Special Products and Factors

A few special binomial products occur so frequently that it is useful to recognize their forms. This will enable you to write their factored forms directly, without trial and error. To prepare for these special products, we first consider the squares of monomials.

Squares of Monomials

Study the squares of monomials in Example 1. Do you see a quick way to find the product?

Example 1a. $(w^5)^2 = w^5 \cdot w^5 = w^{10}$

b. $(4x^3)^2 = 4x^3 \cdot 4x^3 = 4 \cdot 4 \cdot x^3 \cdot x^3 = 16x^6$

Look Closer: In Example 1a, we doubled the exponent and kept the same base. In Example 1b, we squared the numerical coefficient and doubled the exponent.   •

RQ1.   Why do we double the exponent when we square a power?

Example 2   Find a monomial whose square is $36t^8$.

Solution When we square a power, we double the exponent, so t^8 is the square of t^4. Because 36 is the square of 6, the monomial we want is $6t^4$. To check our result, we square $6t^4$ to see that $(6t^4)^2 = 36t^8$.

Squares of Binomials

You can use the distributive law to verify each of the following special products.

$$(a+b)^2 = (a+b)(a+b)$$
$$= a^2 + ab + ab + b^2$$
$$= a^2 + 2ab + b^2$$

$$(a-b)^2 = (a-b)(a-b)$$
$$= a^2 - ab - ab + b^2$$
$$= a^2 - 2ab + b^2$$

Squares of Binomials
1. $(a+b)^2 = a^2 + 2ab + b^2$
2. $(a-b)^2 = a^2 - 2ab + b^2$

RQ2.   Explain why it is NOT true that $(a+b)^2 = a^2 + b^2$.

We can use these results as formulas to compute the square of any binomial.

Example 3   Expand $(2x+3)^2$ as a polynomial.

Solution The formula for the square of a sum says to square the first term, add twice the product of the two terms, then add the square of the second term. We replace a by $2x^3$ and b by $3y$ in the formula.

$$(a+b)^2 = a^2 + 2ab + b^2$$
$$(2x+3)^2 = (2x)^2 + 2(2x)(3) + (3)^2$$

square of   twice their   square of
first term   product   second term

$$= 4x^2 + 12x + 9$$

Of course, you can verify that you will get the same answer for Example 3 if you compute the square by multiplying $(2x+3)(2x+3)$.

Caution!   We cannot square a binomial by squaring each term separately! For example, it is **not** true that

$$(2x+3)^2 = 4x^2 + 9 \qquad \leftarrow \textbf{Incorrect!}$$

We must use the distributive law to multiply the binomial times itself.

RQ3. How do we compute $(a+b)^2$?

Difference of Two Squares

Now consider the product

$$(a+b)(a-b) = a^2 - ab + ab - b^2$$
$$= a^2 - b^2$$

In this product, the two middle terms cancel each other, and we are left with a difference of two squares.

Difference of Two Squares

$$(a+b)(a-b) = a^2 - b^2$$

Example 4   Multiply $(2y+9w)(2y-9w)$

Solution The product has the form $(a+b)(a-b)$, with a replaced by $2y$ and b replaced by $9w$. We use the difference of squares formula to write the product as a polynomial.

$$(a+b)(a-b) = a^2 - b^2$$
$$(2y+9w)(2y-9w) = (2y)^2 - (9w)^2$$

square of   square of
first term   second term

$$= 4y^2 - 81w^2$$

RQ4. Explain the difference between $(a-b)^2$ and $a^2 - b^2$.

Factoring Special Products

The three special products we have just studied are useful as patterns for factoring certain polynomials. For factoring, we view the formulas from right to left.

> **Special Factorizations**
> 1. $a^2 + 2ab + b^2 = (a+b)^2$
> 2. $a^2 - 2ab + b^2 = (a-b)^2$
> 3. $a^2 - b^2 = (a+b)(a-b)$

If we recognize one of the special forms, we can use the formula to factor it. Notice that all three special products involve two squared terms, a^2 and b^2, so we first look for two squared terms in our trinomial.

■ **Example 5**   Factor $x^2 + 24x + 144$

Solution This trinomial has two squared terms, x^2 and 144. These terms are a^2 and b^2, so $a = x$ and $b = 12$. We check whether the middle term is equal to $2ab$.

$$2ab = 2(x)(12) = 24x$$

This is the correct middle term, so our trinomial has the form (1), with $a = x$ and $b = 12$. Thus,

$$a^2 + 2ab + b^2 = (a+b)^2 \qquad \textbf{Replace } a \textbf{ by } x \textbf{ and } b \textbf{ by 12.}$$
$$x^2 + 24x + 144 = (x+12)^2 \qquad ■$$

RQ5.   How can we factor $a^2 - 2ab + b^2$?
RQ6.   How can we factor $a^2 - b^2$?

Caution!   The sum of two squares, $a^2 + b^2$, cannot be factored! For example,

$$x^2 + 16, \qquad 9x^2 + 4y^2, \quad \text{and} \qquad 25y^4 + w^4$$

cannot be factored. You can check, for instance, that $x^2 + 16 \neq (x+4)(x+4)$.   ■

> **Sum of Two Squares**
> The sum of two squares, $a^2 + b^2$, cannot be factored.

As always when factoring, we should check first for common factors.

■ **Example 6**   Factor completely $98 - 28x^4 + 2x^8$

Solution Each term has a factor of 2, so we begin by factoring out 2.

$$98 - 28x^4 + 2x^8 = 2(49 - 14x^4 + x^8)$$

The polynomial in parentheses has the form $(a-b)^2$, with $a = 7$ and $b = x^4$. The middle term is $-2ab = -2(7)(x^4)$. We use equation (2) to write

$$a^2 - 2ab + b^2 = (a-b)^2 \qquad \textbf{Replace } a \textbf{ by 7 and } b \textbf{ by } x^4.$$
$$49 - 14x^4 + x^8 = (7 - x^4)^2$$

Thus, $98 - 28x^4 + 2x^8 = 2(7 - x^4)^2$.

RQ7. What expression involving squares cannot be factored?

Skills Warm-Up

Express each product as a polynomial.

1. $(z-3)^2$ **2.** $(x+4)^2$ **3.** $(3a+5)^2$
4. $(2b-7)^2$ **5.** $(2n-5)(2n+5)$ **6.** $(4m+9)(4m-9)$

Answers: **1.** $z^2 - 6z + 9$ **2.** $x^2 + 8x + 16$ **3.** $9a^2 + 30a + 25$ **4.** $4b^2 - 28b + 49$
5. $4n^2 - 25$ **6.** $16m^2 - 81$

Homework 7.4

Skills Practice

1. Square each monomial.

    a. $(8t^4)^2$ **b.** $(-12a^2)^2$ **c.** $(10h^2k)^2$

2. Find a monomial whose square is given.

    a. $16b^{16}$ **b.** $121z^{22}$ **c.** $36p^6q^{24}$

■ For Problems 3-4, write the area of the square in two different ways:
    a. as the sum of four smaller areas,
    b. as one large square, using the formula Area $= (\text{length})^2$.

3. 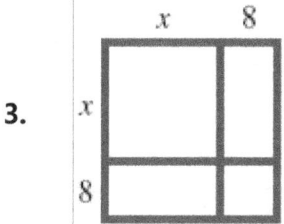   **4.**

■ For Problems 5-7, compute the product.

5. $(n+1)(n+1)$ **6.** $(m-9)(m-9)$ **7.** $(2b+5c)(2b+5c)$

■ For Problems 8-13, use the formulas for squares of binomials to expand the square.

8. $(x+1)^2$ **9.** $(2x-3)^2$ **10.** $(5x+2y)^2$
11. $(3a+b)^2$ **12.** $(7b^3-6)^2$ **13.** $(2h+5k^4)^2$

■ For Problems 14-19, use the formula for difference of two squares to multiply.

14. $(x-4)(x+4)$ **15.** $(x+5z)(x-5z)$ **16.** $(2x-3)(2x+3)$
17. $(3p-4)(3p+4)$ **18.** $(2x^2-1)(2x^2+1)$ **19.** $(h^2+7t)(h^2-7t)$

■ For Problems 20-22, multiply. Write your answer as a polynomial.

20. $-2a(3a-5)^2$ **21.** $4x^2(2x+6y)^2$ **22.** $5mp^2(2m^2-p)(2m^2+p)$

■ For Problems 23-25, factor the squares of binomials.

23. y^2+6y+9 **24.** $m^2-30m+225$ **25.** $a^6-4a^3b+4b^2$

■ For Problems 26-31, factor.

26. z^2-64 **27.** $1-g^2$ **28.** $-225+a^2$
29. x^2-9 **30.** $36-a^2b^2$ **31.** $64y^2-49x^2$

For Problems 32-40, factor completely.

32. $a^4 + 10a^2 + 25$

33. $36y^8 - 49$

34. $16x^6 - 9y^4$

35. $3a^2 - 75$

36. $2a^3 - 12a^2 + 18a$

37. $9x^7 - 81x^3$

38. $12h^2 + 3k^6$

39. $81x^8 - y^4$

40. $162a^4b^8 - 2a^8$

Applications

41. Is $(x - 3)^2$ equivalent to $x^2 - 3^2$? Explain why or why not, and give a numerical example to justify your answer.

42. Use areas to explain why the figure illustrates the product $(a + b)^2 = a^2 + 2ab + b^2$.

43. Use evaluation to decide whether the two expressions $(a + b)^2$ and $a^2 + b^2$ are equivalent.

a	b	$a + b$	$(a + b)^2$	a^2	b^2	$a^2 + b^2$
2	3					
-2	-3					
2	-3					

44. Explain why you can factor $x^2 - 4$, but you cannot factor $x^2 + 4$.

45. a. Expand $(a - b)^3$ by multiplying.
  b. Use your formula from part (a) to expand $(2x - 3)^3$.
  c. Substitute $a = 5$ and $b = 2$ to show that $(a - b)^3$ is not equivalent to $a^3 - b^3$.

46. a. Multiply $(a + b)(a^2 - ab + b^2)$.
  b. Factor $a^3 + b^3$.
  c. Factor $x^3 + 8$.

Chapter 7 Summary and Review

Section 7.1 Polynomials

*   A **polynomial** is a sum of terms, each of which is a power of a variable with a constant coefficient and a whole number exponent.
*   The **degree** of a polynomial in one variable is the largest exponent that appears in any term.
*   **Like terms** are any terms that are exactly alike in their variable factors. The exponents on the variable factors must also match.
*   To add two polynomials, we need only remove parentheses and combine like terms.
*   To subtract polynomials, we change the sign of each term within parentheses, remove the parentheses, and combine like terms.

$$\boxed{\textbf{Profit} = \textbf{Revenue} - \textbf{Cost}}$$

Section 7.2 Products of Polynomials

> **First Law of Exponents**
> To multiply two powers with the same base, we add the exponents and leave the base unchanged. In symbols,
> $$a^m \cdot a^m = a^{m+n}$$

*   We use the distributive law to compute products of two or more polynomials.

Section 7.3 More About Factoring

> **Second Law of Exponents**
> To divide two powers with the same base, we subtract the smaller exponent from the larger one, and keep the same base.
> 1. If the larger exponent occurs in the numerator, put the power in the numerator.
> 2. If the larger exponent occurs in the denominator, put the power in the denominator.
> In symbols,
> $$1. \quad \frac{a^m}{a^n} = a^{m-n} \qquad \text{if} \quad n < m$$
> $$2. \quad \frac{a^m}{a^n} = \frac{1}{a^{n-m}} \qquad \text{if} \quad n > m$$

*   The **greatest common factor** (GCF) is the largest factor that divides evenly into each term of the polynomial: the largest numerical factor and the highest power of each variable.
*   When we represent the product of two binomials by the area of a rectangle, **the products of the entries on the two diagonals are equal.**

> **To Factor $ax^2 + bx + c$ Using the Box Method :**
> 1. Write the quadratic term ax^2 in the upper left sub-rectangle, and the constant term c in the lower right.
> 2. Multiply these two terms to find the diagonal product, D.
> 3. List all possible factors px and qx of D, and choose the pair whose sum is the linear term, bx, of the quadratic trinomial.
> 4. Write the factors px and qx in the remaining sub-rectangles.
> 5. Factor each row of the rectangle, writing the factors on the outside. These are the factors of the quadratic trinomial.

•   We should always begin factoring by checking to see if there is a common factor that can be factored out.

Section 7.4   Special Products and Factors

> **Square of a Binomial**
> 1. $(a+b)^2 = a^2 + 2ab + b^2$
> 2. $(a-b)^2 = a^2 - 2ab + b^2$
>
> **Difference of Squares**
> 3. $(a+b)(a-b) = a^2 - b^2$

> **Special Factorizations**
> 1. $a^2 + 2ab + b^2 = (a+b)^2$
> 2. $a^2 - 2ab + b^2 = (a-b)^2$
> 3. $a^2 - b^2 = (a+b)(a-b)$

> **Sum of Two Squares**
> The sum of two squares, $a^2 + b^2$, cannot be factored.

Review Questions

■ Use complete sentences to answer the questions in Problems 1-10.

1. Explain why $\frac{x}{2} + 3$ is a polynomial but $\frac{2}{x} + 3$ is not a polynomial.
2. A classmate says that $\sqrt{2}x^2 + 3\sqrt{2}x + 1$ is a polynomial, but another classmate disagrees. Who is correct? Explain.
3. If a polynomial is written in descending powers of the variable and the first term has degree 5, what is the degree of the polynomial? Give an example.
4. If a polynomial of degree 3 is added to a polynomial of degree 2, what is the degree of the sum? Give an example.
5. A classmate tells you that, depending on the binomials used, the sum of two binomials can have one, two, three, or four terms. Give an example of each.
6. If a polynomial of degree 3 is multiplied by a polynomial of degree 2, what is the degree of the product? Give an example.
7. How many terms are there in the square of a monomial? Of a binomial? Give examples.
8. Which of the following can be factored? Give examples.
   a. sum of two squares   b. difference of two squares

9. Explain the difference between the sum of two squares and the square of a binomial. Give examples.

10. When you are trying to factor a polynomial, what is the best way to start?

Review Problems

1. Write the polynomial in descending powers of the variable and state the degree of the polynomial.

a. $1 - \dfrac{x^2}{2} + \dfrac{x^4}{24} - \dfrac{x^6}{720}$

b. $10n^4 + 10^6 n + 10^8 n^2$

2. Which of the following are polynomials?

a. $-25y^{10} - 1$

b. $\sqrt{3}\,x + 2$

c. $\dfrac{4}{x^3} - \dfrac{3}{x^2} + \dfrac{1}{x}$

d. $t^4 + 5t^3 - 2\sqrt{t} + 3$

Evaluate each polynomial for the given values of the variables.

3. $-16t^2 + 50t + 5$ for $t = \dfrac{1}{2}$

4. $\dfrac{1}{2}n^2 + \dfrac{1}{2}n$ for $n = 100$

5. $\dfrac{1}{6}z^3 + \dfrac{1}{2}z^2 + z + 1$ for $z = -1$

6. $p^4 + 4p^3 + 6p^2 + 4p + 1$ for $p = -2$

7. $4x^2 - 12xy + 9y^2$ for $x = -3,\ y = -2$

8. $2R^4 S$ for $R = 150,\ S = 0.01$

9. Write as a polynomial: $x(3x[2x(x-2) + 1] - 4)$

10. Write as a polynomial:

$$a^2(a - 3) - 2a(a^2 + 2a^3) + (a - 2)(a - 3)$$

11. Suppose you want to choose four items from a list of n possible items. The number of different ways you can make your choice is given by the polynomial

$$\frac{1}{24}n^4 - \frac{1}{4}n^3 + \frac{11}{24}n^2 - \frac{1}{4}n$$

a. How many different sets of four compact disks can be chosen from a collection with 20 compact disks?

b. Of course you cannot choose four different items from a list of only 3 possible items. What do you get when you evaluate the polynomial for $n = 3$? $n = 2$? $n = 1$?

c. Evaluate the polynomial for $n = 4$. Explain why your answer makes sense in term of what the polynomial represents.

12. The sum of the cubes of the first n counting numbers is given by $\dfrac{n^4}{4} + \dfrac{n^3}{2} + \dfrac{n^2}{4}$. Find the sum of the cubes of the first five counting numbers.

Simplify by combining like terms.

13. $8p^2 q - 3pq^2 - 2pq^2 + p^2 q$

14. $1.7m^3 + 2.6 - 0.3m - 1.4m^2 - 1.2m^3 + 4.5m^2 + 1.1$

■ Add or subtract the polynomials.

15. $(4b^3 + 2b^2 - 3b + 7) - (-2b + 3 - b^3 - 7b)$
16. $(8w^6 - 5w^4 - 3w^2) + (2w^4 - 8w^2 + 4)$

■ Simplify if possible.

17. a. $3x^2 + x^2$
   b. $3x^2 \cdot x^2$

18. a. $5b^3 - b^3$
   b. $5b^3(-b^3)$

19. a. $7a^4 + a^6$
   b. $7a^4 \cdot a^6$

20. a. $2r^4 - r^3$
   b. $2r^4(-r^3)$

21. a. $\dfrac{3b^9}{9b^3}$
   b. $\dfrac{9b^3}{3b^9}$

22. a. $\dfrac{4m^8}{m^8}$
   b. $\dfrac{m^4}{8m^4}$

■ Multiply.

23. $(5m^2n)(-6m^3n^2)$

24. $-7qr^2(2q^4 - 1)$

■ Divide.

25. $\dfrac{-21x^4y^3}{3xy^3}$

26. $\dfrac{3a^2b}{6a^4b^3}$

■ Multiply.

27. $-2y(y - 4)(y + 3)$
28. $6p^4(2p + 1)(p - 4)$
29. $(d - 2)(d^2 - 4d + 4)$
30. $(2k + 1)(k - 2)(k + 3)$
31. $9x^3y(4x - 3y)^2$
32. $(a - 3)^3$

■ Add, subtract, multiply, or divide the polynomials as indicated.

33. $(3a^2 - 4a - 7) - (a^2 - 5a + 2)$
34. $(5ab^2 + 6a^2b) - (3ab^2 + 5ab)$
35. $\dfrac{13c^2d}{26c^3d}$
36. $\dfrac{12m^4 + 4m}{4m}$
37. $7q^2(8 - 7q^2 - q^4)$
38. $2k^2(-3km)(m^3k)$
39. $(9v + 5w)(9v - 5w)$
40. $(x^2 + 1)^2$
41. $-3p^2(p + 2)(p - 5)$
42. $12rs^2(3r - s)(r + 4s)$
43. $(2x - 3)(4x^2 + 6x + 9)$
44. $(3x + 2)(3x + 2)(3x + 2)$

■ For each rectangle, find
  a. its perimeter,   **b.** its area.

45.
$3a^4$ / $2a^4$

46.
$7xy$ / $5xy$

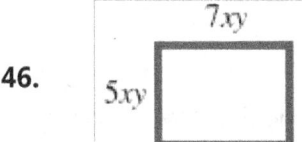

47.
$m + 2n$ / $3m - n$

48.
$4w + 9$ / $4w - 9$

49. Evaluate the expression below for $a = -6.3$, $b = -4.8$, $c = 5.2$.

$$(4a - 3b - 2c) - (a + 6b - 5c) + (3a + 9b - 2c)$$

50. Write a polynomial for the volume of a box whose width is 10 centimeters less than its length, and 2 centimeters more than its height. Let w represent the width of the box.

51. Nova cosmetics sells $140 - 2p$ cans of styling mousse each month if they charge p dollars per can.
   a. Write a polynomial for the company's monthly revenue from mousse.
   b. Find the revenue if each can costs \$4.

52. Newsday magazine surveyed 400 people on the question "Do you think the government is spending too much on defense?" They reported the following results: Of the college-educated respondents, 72% answered yes, and 48% of those without a college education answered yes. Suppose you would like to know how many of the 400 people surveyed answered yes. You will need to know how many of the 400 have a college education. Let x represent this unknown value. Write and simplify expressions in terms of x for each of the following.
   a. How many of the people surveyed do not have a college education?
   b. How many of the college-educated respondents answered yes?
   c. How many of those without a college education answered yes?
   d. How many people total answered yes?

▪ Decide whether each of the following is an equation or a polynomial. If it is an equation, solve it. If it is a polynomial, factor it.

53. $2x^2 + x - 3 = 0$ **54.** $a^2 - 9$ **55.** $2x^2 + x - 3$ **56.** $a^2 = 9$
57. $2x^2 + x = 0$ **58.** $a - 9 = 0$ **59.** $2x + 3 = 0$ **60.** $a^2 = 9a$
61. $2x^3 - 2x$ **62.** $a^4 - 16$ **63.** $p^3 - p = 0$ **64.** $n(n-3)(n+3) = 0$

▪ Factor out the greatest common factor.

65. $12x^5 - 8x^4 + 20x^3$ **66.** $9a^4b^2 + 6a^3b^3 - 3a^2b^4$
67. $30w^9 - 42w^4 + 54w^8$ **68.** $45x^2y^2 + 18x^2y^3 - 27x^3y^3$

▪ Factor out a negative monomial.

69. $-10d^4 + 20d^3 - 5d^2$ **70.** $-6m^3n - 18m^2n + 6mn$ **71.** $-vw^5 - vw^4 + vw^2$

▪ Factor out the common binomial factor.

72. $7q(q-3) - (q-3)$ **73.** $-r^3(3r+2) + 4(3r+2)$ **74.** $5(x-2) - x^2(x-2)$

▪ Factor completely.

75. $3z^3 - 12z$ **76.** $-4x^3y + 8x^2y - 4xy$
77. $a^4 + 10a^2 + 25$ **78.** $4x^8 - 64$
79. $2a^2b^6 + 32a^6b^2$ **80.** $4p^2q^4 + 32p^3q^3 + 64p^4q^2$
81. $-2a^4b - 4a^3b^2 + 30ab^3$ **82.** $15r^3s^2 + 39r^2s^3 - 18rs^4$
83. $2q^4 + 6q^3 - 80q^2$ **84.** $32 - b^4 - 14b^2$
85. $x^2 - 3xy + 2y^2$ **86.** $y^2 - 3by - 28b^2$

87. $3a^2b + 12ab^2 + 9b^3$

88. $80t^4 - 28t^3 - 24t^2$

89. $9x^2y^2 + 3xy - 2$

90. $4a^3x - 2a^2x^2 - 12ax^3$

■ Factor if possible.

91. $h^2 - 24h + 144$

92. $9t^2 - 30tv + 25v^2$

93. $x^2 + 144$

94. $98n^4 - 8n^2$

95. $w^8 + 12w^4 + 36$

96. $q^6 - 14q^3 + 49$

97. $3s^4 - 48$

98. $2y^4 - 2$

Chapter 8 Algebraic Fractions

Lesson 1 Algebraic Fractions

- Evaluate algebraic fractions
- Reduce algebraic fractions

Lesson 2 Operations on Algebraic Fractions

- Multiply algebraic fractions
- Divide algebraic fractions
- Add or subtract like fractions

Lesson 3 Lowest Common Denominator

- Find an LCD
- Add or subtract unlike fractions

Lesson 4 Equations with Fractions

- Solve equations with fractions
- Identify extraneous solutions
- Solve motion problems
- Solve work problems

8.1 Algebraic Fractions

Variables may occur in the numerator or the denominator of a fraction, or both.

■ **Example 1** Choose variables and write a fraction for each situation.
a. The five people in Tom's study group order Chinese food and agree to split the bill equally. How much will Tom's share be?
b. Nurit has $800 to carpet a square bedroom in her house. What price per square foot can she afford?
c. How long does it take an ocean liner to make a trans-Atlantic voyage?

Solutions a. The amount of the bill is unknown. We let b represent the total bill. Then Tom's share is $\dfrac{b}{5}$.
b. The dimensions of the room are unknown. We let L represent the length of bedroom. Then the price per square foot is $\dfrac{800}{s^2}$.
c. The distance and the speed of the liner are unknown. We let D represent the distance traveled, and R the speed, or rate. Then the time for voyage is $\dfrac{D}{R}$. ■

An **algebraic fraction** (or rational expression, as they are sometimes called) is a fraction in which both numerator and denominator are polynomials.

Here are some examples:

$$\frac{3}{x}, \quad \frac{a^2+1}{a-2}, \quad \text{and} \quad \frac{z-1}{2z+3}$$

Look Closer: We can evaluate algebraic fractions just as we do any other algebraic expression. But we cannot evaluate a fraction at any value of the variable that make the denominator equal to zero.   •

■ **Example 2a.**   Evaluate $\dfrac{a^2+1}{a-2}$ for $a = 4$.

b. For what value of a is the fraction undefined?

Solutions a.   Substitute 4 for a in the fraction.

$$\begin{aligned}\frac{a^2+1}{a-2} &= \frac{4^2+1}{4-2} \qquad \text{Simplify numerator and denominator.}\\ &= \frac{17}{2}\end{aligned}$$

b. Because we cannot divide by zero, a fraction is undefined if its denominator is zero. This fraction is undefined for $a = 2$, because $\dfrac{2^2+1}{2-2} = \dfrac{5}{0}$. ■

RQ1.   What is an algebraic fraction? Give an example.
RQ2.   What values of the variable must we exclude when working with algebraic fractions?

Applications of Algebraic Fractions

Envirogreen Technology, Inc. decides to produce a water filter for home use. They spend $5000 for startup costs, and each filter costs $50 to produce. To help her decide on a selling price for the filters, the marketing manager would like to know the average cost per filter if they produce x filters. She first computes the total cost of producing x filters.

$$\text{Total Cost} = \text{Startup Cost} + \text{Cost of } x \text{ Filters}$$
$$= 5000 + 50x$$

Then, to find the average cost per filter, she divides the total cost by the number of filters produced.

$$\text{Average Cost} = \frac{\text{Total Cost}}{\text{Number of Filters}} = \frac{5000 + 50x}{x}$$

This expression is an algebraic fraction.

■ **Example 3**   Sketch a graph of the equation $A = \dfrac{5000 + 50x}{x}$

Solution We evaluate the average cost for various production levels, x, and make a table of values. We plot the points and connect them with a smooth curve to obtain the graph shown below.

x	A
50	150
100	100
200	75
400	62.50
500	60
1000	55
1250	54
2000	52.50

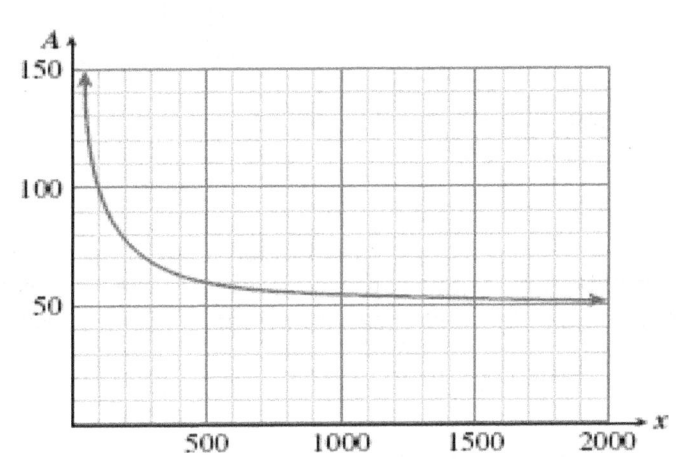

Reducing Fractions

We can **reduce** a fraction if we can divide both numerator and denominator by a common factor. In algebra, it is helpful to think of factoring out the common factor first. For example,

$$\frac{27}{36} = \frac{\cancel{9} \cdot 3}{\cancel{9} \cdot 4} = \frac{3}{4}$$

where we have divided both numerator and denominator by 9. The new fraction has the same value as the old one, namely 0.75, but it is simpler (the numbers are smaller.) Reducing is an application of the fundamental principle of fractions.

Fundamental Principle of Fractions
We can multiply or divide the numerator and denominator of a fraction by the same nonzero factor, and the new fraction will be equivalent to the old one.

$$\frac{a \cdot c}{b \cdot c} = \frac{a}{b} \quad \text{if} \quad b, c \neq 0$$

■ **Example 3**   Reduce the fraction $\dfrac{42ab^2}{35ab^3}$

Solution First we consider the numerical part of the fraction: look for the largest common factor of 42 and 35. This factor is 7, so we write 42 and 35 in factored form:

$$\frac{42ab^2}{35ab^3} = \frac{7 \cdot 6 \; ab^2}{7 \cdot 5 \; ab^3}$$

Next, we write the variable parts of the numerator and denominator in factored form. Remember that b^2 means $b \cdot b$ and b^3 means $b \cdot b \cdot b$.

$$\frac{42ab^2}{35ab^3} = \frac{7 \cdot 6 \; a \, b \, b}{7 \cdot 5 \; a \, b \, b \, b}$$

Finally, we divide any common factors from the numerator and denominator.

$$\frac{42ab^2}{35ab^3} = \frac{\cancel{7} \cdot 6 \; \cancel{a} \cancel{b} \cancel{b}}{\cancel{7} \cdot 5 \; \cancel{a} \cancel{b} \cancel{b} \, b}$$

The reduced fraction is $\dfrac{6}{5b}$.   ■

Look Closer: When we apply the fundamental principle, we often say that we are canceling common factors. In this context, "canceling" means dividing: Because division is the opposite operation for multiplication, we can cancel expressions that are multiplied together.   •

■ **Example 4**   Reduce the fraction $\dfrac{8x + 4}{4}$

Solution We cannot cancel the 4's, because 4 is added to $8x$, not multiplied. Instead, we write the numerator as a product instead of a sum. To do that, we factor the numerator.

$$\frac{8x + 4}{4} = \frac{4(2x + 1)}{4}$$

We see that 4 is a factor of the numerator, so we can divide top and bottom of the fraction by 4 to get

$$\frac{\cancel{4}(2x + 1)}{\cancel{4}} = \frac{1(2x + 1)}{1} = 2x + 1$$ ■

RQ3.   What does "canceling" mean when we reduce a fraction?

Caution!   We can cancel common **factors** (expressions that are multiplied together), but not common **terms** (expressions that are added or subtracted). Use your calculator to decide which calculation in each pair is correct. (In the second pair, choose a value for the variable and evaluate.)

1a. $\dfrac{12}{8} = \dfrac{4 \cdot 3}{4 \cdot 2} \rightarrow \dfrac{3}{2}$ **b.** $\dfrac{7}{6} = \dfrac{4+3}{4+2} \rightarrow \dfrac{3}{2}$

2a. $\dfrac{5x}{8x} \rightarrow \dfrac{5}{8}$ **b.** $\dfrac{x+5}{x+8} \rightarrow \dfrac{5}{8}$

Look Ahead: Now we can apply our method to algebraic fractions. We must be very careful to **cancel only common factors, never common terms**. For this reason, we will think of reducing fractions in two steps: first factor the numerator and denominator completely, then divide by any common factors.

> **To Reduce an Algebraic Fraction:**
> 1. Factor numerator and denominator completely.
> 2. Divide numerator and denominator by any common factors.

Example 5a.   Reduce $\dfrac{3x^2 + 12x}{6x + 24}$

b. Evaluate the fraction for $x = 4$.

Solutions a.   We first factor the numerator and denominator completely.

$$\frac{3x^2 + 12x}{6x + 24} = \frac{3x(x+4)}{2 \cdot 3(x+4)}$$

Then we divide numerator and denominator by any common factors.

$$\frac{3x(x+4)}{2 \cdot 3(x+4)} = \frac{\cancel{3}x(x\cancel{+4})}{2 \cdot \cancel{3}(x\cancel{+4})} = \frac{x}{2}$$

b. If we substitute $x = 4$ into the original fraction and simplify, we find

$$\frac{3(4)^2 + 12(4)}{6(4) + 24} = \frac{48 + 48}{24 + 24} = \frac{96}{48} = 2$$

However, the reduced fraction, $\dfrac{x}{2}$, is equivalent to the original one, so we should get the same answer if we substitute $x = 4$ into the reduced fraction:

$$\frac{x}{2} = \frac{4}{2} = 2$$

Caution!   In Example 4, it would be **incorrect** to "cancel" terms of the numerator and denominator separately, as shown below:

$$\frac{\overset{x}{\cancel{3x^2}} + \overset{x}{\cancel{12x}}}{\underset{2}{\cancel{6x}} + \underset{2}{\cancel{24}}} \rightarrow \frac{2x}{4} \qquad \leftarrow \text{Incorrect!}$$

We must always factor the numerator and denominator before attempting to cancel. ▪

Look Closer: Suppose we'd like to evaluate the fraction in Example 5 for $x = 478$. Instead of substituting 478 into the original fraction, it is musch simpler to use the reduced fraction, and get

$$\frac{x}{2} = \frac{478}{2} = 239$$

This the whole point of reducing a fraction: to get a simpler but equivalent expression.   •

> **RQ4.**   Why is it a good idea to reduce an algebraic fraction when possible?

Negative of a Binomial

How can we deal with negative factors in a fraction? First, we consider some facts about the negative, or **opposite**, of an expression.

> A fraction is a negative number if either its numerator or its denominator is negative, but not both. For example,
>
> $$-\frac{2}{3} = \frac{-2}{3} = \frac{2}{-3}$$
>
> However, the fraction $\frac{-2}{-3}$ is a positive number, because the quotient of two negative numbers is positive.

Look Closer: In algebra, we prefer to write such a fraction with the negative sign in the numerator, so that the standard form for the opposite of $\frac{2}{3}$ is $\frac{-2}{3}$.   •

> **RQ5.**   Does a negative sign in front of a fraction apply to the numerator, the denominator, or both?

To find the opposite, or negative, of a binomial (or any other algebraic expression) we multiply the expression by -1.

> **Negative of a Binomial**
> The opposite of $a - b$ is
> $$-(a - b) = -a + b = b - a$$

Example 6   Find the opposite of each binomial.

a. $2a - 3b$

b. $-x - 1$

Solutions a. The opposite of $2a - 3b$ is

$$-(2a - 3b) = -2a + 3b \text{ or } 3b - 2a$$

b. The opposite of $-x - 1$ is

$$-(-x - 1) = x + 1$$

Look Closer: Recall that any number (except zero) divided by itself is 1, and any number divided by its opposite is -1. For example,

$$\frac{8}{-8} = -1 \qquad \text{and} \qquad \frac{-5}{5} = -1$$

The same is true for binomials and other algebraic expressions, so that

$$\frac{b - a}{a - b} = \frac{-(a - b)}{a - b} = -1$$

Here is an example of how opposites can arise in algebraic fractions.

Example 7   Reduce $\dfrac{2x - 4y}{6y - 3x}$

Solution   First, we factor the numerator and denominator.

$$\frac{2x - 4y}{6y - 3x} = \frac{2(x - 2y)}{3(2y - x)}$$

We see that $x - 2y$ is the opposite of $2y - x$, that is, $x - 2y = -(2y - x)$. Thus,

$$\frac{2(x - 2y)}{3(2y - x)} = \frac{-2(2y - x)}{3(2y - x)} = \frac{-2}{3}$$

RQ6.   What do we get when we divide an expression by its opposite?

Skills Warm-Up

Factor completely.

1. $x^2 - 4$

2. $x^2 - 4x$

3. $x^2 - 4x + 4$

4. $4x^2 - 1$

5. $4x^2 - 4x$

6. $4x^2 + 4x + 1$

Answers: **1.** $(x - 2)(x + 2)$ **2.** $x(x - 4)$ **3.** $(x - 2)^2$ **4.** $(2x - 1)(2x + 1)$
5. $4x(x - 1)$ **6.** $(2x + 1)^2$

Homework 8.1

Skills Practice

■ For Problems 1-2,
   a. Evaluate the fraction for the given values of the variable.
   b. For what values of the variable is the fraction undefined?

1. $\dfrac{x+1}{x-3}$, $x = \dfrac{1}{2}, -4$

2. $\dfrac{2a - a^2}{a^2 + 1}$, $a = 3, -1$

3. a. For what values of the variable is the fraction $\dfrac{2x+1}{x^2-1}$ undefined?

   b. Find a value of x for which the fraction is equal to zero.

■ For Problems 4-11, reduce the fraction.

4. $\dfrac{15}{3x}$

5. $\dfrac{24b}{14}$

6. $\dfrac{-5z}{6z}$

7. $\dfrac{5u}{120uv}$

8. $\dfrac{16ab}{-10ab}$

9. $\dfrac{3a^2}{27a}$

10. $\dfrac{-9y^3z}{42yz}$

11. $\dfrac{8u^3v^2}{12v^2w}$

■ For Problems 12-17, reduce the fraction if possible. If the fraction cannot be reduced, state the reason.

12a. $\dfrac{a+4}{a+5}$

b. $\dfrac{a \cdot 4}{a \cdot 5}$

13a. $\dfrac{2 \cdot m}{4 \cdot n}$

b. $\dfrac{2+m}{4+n}$

14a. $\dfrac{z-3}{z+9}$

b. $\dfrac{z(-3)}{z(+9)}$

15a. $\dfrac{u(-v)}{u(v)}$

b. $\dfrac{u-v}{u+v}$

16a. $\dfrac{3+x+y}{2+x+y}$

b. $\dfrac{3xy}{2xy}$

17a. $\dfrac{x+4}{x+2}$

b. $\dfrac{4x}{2x}$

■ For Problems 18-22, reduce each fraction if possible, and select the correct response, (a) or (b).

18. $\dfrac{x+2}{y+2}$ a. $\dfrac{x}{y}$ b. Cannot be reduced

19. $\dfrac{2x+4}{4}$ a. $\dfrac{x+2}{2}$ b. $2x$

20. $\dfrac{y^2-1}{y-1}$ a. $y+1$ b. y

21. $\dfrac{x^2+z^2}{x+z}$ a. $x+z$ b. Cannot be reduced

22. $\dfrac{n^2+n}{n}$ a. n^2 b. $n+1$

23. $\dfrac{z^2 + 2z + 1}{z + 1}$ **a.** $z + 1$ **b.** $z^2 + 2$

24. $\dfrac{m^2 + 4}{m + 2}$ **a.** $m + 2$ **b.** Cannot be reduced

■ For problems 25-30, reduce the fraction to lowest terms.

25. $\dfrac{3x + 12}{6}$ **26.** $\dfrac{b}{b^2 - b}$ **27.** $\dfrac{n^2 + 4n + 4}{n^2 - 4}$

28. $\dfrac{2a^2}{2a^2 - 6a}$ **29.** $\dfrac{3(a + b)}{4(a + b)}$ **30.** $\dfrac{x - 4}{x^2 - 3x - 4}$

■ For Problems 31-34, decide whether the fraction is equivalent to 1, to −1, or cannot be reduced.

31. $\dfrac{x + 4}{x - 4}$ **32.** $\dfrac{x + 3z}{z + 3x}$ **33.** $\dfrac{x^2 - 4}{4 - x^2}$ **34.** $\dfrac{-(2 - t)}{t - 2}$

■ For Problems 35-38, reduce each fraction if possible.

35. $\dfrac{b - 2}{4 - 2b}$ **36.** $\dfrac{a - b}{a^2 - b^2}$ **37.** $\dfrac{(3x + 2y)^2}{4y^2 - 9x^2}$ **38.** $\dfrac{3a - a^2}{a^2 - 2a - 3}$

Applications
■ For Problems 39-42,
   a. Choose a variable to represent the unknown quantity,
   b. then write an algebraic fraction.

39. There are 84 first-graders at Bonnair Elementary School. What fraction of the first-graders have been immunized against German measles?

40. One hundred thirty-eight of the employees at Digitronics enrolled in a bonus incentive plan. What fraction of the employees enrolled?

41. Gladys filled her car with gas three times while driving to Yellowstone National Park, a distance of 480 miles. What was her car's fuel efficiency during the trip?

42. Write a fraction that gives the length of a rectangle in terms of its area and its width.

■ For Problems 43-45, write algebraic fractions.

43. Sharelle's car still had 4 gallons in the gas tank when she filled up for $18.
   a. If the gas tank holds x gallons, what was the price per gallon of the gasoline?
   b. Evaluate your fraction for $x = 14$. What does your answer mean in this context?

44. Morgan drove across country in h hours, but he estimates that he spent 10 hours stopped for rest and meals.
   a. If he drove a total of 2800 miles, what was his average speed on the road?
   b. Evaluate your fraction for $h = 50$. What does your answer mean in this context?

45. The volume of a test tube is given by its height times the area of its cross-section. A test tube that holds 200 cubic centimeters is $2x - 1$ centimeters long.
  a. What is the area of its cross-section?
  b. Evaluate your fraction for $x = 13$. What does your answer mean in the context of the problem?

46. Ernestine wants to make a trip of 12 miles on her bicycle. If her trip takes a total of t hours, her average speed will be v miles per hour, where v is given by $v = \dfrac{12}{t}$.
  a. Complete the table and graph the equation on the grid at right.
  b. What will be Ernestine's average speed if the trip takes her 2 hours? Locate the corresponding point on your graph.
  c. How long will it take Ernestine to finish the trip if she maintains an average speed of 18 miles per hour? Locate the corresponding point on the graph.

t	v
0.75	
1	
1.2	
1.8	
2	
2.5	
3	
4	

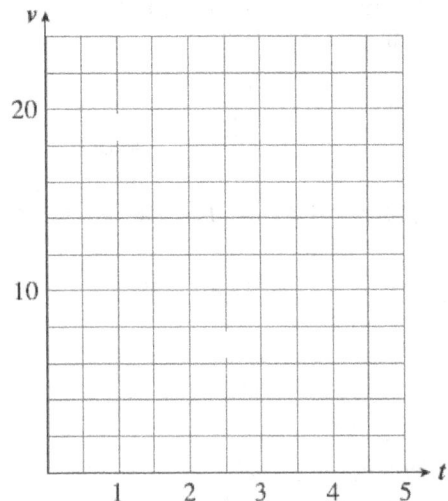

47. The crew team can row at a steady pace of 10 miles per hour in still water. Every afternoon, their training includes a five-mile row upstream on the river. If the current in the river on a given day is v miles per hour, then the time required for this exercise, in minutes, is given by

$$t = \frac{300}{10 - v}$$

Use the graph of this equation shown in the figure to answer questions (a) and (b).
  a. How long does the exercise take if there is no current in the river?
  b. How long will it take if the current is 4 miles per hour?
  c. Find an exact answer for part (b) by using the equation.
  d. If the exercise took $2\frac{1}{2}$ hours, what was the current in the river?
  e. As the speed of the current increases, what happens to the time needed for the exercise?
  f. What happens when the current is 10 miles per hour?

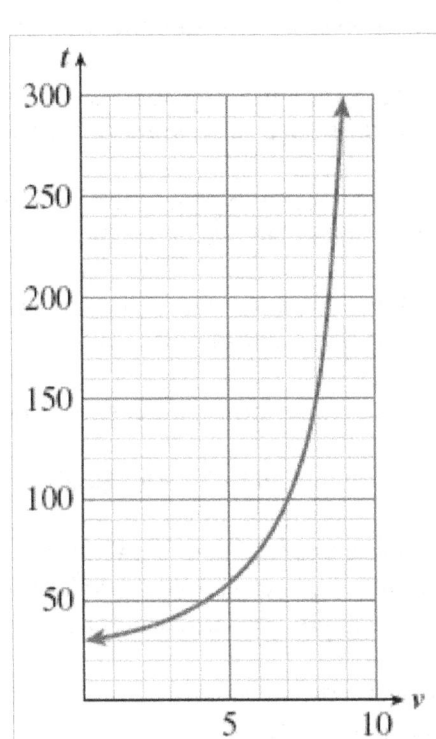

■ Mental Exercise: For Problems 48-50, decide which expressions are equivalent to the given fraction. Do not use pencil, paper, or calculator!

48. $\dfrac{2}{3}$ **a.** $\dfrac{2+4}{3+4}$ **b.** $\dfrac{2 \cdot 4}{3 \cdot 4}$ **c.** $\dfrac{22}{33}$ **d.** $\dfrac{2 \div 4}{3 \div 4}$

49. $\dfrac{3}{5}$ **a.** $\dfrac{3x}{5x}$ **b.** $\dfrac{3-x}{5-x}$ **c.** $\dfrac{30+x}{x+50}$ **d.** $\dfrac{30x}{50x}$

50. $\dfrac{2w-6}{w}$ **a.** -4 **b.** $\dfrac{6-2w}{-w}$ **c.** $2-\dfrac{6}{w}$ **d.** $-\dfrac{2w+6}{w}$

■ For Problems 51-52, explain the error in each calculation.

51. $\dfrac{3x+4}{3} \rightarrow x+4$

52. $\dfrac{2(5z-4)}{5z} \rightarrow \dfrac{2(-4)}{1} = -8$

8.2 Operations on Fractions

Products of Fractions

To multiply two fractions together, we multiply their numerators together and then multiply their denominators together.

Product of Fractions

If $b \neq 0$ and $d \neq 0$, then

$$\frac{a}{b} \cdot \frac{c}{d} = \frac{ac}{bd}$$

To see why this rule works, consider an example. Suppose you make a chocolate cake for the Math Club bake sale, but at the end of the day $\frac{4}{5}$ of the cake is left in the pan. The remaining cake is shown in figure (a).

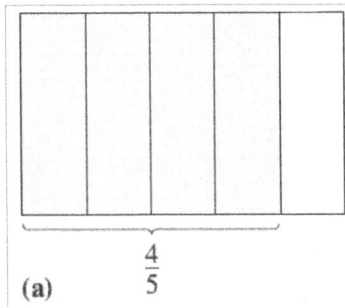

(a) $\frac{4}{5}$

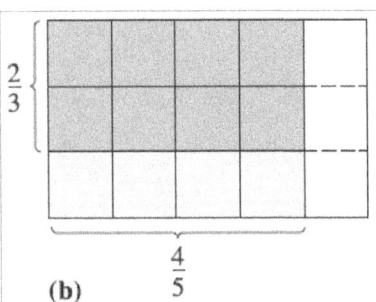

(b) $\frac{4}{5}$

You decide to take $\frac{2}{3}$ of the remaining cake home, and give the rest to your math teacher. You cut the remaining cake into thirds, as shown in figure (b). How much of the original cake are you taking home? Your share is

$$\frac{2}{3} \text{ of } \frac{4}{5} \quad \text{or} \quad \frac{2}{3} \times \frac{4}{5}$$

If you look at figure (b), you can see that you are taking home 8 pieces of cake, and that the original cake would have had 15 pieces of the same size. This means that your share is $\frac{8}{15}$ of the original cake.

We get the same answer if we use the rule above.

$$\frac{2}{3} \times \frac{4}{5} = \frac{2 \times 4}{3 \times 5} = \frac{8}{15}$$
Multiply numerators together.
Multiply denominators together.

Look Closer: There is a shortcut for multiplying fractions that allows us to reduce the answer before we multiply. We do this by dividing out any common factors in a numerator and a denominator. For example, consider the product

$$\frac{12}{35} \cdot \frac{28}{9}$$

Notice that 7 divides evenly into one of the denominators (35) and one of the numerators (28). We write the fractions with the 7's factored out.

$$\frac{12}{7 \cdot 5} \cdot \frac{7 \cdot 4}{9}$$

Also, 12 and 9 are both divisible by 3, so we write them in factored form as well.

$$\frac{3 \cdot 4}{7 \cdot 5} \cdot \frac{7 \cdot 4}{3 \cdot 3}$$

Now we can divide (or "cancel") the common factors from the numerators and denominators.

$$\frac{\cancel{3} \cdot 4}{\cancel{7} \cdot 5} \cdot \frac{\cancel{7} \cdot 4}{\cancel{3} \cdot 3}$$

Finally, we multiply together the remaining factors in the numerator, and multiply the remaining factors in the denominator.

$$\frac{4}{5} \cdot \frac{4}{3} = \frac{16}{15}$$

RQ1. Use the fundamental principle of fractions to explain why the shortcut strategy is valid.

The shortcut strategy also works on algebraic fractions.

■ **Example 1**   Multiply $\dfrac{3a}{4} \cdot \dfrac{5}{6a^2}$

Solution   We begin by factoring each numerator and denominator and then canceling any common factors.

$$\frac{3a}{4} \cdot \frac{5}{6a^2} = \frac{\cancel{3} \cdot \cancel{a}}{4} \cdot \frac{5}{\cancel{3} \cdot 2 \cdot \cancel{a} \cdot a} = \frac{5}{8a} \qquad \text{Cancel before multiplying.}$$

Or we can multiply first, and then reduce the answer.

$$\frac{3a}{4} \cdot \frac{5}{6a^2} = \frac{15a}{24a^2} = \frac{\cancel{3} \cdot 5 \cdot \cancel{a}}{\cancel{3} \cdot 8 \cdot \cancel{a} \cdot a} = \frac{5}{8a} \qquad \text{Cancel after multiplying.}$$

We get the same answer with either method.

When we multiply algebraic fractions, it is usually easier to cancel common factors before multiplying. Otherwise, it may be very difficult to reduce the result!

To Multiply Algebraic Fractions:
1. Factor each numerator and denominator completely.
2. If any factor appears in both a numerator and a denominator, divide out that factor.
3. Multiply the remaining factors of the numerator and the remaining factors of the denominator.
4. Reduce the product if necessary.

So far we have considered fractions whose numerators and denominators are monomials. The next example illustrates how to multiply fractions whose numerators and denominators are binomials.

■ **Example 2**   Multiply: $\dfrac{2x-4}{3x+6} \cdot \dfrac{6x+9}{x-2}$

Solution   We begin by factoring each numerator and denominator. Then we divide numerator and denominator by any common factors.

$$\frac{2x-4}{3x+6} \cdot \frac{6x+9}{x-2} = \frac{2(x-2)}{\cancel{3}(x+2)} \cdot \frac{\cancel{3}(2x+3)}{x-2} \qquad \text{Multiply remaining factors of numerators and of denominators.}$$

$$= \frac{2(2x+3)}{x+2} \quad \text{or} \quad \frac{4x+6}{x+2} \qquad ■$$

Caution!   In Example 2, it would be **incorrect** to try to cancel any terms of numerators and denominators before factoring each binomial. For example, we cannot cancel $3x$ into $6x$, because they are not factors of the fractions. ■

RQ2.   What was the first step in multiplying the two fractions in Example 2?

If we want to multiply a fraction by a whole number, we can write the whole number with a denominator of 1. For example,

$$\frac{2x}{3} \cdot 4 = \frac{2x}{3} \cdot \frac{4}{1} = \frac{8x}{3}$$

Quotients of Fractions

To divide one fraction by another, we multiply the first fraction by the **reciprocal** of the second fraction. For example,

$$\frac{m}{2} \div \frac{2p}{3} = \frac{m}{2} \cdot \frac{3}{2p} = \frac{3m}{4p}$$

Quotient of Fractions
   If $b,\ c,\ d \neq 0$, then

$$\frac{a}{b} \div \frac{c}{d} = \frac{a}{b} \cdot \frac{d}{c}$$

Why does this rule for division work? Consider a simple example, $6 \div \dfrac{1}{3}$. This quotient asks us "how many one-thirds are there in six?" To see the answer, we can draw six whole units as shown below, and split each into thirds.

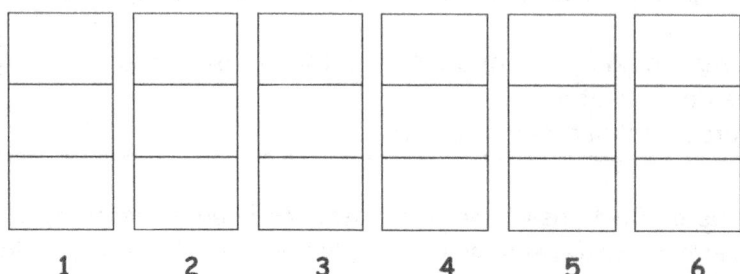

Because there are 3 thirds in each whole unit, we find 6×3, or 18 thirds in six units. The answer is 18. Now, 3 is the reciprocal of $\frac{1}{3}$, so by using the reciprocal rule, we also find

$$6 \div \frac{1}{3} = 6 \times 3 = 18$$

Look Closer: For a more interesting example, you might try using the figure above to show that

$$6 \div \frac{2}{3} = \frac{6}{1} \times \frac{3}{2} = 9$$

(**Hint:** How many groups of two thirds can you make from six whole units?)

> **RQ3.** Explain why dividng a number by 3 is the same as multiplying the number by $\frac{1}{3}$.

■ Example 3   Divide: $\dfrac{15u}{4v} \div \dfrac{10u}{8}$

Solution   We replace the second fraction by its reciprocal, and multiply.

$$\frac{15u}{4v} \div \frac{10u}{8} = \frac{15u}{4v} \cdot \frac{8}{10u}$$

$$= \frac{\overset{3}{\cancel{15u}}}{\underset{v}{\cancel{4v}}} \cdot \frac{\overset{2}{\cancel{8}}}{\underset{2}{\cancel{10u}}} \qquad \begin{array}{l}\text{Divide numerator and denominator by } 5u. \\ \text{Divide numerator and denominator by } 4.\end{array}$$

$$= \frac{3}{v} \cdot \frac{2}{2} = \frac{3}{v} \qquad \text{Multiply numerators; multiply denominators.} \quad ■$$

To divide two algebraic fractions, we take the reciprocal of the divisor and then follow the rules for multiplying algebraic fractions.

> **To Divide One Fraction by Another:**
> 1. Take the reciprocal of the second fraction and change the division to multiplication.
> 2. Follow the rules for multiplication of fractions.

Just as with multiplication, it is important to factor each numerator and denominator before trying to cancel any common factors.

■ Example 4   Divide: $\dfrac{a-2}{6a^2} \div \dfrac{4-a^2}{4a^2-2a}$

Solution   First, we change the operation to multiplication by taking the reciprocal of the divisor.

$$\frac{a-2}{6a^2} \cdot \frac{4a^2-2a}{4-a^2}$$

Now, we follow the rules for multiplication. We begin by factoring each numerator and denominator. Once we have finished factoring, we can look for any common factors to cancel.

$$\frac{a-2}{6a^2} \div \frac{4a^2-2a}{4-a^2} = \frac{a-2}{3\cdot 2\cdot a\cdot a} \cdot \frac{2a(2a-1)}{(2-a)(2+a)}$$ Divide out common factors.

Replace $a-2$ by $-1(2-a)$.

$$= \frac{-1(2\!\!\!\diagup a)}{3\cdot \cancel{2}\cdot \cancel{a}\cdot a} \cdot \frac{\cancel{2a}(2a-1)}{(2\!\!\!\diagup a)(2+a)}$$ Multiply remaining factors.

$$= \frac{-1(2a-1)}{3a(2+a)} = \frac{1-2a}{3a(2+a)}$$ We often leave denominators in factored form.

Caution!   In Example 4, do not try to cancel before factoring the numerators and denominators! For example, it would be **incorrect** to try to cancel a^2 before factoring each binomial.

If a nonfractional expression appears in a quotient, we can write it with a denominator of 1, just as we did for products. For example,

$$\frac{3}{2} \div 6a = \frac{3}{2} \div \frac{6a}{1}$$ Write $6a$ as $\frac{6a}{1}$.

$$= \frac{\cancel{3}}{2} \cdot \frac{1}{2\cdot \cancel{3}a} = \frac{1}{4a}$$

Adding Like Fractions

Fractions with the same denominator are called **like fractions**.

Here are some examples of like fractions:

$$\frac{5}{8} \text{ and } \frac{9}{8}, \qquad \frac{4}{5x} \text{ and } \frac{3}{5x}, \qquad \frac{1}{a-2} \text{ and } \frac{a}{a-2}$$

because they have the same denominators. The following pairs are unlike fractions:

$$\frac{2}{3a} \text{ and } \frac{2}{3a^2}, \qquad \frac{5}{x+1} \text{ and } \frac{2x}{x-1}$$

because they have different denominators.

RQ4. What are like fractions?

We can add or subtract two fractions only if they are like fractions, for the same reason that we can add only like terms. For example,

$$3x + 4x = 7x$$

becuase $3x$ and $4x$ are like terms. Similarly, we can think of the sum $\frac{3}{5} + \frac{4}{5}$ as

$$3\left(\frac{1}{5}\right) + 4\left(\frac{1}{5}\right) = 7\left(\frac{1}{5}\right) = \frac{7}{5}$$

Look Closer: In the example above, the denominators of the fractions tell us what kind of quantity we are adding (namely, fifths). We can add only quantities of the same kind. Note that we added the numerators, $3 + 4$, but kept the same denominator, 5.•

> When we add like fractions, we add their numerators and keep the denominator the same.

Sum or Difference of Like Fractions

If $c \neq 0$, then

$$\frac{a}{c} + \frac{b}{c} = \frac{a+b}{c} \quad \text{and} \quad \frac{a}{c} - \frac{b}{c} = \frac{a-b}{c}$$

The same rules hold true for adding or subtracting algebraic fractions. We can think of the process as combining the numerators over a single denominator. For example,

$$\frac{10}{3x} + \frac{4}{3x} = \frac{10+4}{3x} = \frac{14}{3x}$$ Add the numerators.
Keep the same denominator.

Caution!   In the example above, we do not add the denominators of the fractions; only the numerators. It would be **incorrect** to write $6x$ for the denominator of the sum.

■ **Example 5**   Add: $\dfrac{2x-5}{x+2} + \dfrac{x+4}{x+2}$

Solution We combine the numerators over a single denominator.

$$\frac{2x-5}{x+2} + \frac{x+4}{x+2} = \frac{(2x-5)+(x+4)}{x+2}$$ Add like terms
in the numerator.

$$= \frac{3x-1}{x+2}$$ ■

RQ5.   How do we add like fractions?

Subtracting Like Fractions

We must be careful when subtracting algebraic fractions: A subtraction sign in front of a fraction applies to the entire numerator.

■ **Example 4**   Subtract: $\dfrac{x-3}{x-1} - \dfrac{3x-5}{x-1}$

Solution We combine the numerators over a single denominator. We use parentheses around $3x - 5$ to show that the subtraction applies to the entire numerator.

$$\frac{x-3}{x-1} - \frac{3x-5}{x-1} = \frac{(x-3)-(3x-5)}{x-1} \qquad \text{Remove parentheses;}$$

Remove parentheses; distribute negative sign.

$$= \frac{x-3-3x+5}{x-1} \qquad \text{Combine like terms in the numerator.}$$

$$= \frac{-2x+2}{x-1}$$

We should always check to see whether the fraction can be reduced. Factor the numerator to find

$$\frac{-2x+2}{x-1} = \frac{-2(x \cancel{-} 1)}{x \cancel{-} 1} = -2 \qquad \blacksquare$$

Caution! In Example 4, be careful to **subtract each term** of the numerator of the second fraction. The following calculation is incorrect:

$$\frac{x-3}{x-1} - \frac{3x-5}{x-1} = \frac{x-3-3x-5}{x-1} \qquad \text{Incorrect!}$$

We must change the sign of each term of the second numerator, to get $-3x+5$.   ■

To Add or Subtract Like Fractions
1. Add or subtract the numerators.
2. Keep the same denominator.
3. Reduce the sum or difference if necessary.

RQ6.   When we subtract like fractions, how do we apply the subtraction sign?

Polynomial Division

Not every fraction can be reduced. Consider three improper fractions: $\frac{8}{6}, \frac{8}{4}$, and $\frac{8}{3}$. Can these fractions be simplified?

*   We can reduce the first fraction: $\dfrac{8}{6} = \dfrac{4}{3}$

*   The second fraction reduces to a whole number: $\dfrac{8}{4} = \dfrac{2}{1} = 2$

*   The third fraction does not reduce, but by dividing the denominator into the numerator, we can write it as a whole number plus a proper fraction: $\dfrac{8}{3} = 2\dfrac{2}{3}$

If an algebraic fraction cannot be reduced, we can simplify it by dividing the denominator into the numerator, just as we do with arithmetic fractions. The quotient will be the sum of a polynomial and a simpler algebraic fraction.

If the denominator is a monomial, we can simply divide the monomial into each term of the numerator.

■ **Example 6**   Divide $\dfrac{9x^3 - 6x^2 + 4}{3x}$

Solution We divide $3x$ into each term of the numerator.

$$\frac{9x^3 - 6x^2 + 4}{3x} = \frac{9x^3}{3x} - \frac{6x^2}{3x} + \frac{4}{3x}$$

$$= 3x^2 - 2x + \frac{4}{3x}$$

The quotient is the sum of a polynomial, $3x^2 - 2x$, and an algebraic fraction, $\frac{4}{3x}$. ∎

Skills Warm-Up

▨ Write each expression as a single fraction in lowest terms.

1. $6 \cdot \frac{3}{4}$

2. $6 \div \frac{3}{4}$

3. $\frac{3}{4} \div 6$

4. $6 + \frac{3}{4}$

5. $6 - \frac{3}{4}$

6. $6\frac{3}{4}$

7. Simplify $(x + 2) - (2x - 3)$

8. Reduce $\frac{2x - 4}{3x - 6}$

Answers: **1.** $\frac{9}{2}$ **2.** 8   **3.** $\frac{1}{8}$ **4.** $\frac{27}{4}$ **5.** $\frac{21}{4}$ **6.** $\frac{27}{4}$ **7.** $5 - x$ **8.** $\frac{2}{3}$

Homework 8.2

For Problems 1-4, multiply.

1. $-5c \cdot \dfrac{3}{20}$

2. $\dfrac{5}{6m^2} \cdot 2m$

3. $\dfrac{12s}{5r} \cdot \dfrac{2r}{3s}$

4. $\dfrac{-k^2}{14j} \cdot \dfrac{7j}{2k}$

5. $\dfrac{2}{3x^3} \cdot \dfrac{9x^2}{4}$

6. $\dfrac{21r^2}{4rs} \cdot \dfrac{16s}{5r}$

7. $\dfrac{2}{n} \cdot \dfrac{3}{n+2}$

8. $\dfrac{2b}{3} \cdot \dfrac{4}{b+1}$

9. $\dfrac{a-1}{3} \cdot \dfrac{3}{a^2-1}$

10. $\dfrac{p}{p^2-4} \, (p^2-2p)$

11. $\dfrac{3x-9}{5x-15} \cdot \dfrac{10x-5}{8x-4}$

12. $\dfrac{5a+25}{5a} \cdot \dfrac{10a}{2a+10}$

13. Write each product as a fraction.

    a. $\dfrac{2}{3}x$

    b. $\dfrac{3}{4}(a-b)$

    c. $(n+2) \cdot \dfrac{1}{n-2}$

For Problems 14-15, use the distributive law to multiply.

14. $\dfrac{-2}{t^2} \left(4t^3 - \dfrac{t^2}{8} + \dfrac{3t}{2} \right)$

15. $\dfrac{4}{3}v \left(\dfrac{2}{3}v - \dfrac{6}{v^2} - \dfrac{3}{4v} \right)$

For Problems 16-18, raise to the power.

16. $\left(\dfrac{2}{3z} \right)^2$

17. $\left(\dfrac{-5c}{2d} \right)^2$

18. $\left(\dfrac{-h}{3k} \right)^3$

For Problems 19-28, divide.

19. $\dfrac{24}{5h} \div \dfrac{-8}{5h}$

20. $\dfrac{-9}{2p} \div -36p$

21. $\dfrac{-15}{c^5} \div \dfrac{20}{9c^3}$

22. $\dfrac{12c}{21d} \div \dfrac{24c}{27d}$

23. $\dfrac{2ab^3}{3} \div \left(4a^2b \right)$

24. $1 \div \dfrac{x}{2y}$

25. $\dfrac{a^2-ab}{ab} \div \dfrac{2a-2b}{3ab}$

26. $\dfrac{3xy+x}{y^2-y} \div \dfrac{3y+1}{xy}$

27. $\dfrac{6a^2-12a}{3a+9} \div \dfrac{8a^2-4a^3}{15+5a}$

28. $\dfrac{c^2-6c+5}{c^2+2c-15} \div \dfrac{c^2-3c-10}{c^2+3c+2}$

For Problems 29-34, add or subtract.

29. $\dfrac{5}{2a} + \dfrac{3}{2a}$

30. $\dfrac{3}{x-1} + \dfrac{5}{x-1}$

31. $\dfrac{x-2y}{3x} + \dfrac{x+3y}{3x}$

32. $\dfrac{m^2+1}{m-1} - \dfrac{2m}{m-1}$

33. $\dfrac{z^2-2}{z+2} - \dfrac{z+4}{2+z}$

34. $\dfrac{b+1}{b^2-2b+1} - \dfrac{5-3b}{b^2-2b+1}$

For Problems 35-37, divide the monomial into each term of the polynomial. Write your answer as the sum of a polynomial and an algebraic fraction.

35. $\dfrac{9x^5-6x^2-2}{3x^2}$

36. $\dfrac{3n^3-3n^2+2n-3}{3n^2}$

37. $\dfrac{2x^2y^2-4xy^2+6xy}{2xy^2}$

For Problems 38-40, simplify.

38. a. $\dfrac{2}{x} + \dfrac{5}{x}$ **b.** $\dfrac{2}{x} - \dfrac{5}{x}$ **c.** $\dfrac{2}{x} \cdot \dfrac{5}{x}$ **d.** $\dfrac{2}{x} \div \dfrac{5}{x}$

39. a. $\dfrac{p}{q+2} + \dfrac{r-1}{q+2}$ **b.** $\dfrac{p}{q+2} - \dfrac{r-1}{q+2}$ **c.** $\dfrac{p}{q+2} \cdot \dfrac{r-1}{q+2}$ **d.** $\dfrac{p}{q+2} \div \dfrac{r-1}{q+2}$

40. a. $\left(\dfrac{1}{c} \cdot \dfrac{1}{5}\right) \div \dfrac{2}{5}$ **b.** $\dfrac{1}{c} \cdot \left(\dfrac{1}{5} \div \dfrac{2}{5}\right)$

   c. $\dfrac{1}{c} \div \left(\dfrac{1}{5} \div \dfrac{2}{5}\right)$ **d.** $\left(\dfrac{1}{c} \div \dfrac{1}{5}\right) \div \dfrac{2}{5}$

For Problems 41-46, simplify.

41. $2A^2 \div \dfrac{6A}{B^2}$ **42.** $\dfrac{3T^2}{4K^3} \div 9T^3$ **43.** $\dfrac{1}{5GH} \div \dfrac{G^2}{H^2}$

44. $\dfrac{4V}{D} \cdot \dfrac{LR}{DV}$ **45.** $\dfrac{2L}{c}\left(1 + \dfrac{V^2}{c^2}\right)$ **46.** $\dfrac{q}{8\pi}\left(\dfrac{3}{R} - \dfrac{a^2}{R^3}\right)$

47. a. Multiply: $\dfrac{2m^2 - 8}{3m^2 - 3} \cdot \dfrac{6 - 6m}{2m^2 + 4m}$

   b. Evaluate the expression in (a) for $m = 3$.

48. Multiply: $2x(x+2)\left(\dfrac{1}{2x} + \dfrac{x}{x+2} - 1\right)$

49. Write an algebraic expression for each phrase, and simplify.
   a. One-half of x.
   b. x divided by one-half.
   c. One-half divided by x.
   d. The reciprocal of $a + b$.
   e. Three-fourths of the reciprocal of $a + b$.
   f. The reciprocal of $a + b$ divided by three-fourths.

50. Write an equation that expresses the second variable in terms of the first variable.

a.

m	K
-9	6
-6	4
3	-2
6	-4

b.

t	H
-10	-4
-5	-2
10	4
15	6

8.3 Lowest Common Denominator

Adding and Subtracting Unlike Fractions

To add or subtract fractions with unlike denominators, we first convert the fractions to equivalent forms with the same denominator. For example, to compute the sum

$$\frac{2}{3} + \frac{3}{4}$$

we first find the **lowest common denominator** for the two fractions: The smallest number that is a multiple of both 3 and 4 is 12, so the LCD is 12.

Next, we **build** each fraction to an equivalent fraction with denominator 12. First, we multiply $\frac{2}{3}$ by $\frac{4}{4}$. Because $\frac{4}{4}$ is equal to 1, we have not changed the value of the fraction. 4 is called the **building factor** for $\frac{2}{3}$. Similarly, we multiply $\frac{3}{4}$ by $\frac{3}{3}$.

$$\frac{2}{3} = \frac{2 \cdot 4}{3 \cdot 4} = \frac{8}{12} \qquad \text{The building factor is 4.}$$
$$\frac{3}{4} = \frac{3 \cdot 3}{4 \cdot 3} = \frac{9}{12} \qquad \text{The building factor is 3.}$$

RQ1.   What does it mean to build a fraction?

Look Closer: We can visualize the building process with diagrams. Here are pictures of the original fractions, $\frac{2}{3}$ and $\frac{3}{4}$.

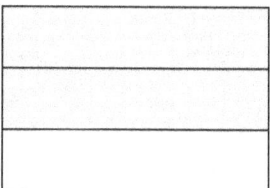   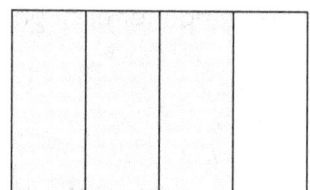

Thirds and fourths are not like fractions -- they are not pieces of the same size. But if we can break up one-third and one-fourth into smaller pieces that *are* the same size, then we can combine all the pieces. This is what the LCD is for. We can break up both thirds and fourths into twelfths, as shown below.

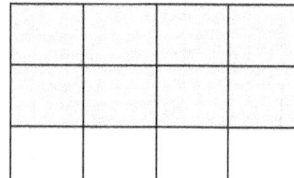   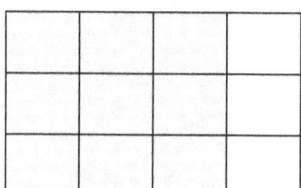

We can see from the figure that $\frac{2}{3} = \frac{8}{12}$ and $\frac{3}{4} = \frac{9}{12}$. Once we have written the fractions with the same denominator, they are like fractions and we can add them. This is what building fractions is all about.   ●

The new fractions, $\dfrac{8}{12}$ and $\dfrac{9}{12}$, are like fractions, so we add them by combining their numerators.

$$\frac{2}{3} + \frac{3}{4} = \frac{8}{12} + \frac{9}{12} = \frac{8+9}{12} = \frac{17}{12}$$

Look Closer:

Building is an application of the fundamental principle of fractions,

$$\frac{a}{b} = \frac{a \cdot c}{b \cdot c}, \qquad \text{if} \quad c \neq 0$$

We can multiply numerator and denominator of a fraction by the same (nonzero) quantity without changing the value of the fraction.

We add or subtract algebraic fractions in the same way.

> **To Add or Subtract Algebraic Fractions**
> 1. Find the lowest common denominator (LCD) for the fractions.
> 2. Build each fraction to an equivalent one with the LCD as denominator.
> 3. Add or subtract the resulting like fractions: Add or subtract their numerators, and keep the same denominator.
> 4. Reduce the sum or difference if necessary.

■ **Example 1**   Subtract: $\dfrac{x+2}{6} - \dfrac{x-1}{15}$

Solution Step 1   We find the LCD: The smallest common multiple of 6 and 15 is 30.
Step 2   We build each fraction to an equivalent one with denominator 30. The building factor for the first fraction is 5, and for the second fraction the building factor is 2.

$$\frac{x+2}{6} = \frac{(x+2) \cdot 5}{6 \cdot 5} = \frac{5x+10}{30}$$
$$\frac{x-1}{15} = \frac{(x-1) \cdot 2}{15 \cdot 2} = \frac{2x-2}{30}$$

Step 3   We subtract the resulting like fractions to obtain

$$\frac{x+2}{6} - \frac{x-1}{15} = \frac{5x+10}{30} - \frac{2x-2}{30} \qquad \text{Combine the numerators; keep the same denominator.}$$

$$= \frac{(5x+10) - (2x-2)}{30} \qquad \text{Simplify the numerator.}$$

$$= \frac{5x+10-2x+2}{30} = \frac{3x+12}{30}$$

Step 4   We reduce the fraction.

$$\frac{3x+12}{30} = \frac{\cancel{3}(x+4)}{\cancel{3} \cdot 10} = \frac{x+4}{10}$$

RQ2.   What is the first step in adding or subtracting unlike fractions?

Caution!   In Example 1, be careful to subtract **each term** of the numerator in the second fraction.

$$\frac{5x + 10}{30} - \frac{2x - 2}{30} = \frac{3x + 12}{30}$$

because $(5x + 10) - (2x - 2) = 5x + 10 - 2x + 2 = 3x + 12.$

Finding the Lowest Common Denominator

How can we find an LCD for fractions whose denominators are algebraic expressions?

> The **lowest common denominator** for two or more algebraic fractions is the simplest algebraic expression that is a multiple of each denominator.

The easiest case occurs when neither denominator can be factored. In that case, their LCD is just the product of the two expressions. For example, the LCD for the fractions $\frac{6}{x}$ and $\frac{x}{x - 2}$ is $x(x - 2)$.

■ **Example 2**   Add: $\frac{6}{x} + \frac{x}{x - 2}$

Solution **Step 1: Find the LCD.**   The LCD is $x(x - 2)$.
Step 2: Build.   We build each fraction to an equivalent one with denominator $x(x - 2)$. To find the building factors, we compare the given denominator with the desired LCD. The building factor for each fraction is the missing factor. Thus, the building factor for $\frac{6}{x}$ is $x - 2$, and the building factor for $\frac{x}{x - 2}$ is x.

$$\frac{6}{x} = \frac{6(x - 2)}{x(x - 2)} = \frac{6x - 12}{x(x - 2)}$$

$$\frac{x}{x - 2} = \frac{x(x)}{(x - 2)(x)} = \frac{x^2}{x(x - 2)}$$

Step 3: Combine.   We combine the resulting like fractions.

$$\frac{6}{x} + \frac{x}{x - 2} = \frac{6x - 12}{x(x - 2)} + \frac{x^2}{x(x - 2)} \qquad \text{Combine the numerators;}$$
$$\text{keep the same denominator.}$$

$$= \frac{x^2 + 6x - 12}{x(x - 2)}$$

Step 4: Reduce.   The numerator of this fraction cannot be factored, so the sum cannot be reduced. ■

RQ3.   When can we multiply two denominators together to find their LCD?
RQ4.   How do we find the building factor for each fraction?

However, if the denominators contain any common factors, the simplest common denominator is not just the product of the two denominators. In this case we must first factor each denominator to find their LCD.

Example 3   Find the LCD for $\dfrac{5}{12a^2} + \dfrac{7}{18a}$

Solution The LCD is not $12a^3(18a) = 216a^4$. It is true that 216 is a multiple of both 12 and 18, but it is not the smallest one! We can find a smaller common denominator by factoring each denominator.

$$12a^3 = \mathbf{2 \cdot 2 \cdot 3} \cdot \mathbf{a \cdot a \cdot a}$$
$$18a = 2 \cdot \mathbf{3 \cdot 3} \cdot a$$

To find an expression that both $12a^3$ and $18a$ divide into evenly, we need only enough factors to "cover" each of them separately. In this case two 2's, two 3's and three a's are sufficient, so the LCD is

$$\text{LCD} = \mathbf{2 \cdot 2 \cdot 3 \cdot 3 \cdot a \cdot a \cdot a} = 36a^3$$

You can check that $36a^3$ is a multiple of both $12a^3$ and $18a$. ■

Look Closer: To find the LCD in Example 3, notice that for each factor, we chose the denominator with the most copies of that factor, and included them in the LCD. For example, $12a^3$ has only one factor of 3, but $18a$ has two factors of 3, so we used two factors of 3 in the LCD.

To Find the LCD:
1. Factor each denominator completely, and arrange the factors in order.
2. For each factor,
   a. Which denominator has the most copies of that factor? Circle them. (If there is a tie, either denominator will do.)
   b. Include all the circled factors in the LCD.
3. Multiply together the factors of the LCD.

RQ5.   We would like to add two unlike fractions. If each denominator contains one factor of x, how many factors of x do we include in the LCD?

Now let us see how to incorporate this new skill into the steps for adding fractions.

Example 4   Add: $\dfrac{x-4}{x^2-2x} + \dfrac{4}{x^2-4}$

Solution Step 1   To find the **LCD**, we first factor each denominator completely.

$$x^2 - 2x = x(x-2)$$
$$x^2 - 4 = (x-2)(x+2)$$

The LCD is $x(x-2)(x+2)$.
Step 2   We **build** each fraction to an equivalent one with the LCD as its denominator.

$$\frac{x-4}{x(x-2)} = \frac{(x-4)(x+2)}{x(x-2)(x+2)} = \frac{x^2-2x-8}{x(x-2)(x+2)}$$

$$\frac{4}{(x-2)(x+2)} = \frac{4x}{(x-2)(x+2)x}$$

Step 3   Because the fractions are now like fractions, we can **combine** their numerators.

$$\frac{x-4}{x^2-2x} + \frac{4}{x^2-4} = \frac{x^2-2x-8}{x(x-2)(x+2)} + \frac{4x}{x(x-2)(x+2)}$$

$$= \frac{x^2+2x-8}{x(x-2)(x+2)}$$

Step 4   We **reduce** the fraction by first factoring numerator and denominator, then dividing out any common factors.

$$\frac{x^2+2x-8}{x(x-2)(x+2)} = \frac{(x+4)(x-2)}{x(x-2)(x-2)} = \frac{x+4}{x(x+2)}$$

RQ6.   State the four steps for adding or subtracting fractions.

Skills Warm-Up

Simplify.

1. $2x + 5 - (3x + 2)$ **2.** $4(x-3) - 3(2x-5)$ **3.** $2(3x - x^2) + x(2x - 4)$

4. $(x-3)2x + (x+2)4$ **5.** $x^2(x-1) - 2x(x+2)$ **6.** $3x(x+4) - 2(x^2 - 16) - 4x$

Answers: **1.** $-x + 3$ **2.** $-4x + 3$ **3.** $2x$ **4.** $2x^2 - 2x + 8$ **5.** $x^3 - 3x^2 - 4x$

6. $x^2 + 8x + 32$

Homework 8.3

Skills Practice

■ For Problems 1-12, find the LCD, then add or subtract.

1. $\dfrac{3}{x} - \dfrac{4}{y}$

2. $\dfrac{-3}{a} - \dfrac{2a}{b}$

3. $1 - \dfrac{3}{a}$

4. $\dfrac{1}{x} + \dfrac{x}{y} + \dfrac{x}{z}$

5. $\dfrac{5}{xy} - \dfrac{x}{y}$

6. $\dfrac{2}{z} - \dfrac{1}{2z}$

7. $\dfrac{u-4}{3} + \dfrac{6}{v}$

8. $\dfrac{3}{x} - \dfrac{2}{x+1}$

9. $\dfrac{3}{n+3} + \dfrac{4n}{n-3}$

10. $h - \dfrac{3}{h+2}$

11. $\dfrac{v+1}{v} + \dfrac{1}{v-1}$

12. $\dfrac{2}{x} + \dfrac{x}{x-2} - 2$

■ For Problems 13-24,
  a. Find the lowest common denominator for the fractions.
  b. Add or subtract.

13. $\dfrac{-2}{5n} + \dfrac{q}{4n}$

14. $\dfrac{-6}{s} + \dfrac{3}{s^2}$

15. $\dfrac{z}{4x^2} - \dfrac{1}{6xz}$

16. $\dfrac{5}{2x} + \dfrac{3}{4x^2}$

17. $\dfrac{2z-3}{8z} + \dfrac{z-2}{6z}$

18. $\dfrac{3}{2a-b} + \dfrac{1}{8a-4b}$

19. $\dfrac{1}{x} + \dfrac{1}{d-x}$

20. $a + \dfrac{N-a^2}{2a}$

21. $\dfrac{1}{2} \cdot \dfrac{a}{t} - \dfrac{m}{a}$

22. $\dfrac{q}{4\pi r} + \dfrac{qa}{2\pi r^2}$

23. $\dfrac{L}{c-V} + \dfrac{L}{c+V}$

24. $\dfrac{q}{r-a} - \dfrac{2q}{r} + \dfrac{q}{r+a}$

■ For Problems 25-30,
  a. Find the lowest common denominator for the fractions.
  b. Add or subtract.

25. $\dfrac{2x+3}{x-1} + \dfrac{2x-5}{1-x}$

26. $\dfrac{5}{2p-4} - \dfrac{2}{6-3p}$

27. $\dfrac{-2}{m^2+3m} + \dfrac{1}{m^2-9}$

28. $\dfrac{4}{k^2-3k} + \dfrac{1}{k^2+k}$

29. $\dfrac{x-1}{x^2+3x} + \dfrac{x}{x^2+6x+9}$

30. $\dfrac{n+2}{2n^2-3n} - \dfrac{7}{6n-9}$

■ For Problems 31-33, combine the fractions.

31a. $\dfrac{2b}{a} \cdot \dfrac{4}{b}$
  b. $\dfrac{2b}{a} - \dfrac{4}{b}$
  c. $\dfrac{2b}{a} \div \dfrac{4}{b}$
  d. $\dfrac{2b}{a} + \dfrac{4}{b}$

32a. $\dfrac{2y}{3x^2} \div \dfrac{6}{xy}$
  b. $\dfrac{2y}{3x^2} + \dfrac{6}{xy}$
  c. $\dfrac{2y}{3x^2} \cdot \dfrac{6}{xy}$
  d. $\dfrac{2y}{3x^2} - \dfrac{6}{xy}$

33a. $\dfrac{3}{2+a} - \dfrac{2}{a^2-4}$
  b. $\dfrac{3}{2+a} \cdot \dfrac{2}{a^2-4}$
  c. $\dfrac{3}{2+a} + \dfrac{2}{a^2-4}$
  d. $\dfrac{3}{2+a} \div \dfrac{2}{a^2-4}$

Applications

■ For Problems 34-35, write algebraic fractions in simplest form.

34. The dimensions of a rectangular rug are $\dfrac{12}{x}$ feet and $\dfrac{12}{x-2}$ feet.
   a. Write an expression for the area of the rug.
   b. Write an expression for the perimeter of the rug.

35. Colonial Airline has a commuter flight between Richmond and Washington, a distance of 100 miles. The plane flies at x miles per hour in still air. Today there is a steady wind from the north at 10 miles per hour.
   a. How long will the flight from Richmond to Washington take?
   b. How long will the flight from Washington to Richmond take?
   c. How long will a round trip take?
   d. Evaluate your fractions in parts (a)-(c) for $x = 150$.

36. Francine's cocker spaniel eats a large bag of dog food in d days, and Delbert's sheep dog takes 5 days less to eat the same size bag.
   a. What fraction of a bag of dog food does Francine's cocker spaniel eat in one day?
   b. What fraction of a bag of dog food does Delbert's sheep dog eat in one day?
   c. If Delbert and Francine get married, what fraction of a bag of dog food will their dogs eat in one day?
   d. If $d = 25$, how soon will Delbert and Francine have to buy more dog food?

8.4 Equations with Fractions

How can we solve an equation that involves algebraic fractions? Recall that to solve the equation

$$\frac{3x}{4} = 9$$

we multiply both sides by 4 (the denominator of the fraction) to clear the fraction.

$$4\left(\frac{3x}{4}\right) = (9)\,4$$
$$3x = 36$$

We can now finish the solution to get $x = 12$.
    We use the same idea to solve an equation with algebraic fractions.

■ **Example 1** Solve $\dfrac{5000 + 50x}{x} = 75$

Solution We multiply both sides of the equation by x, the denominator of the fraction, to get

$$x\left(\frac{5000 + 50x}{x}\right) = (75)\,x$$
$$5000 + 50x = 75x$$

Then we proceed as usual to finish the solution. Subtract $50x$ from both sides to find

$$5000 = 25x$$
$$200 = x$$

You can check that $x = 200$ does satisfy the equation. ■

Using an LCD to Clear Fractions

If an equation contains more than one fraction, we multiply both sides of the equation by the LCD of all the fractions. This will clear all the denominators at once.

■ **Example 2** Solve $\dfrac{x}{3} - 2 = \dfrac{4}{5} + \dfrac{x}{5}$

Solution The LCD of $\dfrac{x}{3}$, $\dfrac{4}{5}$, and $\dfrac{x}{5}$ is 15. We multiply both sides of the equation by 15.

$$15\left(\frac{x}{3} - 2\right) = \left(\frac{4}{5} + \frac{x}{5}\right)15 \qquad \text{**Apply the distributive law.**}$$

$$15\left(\frac{x}{3}\right) - 15(2) = 15\left(\frac{4}{5}\right) + 15\left(\frac{x}{5}\right) \qquad \text{**Simplify each product.**}$$
$$5x - 30 = 12 + 3x$$

Now we can proceed as usual to complete the solution.

$$5x - 30 = 12 + 3x$$

Subtract $3x$ from both sides; add 30 to both sides.

$$2x = 42$$

Divide both sides by 2.

$$x = 21$$ ◼

Caution!   In Example 2, we multiplied each term by the LCD, 15, including terms that are not fractions, namely –2 in this example. Be sure to multiply **each** term of the equation by the LCD. ◼

RQ1.   When clearing fractions, which terms of the equation should we multiply by the LCD?

Variables in the Denominator

Equations that involve algebraic fractions can also be solved using an LCD.

◼ **Example 3**   Solve $\dfrac{3}{4} = 8 - \dfrac{2x + 11}{x - 5}$

Solution   The LCD for the two fractions in the equation is $4(x - 5)$. We multiply both sides of the equation by $4(x - 5)$.

$$4(x - 5)\left(\frac{3}{4}\right) = \left(8 - \frac{2x + 11}{x - 5}\right) \cdot 4(x - 5) \qquad \textbf{Apply the distributive law.}$$

$$\cancel{4}(x - 5)\left(\frac{3}{\cancel{4}}\right) = 4(x - 5)(8) - 4(x \cancel{- 5})\left(\frac{2x + 11}{x \cancel{- 5}}\right)$$

$$3(x - 5) = 32(x - 5) - 4(2x + 11)$$

We proceed as usual to complete the solution. First we use the distributive law to remove parentheses.

$$3x - 15 = 32x - 160 - 8x - 44 \qquad \textbf{Combine like terms.}$$
$$3x - 15 = 24x - 204$$
$$-21x = -189$$
$$x = 9$$

The solution is $x = 9$. You can check that $x = 9$ satisfies the original equation. ◼

RQ2.   How do we "clear" the fractions from an equation?

Look Ahead: Remember that we do not obtain an equivalent equation if we multiply both sides of an equation by zero. In Example 3 we multiplied by $4(x - 5)$. Is it possible that $4(x - 5)$ equals zero? After solving, we found that $x = 9$, so $4(x - 5) = 16$, which is not zero, and the multiplication step was valid. The next example illustrates what can go wrong if we multiply by zero. •

Extraneous Solutions

Example 4   Solve $6 + \dfrac{4}{x-3} = \dfrac{x+1}{x-3}$

Solution   We multiply both sides of the equation by the LCD, $x-3$, to clear the fractions.

$$(x-3)\left(6 + \frac{4}{x-3}\right) = \left(\frac{x+1}{x-3}\right)(x-3) \quad \textbf{Apply the distributive law.}$$

$$(x-3)(6) + (x\!\!\!\not/3)\left(\frac{4}{x\!\!\!\not/3}\right) = \left(\frac{x+1}{x\!\!\!\not/3}\right)(x\!\!\!\not/3)$$

$$6(x-3) + 4 = x + 1$$

We complete the solution as usual.

$$6x - 18 + 4 = x + 1$$
$$6x - 14 = x + 1$$
$$5x = 15$$
$$x = 3$$

The solution appears to be $x = 3$. But we have a problem, because the LCD, $x - 3$, equals zero when $x = 3$. We have multiplied both sides of the equation by zero. When we try to check the solution we find

$$6 + \frac{4}{3-3} = \frac{3+1}{3-3} \quad \textbf{Simplify each term.}$$
$$6 + \frac{4}{0} = \frac{4}{0}$$

Because division by zero is undefined, 3 is not a solution after all. The original equation does not have a solution.

In Example 4, when we multiplied both sides of the equation by zero we found a false solution for the equation. Such solutions are called **extraneous solutions.** There is always a danger that an extraneous solution may be introduced when we multiply by an expression that contains the variable.

> **RQ3.**   When might an extraneous solution be introduced?

> We should always check for extraneous solutions when solving equations that involve algebraic fractions.

To check a solution, we substitute it into the original equation. If a possible solution causes any of the denominators in the equation to equal zero, then that solution is extraneous.

> **RQ4.**   How do we check for extraneous solutions?

Formulas

We can also solve formulas that involve algebraic fractions.

■ **Example 5**   Solve for P: $\dfrac{1}{T} = \dfrac{PR}{A - P}$

Solution The LCD for the two fractions in the equation is $T(A - P)$. We multiply both sides of the equation by the LCD to obtain

$$\cancel{T}(A - P)\frac{1}{\cancel{T}} = \frac{PR}{\cancel{A - P}}\,T(\cancel{A - P})$$
$$A - P = PRT$$

Next, we get all the terms containing the desired variable, P, on one side of the equation. We add P to both sides to get

$$A = P + PRT$$

We now have two unlike terms that contain the desired variable. To proceed, we factor out this variable, and then divide both sides by the remaining factor.

$$A = P(1 + RT) \qquad \textbf{Divide both sides by } 1 + RT.$$

$$\frac{A}{1 + RT} = \frac{P(1 + \cancel{RT})}{1 + \cancel{RT}}$$

Thus, the new version of the formula is

$$P = \frac{A}{1 + RT}$$

■

RQ5.   When solving a formula, what should we do if there are two terms that contain the variable?

Applications

Recall the formula $d = rt$, which is useful in solving problems about motion. By solving for r or t, we can write the equation in the forms

$$r = \frac{d}{t} \qquad \text{or} \qquad t = \frac{d}{r}$$

if either of these is more useful for the problem.

■ **Example 6**   A cruise boat travels 18 miles downstream and back in $4\frac{1}{2}$ hours. If the speed of the current is 3 miles per hour, what is the speed of the boat in still water?

Solution We let x stand for unknown quantity, the speed of the boat in still water. Then we make a table showing the distance, rate, and time for each part of the trip. We begin by filling in the information given in the problem.

	Distance	Rate	Time
Downstream trip	18	$x + 3$	
Upstream trip	18	$x - 3$	

We use the formula $t = \dfrac{d}{r}$ to fill in the last column of the table.

	Distance	Rate	Time
Downstream trip	18	$x + 3$	$\dfrac{18}{x + 3}$
Upstream trip	18	$x - 3$	$\dfrac{18}{x - 3}$

Notice that we did not use the $4\frac{1}{2}$ hours in the table, because it was not the trip upstream or the trip downstream that took $4\frac{1}{2}$ hours, but the total trip. We use the $4\frac{1}{2}$ to write an equation: the sum of the times for the upstream and downstream trips was $4\frac{1}{2}$ or $\frac{9}{2}$ hours:

$$\frac{18}{x + 3} + \frac{18}{x - 3} = \frac{9}{2}$$

To solve the equation, we multiply both sides by the LCD, $2(x - 3)(x + 3)$.

$$2(x - 3)(x + 3)\left(\frac{18}{x + 3} + \frac{18}{x - 3}\right) = \left(\frac{9}{2}\right)2(x - 3)(x + 3)$$

Next, we apply the distributive law to multiply each term of the equation by the LCD.

$$2(x - 3)(x \cancel{+3})\left(\frac{18}{x\cancel{+3}}\right) + 2(x - 3)(x\cancel{+3})\left(\frac{18}{x\cancel{-3}}\right) = \left(\frac{9}{\cancel{2}}\right)\cancel{2}(x - 3)(x + 3)$$

$$36(x - 3) + 36(x + 3) = 9(x - 3)(x + 3)$$

We simplify each side of the equation and write it in standard form.

$$36x - 108 + 36x + 108 = 9x^2 - 81 \quad \textbf{Combine like terms.}$$
$$9x^2 - 72x - 81 = 0$$

The equation is quadratic, and we can solve it by factoring.

$$9(x^2 - 8x - 9) = 0$$
$$9(x - 9)(x + 1) = 0 \qquad \textbf{Set each factor equal to zero.}$$
$$x - 9 = 0 \qquad x + 1 = 0$$
$$x = 9 \qquad\quad x = -1$$

Finally, we check each solution by substituting into the original equation. Neither solution is extraneous. However, the speed of the boat is not a negative number, so we discard the solution $x = -1$. The boat travels at 9 miles per hour in still water. ■

Look Closer: Problems involving other types of rates can be solved with similar techniques. Suppose it takes you 8 hours to type a term paper for your history class. If you work at a constant rate, in 1 hour you would complete $\frac{1}{8}$ of your task. The rate at which you work, or your work rate, is one-eighth job per hour. In 3 hours you would complete

$$3 \cdot \frac{1}{8} \quad \text{or} \quad \frac{3}{8}$$

of the job. In t hours you would complete

$$t \cdot \frac{1}{8} \quad \text{or} \quad \frac{t}{8}$$

of the job.

In general, the amount of work done, expressed as a fraction of 1 whole job, is given by the following formula.

Work Formula

$$\textbf{work rate} \times \textbf{time} = \textbf{work completed}$$

$$rt = w$$

Skills Warm-Up

Write each expression as a single fraction in simplest form.

1. $\dfrac{1}{x-2} + \dfrac{2}{x}$

2. $\dfrac{1}{x-2} - \dfrac{2}{x}$

3. $\dfrac{1}{x-2} \cdot \dfrac{2}{x}$

4. $\dfrac{1}{x-2} \div \dfrac{2}{x}$

Answers: 1. $\dfrac{3x-4}{x^2-2x}$ 2. $\dfrac{4-x}{x^2-2x}$ 3. $\dfrac{2}{x^2-2x}$ 4. $\dfrac{x}{2x-4}$

Homework 8.4

Skills Practice

For Problems 1-10, solve the equation.

1. $\dfrac{5x}{2} - 1 = x + \dfrac{1}{2}$

2. $\dfrac{t}{6} - \dfrac{7}{3} = \dfrac{2t}{9} = \dfrac{t}{4}$

3. $\dfrac{2}{3}(x - 1) + x = 6$

4. $\dfrac{3x^2}{2} - \dfrac{x}{4} = \dfrac{1}{2}$

5. $2 + \dfrac{5}{2x} = \dfrac{3}{x} + \dfrac{3}{2}$

6. $1 + \dfrac{1}{x(x - 1)} = \dfrac{3}{x}$

7. $\dfrac{4}{x - 1} - \dfrac{4}{x + 2} = \dfrac{3}{7}$

8. $\dfrac{1}{x - 2} + \dfrac{1}{x + 2} = \dfrac{4}{x^2 - 4}$

9. $\dfrac{15x}{1 + x^2} = 6$

10. $\dfrac{3x + 2}{x} = \dfrac{x + 9}{x + 6}$

For Problems 11-18, solve the formula for the specified variable.

11. $V = \dfrac{hT}{P}$ for T

12. $m = \dfrac{2E}{v^2}$ for E

13. $m = \dfrac{y - k}{x - h}$ for x

14. $a = \dfrac{F}{m + M}$ for m

15. $\dfrac{1}{R} = \dfrac{1}{A} + \dfrac{1}{B}$ for A

16. $I = \dfrac{E}{R + \dfrac{r}{n}}$ for R

17. $r = \dfrac{dc}{1 - ec}$ for e

18. $w = 0.622 \dfrac{e}{P - e}$ for e

For Problems 19-20, solve the proportion.

19. $\dfrac{p - 3}{2} = \dfrac{7}{p + 2}$

20. $\dfrac{m - 4}{2m + 1} = \dfrac{m + 1}{2m - 1}$

Applications

21. A small lake in a state park has become polluted by runoff from a factory upstream. The cost for removing p percent of the pollution from the lake is given, in thousands of dollars, by

$$C = \dfrac{25p}{100 - p}$$

How much of the pollution can be removed for $25,000?

22. During the baseball season so far this year, Pete got hits 44 times out of 164 times at bat.
   a. What is Pete's batting average so far? (Batting average is the fraction of at-bats that resulted in hits.)
   b. If Pete gets hits on every one of his next x at-bats, write an expression for his new batting average.
   c. How many consecutive hits does Pete need to raise his batting average to 0.350?

23. The rectangle $ABCD$ in the figure is divided into a square and a smaller rectangle, $CDEF$. The two rectangles $ABCD$ and $CDEF$ are **similar**, which means that their corresponding sides are proportional. A rectangle with this property is called a golden rectangle, and the ratio of its length to its width is called the golden ratio. The golden ratio appears frequently in art and

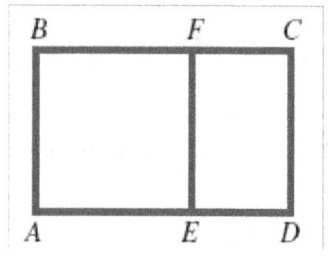

nature, and is considered to give the most pleasing proportions to many figures. We'll compute the golden ratio as follows.
   a. Let $AB = 1$ and $AD = x$. What are the lengths of AE, ED, and CD?
   b. Write a proportion in terms of x for the similarity of rectangles $ABCD$ and $CDEF$. Be careful to match up the corresponding sides.
   c. Solve your proportion for x. Find the golden ratio, $\dfrac{AD}{AB} = \dfrac{x}{1}$.

24. Compare the procedures for adding fractions and for solving fractional equations in the problems below. Explain how the LCD is used differently for adding and for solving equations.

   a. Add: $\dfrac{x-1}{4} + \dfrac{3x}{5}$ b. Solve: $\dfrac{x-1}{4} + \dfrac{3x}{5} = 1$

▆ For Problems 25-28,
   a. Write an equation to model the problem, and
   b. Solve your equation and answer the question posed in the problem.

25. An express train travels 180 miles in the same time that a freight train travels 120 miles. If the express train travels 20 miles per hour faster than the freight train, find the speed of each.

26. Sam Scholarship and Reginald Privilege each travel the 360 miles to Fort Lauderdale on spring break, but Reginald drives his Porsche while Sam hitches a ride on a vegetable truck. Reginald travels 20 miles per hour faster than Sam does and arrives in 3 hours less time. How fast did each travel?

27. Periwinkle Publishing can print the first run of a volume of poems in 10 hours on their new press; the same job takes 18 hours on their old machine. The new press is finishing another job, so the press manager starts the old machine 4 hours ahead of the new one. How many hours are needed with both presses running to finish the printing?

28. It takes 30 minutes to fill a large water tank. However, the tank has a small leak that would completely drain it in 4 hours. How long will it take to fill the tank if the leak is not plugged?

29. Find the error in the following "proof" that $1 = 0$:
Start by letting $x = 1$.

$$x = 1 \qquad \text{Multiply both sides by } x.$$
$$x^2 = x \qquad \text{Subtract 1 from both sides.}$$
$$x^2 - 1 = x - 1 \qquad \text{Factor the left side.}$$
$$(x - 1)(x + 1) = x - 1 \qquad \text{Divide both sides by } x - 1.$$
$$\frac{(x - 1)(x + 1)}{x - 1} = \frac{x - 1}{x - 1} \qquad \text{Simplify both sides.}$$
$$x + 1 = 1 \qquad \text{Subtract 1 from both sides.}$$
$$x = 0$$

Because $x = 1$ and $x = 0$, we have "proved" that $1 = 0$.

For Problems 30-31, find the error in the "solution" and correct it.

30.
$$\frac{a}{2} - \frac{a}{3} = 1$$
$$6\left(\frac{a}{2} - \frac{a}{3}\right) = 1$$
$$3a - 2a = 1$$
$$a = 1$$

31.
$$m^2 - 6m = 0$$
$$m^2 = 6m$$
$$\frac{m^2}{m} = \frac{6m}{m}$$
$$m = 6$$

32. If two hens lay two eggs in 2 days, how long will it take six hens to lay six eggs?

Chapter 8 Summary and Review

Section 8.1 Algebraic Fractions

• An **algebraic fraction** (or *rational expression*, as they are sometimes called) is a fraction in which both numerator and denominator are polynomials.

• An algebraic fraction is undefined at any values of the variable that make the denominator equal to zero.

• We use the fundamental principle of fractions to reduce or build fractions.

Fundamental Principle of Fractions
We can multiply or divide the numerator and denominator of a fraction by the same nonzero factor, and the new fraction will be equivalent to the old one.

$$\frac{a \cdot c}{b \cdot c} = \frac{a}{b} \quad \text{if} \quad b, c \neq 0$$

To Reduce an Algebraic Fraction:
1. Factor the numerator and denominator.
2. Divide the numerator and denominator by any common factors.

• We can cancel common **factors** (expressions that are multiplied together), but not common **terms** (expressions that are added or subtracted).

• A fraction is a negative number if either its numerator or its denominator is negative, but not both.

Negative of a Binomial
The opposite of $a - b$ is

$$-(a - b) = -a + b = b - a$$

Section 8.2 Operations on Algebraic Fractions

Product of Fractions
If $b \neq 0$ and $d \neq 0$, then

$$\frac{a}{b} \cdot \frac{c}{d} = \frac{ac}{bd}$$

To Multiply Algebraic Fractions:
1. Factor each numerator and denominator completely.
2. If any factor appears in both a numerator and a denominator, divide out that factor.
3. Multiply the remaining factors of the numerator and the remaining factors of the denominator.
4. Reduce the product if necessary.

Quotient of Fractions
If $b, c, d \neq 0,$ then

$$\frac{a}{b} \div \frac{c}{d} = \frac{a}{b} \cdot \frac{d}{c}$$

To Divide Algebraic Fractions:
1. Take the reciprocal of the second fraction and change the operation to multiplication.
2. Follow the rules for multiplication of fractions.

- Fractions with the same denominator are called **like** fractions.

Sum or Difference of Like Fractions
If $c \neq 0,$ then

$$\frac{a}{c} + \frac{b}{c} = \frac{a+b}{c} \quad \text{and} \quad \frac{a}{c} - \frac{b}{c} = \frac{a-b}{c}$$

To Add or Subtract Like Fractions
1. Add or subtract the numerators.
2. Keep the same denominator.
3. Reduce the sum or difference if necessary.

Section 8.3 Lowest Common Denominator

- To add or subtract fractions with unlike denominators, we must first convert the fractions to equivalent forms with the same denominator.
- The **lowest common denominator** for two or more algebraic fractions is the simplest algebraic expression that is a multiple of each denominator.

To Find the LCD:
1. Factor each denominator completely.
2. Include each different factor in the LCD as many times as it occurs in any *one* of the given denominators.

To Add or Subtract Algebraic Fractions
1. Find the lowest common denominator (LCD) for the fractions.
2. Build each fraction to an equivalent one with the same denominator.
3. Add or subtract the resulting like fractions: Add or subtract their numerators, and keep the same denominator.
4. Reduce the sum or difference if necessary.

Section 8.4 Equations with Fractions

- To solve an equation that contains algebraic fractions, we first clear the denominators by multiplying both sides of the equation by the LCD of the fractions.
- When clearing fractions from am equation, we must be sure to multiply **each** term of the equation by the LCD.

- Whenever we multiply an equation by an expression containing the variable, we should check for extraneous solutions.

> **Work Formula**
>
> **work rate × time = work completed**
>
> $$rt = w$$

Review Questions

■ For Problems 1-10, use complete sentences to answer the questions.

1. When reducing an algebraic fraction, we should always _____ before we _____ .
2. The fundamental principle of fractions says that we can cancel _____ , but not _____ .
3. Delbert says that to multiply two algebraic fractions, we just multiply the numerators together and multiply the denominators together. Comment on Delbert's method.
4. To divide by a fraction is the same as to _____ by its _____ .
5. Describe how to add or subtract unlike fractions in three steps.
6. Why do we need to find an LCD when adding unlike fractions?
7. Francine says that to solve an equation containing algebraic fractions, we build each fraction to an equivalent one with the LCD. Comment on Francine's method.
8. When might you expect to encounter extraneous solutions?
9. Which of the operations listed below use an LCD?

    add fractions   multiply fractions   solve an equation with fractions
    subtract fractions   divide fractions

10. Which of the operations listed above use building factors?

Review Problems

1. **a.** Evaluate the fraction $\dfrac{s^2 - s}{s^2 + 3s - 10}$ for $s = -2$
    b. For what value(s) of s is the fraction undefined?

2. Ed's Diner uses a package of coffee filters every $x + 5$ days.
    a. What fraction of a package does Ed use every day?
    b. What fraction of a package does Ed use in one week?

■ For Problems 3-10, reduce the fraction if possible.

3. $\dfrac{a + 3}{b + 3}$

4. $\dfrac{5x + 7}{5x}$

5. $\dfrac{10 + 2y}{2y}$

6. $\dfrac{3x^2 - 1}{1 - 3x^2}$

7. $\dfrac{v - 2}{v^2 - 4}$

8. $\dfrac{q^5 - q^4}{q^4}$

9. $\dfrac{-3x}{6x^2 + 9x}$

10. $\dfrac{x^2 + 5x + 6}{x^2 - 4}$

■ For Problems 11-16,
    a. add the fractions,
    b. multiply the fractions.

11. $\dfrac{3}{8}, \dfrac{5}{12}$

12. $\dfrac{2}{x}, \dfrac{1}{x + 2}$

13. $\dfrac{3x}{2x + 2}, \dfrac{x + 1}{6x}$

14. $\dfrac{x+1}{x-1}, \dfrac{1}{x^2-1}$

15. $2, \dfrac{1}{x}$

16. $x, \dfrac{1}{x+2}$

■ For Problems 17-22, write the expression as a single fraction in lowest terms.

17. $\dfrac{4c^2 d}{3} \div (6cd^2)$

18. $\dfrac{u^2 - 2uv}{uv} \div \dfrac{3u - 6v}{2uv}$

19. $\dfrac{2m^2 - m - 1}{m+1} - \dfrac{m^2 - m}{m+1}$

20. $\dfrac{3}{2p} + \dfrac{7}{6p^2}$

21. $\dfrac{5q}{q-3} - \dfrac{7}{q} + 3$

22. $\dfrac{2w}{w^2 - 4} + \dfrac{4}{w^2 + 4w + 4}$

23. On Saturday mornings, Olive takes her motorboat 5 miles upstream to the general store for supplies and then returns home. The current in the river is 2 miles per hour. Let x represent Olive's speed in still water, and write algebraic fractions to answer each question.
   a. How long does it take Olive to get to the store?
   b. How long does the return trip take?
   c. How long does the round trip take?

24. On spring break, Johann and Sebastian both walk from the university to the next town. Johann leaves at noon and Sebastian leaves one hour later, but Sebastian walks 1 mile per hour faster. Let r stand for Johann's walking speed. Write polynomials to answer the following questions.
   a. What is Sebastian's walking speed?
   b. How far has Sebastian walked at 3 pm?
   c. How far has Johann walked at 3 pm?
   d. How far apart are Johann and Sebastian at 3 pm?

■ For Problems 25-28, solve the equation.

25. $q - \dfrac{16}{q} = 6$

26. $\dfrac{2-x}{5x} = \dfrac{4}{15x} - \dfrac{1}{6}$

27. $\dfrac{9}{m+2} + \dfrac{2}{m} = 2$

28. $\dfrac{15}{x^2 - 3x} + \dfrac{4}{x} = \dfrac{5}{x-3}$

■ For Problems 29-30, solve for the indicated variable.

29. $\dfrac{1}{x} + \dfrac{1}{y} = \dfrac{1}{z}$ for x

30. $y = \dfrac{2x+3}{1-x}$ for x

■ For Problems 31-32,
   a. Write an equation to model the problem, and
   b. Solve your equation and answer the question posed in the problem.

31. On a walking tour, Nora walks uphill 5 miles to an inn where she has lunch. After lunch, she increases her speed by 2 miles per hour and walks for 8 more miles. If she walked for 1 hour longer before lunch than after lunch, what was her speed before lunch?

32. Brenda can fill her pool in 30 hours using the normal intake pipe. She can instead fill the pool in 45 hours using the garden hose. How long will it take to fill the empty pool if both the pipe and garden hose are running?

Chapter 9   More about Exponents and Roots

Lesson 1   Laws of Exponents

- Apply the laws of exponents
- Simplify expressions involving powers

Lesson 2   Negative Exponents and Scientific Notation

- Write expressions using negative exponents
- Simplify expressions using the laws of exponents
- Convert between standard and scientific notation
- Perform computations using scientific notation

Lesson 3   Properties of Radicals

- Apply the product and quotient rules for radicals
- Simplify square roots
- Add or subtract like radicals

Lesson 4   Operations on Radicals

- Multiply radical expressions
- Divide radicals
- Simplify fractions involving radicals
- Rationalize denominators

Lesson 5   Equations with Radicals

- Solve equations involving radicals
- Identify extraneous solutions

9.1 Laws of Exponents

Products and Quotients

In Chapter 7 we learned the first and second laws of exponents, which we use to compute products and quotients of powers.

■ **Example 1**   Simplify each product or quotient.

   a. $5^2 \cdot 5^6$ **b.** $\dfrac{3^7}{3^2}$ **c.** $\dfrac{2^3}{2^5}$

Solutions a. To multiply two powers with the same base, we add the exponents and leave the base unchanged.

$$5^2 \cdot 5^6 = 5^8$$

b. To divide two powers with the same base, we subtract the smaller exponent from the larger. If the larger exponent occurs in the numerator, we put the power in the numerator.

$$\frac{3^7}{3^2} = 3^5 \qquad \text{Larger exponent is in the numerator.}$$

c. If the larger exponent occurs in the denominator, we put the power in the denominator.

$$\frac{2^3}{2^5} = \frac{1}{2^2} \qquad \text{Larger exponent is in the denominator.} \qquad ■$$

Caution! In Example 1a, it is **not** correct to multiply the bases:

$$5^2 \cdot 5^6 \rightarrow 25^8 \qquad \textbf{Incorrect!}$$

The exponents tell us how many copies of the base to multiply together, but the base does not change when we apply the first law of exponents.

RQ1.   Why do we add exponents when we multiply powers with the same base?

These laws are expressed in symbols below.

First Law of Exponents

$$a^m \cdot a^n = a^{m+n}$$

Second Law of Exponents

$$\frac{a^m}{a^n} = a^{m-n} \quad (n < m)$$

$$\frac{a^m}{a^n} = \frac{1}{a^{n-m}} \quad (n > m)$$

In this Lesson we'll study three more laws of exponents.

Power of a Power

Consider the expression $(x^4)^3$, or the cube of x^4. We can simplify this expression as follows.

$$(x^4)^3 = (x^4)(x^4)(x^4) = x^{4+4+4} = x^{12} \quad \textbf{Add exponents.}$$

We wrote $(x^4)^3$ as a repeated product and applied the first law of exponents to add the exponents. Of course, because repeated addition is actually multiplication, we can just multiply the exponents together: $3(4) = 12$. This gives us another rule.

> **Third Law of Exponents**
> To raise a power to a power, keep the same base and multiply the exponents. In symbols,
>
> $$(a^m)^n = a^{mn}$$

■ **Example 2a.** $(b^2)^4 = b^{2 \cdot 4} = b^8$ **Multiply exponents.**
b. $(y^2)^5 = y^{2 \cdot 5} = y^{10}$ ■

Caution! Note carefully the difference between the two expressions $(x^3)(x^4)$ and $(x^3)^4$:

$$(x^3)(x^4) = x^{3+4} = x^7 \quad \textbf{Add the exponents.}$$
$$(x^3)^4 = x^{3 \cdot 4} = x^{12} \quad \textbf{Multiply the exponents.}$$

The first expression is a product, so we add the exponents. The second expression raises a power to a power, so we multiply the exponents. ■

RQ2. Which is larger, $(2^3)^3$ or $2^3 \cdot 2^3$? Why?

Power of a Product

To simplify the expression $(2x)^3$, we can use the commutative and associative properties of multiplication to write

$$\begin{aligned}(2x)^3 &= (2x)(2x)(2x) \\ &= 2 \cdot 2 \cdot 2 \cdot x \cdot x \cdot x \\ &= 2^3 x^3\end{aligned}$$

Thus, to raise a product to a power we can raise each factor to the power. This example illustrates the following rule.

> **Fourth Law of Exponents**
> To raise a product to a power, raise each factor to the power. In symbols,
>
> $$(ab)^n = a^n b^n$$

■ **Example 3a.** $(5ab)^2 = 5^2 a^2 b^2 = 25a^2b^2$ **Square each factor.**
b. $(-xy^3)^4 = (-x)^4(y^3)^4$ Raise each factor to the fourth power;
       $= x^4 y^{12}$ apply the third law of exponents. ■

Caution!   Note the difference between the two expressions $3a^2$ and $(3a)^2$:

$$3a^2 \quad \text{cannot be simplified, but}$$

$$(3a)^2 = 3^2 a^2 = 9a^2$$

In the expression $3a^2$, only the factor a is squared, but in $(3a)^2$ both 3 and a are squared. ■

RQ3. Explain why we cannot use the fourth law of exponents to simpify $(a+b)^3$ as $a^3 + b^3$.

Power of a Quotient

To simplify the expression $\left(\dfrac{x}{2}\right)^4$, we multiply together four copies of the fraction $\dfrac{x}{2}$. That is,

$$\left(\frac{x}{2}\right)^4 = \frac{x}{2} \cdot \frac{x}{2} \cdot \frac{x}{2} \cdot \frac{x}{2} = \frac{x \cdot x \cdot x \cdot x}{2 \cdot 2 \cdot 2 \cdot 2}$$
$$= \frac{x^4}{2^4} = \frac{x^4}{16}$$

This example suggests the following rule.

> **Fifth Law of Exponents**
> To raise a quotient to a power, raise both the numerator and the denominator to the power. In symbols,
> $$\left(\frac{a}{b}\right)^n = \frac{a^n}{b^n}$$

■ **Example 4a.** $\left(\dfrac{x}{y}\right)^5 = \dfrac{x^5}{y^5}$ **Raise top and bottom to the fifth power.**

b. $\left(\dfrac{2}{y^2}\right)^3 = \dfrac{2^3}{(y^2)^3}$ **Raise top and bottom to the third power.**

$= \dfrac{2^3}{y^{2 \cdot 3}} = \dfrac{8}{y^6}$ **Apply the third law of exponents to the denominator.** ■

RQ4. Why does $\left(\dfrac{1}{x}\right)^4 = \dfrac{1}{x^4}$?

Using the Laws of Exponents

We can use the laws of exponents along with the order of operations to simplify algebraic expressions. The five laws of exponents are stated together below. All of the laws are valid when the base a is not equal to zero and when the exponents m and n are positive integers.

Laws of Exponents

1. $a^m \cdot a^n = a^{m+n}$

2. $\dfrac{a^m}{a^n} = a^{m-n}$ $(n < m)$

   $\dfrac{a^m}{a^n} = \dfrac{1}{a^{n-m}}$ $(n > m)$

3. $(a^m)^n = a^{mn}$

4. $(ab)^n = a^n b^n$

5. $\left(\dfrac{a}{b}\right)^n = \dfrac{a^n}{b^n}$

Example 5 Simplify $2x^2 y(3xy^2)^4$

Solution According to the order of operations, we should perform powers before products, so we first simplify the factor $(3xy^2)^4$.

$$(3xy^2)^4 = 3^4 x^4 (y^2)^4 \quad \textbf{Fourth law: raise each factor to the power.}$$
$$= 81 x^4 y^8 \quad \textbf{Third law: } (y^2)^4 = y^{2 \cdot 4}$$

Now multiply the result by $2x^2 y$, and apply the first law to obtain

$$2x^2 y(81 x^4 y^8) = 2 \cdot 81 x^2 \cdot x^4 \cdot y \cdot y^8$$
$$= 162 x^6 y^9$$

RQ5. In Example 5, why did we start by simplifying $(3xy^2)^4$?

Example 6 Simplify $(-x)^4 (-xy)^3$

Solution Each power should be simplified before we compute their product. Because 4 is an even exponent, $(-x)^4$ is positive, so

$$(-x)^4 = x^4$$

To simplify $(-xy)^3$, we apply the fourth law of exponents:

$$(-xy)^3 = (-x)^3 y^3 \quad \textbf{3 is an odd power, so } (-x)^3 = -x^3.$$
$$= -x^3 y^3$$

Finally, we multiply the powers together to get

$$(-x)^4 (-xy)^3 = x^4 (-x^3 y^3) \quad \textbf{Apply the first law.}$$
$$= -x^7 y^3$$

RQ6. In the expression $-x^3 y^3$, does the negtive sign apply to both factors, or only one of them?

Skills Warm-Up

Simplify each expression according to the order of operations.

1. $10 - 6^2$
2. $10(-6)^2$
3. $(10 - 6)^2$
4. $2 \cdot 5^2 - 3^2$
5. $-2 \cdot 5^2 (-3^2)$
6. $-2^2 - (5 - 3)^2$

Answers: **1.** -26 **2.** 360 **3.** 16 **4.** 41 **5.** 450 **6.** -8

Homework 9.1

1. Simplify each power by using the third law of exponents.
   a. $(t^3)^5$ b. $(b^4)^2$ c. $(w^{12})^{12}$

2. Simplify each power by using the fourth law of exponents.
   a. $(5x)^3$ b. $(-3wz)^4$ c. $(-ab)^5$

3. Simplify each power by using the fifth law of exponents.
   a. $\left(\dfrac{w}{2}\right)^6$ b. $\left(\dfrac{5}{v}\right)^4$ c. $\left(\dfrac{-m}{p}\right)^3$

4. Simplify each expression.
   a. $x^3 \cdot x^6$ b. $(x^3)^6$ c. $\dfrac{x^3}{x^6}$ d. $\dfrac{x^6}{x^3}$

5. Explain the difference between the first and third laws of exponents. Use examples.

6. Find and correct the error in each calculation.
   a. $2 \cdot 3^2 \rightarrow 36$ b. $-10^2 \rightarrow 100$ c. $a^4 \cdot a^3 \rightarrow a^{12}$

For Problems 7-9, use the laws of exponents to simplify each expression.

7. $(2p^3)^5$ 8. $\left(\dfrac{-3}{q^4}\right)^5$ 9. $\left(\dfrac{-2h^2}{m^3}\right)^4$

For Problems 10-15, simplify.

10. $x^3(x^2)^5$ 11. $(2x^3y)^2(xy^3)^4$ 12. $[ab^2(a^2b)^3]^3$
13. $-a^2(-a)^2$ 14. $-(-xy)^2(xy^2)$ 15. $-4p(-p^2q^2)^2(-q^3)^2$

For Problems 16-18, simplify.

16. $2y(y^3)^2 - 2y^4(3y)^3$ 17. $2a(a^2)^4 + 3a^2(a^6) - a^2(a^2)^3$
18. $-3v^2(2v^3 - v^2) + v(-4v)^2$

For Problems 19-23, simplify each pair of expressions as much as possible.

19a. $4x^2 + 2x^4$ b. $4x^2(2x^4)$
20a. $(-x)^3x^4$ b. $[(-x^3)(-x)]^4$
21a. $(3x^2)^4(2x^4)^2$ b. $(3x^2)^4 - (2x^4)^2$
22a. $6x^3 - 3x^6$ b. $6x^3(-3x^6)$
23a. $6x^3 - 3x^3(x^3)$ b. $(6x^3 - 3x^3)x^3$

For Problems 24-26, find the value of n.

24. $b^3 \cdot b^n = b^9$ 25. $\dfrac{c^8}{c^n} = c^2$ 26. $\dfrac{n^3}{3^3} = 8$

For Problems 27-28, factor.

27. $x^4 + x^6 = x^2(\quad)$ 28. $4m^4 - 4m^8 + 8m^{16} = 4m^4(\quad)$

■ Mental exercise: For Problems 29-34, replace the comma with the appropriate symbol, $<$, $>$, or $=$. Do not use pencil, paper, or calculator.

29. -8^2, 64

30. $(-3)^5$, -3^5

31. $\left(\dfrac{-7}{4}\right)^{11}$, 0

32. $6^{10} \cdot 4^{10}$, 24^{10}

33. $(8-2)^3$, $8-2^3$

34. $(17^4)^5$, $(17^5)^4$

9.2 Negative Exponents and Scientific Notation

In this Lesson consider negative exponents. Negative exponents simplify many calculations and give us a useful form for writing very large or very small numbers.

First, what do negative exponents mean? In order for them to be useful, their properties must fit with what we already know about exponents.

Zero as an Exponent

We know that a^n means the product of n factors of a. For example, a^3 means $a \cdot a \cdot a$. Is there a reasonable meaning for the power a^0?

Consider the quotient $\dfrac{a^4}{a^4}$. If a is not zero, this quotient equals 1, because any nonzero number divided by itself is 1. On the other hand, we can also think of $\dfrac{a^4}{a^4}$ as a quotient of powers and subtract the exponents. If we extend the second law of exponents to include the case $m = n$, we have

$$\frac{a^4}{a^4} = a^{4-4} = a^0$$

Thus, it seems reasonable to make the following definition.

> **Zero as an Exponent**
>
> $$a^0 = 1, \quad \text{if} \quad a \neq 0$$

For example,

$$3^0 = 1, \quad (-427)^0 = 1, \quad \text{and} \quad (5xy)^0 = 1 \quad (\text{if } x, y \neq 0)$$

> **RQ1.**   Why does it make sense that $a^0 = 1$ (as long as $a \neq 0$) ?

Negative Exponents

We can also use the second law of exponents to give meaning to negative exponents. Consider the quotient $\dfrac{a^4}{a^7}$. According to the second law of exponents,

$$\frac{a^4}{a^7} = \frac{1}{a^{7-4}} = \frac{1}{a^3}$$

However, if we allow negative numbers as exponents, we can apply the first half of the second law to obtain

$$\frac{a^4}{a^7} = a^{4-7} = a^{-3}$$

We therefore define a^{-3} to mean $\dfrac{1}{a^3}$. In general, for $a \neq 0$, we make the following definition.

Negative Exponents

$$a^{-n} = \frac{1}{a^n}, \quad \text{if } a \neq 0$$

For example,

$$2^{-4} = \frac{1}{2^4} \quad \text{and} \quad x^{-5} = \frac{1}{x^5}$$

A negative power is the reciprocal of the corresponding positive power.

■ **Example 1** Write each expression without exponents.

a. 10^{-4}
b. $\left(\dfrac{1}{4}\right)^{-3}$
c. $\left(\dfrac{3}{5}\right)^{-2}$

Solutions a. $10^{-4} = \dfrac{1}{10^4} = \dfrac{1}{10,000}$, or 0.0001

b. To compute a negative power of a fraction, we compute the corresponding positive power of its reciprocal. (This is because of the fifth law of exponents.)

$$\left(\frac{1}{4}\right)^{-3} = 4^3 = 64$$

c. As in part (b), we compute the corresponding positive power of the reciprocal of $\frac{3}{5}$.

$$\left(\frac{3}{5}\right)^{-2} = \left(\frac{5}{3}\right)^2 = \frac{25}{9}$$ ■

RQ2. What does a negative exponent mean?

Caution!   A negative exponent does **not** mean that the power is negative.
For example,

$$2^{-4} \neq -2^4$$ ■

We can also rewrite fractions using negative exponents.

■ **Example 2** Write each expression using negative exponents.

a. $\dfrac{1}{3^4}$
b. $\dfrac{7}{10^2}$
c. $\dfrac{8}{x}$

Solutions a. $\dfrac{1}{3^4} = 3^{-4}$
b. $\dfrac{7}{10^2} = 7 \cdot \dfrac{1}{10^2} = 7 \cdot 10^{-2}$
c. $\dfrac{8}{x} = 8 \cdot \dfrac{1}{x} = 8x^{-1}$ ■

Caution!   An exponent applies only to its base. For example, the exponent -2 in the expression $3x^{-2}$ applies only to x, but in $(3x)^{-2}$, the exponent applies to $3x$. Thus,

$$3x^{-2} = 3 \cdot \frac{1}{x^2} = \frac{3}{x^2} \quad \text{but} \quad (3x)^{-2} = \frac{1}{(3x)^2} = \frac{1}{9x^2}$$ ■

Laws of Exponents

The laws of exponents apply to negative exponents. In particular, if we allow negative exponents, we can write the second law as a single rule.

Laws of Exponents

1. $a^m \cdot a^n = a^{m+n}$ 4. $(ab)^n = a^n b^n$

2. $\dfrac{a^m}{a^n} = a^{m-n}, \quad a \neq 0$ 5. $\left(\dfrac{a}{b}\right)^n = \dfrac{a^n}{b^n}, \quad b \neq 0$

3. $(a^m)^n = a^{mn}$

■ **Example 3**   Simplify by using the first or second law of exponents.

a. $x^5 \cdot x^{-8}$
                                              b. $\dfrac{5^2}{5^{-6}}$

Solutions a.   Apply the first law: add the exponents.

$$x^5 \cdot x^{-8} = x^{5-8} = x^{-3}$$

b. Apply the second law: subtract the exponents.

$$\frac{5^2}{5^{-6}} = 5^{2-(-6)} = 5^8$$ ■

Look Closer: As a special case of the second law, note that $1 = a^0$, so we can write

$$\frac{1}{a^{-n}} = \frac{a^0}{a^{-n}} = a^{0-(-n)} = a^n$$ •

Thus,

$$\frac{1}{a^{-n}} = a^n \quad \text{and} \quad \frac{b}{a^{-n}} = ba^n, \quad a \neq 0$$

■ **Example 4a.** $\dfrac{1}{2^{-3}} = 2^3$ b. $\dfrac{8}{x^{-5}} = 8x^5$ ■

Here are some examples using the laws of exponents.

■ **Example 5**   Simplify by using the third or fourth law of exponents.

a. $(2^{-3})^{-3}$ b. $(ab)^{-3}$

Solutions a.   Apply the third law: multiply the exponents.

$$(2^{-3})^{-3} = 2^{-3(-3)} = 2^9$$

b. Apply the fourth law: raise each factor to the power.

$$(ab)^{-3} = a^{-3}b^{-3}$$
■

RQ3.   What is the difference between 2^{-3} and -2^3?

Powers of Ten

Scientists and engineers often encounter very large numbers such as

$$5,980,000,000,000,000,000,000,000$$

(the mass of the earth in kilograms) and very small numbers such as

$$0.000\ 000\ 000\ 000\ 000\ 000\ 000\ 00167$$

(the mass of a hydrogen atom in grams) in their work. These numbers can be written in a more compact and useful form using powers of 10.

Recall the following facts about our base-10 number system.

■ **Example 6a.** Multiplying a number by a positive power of 10 has the effect of moving the decimal point k places to the right, where k is the exponent on 10. For example,

$$2.358 \times 10^2 = 235.8 \qquad \text{and} \qquad 17 \times 10^4 = 170,000$$

b. Multiplying a number by a negative power of 10 has the effect of moving the decimal point to the left. For example,

$$5452 \times 10^{-3} = 5.452 \qquad \text{and} \qquad 2.3 \times 10^{-5} = 0.000\ 023$$
■

RQ4.   What happens to a number when we multiply it by a power of 10?

We can also reverse the process above to write a number in factored form, where one factor is a power of 10.

■ **Example 7**   Fill in the correct power of 10 for each factored form.

a. $38,400 = 3.84 \times$ _____   **b.** $0.0057 = 5.7 \times$ _____

Solutions a.   To recover $38,400$ from 3.84, we must move the decimal point four places to the right, so we multiply by 10^4. Thus,

$$38,400 = 3.84 \times 10^4$$

b. To recover 0.0057 from 5.7, we must move the decimal point three places to the left, so we multiply by 10^{-3}. Thus,

$$0.0057 = 5.7 \times 10^{-3}$$
■

Scientific Notation

If there is just one digit to the left of the decimal point in the factored form, we say that the number is written in **scientific notation**.

> A number is in **scientific notation** if it is expressed as the product of a number between 1 and 10 and a power of 10.

For example,

$$4.18 \times 10^{12}, \quad 2.9 \times 10^{-8}, \quad \text{and} \quad 4 \times 10^{1}$$

are written in scientific notation.

To write a number in scientific notation, we first position the decimal point and then determine the correct power of 10.

> **To Write a Number in Scientific Notation:**
> 1. Locate the decimal point so that there is exactly one nonzero digit to its left.
> 2. Count the number of places you moved the decimal point: this determines the power of 10.
>   a. If the original number is greater than 10, the exponent is positive.
>   b. If the original number is less than 1, the exponent is negative.

Example 8   Write each number in scientific notation.
a. 62,000,000   b. 0.000431

Solutions a.   We position the decimal point so that there is just one nonzero digit to the left of the decimal.

$$62,000,000 = 6.2 \times \underline{\hspace{2cm}}$$

To recover 62,000,000 from 6.2, we move the decimal point seven places to the right. Therefore, we multiply 6.2 by 10^{7}.

$$62,000,000 = 6.2 \times 10^{7}$$

b. We position the decimal point so that there is just one nonzero digit to the left of the decimal.

$$0.000431 = 4.31 \times \underline{\hspace{2cm}}$$

To recover 0.000431 from 4.31, we move the decimal point four places to the left. Therefore, we multiply 4.31 by 10^{-4}.

$$0.000431 = 4.31 \times 10^{-4}$$

Look Closer: The easiest way to remember which way to move the decimal point is to note whether the number is large (greater than 10) or small (less than 1). For example,

- 62,000,000 is a large number, so the exponent on 10 in its scientific form is positive. On the other hand,
- 0.000431 is a decimal fraction less than 1, so the exponent on 10 in its scientific form is negative.

> **RQ5.** How can you tell whether the exponent in scientific notation should be
> positive or negative?

Calculators and Scientific Notation

Your calculator displays numbers in scientific notation if they have too many digits to
fit in the display screen.

Example 9   Use a calculator to compute the square of 12,345,678.

Solution We enter

$$12345678 \;\boxed{x^2}$$

(and press enter on a graphing calculator.) A scientific calculator displays the result as

$$\boxed{1.524157653 \qquad 14}$$

and a graphing calculator displays

$$1.524157653\ \mathbf{E}\ 14$$

(Some calculators may round the result to fewer decimal places.) Both of these displays
represent the number $1.524157653 \times 10^{14}$.

Because scientific notation always involves a power of 10, most calculators do not
display the 10, but only its exponent. If you now press the $\boxed{1/x}$ key, your calculator will
display the reciprocal of $1.524157653 \times 10^{14}$ as

$$\boxed{6.56100108 \qquad -15} \qquad \text{or} \qquad 6.56100108\ \mathbf{E}\ -15$$

This is how your calculator displays the number $6.56100108 \times 10^{-15}$.

Caution!   Do not use your calculator's notation when giving an answer in scientific
notation. For example, if your calculator reports a result as 3.47 **E** 8, you should write
3.47×10^8.

> **RQ6.** How does your calculator display scientific notation?

Your calculator has a special key for entering numbers in scientific notation.

To enter a number in scientific notation into the calculator, we use the key
marked either $\boxed{\textbf{EXP}}$ or $\boxed{\textbf{EE}}$.

For instance, to enter 6.02×10^{23}, we key in the sequence

$$6.02 \;\boxed{\textbf{EXP}}\; 23$$

To enter a number with a negative exponent, such as 1.66×10^{-27}, on a scientific
calculator we key in

$$1.66 \;\boxed{\textbf{EXP}}\; 27 \;\boxed{+/-}$$

On a graphing calculator we key in

$$1.66 \boxed{\text{EE}} \boxed{(-)} 27$$

■ **Example 10**   The People's Republic of China encompasses about $2,317,400,000$ acres of land and has a population of $1,335,300,000$ people. How many acres of land per person are there in China?

Solution We divide the number of acres of land by the number of people. To do this, we first write each number in scientific notation.

$$2,317,400,000 = 2.3174 \times 10^9$$
$$1,335,300,000 = 1.3353 \times 10^9$$

We enter the division into the calculator as follows.

$$2.3174 \boxed{\text{EXP}} 9 \boxed{\div} 1.3353 \boxed{\text{EXP}} 9 \boxed{=}$$

The calculator displays the quotient, 1.735. Thus, there are about 1.735 acres of land per person in China.   ■

Mental Calculation with Scientific Notation

We can often obtain a quick estimate for a calculation by converting each figure to scientific notation and rounding to just one or two digits.

■ **Example 11**   Use scientific notation to find the product

$$(6,200,000,000)(0.000\,000\,3)$$

Solution We convert each number to scientific notation.

$$(6,200,000,000)(0.000\,000\,3) = (6.2 \times 10^9)(3 \times 10^{-7})$$

We can use a calculator (if necessary) to combine the decimal numbers, and the first law of exponents to combine the powers of 10.

$$
\begin{aligned}
(6,200,000,000)(0.000\,000\,3) &= (6.2 \times 10^9)(3 \times 10^{-7}) & \text{Rearrange the factors.} \\
&= (6.2 \times 3) \times (10^9 \times 10^{-7}) & 10^{9-7} = 10^2 \\
&= 18.6 \times 10^2 = 1860 &
\end{aligned}
$$
■

Skills Warm-Up

■ Decide whether each statement is true or false.

1. The reciprocal of $x + 2$ is $\dfrac{1}{x} + \dfrac{1}{2}$.

2. The reciprocal of $\dfrac{1}{3} + \dfrac{1}{4}$ is 7.

3. $\dfrac{3}{\frac{1}{5}} = 15$

4. $\dfrac{3}{\frac{1}{5} + \frac{1}{2}} = 15 + 6$

5. The reciprocal of $\dfrac{1}{x}$ is x.

6. The reciprocal of $\dfrac{1}{x+y}$ is $x + y$.

Answers: 1. False   **2.** False   **3.** True   **4.** False   **5.** True   **6.** True

Homework 9.2

Skills Practice

■ For Problems 1-2, write without using zero or negative exponents and simplify.

1. a. 5^{-2} **b.** x^{-6} **c.** $(8x)^0$ **d.** $\left(\dfrac{3}{4}\right)^{-3}$

2. a. $\left(\dfrac{b}{3}\right)^{-4}$ **b.** $(2q)^{-5}$ **c.** $3 \cdot 4^{-3}$ **d.** $4x^{-2}$

■ For Problems 3-4, write each expression using negative exponents.

3. a. $\dfrac{1}{2^3}$ **b.** $\dfrac{3}{5^2}$ **c.** $\dfrac{1}{27}$

4. a. $\dfrac{x}{625}$ **b.** $\dfrac{2}{z^2}$ **c.** $\left(\dfrac{z}{10}\right)^5$

5. Find and correct the error in each calculation.

   a. $x^0 \rightarrow 0$ **b.** $w^{-3} \rightarrow -w^3$ **c.** $2x^{-4} \rightarrow \dfrac{1}{2x^4}$

6. Simplify by using the first law of exponents.
   a. $x^{-3} \cdot x^8$ **b.** $5^{-4} \cdot 5^{-3}$ **c.** $(3b^{-5})(5b^2)$

7. Simplify by using the second law of exponents.
   a. $\dfrac{c^{-7}}{c^{-4}}$ **b.** $\dfrac{8b^{-4}}{4b^{-8}}$ **c.** $\dfrac{6^6}{6^{-2}}$

8. Simplify.
   a. $\dfrac{1}{6^{-3}}$ **b.** $\dfrac{3}{2^{-6}}$ **c.** $\dfrac{8x^3}{y^{-5}}$

9. Simplify by using the third law of exponents.
   a. $(8^{-2})^5$ **b.** $(w^{-6})^{-3}$ **c.** $(d^6)^{-4}$

10. Simplify by using the fourth law of exponents.
   a. $(pq)^{-5}$ **b.** $(3x)^{-2}$ **c.** $5(2r)^{-3}$

■ For Problems 11-14, simplify.

11. $(a^{-4}c^2)^{-3}$ **12.** $(2u^2)^{-3}(u^{-4})^2$

13. $\dfrac{5k^{-3}(k^4)^{-3}}{6k^{-5}}$ **14.** $\left(\dfrac{2p^{-3}}{p^2}\right)^{-2}$

■ Mental exercise: For Problems 15-16, simplify each quotient without using pencil, paper, or calculator. Write your answer as a power of 10.

15a. $\dfrac{10^3}{10^{-2}}$ **b.** $\dfrac{10^{-5}}{10^{-3}}$ **c.** $\dfrac{10}{10^{-1}}$

16a. $\dfrac{10^{-3} \times 10^{-2}}{10^{-6}}$ **b.** $\dfrac{10^{-5}}{10^{-4} \times 10^7}$

17. Compute each product.

    a. 4.3×10^4 **b.** 8×10^{-6} **c.** 0.002×10^{-2}

18. Complete each factored form.

    a. $234 = 2.34 \times$ _____   **b.** $0.92 = 9.2 \times$ _____

    c. $1,720,000 = 1.72 \times$ _____

■ For Problems 19-20, write in scientific notation.

19a.   4834   **b.** 0.072 **c.** $0.000\,007$

20a.   685,000,000   **b.** 56.74×10^4 **c.** 385×10^{-3}

■ For Problems 21-24, use scientific notation to compute.

21. $(2,000,000)(0.000\,07)$ **22.** $0.000\,036 \div 0.000\,9$

23. $\dfrac{(80,000,000,000)(0.000\,6)}{20,000}$ **24.** $\dfrac{(0.000\,000\,25)(90,000,000)}{(1,500,000)(0.000\,000\,2)}$

Applications

■ For Problems 25-28, suppose that each of the following numbers is written in scientific notation. Decide whether the exponent on 10 is positive or negative.

25. The population of Los Angeles County.

26. The length of a football field in miles.

27. The number of seconds in a year.

28. The speed at which your hair grows in miles per hour.

29. Write each number in scientific notation.

    a. The height of Mount Everest is 29,141 feet.

    b. The wavelength of red light is 0.000 076 centimeters.

30. Write each number in standard notation.

    a. An amoeba weighs about 5×10^{-6} gram.

    b. Our solar system has existed for about 5×10^9 years.

■ For Problems 31-34, give your answers in scientific notation rounded to two decimal places.

31. A light year is the distance that light can travel in one year. Light travels at approximately 1.86×10^5 miles per second. How many miles are there in one light year? (One year is approximately 3.16×10^7 seconds.)

32. A 1-foot high stack of dollar bills contains 3.6×10^3 bills.

    a. In January, 2013 the U. S. federal debt was 1.644×10^{13} dollars. How tall a stack of one-dollar bills, in feet, would be needed to pay off the federal debt?

    b. Express the height of the stack of bills in part (a) in miles. (One mile equals 5280 feet.)

33. There are about 200 million insects for every person on earth. In 2013, the world's population is about 7.1 billion people.

   a. How many insects are there on earth?

   b. If the average insect weighs 3×10^{-4} gram, how much do all the insects on earth weigh?

34. There are 5,800,000 cubic miles of fresh water on earth. Each cubic mile is equal to 110,000,000,000 gallons of water. If the population of earth was about 7.1 billion people in 2013, how many gallons of fresh water were there for each person?

9.3 Properties of Radicals

Radicals with the same value may be written in different forms. Consider the following calculation:

$$(2\sqrt{2})^2 = 2^2(\sqrt{2})^2 \qquad \text{By the fourth law of exponents.}$$
$$= 4(2) = 8$$

This calculation shows that $2\sqrt{2}$ is equal to $\sqrt{8}$, because the square of $2\sqrt{2}$ is 8. You can verify on your calculator that $2\sqrt{2}$ and $\sqrt{8}$ have the same decimal approximation, 2.828.

■ **Example 1**   Show that $\sqrt{45} = 3\sqrt{5}$.

Solution $\sqrt{45}$ is a number whose square is 45. We can square $3\sqrt{5}$ as follows:

$$(3\sqrt{5})^2 = 3^2(\sqrt{5})^2$$
$$= 9(5) = 45$$

Because the square of $3\sqrt{5}$ is equal to 45, it is the case that $3\sqrt{5} = \sqrt{45}$.   ■

It is usually helpful to write a radical expression as simply as possible. The expression $2\sqrt{2}$ is considered simpler than $\sqrt{8}$, because the radicand is a smaller number. Similarly, $3\sqrt{5}$ is simpler than $\sqrt{45}$. In this section we discover properties of radicals that help us simplify radical expressions.

Look Ahead: In the Activities we will verify the following properties of radicals.

Product Rule for Radicals

$$\text{If} \quad a, b \geq 0, \quad \text{then} \quad \sqrt{ab} = \sqrt{a}\sqrt{b}$$

Quotient Rule for Radicals

$$\text{If} \quad a \geq 0, \, b > 0, \quad \text{then} \quad \sqrt{\frac{a}{b}} = \frac{\sqrt{a}}{\sqrt{b}}$$

Caution!   It is just as important to remember that we do not have a sum or difference rule for radicals. That is, in general,

$$\sqrt{a+b} \neq \sqrt{a} + \sqrt{b}$$

$$\sqrt{a-b} \neq \sqrt{a} - \sqrt{b}$$
■

■ **Example 2**   Decide which statement is true and which is false.
a. $\sqrt{4} + \sqrt{9} = \sqrt{13}$?   **b.** $\sqrt{4}\sqrt{9} = \sqrt{36}$?

Solutions a. $\sqrt{4} + \sqrt{9} = 2 + 3 = 5$, but $\sqrt{13} \neq 5$, so the first statement is false.
b. $\sqrt{4}\sqrt{9} = 2(3) = 6$, and $\sqrt{36} = 6$, so the second statement is true.   ■

Simplifying Square Roots

We can use the product rule for radicals to write $\sqrt{12} = \sqrt{4}\sqrt{3}$. Now, $\sqrt{4} = 2$, so we can simplify $\sqrt{12}$ as

$$\sqrt{12} = \sqrt{4}\sqrt{3} = 2\sqrt{3}$$

The expression $2\sqrt{3}$ is a simplified form for $\sqrt{12}$. The factor of 4, which is a perfect square, has been removed from the radical. This example illustrates a strategy for simplifying radicals.

To Simplify a Square Root:
1. Factor any perfect squares from the radicand.
2. Use the product rule to write the radical as a product of two square roots.
3. Simplify the square root of the perfect square.

■ **Example 3**   Simplify $\sqrt{45}$

Solution We look for a perfect square that divides evenly into 45. The largest perfect square that divides 45 is 9, so we factor 45 as $9 \cdot 5$. Then we use the product rule to write

$$\sqrt{45} = \sqrt{9 \cdot 5} = \sqrt{9}\sqrt{5}$$

Finally, we simplify $\sqrt{9}$ to get

$$\sqrt{45} = \sqrt{9}\sqrt{5} = 3\sqrt{5} \qquad ■$$

Caution!   Finding a decimal approximation for a radical is not the same as simplifying the radical. In Example 1, we can use a calculator to find

$$\sqrt{45} \approx 6.708$$

but 6.708 is not the exact value for $\sqrt{45}$. For long calculations, too much accuracy may be lost by approximating each radical. However, $3\sqrt{5}$ is equivalent to $\sqrt{45}$, which means that their values are exactly the same. We can replace one expression by the other without losing accuracy. ■

RQ1.   What is the difference between simplifying a square root and approximating a square root?

Square Root of a Variable Expression

To square a power of a variable we double the exponent. For example, the square of x^5 is

$$(x^5)^2 = x^{5 \cdot 2} = x^{10}$$

Because taking the square root of a number is the opposite of squaring a number, to take the square root of an even power we divide the exponent by 2.

Example 4   If x is a nonnegative number,

$$\sqrt{x^{10}} = x^5 \quad \text{because} \quad (x^5)^2 = x^{10}$$

Similarly, if a is not negative,

$$\sqrt{a^6} = a^3 \quad \text{because} \quad (a^3)^2 = a^6$$

RQ2. How do we simplify the square root of a variable raised to an even exponent?

For the rest of this section we'll assume that all variables are nonnegative.

Caution!   Note that $\sqrt{a^{16}}$ is not equal to a^4. Compare the two radicals:

$$\sqrt{16} = 4 \quad \text{but} \quad \sqrt{a^{16}} = a^8$$

Look Closer: How can we simplify the square root of an odd power? We write the power as a product of two factors, one having an even exponent and one having exponent 1.

Example 5   Simplify $\sqrt{x^7}$

Solution We factor x^7 as $x^6 \cdot x$. Then we use the product rule to write

$$\sqrt{x^7} = \sqrt{x^6 \cdot x} = \sqrt{x^6} \cdot \sqrt{x}$$

Finally, we simplify the square root of x^6 to get

$$\sqrt{x^7} = \sqrt{x^6}\sqrt{x} = x^3\sqrt{x}$$

RQ3. How do we simplify the square root of a variable raised to an odd exponent?

If the radicand contains more than one variable or a coefficient, we consider the constants and each variable separately. We try to remove the largest factors possible from the radicand.

Example 6   Simplify $\sqrt{20x^2y^3}$

Solution We look for the largest perfect square that divides 20; it is 4. We write the radicand as the product of two factors, one of which contains the perfect square and even powers of the variables. That is,

$$20x^2y^3 = 4x^2y^2 \cdot 5y \quad \textbf{Factor into perfect squares and "leftovers."}$$

Now we write the radical as a product.

$$\sqrt{20x^2y^3} = \sqrt{4x^2y^2 \cdot 5y} = \sqrt{4x^2y^2}\sqrt{5y}$$

Finally, we simplify the first of the two factors to find

$$\sqrt{20x^2y^3} = \sqrt{4x^2y^2}\sqrt{5y} \quad \text{Take square root of the first factor.}$$
$$= 2xy\sqrt{5y}$$

RQ4. In Example 5, what will we get if we square $2xy\sqrt{5y}$?

Sums and Differences

What about sums and differences of radicals? We cannot add or subtract radicals by combining their radicands. That is,

$$\sqrt{a} + \sqrt{b} \neq \sqrt{a+b}$$

You can verify some examples on your calculator:

1. $\sqrt{3} + \sqrt{5} \neq \sqrt{8}$
2. $\sqrt{4} + \sqrt{16} \neq \sqrt{20}$
3. $\sqrt{7} + \sqrt{7} \neq \sqrt{14}$

So, we cannot simplify a sum or difference if the expressions under the radical are different. However, we can combine radicals with the same radicand. For example, we can write

$$\sqrt{7} + \sqrt{7} = 2\sqrt{7}$$

just as we write

$$x + x = 2x$$

Square roots with identical radicands are called **like radicals**.

Look Closer: We can add or subtract like radicals in the same way that we add or subtract like terms, namely by adding or subtracting their coefficients. For example,

$$2r + 3r = 5r$$

where r is a variable that can stand for any real number. In particular, if $r = \sqrt{2}$, we have

$$2\sqrt{2} + 3\sqrt{2} = 5\sqrt{2}$$

Thus, we may add like radicals by adding their coefficients. The same idea applies to subtraction.

RQ5.   What are like radicals?

Example 7   Simplify if possible.
a. $7\sqrt{3} - 2\sqrt{3}$ **b.** $3\sqrt{2} + 4\sqrt{3}$

Solutions a. Because $7\sqrt{3}$ and $2\sqrt{3}$ are like radicals, we can combine them as

$$7\sqrt{3} - 2\sqrt{3} = 5\sqrt{3}$$

b. However, $3\sqrt{2}$ and $4\sqrt{3}$ are not like radicals. We cannot simplify sums or differences of unlike radicals. Thus, $3\sqrt{2}+4\sqrt{3}$ cannot be combined into a single term. ■

Caution!   When adding or subtracting like radicals, we do **not** add or subtract the radicands. For example,

$$3\sqrt{5}+4\sqrt{5}=7\sqrt{5}$$

not $7\sqrt{10}$. ■

> **RQ6.**   How do we add or subtract like radicals?
> **RQ7.**   How do we add or subtract unlike radicals?

Sometimes we must simplify the square roots in a sum or difference before we can recognize like radicals.

■ **Example 8**   Simplify $\sqrt{20}-3\sqrt{50}+2\sqrt{45}$

Solution We simplify each square root by removing perfect squares from the radicals.

$$\sqrt{20}-3\sqrt{50}+2\sqrt{45}=\sqrt{4\cdot5}-3\sqrt{25\cdot2}+2\sqrt{9\cdot5}$$
$$=2\sqrt{5}-3\cdot5\sqrt{2}+2\cdot3\sqrt{5}$$
$$=2\sqrt{5}-15\sqrt{2}+6\sqrt{5}$$

Then we combine the like radicals $2\sqrt{5}$ and $6\sqrt{5}$ to get

$$\sqrt{20}-3\sqrt{50}+2\sqrt{45}=8\sqrt{5}-15\sqrt{2}$$ ■

Skills Warm-Up

Find the missing factor.

1. $60x^9=3x^3\cdot?$ 　　　　　**2.** $16z^{16}=4z^4\cdot?$ 　　　　　**3.** $108a^5b^2=36a^4b^2\cdot?$

4. $\dfrac{20}{7}m^7=4m^6\cdot?$ 　　　　**5.** $\dfrac{5k^5}{9n}=\dfrac{k^4}{9}\cdot?$ 　　　　**6.** $\dfrac{a^2+4a^4}{8}=\dfrac{a^2}{4}\cdot?$

Answers: 1. $20x^6$ 　　**2.** $4z^8$ 　　**3.** $3a$ 　　**4.** $\dfrac{5m}{7}$ 　　**5.** $\dfrac{5k}{n}$ 　　**6.** $\dfrac{1+4a^2}{2}$

Homework 9.3

■ For Problems 1-3, decide whether the statement is true or false. Then use a calculator to verify your answer.

1. $\sqrt{6} = \sqrt{2}\sqrt{3}$ **2.** $\sqrt{16} = \sqrt{18} - \sqrt{2}$ **3.** $\sqrt{5} + \sqrt{5} = \sqrt{10}$

■ For Problems 4-9, find the square root.

4. $\sqrt{y^8}$ **5.** $\sqrt{n^{36}}$ **6.** $\pm\sqrt{16x^4}$

7. $-\sqrt{121a^2b^6}$ **8.** $\sqrt{9(x+y)^2}$ **9.** $-\sqrt{\dfrac{64}{b^6}}$

■ For Problems 10-15, simplify the square root.

10. $\sqrt{8}$ **11.** $-\sqrt{20}$ **12.** $\sqrt{125}$

13. $\sqrt{x^3}$ **14.** $-\sqrt{b^{11}}$ **15.** $\sqrt{p^{25}}$

■ For Problems 16-24, simplify the square root.

16. $\sqrt{8a^3}$ **17.** $\pm\sqrt{72m^9}$ **18.** $\sqrt{\dfrac{x^8}{27}}$

19. $\sqrt{48c^6d}$ **20.** $-\sqrt{\dfrac{45}{4}b^2d^3}$ **21.** $\sqrt{\dfrac{9w^3}{28z}}$

22. $3\sqrt{4x^3}$ **23.** $-2a\sqrt{50a^3b^2}$ **24.** $-\dfrac{2}{3k}\sqrt{9b^3k^5}$

■ For Problems 25-31, simplify if possible.

25. $\sqrt{3} + 2\sqrt{3}$ **26.** $3\sqrt{5} - 3\sqrt{7}$ **27.** $2\sqrt{6} - 9\sqrt{6}$

28. $\sqrt{20} + \sqrt{45} - 2\sqrt{80}$ **29.** $\sqrt{3} - 2\sqrt{12} - \sqrt{18}$

30. $\sqrt{8a} + \sqrt{18a} - 7\sqrt{2a}$ **31.** $2\sqrt{5x^3} - x\sqrt{125x} - 3\sqrt{20x^2}$

■ For Problems 32-34, find and correct the error.

32. $\sqrt{36 + 64} \rightarrow 6 + 8$ **33.** $\sqrt{3} + \sqrt{3} \rightarrow \sqrt{6}$ **34.** $\sqrt{9 + x^2} \rightarrow 3 + x$

■ Mental exercise: For Problems 35-37, choose the best approximation for the square root. Do not use pencil, paper, or calculator.

35. $\sqrt{13}$ **a.** 6   **b.** 6.5   **c.** 3.5   **c.** 4

36. $\sqrt{72}$ **a.** 64   **b.** 9   **c.** 36   **d.** 81

37. $\sqrt{125.6}$ **a.** 11   **b.** 12   **c.** 15   **d.** 25

■ For Problems 38-40, write the expression as the square root of an integer. **Hint:** square the expression.

38. $2\sqrt{5}$ **39.** $2\sqrt{3}$ **40.** $3\sqrt{6}$

9.4 Operations on Radicals

Products

We have used the product rule for radicals,

$$\sqrt{ab} = \sqrt{a}\sqrt{b}$$

to simplify square roots. We can also use the product rule to multiply radicals together.

> **Product Rule**
>
> $$\text{If } a, b \geq 0, \text{ then } \quad \sqrt{a}\sqrt{b} = \sqrt{ab}$$

For example,

$$\sqrt{2} \cdot \sqrt{3} = \sqrt{6} \qquad \text{and} \qquad \sqrt{5} \cdot \sqrt{x} = \sqrt{5x}$$

Example 1 Find the product and simplify $\sqrt{2x}\sqrt{10xy}$

Solution First, we apply the product rule.

$$\sqrt{2x}\sqrt{10xy} = \sqrt{20x^2y} \qquad \text{Factor out perfect squares.}$$
$$= \sqrt{4x^2}\sqrt{5y} = 2x\sqrt{5y}$$

RQ1. Which of these statements is correct:

    a. $\sqrt{x^2 + 4} = \sqrt{x^2} + \sqrt{4}$ **b.** $\sqrt{4x^2} = \sqrt{4}\sqrt{x^2}$

    c. $\sqrt{x^2 - 4} = \sqrt{x^2} - \sqrt{4}$ **d.** $\sqrt{\dfrac{x^2}{4}} = \dfrac{\sqrt{x^2}}{\sqrt{4}}$

We use the distributive law to remove parentheses from products involving radicals. Compare the product involving radicals on the left with the more familiar use of the distributive law on the right.

$$6(\sqrt{3} + \sqrt{2}) = 6\sqrt{3} + 6\sqrt{2} \qquad\qquad 6(x + y) = 6x + 6y$$

Example 2 Multiply $5(3x + 4\sqrt{2})$

Solution Apply the distributive law to find

$$5(3x + 4\sqrt{2}) = 5 \cdot 3x + 5 \cdot 4\sqrt{2} = 15x + 20\sqrt{2}$$

RQ2. Explain why this statement is **incorrect:** $5 \cdot 4\sqrt{2} = 20 \cdot 5\sqrt{2}$.

In the next example, we use the distributive law along with the product rule to simplify the product.

Example 3   Multiply $2\sqrt{5}\,(3+\sqrt{3})$

Solution Apply the distributive law to obtain

$$2\sqrt{5}\,(3+\sqrt{3}) = 2\sqrt{5}\cdot 3 + 2\sqrt{5}\cdot\sqrt{3} \qquad \text{Apply the product rule to the second term.}$$
$$= 6\sqrt{5} + 2\sqrt{15}$$

Caution!   Note the difference between the products

$$\sqrt{5}\cdot 3 = 3\sqrt{5} \qquad \text{and} \qquad \sqrt{5}\cdot\sqrt{3} = \sqrt{15}$$

The rule $\sqrt{a}\sqrt{b} = \sqrt{ab}$ applies to the second product, but not the first.

RQ3.   Explain how to multiply $\left(3\sqrt{2}\right)\left(4\sqrt{5}\right)$.

To multiply binomials, we use the FOIL method.

Example 4   Multiply $(2+\sqrt{3})(1-2\sqrt{3})$, then simplify.

Solution $(2+\sqrt{3})(1-2\sqrt{3}) = 2\cdot 1 - 2\cdot 2\sqrt{3} + \sqrt{3}\cdot 1 - \sqrt{3}\cdot 2\sqrt{3}$

$\qquad\qquad\qquad\qquad\qquad\quad$ F $\qquad$ O $\qquad\quad$ I $\qquad\quad$ L

$$= 2 - 4\sqrt{3} + \sqrt{3} - 2\cdot 3 \qquad \text{Note that } \sqrt{3}\cdot\sqrt{3} = 3.$$
$$= 2 - 4\sqrt{3} + \sqrt{3} - 6 \qquad \text{Combine like terms.}$$
$$= -4 - 3\sqrt{3}$$

RQ4.   Explain why $\sqrt{b}\sqrt{b} = b$.

Quotients

We use the quotient rule to simplify square roots of fractions by writing $\sqrt{\dfrac{a}{b}} = \dfrac{\sqrt{a}}{\sqrt{b}}$.

We can also use the quotient rule to simplify quotients of square roots.

Quotient Rule

$$\text{If } a \geq 0 \text{ and } b > 0, \quad \text{then} \quad \frac{\sqrt{a}}{\sqrt{b}} = \sqrt{\frac{a}{b}}$$

Example 5   Simplify $\dfrac{2\sqrt{30a}}{\sqrt{6a}}$

Solution We use the quotient rule to write the expression as a single radical. Then we simplify the fraction inside the radical.

$$\frac{2\sqrt{30a}}{\sqrt{6a}} = 2\sqrt{\frac{30a}{6a}} \qquad \text{Reduce the fraction.}$$
$$= 2\sqrt{5}$$

> **RQ5.**   Is it true that we can only simplify a product or quotient of radicals if they are like radicals?

Fractions

When we solve quadratic equations, the solutions are often fractions that involve square roots. Many such fractions can be simplified using properties of radicals.

■ **Example 6**   One of the solutions of the equation $4x^2 - 8x = 1$ is $\dfrac{8 + \sqrt{80}}{8}$.
Simplify this radical expression.

Solution First, simplify the square root: $\sqrt{80} = \sqrt{16 \cdot 5} = 4\sqrt{5}$. Thus,

$$\frac{8 + \sqrt{80}}{8} = \frac{8 + 4\sqrt{5}}{8} \qquad \text{Factor numerator and denominator.}$$
$$= \frac{\cancel{4}(2 + \sqrt{5})}{\cancel{4} \cdot 2} \qquad \text{Divide out common factors.}$$
$$= \frac{2 + \sqrt{5}}{2} \qquad\qquad\qquad ■$$

Caution!   In Example 6, note that

$$8 + 4\sqrt{5} \neq 12\sqrt{5}$$

because 8 and $4\sqrt{5}$ are not like terms. Also note that

$$\frac{8 + \sqrt{80}}{8} \neq \sqrt{80}$$

Because 8 is a term of the numerator, not a factor, we cannot cancel the 8's.   ■

> **RQ6.**   Which of these can be reduced: $\dfrac{4 + \sqrt{6}}{2}$ or $\dfrac{4\sqrt{6}}{2}$? Why?

We can add or subtract fractions involving radicals.

■ **Example 7**   Subtract $\dfrac{1}{2} - \dfrac{\sqrt{3}}{3}$

Solution The LCD for the two fractions is 6. We build each fraction to an equivalent one with denominator 6, then combine the numerators.

$$\frac{1}{2} - \frac{\sqrt{3}}{3} = \frac{1 \cdot 3}{2 \cdot 3} - \frac{\sqrt{3} \cdot 2}{3 \cdot 2}$$
$$= \frac{3}{6} - \frac{2\sqrt{3}}{6} = \frac{3 - 2\sqrt{3}}{6} \qquad ■$$

Rationalizing the Denominator

For some applications, it is easier to work with expressions that do not have radicals in the denominators of fractions. For example, the fraction $\dfrac{\sqrt{2}}{\sqrt{3}}$ is equivalent to

$$\frac{\sqrt{2}\cdot\sqrt{3}}{\sqrt{3}\cdot\sqrt{3}} = \frac{\sqrt{6}}{3}$$

We multiplied numerator and denominator by $\sqrt{3}$, the same root that appeared in the denominator originally, and thus eliminated the radical from the denominator. This process is called **rationalizing the denominator**.

■ **Example 9**   Rationalize the denominator of $\dfrac{10}{\sqrt{50}}$

Solution We simplify the radical before attempting to rationalize the denominator.

$$\frac{10}{\sqrt{50}} = \frac{10}{5\sqrt{2}} = \frac{2}{\sqrt{2}}$$

Now we multiply top and bottom of the fraction by $\sqrt{2}$.

$$\frac{2\cdot\sqrt{2}}{\sqrt{2}\cdot\sqrt{2}} = \frac{2\sqrt{2}}{2} = \sqrt{2} \qquad ■$$

RQ7.   What does it mean to rationalize the denominator?
RQ8.   What should you do before trying to rationalize a denominator?

Skills Warm-Up

■ Each simplification is incorrect. Give the correct simplification, or write c.b.s. if the expression cannot be simplified.

1. $2+4\sqrt{5} \rightarrow 6\sqrt{5}$

2. $\dfrac{2+3\sqrt{5}}{2} \rightarrow 3\sqrt{5}$

3. $3\sqrt{5}+3\sqrt{2} \rightarrow 3\sqrt{7}$

4. $\dfrac{4+\sqrt{6}}{2} \rightarrow 2+\sqrt{3}$

5. $2(3\sqrt{5}) \rightarrow 6\sqrt{10}$

6. $2(3+\sqrt{5}) \rightarrow 6+\sqrt{10}$

7. $(3\sqrt{5})(4\sqrt{5}) \rightarrow 12\sqrt{5}$

8. $3\sqrt{5}+4\sqrt{5} = 7\sqrt{10}$

Answers: 1. c.b.s.   **2.** c.b.s.   **3.** c.b.s.   **4.** c.b.s.   **5.** $6\sqrt{5}$ **6.** $6+2\sqrt{5}$ **7.** 60
8. $7\sqrt{5}$

Homework 9.4

Skills Practice

■ For Problems 1-3, simplify the product.

1. $\sqrt{8}\sqrt{2}$

2. $\sqrt{2x}\sqrt{3x}$

3. $(3\sqrt{8a})(a\sqrt{18a})$

■ For problems 4-6, multiply.

4. $\sqrt{2}\,(3+\sqrt{3})$

5. $\sqrt{5}\,(4+2\sqrt{15})$

6. $2\sqrt{p}\,(\sqrt{2p}-p\sqrt{2})$

■ For Problems 7-10, multiply.

7. $(3+\sqrt{2})(1-\sqrt{2})$

8. $(4-\sqrt{a})(4+\sqrt{a})$

9. $(2+\sqrt{3})^2$

10. $(2\sqrt{w}+\sqrt{5})(\sqrt{w}-2\sqrt{5})$

■ For Problems 11-13, simplify the quotient.

11. $\dfrac{\sqrt{18}}{\sqrt{2}}$

12. $\dfrac{\sqrt{75x^3}}{\sqrt{3x}}$

13. $\dfrac{\sqrt{48b}}{\sqrt{27b}}$

■ For Problems 14-16, reduce if possible.

14. $\dfrac{9-3\sqrt{5}}{3}$

15. $\dfrac{-8+\sqrt{8}}{4}$

16. $\dfrac{6a-\sqrt{18}}{6a}$

■ For Problems 17-20, write the expression as a single fraction in simplest form.

17. $\dfrac{5}{4}+\dfrac{3\sqrt{2}}{2}$

18. $\dfrac{3}{2a}+\dfrac{\sqrt{3}}{6a}$

19. $\dfrac{3\sqrt{3}}{2}+3$

20. $\dfrac{3}{4}-2\sqrt{y}$

■ For Problems 21-24, find and correct the error in the calculation.

21. $8\sqrt{7}-2\sqrt{5}\to 6\sqrt{2}$

22. $6\sqrt{8}\to\sqrt{48}$

23. $\dfrac{5-10\sqrt{3}}{5}\to -10\sqrt{3}$

24. $\sqrt{x^2+16}\to x+4$

■ For Problems 25-28, rationalize the denominator.

25. $\dfrac{5}{\sqrt{2}}$

26. $\dfrac{a\sqrt{2}}{\sqrt{a}}$

27. $\dfrac{b\sqrt{21}}{\sqrt{3b}}$

28. $\sqrt{\dfrac{7x}{12}}$

Applications

29. a. Write an expression for the height of an equilateral triangle of side w. (See the figure below left.)

    b. Write an expression for the area of the triangle in part (a).

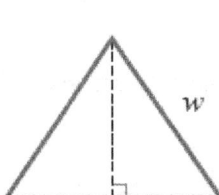 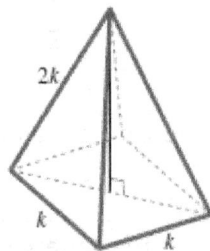

30. a. Write an expression for the height of the pyramid shown above right in terms of k.

    b. Write an expression for the volume of the pyramid in terms of k.

For Problems 31-33, verify by substitution that the given value is a solution of the equation.

31. $t^2 - 2\sqrt{3}\,t + 3 = 0,$ $t = \sqrt{3}$

32. $s^2 + 1 = 4s,$ $s = \sqrt{3} + 2$

33. $x^2 + 3x + 1 = 0,$ $x = \dfrac{-3 + \sqrt{5}}{2}$

9.5 Equations with Radicals

When an object falls from a height of h meters, the time it takes to reach the ground is given in seconds by

$$t = \sqrt{\frac{h}{4.9}}$$

This equation is called a radical equation, because an unknown value, h, appears under a radical sign.

> A **radical equation** is one in which the variable appears under a radical.

■ **Example 1**   If you drop a penny from the top of the Sears Tower in Chicago, it will take 9.5 seconds to reach the ground. Find the height of the Sears Tower by solving the equation

$$9.5 = \sqrt{\frac{h}{4.9}}$$

Solution   To solve the equation, we undo the operations performed on h. To undo the operation of taking a square root, we can square both sides of the equation. Remember that we undo the operations in the opposite order, as follows.

Operations performed on r
1.   Divide by 4.9
2.   Take square root

Steps for solution
1.   Square both sides
2.   Multiply by 4.9

Here is the solution.

$$9.5^2 = \left(\sqrt{\frac{h}{4.9}}\right)^2 \qquad \textbf{Square both sides of the equation.}$$

$$90.25 = \frac{h}{4.9} \qquad \textbf{Multiply both sides by 4.9.}$$

$$442.225 = h$$

The Sears Tower is about 442 meters tall.   ■

> **RQ1.**   What is a radical equation?

■ **Example 2**   Solve the radical equation $\sqrt{x - 3} = 4$

We square both sides of the equation to produce an equation without radicals.

$$(\sqrt{x - 3})^2 = 4^2$$
$$x - 3 = 16$$
$$x = 19$$

You can check that $x = 19$ satisfies the original equation.   ■

Extraneous Solutions

The technique of squaring both sides may introduce **extraneous solutions**. (Recall that an extraneous solution is a value that is not a solution to the original equation.) For example, consider the equation

$$\sqrt{x + 2} = -3$$

Squaring both sides gives us

$$(\sqrt{x + 2})^2 = (-3)^2$$
$$x + 2 = 9$$
$$x = 7$$

However, if we substitute $x = 7$ into the original equation, we see that it is not a solution:

$$\sqrt{7 + 2} \overset{?}{=} -3$$
$$\sqrt{9} \overset{?}{=} -3$$
$$3 \neq -3$$

The value $x = 7$ is a solution to the squared equation, but not to the original equation. In this case, the original equation has no solution.

RQ2. When we solve a radical equation, what causes extraneous solutions to be introduced?

Whenever we square both sides of an equation, we must check the solutions in the original equation.

If a radical equation involves several terms, it is easiest to isolate the radical term on one side of the equation before squaring both sides.

■ **Example 3**   Solve $4 + \sqrt{8 - 2x} = x$

Solution We first isolate the radical by subtracting 4 from both sides to get

$$\sqrt{8 - 2x} = x - 4$$

Now we square both sides to obtain

$$(\sqrt{8 - 2x})^2 = (x - 4)^2$$
$$8 - 2x = x^2 - 8x + 16$$
$$0 = x^2 - 6x + 8$$
$$0 = (x - 4)(x - 2)$$

Thus, the possible solutions are $x = 4$ and $x = 2$. We check both of these in the original equation.

Check: For $x = 4$,   For $x = 2$,

$$4 + \sqrt{8 - 2(4)} \overset{?}{=} 4 \qquad\qquad 4 + \sqrt{8 - 2(2)} \overset{?}{=} 2$$
$$4 + \sqrt{0} \overset{?}{=} 4 \qquad\qquad\quad 4 + \sqrt{4} \overset{?}{=} 2$$
$$4 = 4 \qquad\qquad\qquad\quad 6 \neq 2$$

Thus, $x = 2$ is an extraneous solution. The only solution to the original equation is $x = 4$.

> **Caution!**   When squaring both sides of an equation, we must be careful to square the entire expression on either side of the equal sign. It is incorrect to square each term separately. Thus, in Example 1, it would not be correct to write
>
> $$(\sqrt{8-2x})^2 = x^2 - 4^2 \qquad \leftarrow \text{Incorrect!} \qquad \blacksquare$$

> **RQ3.**   True or false: to solve a radical equation, we should square each term of the equation. Explain.

Equations with Cube Roots

We can also solve equations in which the variable appears under a cube root.

> **Example 4**   Solve the equation $15 - 2\sqrt[3]{x-4} = 9$
>
> **Solution**   We first isolate the cube root.
>
> $$\begin{aligned} 15 - 2\sqrt[3]{x-4} &= 9 && \textbf{Subtract 15 from both sides.} \\ -2\sqrt[3]{x-4} &= -6 && \textbf{Divide both sides by } -2. \\ \sqrt[3]{x-4} &= 3 \end{aligned}$$
>
> Next, we undo the cube root by cubing both sides of the equation.
>
> $$\begin{aligned} \left(\sqrt[3]{x-4}\right)^3 &= 3^3 \\ x - 4 &= 27 \end{aligned}$$
>
> Finally, we add 4 to both sides to find the solution, $x = 31$. We do not have to check for extraneous solutions when we cube both sides of an equation, but it is a good idea to check the solution for accuracy anyway.
>
> **Check:**   Substitute **31** for x into the left side of the equation.
>
> $$\begin{aligned} 15 - 2\sqrt[3]{31-4} &= 15 - 2\sqrt[3]{27} \\ &= 15 - 2(3) \\ &= 15 - 6 = 9 && \textbf{The solution checks.} \qquad \blacksquare \end{aligned}$$

> **RQ4.**   How can we solve an equation if the variable appears under a cube root?

Extraction of Roots

Now we'll compare solving radical equations, where we square both sides, with quadratic equations, where we may be able to take square roots of both sides. We may encounter fractions when we solve quadratic equations by extraction of roots.

> **Example 5**   Solve by extraction of roots: $2(3x-1)^2 = 36$
>
> **Solution**   We first isolate the squared expression: we divide both sides by 2.

$$(3x - 1)^2 = 18 \qquad \text{Extract square roots.}$$
$$3x - 1 = \pm\sqrt{18} \qquad \text{Simplify the radical.}$$
$$3x - 1 = \pm 3\sqrt{2} \qquad \text{Write as two equations.}$$

Now we solve each equation, to find two solutions.

$$3x - 1 = 3\sqrt{2} \qquad\qquad 3x - 1 = -3\sqrt{2}$$
$$3x = 1 + 3\sqrt{2} \qquad\qquad 3x = 1 - 3\sqrt{2}$$
$$x = \frac{1 + 3\sqrt{2}}{3} \qquad\qquad x = \frac{1 - 3\sqrt{2}}{3}$$

We write the solutions as $\dfrac{1 \pm 3\sqrt{2}}{3}$.

RQ5. Is it true that $1 + 3\sqrt{2} = 4\sqrt{2}$? Why or why not?

Skills Warm-Up

Square each expression.

1. **a.** $3x$

   b. $3 + x$

2. **a.** $\sqrt{3x}$

   b. $3\sqrt{x}$

3. **a.** $\sqrt{x + 3}$

   b. $\sqrt{x} + \sqrt{3}$

4. **a.** $x - \sqrt{3}$

   b. $3 - \sqrt{x}$

5. **a.** $2 - 3\sqrt{x}$

   b. $2\sqrt{2x - 3}$

6. **a.** $2 + \sqrt{x - 3}$

   b. $3 + 2\sqrt{x - 3}$

Answers: 1a. $9x^2$ **b.** $9 + 6x + x^2$ **2a.** $3x$ **b.** $9x$ **3a.** $x + 3$ **b.** $x + 2\sqrt{3x} + 3$
4a. $x^2 - 2\sqrt{3}\,x + 3$ **b.** $9 - 6\sqrt{x} + x$ **5a.** $4 - 12\sqrt{x} + 9x$ **b.** $8x - 24$
6a. $x + 1 + 4\sqrt{x - 3}$ **b.** $4x - 3 + 12\sqrt{x - 3}$

Homework 9.5

Skills Practice

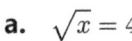 For Problems 1-5, solve and check.

1. $\sqrt{x+4} = 5$ **2.** $\sqrt{x} - 4 = 5$ **3.** $6 - \sqrt{x} = 8$

4. $2 + 3\sqrt{x-1} = 8$ **5.** $2\sqrt{3x+1} - 3 = 5$

6. Use the graph of $y = \sqrt{x}$ shown below to solve the equations. (You may have to estimate some of the solutions.) Check your answers algebraically.

a. $\sqrt{x} = 4$
b. $\sqrt{x} = 2.5$
c. $\sqrt{x} = -2$
d. $\sqrt{x} = 5.3$

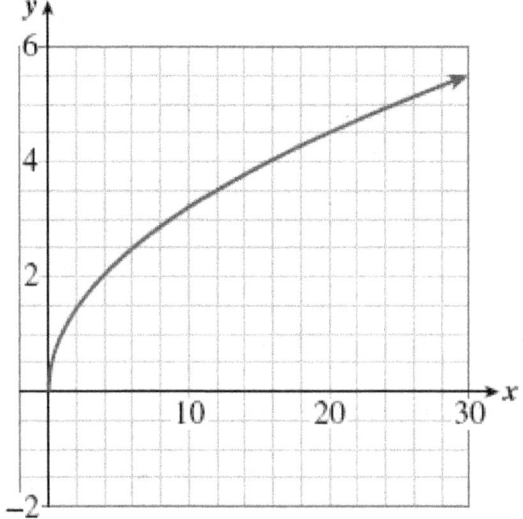

7. The equation for the semicircle shown is

$$y = \sqrt{9 - x^2}$$

Find the x-coordinates of two different points on the semicircle that have y-coordinate 2.

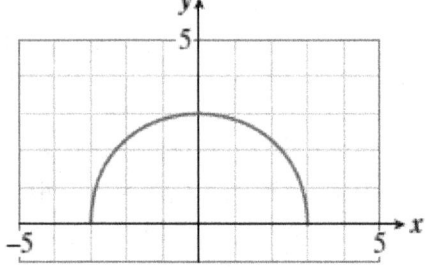

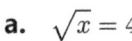 For Problems 8-12, solve, and check for extraneous solutions.

8. $\sqrt{x} = 3 - 2x$ **9.** $\sqrt{x+4} + 2 = x$ **10.** $x + \sqrt{2x+7} = -2$

11. $\sqrt{x+7} = 2x + 4$ **12.** $6 + \sqrt{5x-4} - x = 4$

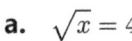 For Problems 13-15, solve.

13. $2\sqrt[3]{x} + 15 = 5$ **14.** $\sqrt[3]{2x-5} - 1 = 2$ **15.** $2 = 8 - 3\sqrt[3]{x^3 + 1}$

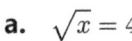 For Problems 16-18, solve by extraction of roots, and simplify your answer.

16. $(2x+1)^2 = 8$ **17.** $3(2x-8)^2 = 60$ **18.** $\dfrac{4}{3}(x+3)^2 = 24$

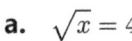 For Problems 19-20, solve the formula for the indicated variable.

19. $x^2 + a^2 = b^2$ for x **20.** $\dfrac{x^2}{4} - y^2 = 1$ for x

Applications

21. a. Complete the table of values and graph $y = \sqrt{x-4}$.

x	y
4	
5	
6	
10	
16	
19	
24	

b. Solve $\sqrt{x-4} = 3$ graphically and algebraically. Do your answers agree?

22. a. Complete the table of values and graph $y = 4 - \sqrt{x+3}$.

x	y
−3	
−2	
0	
1	
4	
8	
16	

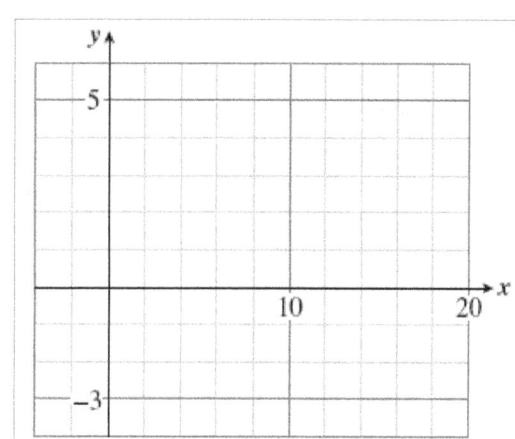

b. Solve $4 - \sqrt{x+3} = 1$ graphically and algebraically. Do your answers agree?

23. The higher your altitude, the farther you can see to the horizon, if nothing blocks your line of sight. From a height of h meters, the distance d to the horizon in kilometers is given by

$$d = \sqrt{12h}$$

a. Mt. Wilson is part of the San Gabriel mountains north of Los Angeles, and it has an elevation of 1740 meters. How far can you see from Mt. Wilson?

b. The new Getty Center is built on the hills above Sunset Boulevard in Los Angeles, and from the patio on a clear day you can see the city of Long Beach, 44 kilometers away. What does this tell you about the elevation at the Getty Center?

24. The speed of a tsunami is given, in miles per hour, by

$$s = 3.9\sqrt{d}$$

where d is the depth of the ocean beneath the wave, in feet. A tsunami traveling along the Aleutian Trench off the coast of Alaska is moving at a speed of over 615 miles per hour. Find the depth of Aleutian Trench.

25. The height of a cylindrical storage tank is four times its radius. If the tank holds V cubic inches of liquid, its radius in inches is

$$r = \sqrt[3]{\frac{V}{12.57}}$$

a. If the tank should hold 340 cubic inches, what must its radius be?
b. If the radius of the tank is 5.5 inches, how much liquid can it hold?

Chapter 9 Summary and Review

Section 9.1 Laws of Exponents

First Law of Exponents

$$a^m \cdot a^n = a^{m+n}$$

Second Law of Exponents

$$\frac{a^m}{a^n} = a^{m-n} \quad (n < m)$$

$$\frac{a^m}{a^n} = \frac{1}{a^{n-m}} \quad (n > m)$$

Third Law of Exponents

To raise a power to a power, keep the same base and multiply the exponents. In symbols,

$$(a^m)^n = a^{mn}$$

Fourth Law of Exponents

To raise a product to a power, raise each factor to the power. In symbols,

$$(ab)^n = a^n b^n$$

Fifth Law of Exponents

To raise a quotient to a power, raise both the numerator and the denominator to the power. In symbols,

$$\left(\frac{a}{b}\right)^n = \frac{a^n}{b^n}$$

Section 9.2 Negative Exponents and Scientific Notation

Zero as an Exponent

$$a^0 = 1, \quad \text{if} \quad a \neq 0$$

Negative Exponents

$$a^{-n} = \frac{1}{a^n}, \quad \text{if} \quad a \neq 0$$

• A negative power is the reciprocal of the corresponding positive power. A negative exponent does **not** mean that the power is negative.

- The laws of exponents apply to negative exponents. In particular, if we allow negative exponents, we can write the second law as a single rule.

$$\frac{a^m}{a^n} = a^{m-n}, \quad a \neq 0$$

- A number is in **scientific notation** if it is expressed as the product of a number between 1 and 10 and a power of 10.
- To write a number in scientific notation, we first position the decimal point and then determine the correct power of 10.
- We can often obtain a quick estimate for a calculation by converting each figure to scientific notation and rounding to just one or two digits.

Section 9.3 Properties of Radicals

Product Rule for Radicals

$$\text{If} \quad a, b \geq 0, \quad \text{then} \quad \sqrt{ab} = \sqrt{a}\sqrt{b}$$

Quotient Rule for Radicals

$$\text{If} \quad a \geq 0, \, b > 0, \quad \text{then} \quad \sqrt{\frac{a}{b}} = \frac{\sqrt{a}}{\sqrt{b}}$$

- It is just as important to remember that we do not have a sum or difference rule for radicals. That is, in general,

$$\sqrt{a+b} \neq \sqrt{a} + \sqrt{b}$$

$$\sqrt{a-b} \neq \sqrt{a} - \sqrt{b}$$

To Simplify a Square Root:
1. Factor any perfect squares from the radicand.
2. Use the product rule to write the radical as a product of two square roots.
3. Simplify the square root of the perfect square.

- Finding a decimal approximation for a radical is not the same as simplifying the radical.
- To take the square root of an even power we divide the exponent by 2.
- Square roots with identical radicands are called **like radicals**.
- We can add or subtract like radicals in the same way that we add or subtract like terms, namely by adding or subtracting their coefficients.

Section 9.4 Operations on Radicals

- We can use the product rule to multiply radicals together: $\sqrt{a}\sqrt{b} = \sqrt{ab}$
- We can use the quotient rule to simplify quotients of square roots: $\frac{\sqrt{a}}{\sqrt{b}} = \sqrt{\frac{a}{b}}$

• We can multiply numerator and denominator of a fraction by the same root that appears in the denominator to eliminate the radical from the denominator. This process is called **rationalizing the denominator**.

Section 9.5 Equations with Radicals
• A **radical equation** is one in which the variable appears under a radical.

To Solve a Radical Equation:
1. Isolate the radical on one side of the equation.
2. Square both sides of the equation.
3. Continue as usual to solve for the variable.

• The technique of squaring both sides may introduce **extraneous solutions**.

• If a radical equation involves several terms, it is easiest to isolate the radical term on one side of the equation before squaring both sides.

• To solve an equation in which the variable appears under a cube root, we isolate the cube root, then cube both sides of the equation.

• We do not have to check for extraneous solutions when we cube both sides of an equation.

Review Questions

▇ Use complete sentences to answer the questions in Problems 1-12.

1. State the third law of exponents, and compare with the first law.
2. State the fourth and fifth laws of exponents and give examples.
3. Give the definition of a^{-n}, and give an example.
4. State the second law of exponents.
5. Explain how the quotient $\frac{2^5}{2^5} = 1$ illustrates the definition of a^0.
6. Describe the form of a number written in scientific notation.
7. State two properties of radicals that are useful in simplifying radical expressions.
8. State two similar "rules" for radicals that are false.
9. Explain how to simplify a square root to a classmate who was absent that day.
10. When can you simplify a sum or difference of square roots? How?
11. Explain how to rationalize the denominator of a fraction.
12. How do we solve a radical equation?

Review Problems

▇ For Problems 1-8, simplify the expression.

1a. $a^4 \cdot a^6$ **b.** $(a^4)^6$ **2a.** $\dfrac{a^4}{a^6}$ **b.** $\dfrac{a^6}{a^4}$

3a. $(2a^2)^3$ **b.** $2a^2(a^2)^3$ **4a.** $\left(\dfrac{-3u}{v^2}\right)^4$ **b.** $\dfrac{-3u^4}{v^2(v^4)}$

5. $-4x(-2x^2)^3$ **6.** $-3w^2(-w^3)^2$

7. $4t^2(t^2)^3 - (6t^4)^2$ **8.** $(3v)^3(-v^3) - (2v)^2(-v^4)$

■ For Problems 9-18, simplify and write without negative exponents.

9a. $3x^{-2}$ **b.** $(3x)^{-2}$ **10a.** $(4y)^0$ **b.** $4y^0$

11a. $\left(\dfrac{5}{z}\right)^{-2}$ **b.** $\dfrac{5}{z^{-2}}$ **12a.** $\dfrac{16c^{-4}}{8c^{-8}}$ **b.** $\dfrac{16c^{-4}}{-8c^8}$

13. $3p^{-4}(2p^{-3})$ **14.** $2q^{-4}(2q)^{-3}$ **15.** $\dfrac{(4k^{-3})^2}{2k^{-5}}$

16. $\dfrac{6h^{-4}(2h^{-2})}{3h^{-3}}$ **17.** $5g^{-6}(g^{-3})^{-2}$ **18.** $(8n)^{-2}(n^{-3})^{-4}$

■ For Problems 19-24, write in scientific notation.

19. 586,000 **20.** 12,400,000 **21.** 0.0007
22. 0.000 009 **23.** 483×10^3 **24.** 0.0035×10^2

■ For Problems 25-28, use scientific notation to compute.

25. $(48,000,000)(380,000,000)$ **26.** $(0.000\,002\,4)(1,900,000,000)$

27. $\dfrac{0.000\,000\,005}{0.000\,2}$ **28.** $\dfrac{38,500,000}{(0.000\,8)(0.001\,7)}$

29. One atomic unit is equal to 1.66×10^{-27} kilogram. What is the mass of 6.02×10^{23} atomic units?

30. The mass of an electron is 9.11×10^{-31} kilogram, and the mass of a proton is 1.67×10^{-27} kilogram. How many electrons would you need to match the mass of one proton?

■ For Problems 31-32, simplify the radical if possible.

31a. $\sqrt{4x^6}$ **b.** $\sqrt{4 + x^6}$ **c.** $\sqrt{(4 + x)^6}$
32a. $\sqrt{1 - w^9}$ **b.** $\sqrt{1 - w^8}$ **c.** $\sqrt{-w^8}$

33. $-\sqrt{27m^5}$ **34.** $\pm\sqrt{98q^{99}}$ **35.** $\sqrt{\dfrac{a^3c}{16}}$

36. $\sqrt{\dfrac{50b^7}{2g^4}}$ **37.** $\dfrac{2}{3}b\sqrt{12b^3}$ **38.** $\dfrac{4}{3a^2}\sqrt{45a^3}$

■ For Problems 39-46, simplify the expression.

39. $3\sqrt{24} + 2\sqrt{18} - 5\sqrt{6}$ **40.** $2x\sqrt{x} - 3\sqrt{x^3} - 6\sqrt{x}$

41. $\dfrac{\sqrt{54w^{12}}}{\sqrt{9w^6}}$ **42.** $\dfrac{\sqrt{24n^3}}{\sqrt{6n^5}}$ **43.** $\dfrac{6 - 3\sqrt{12}}{3}$

44. $\dfrac{\sqrt{8} - \sqrt{12}}{6}$ **45.** $\dfrac{2}{3} - \dfrac{\sqrt{3}}{2}$ **46.** $\dfrac{2\sqrt{3}}{5} - 1$

■ For Problems 47-48, solve by extraction of roots.

47. $(3a - 2)^2 = 24$ **48.** $5(2d + 1)^2 = 90$

For Problems 49-50, solve for the variable indicated.

49. $2a^2 + 4b^2 = c^2$ for b

50. $25w^2 - k = 16m$ for w

For Problems 51-56, multiply and simplify.

51. $\sqrt{3}(\sqrt{2} - \sqrt{6})$

52. $3\sqrt{2}(8\sqrt{6} - \sqrt{12})$

53. $(2 - \sqrt{d})(2 + \sqrt{d})$

54. $(5 - 3\sqrt{2})(3 + \sqrt{2})$

55. $(\sqrt{7} + 3)^2$

56. $(3\sqrt{t} + 1)^2$

For Problems 57-62, simplify the expression and rationalize the denominator if necessary.

57. $\dfrac{2}{\sqrt{x}}$

58. $\sqrt{\dfrac{3a}{b}}$

59. $\dfrac{2\sqrt{5}}{\sqrt{8}}$

60. $\dfrac{a\sqrt{32}}{\sqrt{2a}}$

61. $\dfrac{2}{\sqrt{7}} + \dfrac{3\sqrt{7}}{7}$

62. $\dfrac{1}{2\sqrt{3}} - \dfrac{1}{3\sqrt{2}}$

For Problems 63-64, verify by substitution that the given value is a solution of the equation.

63. $2x^2 - 2x - 3 = 0, \quad x = \dfrac{1 + \sqrt{7}}{2}$

64. $x^2 + 4x - 1 = 0, \quad x = 2 - \sqrt{5}$

For Problems 65-66, find the length of the third side of the right triangle.

65.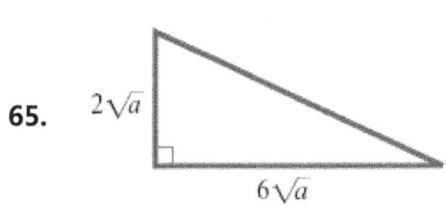

66.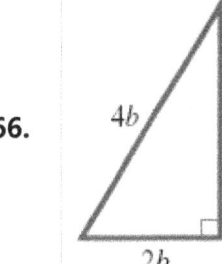

For Problems 67-72, solve.

67. $3\sqrt{x + 2} - 4 = 5$

68. $\sqrt{x - 3} + 4 = 2$

69. $\sqrt{2x + 1} = x - 7$

70. $4\sqrt{4x + 1} = 5x + 2$

71. $\sqrt[3]{3x + 2} - 4 = 1$

72. $9 - 4\sqrt[3]{1 - 2x} = 17$

73. The time it takes for a pendulum to complete one full swing, from right to left and back again, is given in seconds by the formula

$$T = 2\pi\sqrt{\dfrac{L}{32}}$$

where L is the length of the pendulum in feet. The longest pendulum in the world is a reconstruction of Foucault's pendulum in the Convention Center in Portland, Oregon. The pendulum weighs 900 pounds and takes 10.54 seconds to complete one full swing. To the nearest foot, how long is the pendulum?

74. The velocity, v, of a satellite orbitting the earth is given in miles per hour by

$$v = \sqrt{\frac{1.24 \times 10^{12}}{R + h}}$$

where h is the altitude of the satellite in miles, and R is the radius of the earth, about 3960 miles. The Russian space station Mir has an orbital velocity of 17,187 miles per hour. What is its altitude?

Answers to Homework Problems

Homework 1.1

1.

m	g
2	5
3	6
5	8
10	13
12	15
16	19
18	21
m	$m+3$

$g = m + 3$

2.

t	w
0	20
2	18
4	16
5	15
6	14
10	10
12	8
t	$20-t$

$w = 20 - t$

3.

b	x
0	0
2	1
4	2
5	2.5
6	3
8	4
9	4.5
b	$\frac{b}{2}$

$x = \frac{b}{2}$

4.

z	3	6	8	12	15	18	20	z
r	2	4	$\frac{16}{3}$	8	10	12	$\frac{40}{3}$	$\frac{2}{3}z$

$r = \frac{2}{3}z$

5. Answers will vary.

Sample Answer:

n	0	5	10	15	20
W	0	6	12	18	24

6. Answers will vary.

Sample Answer:

x	0	2	4	6	8
M	0	3	6	9	12

7. a, b, and c.

x	y
4	0.5
8	1
10	1.25
16	2

The three tables have identical values.

8. a, b, and c.

x	y
5	3
10	6
12	7.2
1	0.6

The three tables have identical values.

9.

Ridgecrest	70	75	82	86	90	R
Calculation	$70+15$	$75+15$	$82+15$	$86+15$	$90+15$	$R+15$
Sunnyvale	85	90	97	101	105	$R+15$

a. Add 15° to Ridgecrest's temperature to find Sunnyvale's temperature.

b. Temperature in Sunnyvale = Temperature in Ridgecrest + 15

c. $S = R + 15$

d.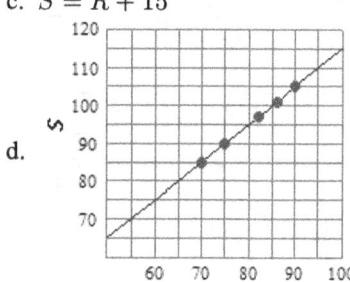

10.

driven	40	60	90	120
Calculation	$200-40$	$200-60$	$200-90$	$200-120$
remaining	160	140	110	80

driven	140	170	d
Calculation	$200-140$	$200-170$	$200-d$
remaining	60	30	$200-d$

a. Subtract the miles driven from 200 to get Jerome's remaining miles.

b. Miles remaining = 200 − miles driven

c. $r = 200 - d$

d.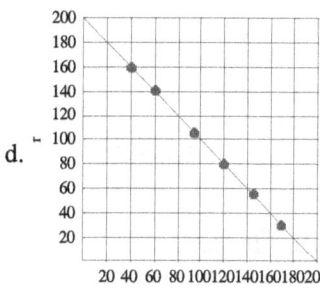

11.

Total	15	30	45	60	75	81	b
Calculation	$\frac{15}{3}$	$\frac{30}{3}$	$\frac{45}{3}$	$\frac{60}{3}$	$\frac{75}{3}$	$\frac{81}{3}$	$\frac{b}{3}$
share	5	10	15	20	25	27	$\frac{b}{3}$

a. Divide the total bill by 3 to get Milton's share.

b. Milton's share $= \frac{\text{total bill}}{3}$

c. $s = \frac{b}{3}$

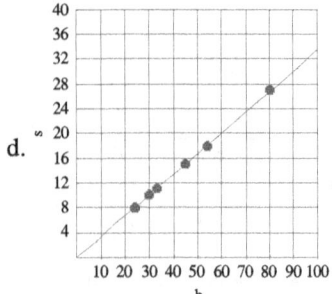

d.

12.

Total Calories	1000	1500	2000
Calculation	0.30×1000	0.30×1500	0.30×2000
Fat Calories	300	450	600

Total Calories	2500	3000	C
Calculation	0.30×2500	0.30×3000	$0.30 \times C$
remaining	750	900	$0.30 \times C$

a. Multiply the Total Calories by 0.30
b. Fat Calories = $0.30 \times$ Total Calories
c. $F = 0.30 \times C$

d.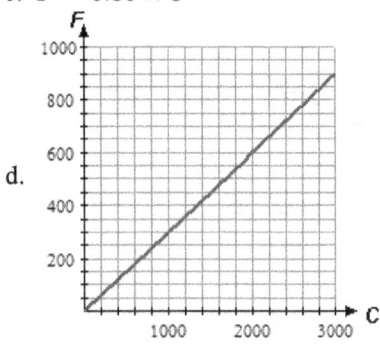

13. a. Horizontal: Number of years,
       Vertical: Amount in account
    b. About \$3000
    c. Approximately after 13.75 years
    d. About \$200   e. About \$400
14. a. Horizontal: Number of minutes,
       Vertical: Temperature of soup
    b. After about 3.5 minutes   c. 70 degrees
    d. About 55 degrees   e. 60 degrees
15. Graph C
16. a. I. Graph C II. Graph D III. Gaph A
    b. Answers will vary. Sample Answer:
       Delbert walks toward school, about halfway there
       he realizes that he left a book at home. He finds
       and retrieves the book, then walks back to school.

17. a.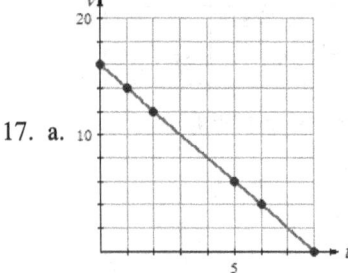

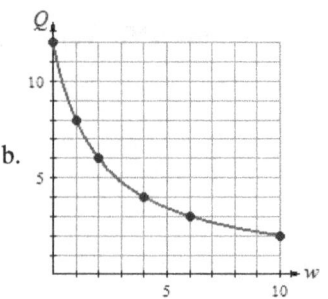

b.

18. a.
x	0	10	30	40	60	70
y	70	60	40	30	10	0

   b. $y = 70 - x$
19. a. $y = 2 - x$
   b.
x	0	0.2	0.3	0.5	0.6	0.7
y	2	1.8	1.7	1.5	1.4	1.3

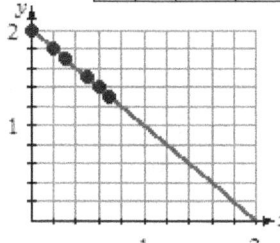

20. a. $y = \frac{120}{x}$
   b.
x	1	2	3	4	6	8
y	120	60	40	30	20	15

Homework 1.2

1. $4y$
2. $2b$
3. $1.15g$
4. $t - 5$
5. $\frac{7}{w}$
6. $\frac{B}{4}$
7. $T + 20$
8. $16 - p$
9. $\frac{15}{M}$
10. $R - 3.5$
11. $n - 6$
12. $6n$
13. $6 - n$
14. $\frac{6}{n}$
15. $2x$
16. $4b$
17. $2v$
18. a. $\frac{x}{6}, \frac{1}{6}x$ b. $0.06, \frac{6}{100}x$
19. a. $\frac{1}{5}m$
    b.
Spending Money	m	20	25	60	80
Amount Saved	$\frac{1}{5}m$	4	5	12	16

20. a. $\frac{1}{4}W$
    b.
Width of photo	W	6	10	24	30
Height of photo	$\frac{1}{4}W$	1.5	2.5	6	7.5

21. a. $\frac{2}{3}p$

b.

Members present	p	90	96	120	129
Votes Needed	$\frac{2}{3}p$	60	64	80	86

22. a. $\frac{3}{5}D$

b.

Stock Dividend	D	40	85	115	170
Maria's Share	$\frac{3}{5}D$	24	51	69	102

23. Ticket Price: p; $15p$

24. Cost of a light bulb: b; $3b$

25. Savings account balance: s; $\frac{3}{5}s$

26. Price of the pizza : p; $\frac{p}{6}$

27. Weight of copper : w; $\frac{w}{16}$

28. School buses : b, $0.09b$

29. Cost of vaccine : v, $v - 16$

30. Cost of each computer : c, $32c$

31. a. $480 - x$ b. $500 - x$ c. $30 - x$

32. a. $x - 15$ b. $x - 12$ c. $x - 18$

33. Amount of milk : m, $0.70m$

34. Number of days : d, $\frac{1200}{d}$

35. Votes opponent received : v, $v - 432$

36. Budget : B, $B + 2000$

37. a. $d = 180t$

b. 360 miles; 630 miles; 2160 miles

38. a. $C = 0.0002g$

b. 0.08 gal, 1 gal, 10 gal

39. a. $B = 0.085R$

b. $8500, $42500, $170,000

40. a.. $P = R - 4000$

b. $6000, $2500, -1500

Homework 1.3

1. a. no b. yes
   c. no d. no
2. a. yes b. yes
   c. no d. no
3. a. yes b. no
   c. no d. yes
4. a. no b. yes
   c. yes d. no
5. a. no b. no
   c. yes d. yes
6. a. yes b. yes
   c. no d. no
7. horizontal: 0.25; vertical: 4
8. horizontal: 50; vertical: 0.5
9. a.

x	0	2	5	6	10	12	16
y	16	14	11	10	6	4	0

b. $y = 16 - x$

10. a.

x	0	1	4	5	7	8	10
y	0	4	16	20	28	32	40

b. $y = 4x$

11. f 12. g 13. c, e 14. b

15. a. $35

b.

w	12	16	20	30	36	40
p	15	20	25	37.50	45	50

c.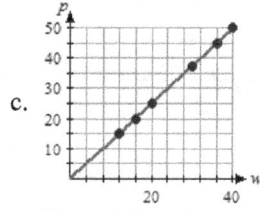

16. a. $D = 1000 - L$

b.

L	250	300	450	600
D	750	700	550	400

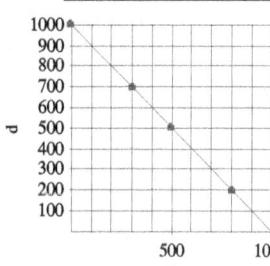

17. a. $S = \dfrac{3600}{m}$

b.

m	20	40	60	100
S	180	90	60	36

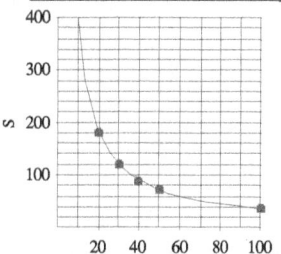

18. a. $p = 110$ b. $p = 60$

19. a. $g \approx 13.9$ b. $g \approx 11.7$

20 a. $(50, 4)$

b. The tax on a $50 item is $4.

21 a. $(4, 16)$

b. A 4-inch tall bunny weighs 16 ounces.

22. a $I = 80t$

b.

t	2	4	7
I	$160	$320	$560

c.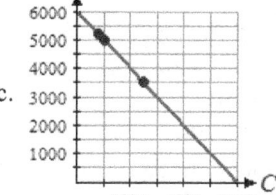

23. a $P = 6000 - C$

b.

C	$800	$1000	$2500
P	$5200	$5000	$3500

c.

24. a $I = 15000r$

b.

r	8%	$10\frac{1}{4}$%	12%
I	$1200	$1537.50	$1800

c.

25. a. $S = \frac{540}{n}$

b.

n	20	25	30
S	27	21.6	18

c.

Homework 1.4

1.

t	q
2	11
4	13
6	15
9	18
21	30
30	39

To find a value for q, add 9 to the value of t. To find a value for t, subtract 9 from the value of q.

2.

n	p
0	0
2	10
4	20
5	25
7	35
11	55

To find a value for p, multiply the value of n by 5. To find a value for n, divide the value of p by 5.

3. a. yes   b. no   c. yes

4. $x = 14$ 5. $y = 2.8$ 6. $y = 36$

7. $b = 12$ 8. $a = 3.9$ 9. $x = 4$

10. $x = \frac{106}{17}$ 11. $z = \frac{10}{3}$ 12. $k = 0$

13. a. $I = Prt$ becomes $75 = P \cdot 0.03 \cdot 1$

   b. $P = 2500$; Clive loaned his brother $2500.

14. a. $A = \frac{S}{n}$ becomes $38.25 = \frac{S}{8}$

   b. $S = 306$; Andy had 306 points.

15. a. $d = rt$: $234 = 13t$

   b. $t = 18$; It takes 18 hours.

16. a. $A = lw$: $400 = l \cdot 16$

   b. $l = 25$; The roll is 25 feet long.

17. $x + 7 = 26$ 18. $x + 7 = 26$

19. $\frac{x}{7} = 26$ 20. $7x = 26$

21. $\frac{x}{26} = 7$ 22. $x - 7 = 26$

23. a. B = Amount before Craft Fair

   b. $B - 24 = 39$ c. $63

24. a. B = Brenda's weight

b. $B + 32 = 157$ c. 125 pounds

25. a. w = Hourly wage

   b. $20w = 136$ c. $6.80

26. a. P = Total Profit

   b. $\frac{P}{8} = 64$ c. $512

27. $89 = p - 26$

   Her mother paid $115.

28. $360 = 0.40I$

   Emily makes $900 per month.

29. $1978 = B - 2378$

   There wase $4356 in her account.

30. a. price of new car : n; price of used car : u;

   $u = n - 3400$

   b. $11,100   c. $12,600

31. a. number of games played: p;

     number of games won: w; $w = 0.60p$

   b. 72   c. 160

32. a. selling price : s;   profit : P;

    $P = 0.18s$

   b. $10.80   c. $40

Homework 1.5

1. 14   2. 3   3. 72   4. 18

5. $\frac{3}{2}$ 6. 23   7. 0   8. 9

9. 4   10. 9   11. 3   12. 1

13. 0.859   14. 0.729

15. 42.705   16. 2204.533

17. a. 18   b. 50   18a. 3   b. 18

19a. 2   b. 8   20a. 42   b. 12

21a. 2   b. 18   22a. 125   b. 5

23.

z	$5z$	$5z - 3$
2	10	7
4	20	17
5	25	22

24.

Q	$12 + Q$	$2(12 + Q)$
0	12	24
4	16	32
8	20	40

25. 26   26. 53   27. $\frac{7}{12}$

28. $\frac{19}{18}$ 29. $\frac{15}{2}$ 30. $\frac{13}{10}$

31. a. $(20 - 2) \cdot 8 + 1 = 145$

   b. $20 - 2 \cdot (8 + 1) = 2$

   c. $20 - 2 \cdot 8 + 1 = 5$

32. $4(5 + 7) \div (10 - 8)$

33. a. $\dfrac{20 - 8}{6 + 4 + 10}$ b. $(20 - 8) \div (6 + 4 + 10)$

34. First multiply 32 and 12, then divide the result by 4, then subtract the result from 825, and finally add the result to 2.

35. $\dfrac{12}{m - 3}$ 36. $\dfrac{m}{12} - 3$

37. $\dfrac{m - 3}{12}$ 38. $\dfrac{12}{m} - 3$

39. Subtracting $9 - 3$ is incorrect. Should be

$$9 - 3(2) = 9 - 6 = 3$$

40. $(2.3 + 5.7)6 - (1.2 + 3.3)2 = 39$ sq cm

Chapter 1 Review

1.

Gallons	3	5	8	10	11	12
Miles	66	110	176	220	242	264

b. Multiply 22 by the number of gallons.

c. $m = 22g$

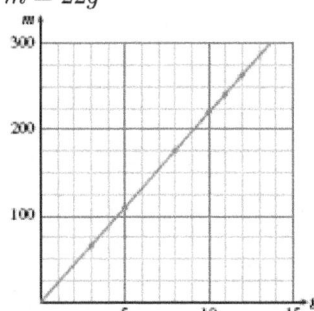

d.

3.

x	.5	1	1.5	2	4	6	7.5
y	.125	.25	.375	.5	1	1.5	1.875

$y = \frac{x}{4}$

5. $z + 5$ 7. $f - 60$

9. $y + 6$ 11. $0.08b$

13a. $d = 88t$ b. 440 ft, 44 ft, 2640 ft

15a. $I = 35t$ b. \$35, \$70, \$175

17a. $n = b - 60$

b.

b	100	120	150	180	200
n	40	60	90	120	140

19a. $n = 2s$

b.

s	2	3	5	8
n	4	6	10	16

21. a, c, d

23a. yes   b. no

25. 9   27. 0.4   29. 0

31a. 32   b. 24

33. $171 = 0.095P$; \$1800

35. $106,000 = 0.53W$; $200,000$

37. $\frac{w}{85} = 0.7$; 59.5 lb

39. 3   41. 6

43. $\frac{1}{2}d + 4$ 45. $3(l + 5.6)$

47. 4   49. 5

51. \$1300

Homework 2.1

1a. 2   b. -2 c. -8 2a. -35 b. -5 c. 5

3a. -25 b. 4.1   c. $\frac{-1}{6}$ 4a. -49 b. 2.4   c. $\frac{-9}{8}$

5a. $4 + (-8) = -4$ b. $3 + 9 = 12$ c. $-8 + 6 = -2$

d. $-6 + (-5) = -11$

6a. 6   b. 10   c. -10 d. 2

7a. -11 b. -9 c. 0   d. -8

8a. 34   b. 6   c. -2 d. -10

9a. 10   b. 10   c. 10

10a. -8 b. -8 c. -8

11a. -5 b. -9 c. 5   d. 9   e. 5

12a. 8   b. -18 c. -8 d. 18   e. -8

13a. -6 b. -18 14a. -13 b. -18

15a. -2 b. 12   c. -39 16a. 4   b. -6 c. -9

17a. 32   b. -3 c. 4

18a. 18   b. undefined   c. 0

19a. 0   b. -8 c. undefined

20a. $\frac{-3}{8}$ b. 2.6   c. $\frac{2}{3}$

21a. -16 b. 48   c. -8 d. -16 e. -3

22a. -6 b. 9   c. 0   d. -6 e. 1

23a. $5, -5$ b. $5, 5$ c. $0, 5$ d. $0, 5$

24a. $5, \frac{1}{5}$ b. $-5, \frac{1}{5}$ c. $5, 5$ d. $5, -5$

25. $x = 5$ 26. $z = \frac{-4}{3}$ 27. $a = -32$

28. $x = 6$ 29. $t = -13$ 30. $b = \frac{9}{2}$

31a. $x = 3$ b. $x = -6$ 32. answers may vary

33. \$500   34. 40 feet   35. $-6°$

36. $- \$695$ 37. 14,776 feet   38. $-\$35.40$

39. -48 meters   40. $-28.75°$ per hour

41. $n + 12 = 5$ 42. $-12 + n = -5$

43. $-12 + n = -5$ 44. $-12 + n = 5$

Homework 2.2

1. 29.8 meters   2. 24 square cm   3. \$5400

4. $37°C$ 5. -18 6. 2

7. -15 8. -15

9a. -24 b. 24   c. -24 d. 48   e. -48 f. 48

10a. negative   b. positive

11a. 2   b. -2 12a. 7   b. -22

13a. -6 b. -2 14a. 12.96   b. -19.2

15a. -60 b. 7   c. 27   d. 6   e. -23 f. 35

16a. 2   b. -3 c. -6 d. 27   e. 2   f. -8

17. 13   18. -12 19. -18 20. $\frac{1}{13}$

21. $-\frac{1}{9}$ 22. $\frac{7}{6}$ 23a. i. $-\frac{15}{4}$ ii. $-\frac{15}{4}$ iii. -3.75

b. No   24. a, d   25. $2x + 10$ 26. $7x + 42$

27a.   \$4400 ;\$2600 b.   Multiply the number of weeks by \$200, then subtract from \$5000.

c.

weeks	2	4	5	6	10	15	20
left	4600	4200	4000	3800	3000	2000	1000

d. $5000 - 200w$ e. $S = 5000 - 200w$

28.a. \$720; \$960   b. Subtract \$2000 from her income, then multiply the result by 0.12

c.

I	5000	7,000	12,000	15,000	20,000	24,000	30,000
T	360	600	1200	1560	2160	2640	3360

d. $0.12(I - 2000)$ e. $T = 0.12(I - 2000)$

29a. $w =$ width (inches): $2w - 3$ b. 23 inches

30a. $P =$ principle (\$): $0.40P + 20$ b. \$220

31a. $t =$ number of childrens tickets; $8(150 - t)$

b. \$536

32a. $p =$ profit; $\frac{1}{3}(p - 50)$ b. \$150

33. $65 + 30t$ 34. $800 + \frac{a}{2}$ 35. $\frac{80 - M}{4}$

36. $50 + \frac{1}{3}I$ 37. $200 - 15w$ 38. $600 + 80t$

39. $\frac{c - 50}{10}$ 40. $35(x + 3)$

41a. $\frac{x}{4 + x}$ b. Answers may vary

42a. $(a + 5)(a - 5)$ b. Answers may vary

Homework 2.3

1a. No b. Yes c. No   d. Yes

2a. Yes b. Yes   c. No   d. No

3a. $(-6, 2)$ b. $(-5, -7)$ c. $(8, -4)$

d. $(1, 7)$ e. $(0, -9)$ f. $(-2, 0)$

4a. $(-10, 0)$ b. $(-2, -4)$ c. $(0, -5)$

d. $(2, -6)$ e. $(10, -10)$

5a. $(0, -600)$ **b.** $(10, -200)$
c. $(15, 0)$ **d.** $(20, 200)$
6a. $P = 100 - 6d$
b.

d	0	2	5	10	15
P	100	88	70	40	10

c.
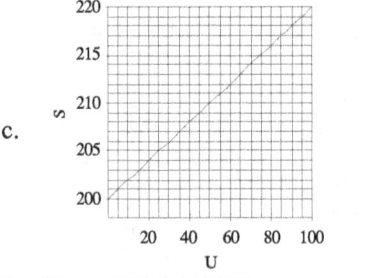

7a. $M = 200 + \frac{1}{5}U$
b.

U	20	40	80	100	200
M	204	208	216	220	240

c.
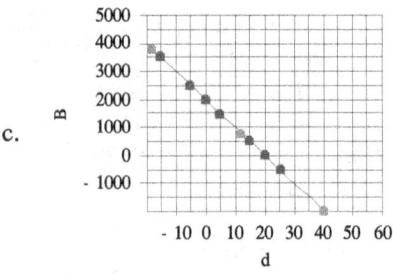

8a. $B = -100d + 2000$
b.

d	B
-15	3500
-5	2500
0	2000
5	1500
15	500
20	0

c.
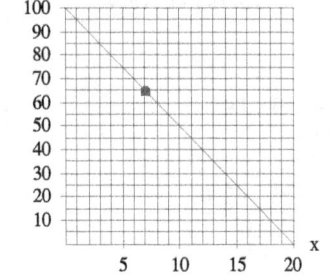

d. -27 feet **e.** floor 8
9a. \$540 **b.** \$30 **c.** \$360 **d.** 18
e. $B = 540 - 30m$
10a. $s = 100 - 5x$
b.

x	2	5	6	12
s	90	75	70	40

c. $65 = 100 - 5x;\ 7$
11a. $d = 120 - \frac{1}{2}w$
b.

w	24	36	84	120
d	108	102	78	60

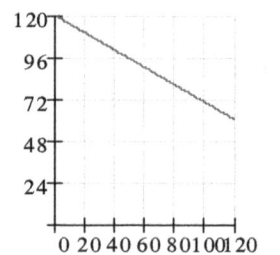

c. 48 weeks; $96 = 120 - \frac{1}{2}w$

12.
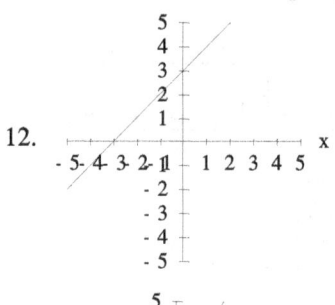

13.
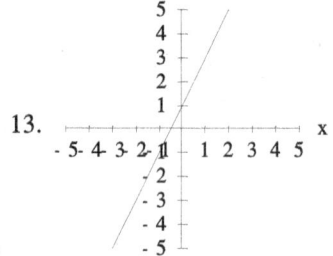

14.
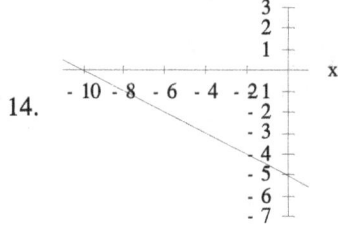

15.
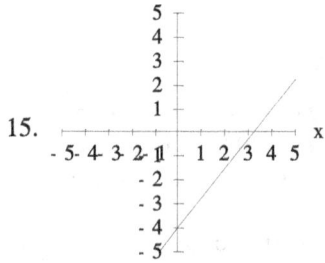

16a. -4 **b.** -5 **c.** $-4 = 2(-5) + 6$
d. 1 **e.** $-4.9, -4$, etc.
17a. $x = -16$ **b.** $x = -40$
18a. $x = 6$ **b.** $x = 12$
19. $t \approx 1.5$ **20.** $x \approx -2.5$

Homework 2.4

1.
x	$2x$	$2x+4$
3	6	10
6	12	16
5	10	14
8	16	20

2.
q	$q-3$	$5(q-3)$
3	0	0
5	2	10
4	1	5
7	4	20

3. $x = 3$ 4. $a = 20$ 5. $x = 4$
6. $x = 0$ 7. $p = 13$ 8. $z = 0$
9. $k = 41$ 10. $x = 18$ 11. $b = 6$
12. $c = -2$ 13. $t = 1$ 14. $h = 5$
15. $b = 18$ 16. $y = 10$ 17. $x = -4$
18. $w = 13.4$ 19. $x = 3.4$ 20a. $x \leq -15$

b. (number line: -20 -15 -10 -5 0 5 10 15)

c. $x = -15.1$ is a solution, $x = -14.9$ is not.
21a. $y > -5$

b. (number line: -10 -5 0 5 10)

c. $y = -4.9$ is a solution, $y = -5$ is not.
22a. $x \leq 12$

b. (number line: 0 10)

c. $x = 12$ is a solution, $x = 13$ is not.
23. $x > 2$ 24. $x \geq -3$ 25. $x < -6$
26. $-1 \leq x \leq 4$ 27. $-7 < b \leq -2$
28. $-7 \leq w < -3$ 29. 7
30. 6   31. -3
32a. 8.6 lbs   b. 36 weeks   c. $w = 3.8 + 0.6t$
d. $26 = 3.8 + 0.6t$; $t = 37$ weeks
33a. $P \approx 20$ b. $m \approx 80$ minutes
c. $P = 13 + 0.15m$
d. $25 = 13 + 0.15m$; $80 = m$
34a. $m =$ each installment ($)
b. $10200 = 1200 + 36m$
c. Each installment is $250.
35a. $d =$ assigned distance
b. $5(d + 0.5) = 22.5$
c. She is assigned 4 miles.
36a. $p =$ number of planted seeds
b. $112 = 0.6p - 38$
c. 250 seeds were planted.
37a. $c =$ average temperature change per day
b. $26 = -6 + 4c$
c. The average change was 8 degrees per day.
38a. $w =$ time it takes (weeks) b. $162 = 196 - 4w$
c. It will take 8.5 weeks.
39. 175 yards   40. 6 meters
41. $4x + 8 = 36$; $x = 7$
42. $\frac{1}{2}(2)(3 + x) = 20$; $x = 17$

Homework 2.5

1. $-4x$ 2. $-12.8a$ 3. t 4. $-bc$
5. $12y - 4$ 6. $-8st + 9s - 2$ 7. $9x - 2$
8. $6y - 10$ 9. $2a$ 10. $-5b + 2$ 11. 4
12. $\frac{-9}{2}$ 13. 0   14. $x > -2$ 15. $\frac{8}{3}$
16. $y \leq -4$ 17. $w \geq 21$ 18. $p = -1$
19. a. -21 b. $-6y - 6$ c. -21 20. -170
21. $x = 0$: $2 + 7x = 2 + 7(0) = 2$;
          $9x = 9(0) = 0$
22. $a = 0$: $-(a - 3) = -(0 - 3) = 3$;

$-a - 3 = -0 - 3 = -3$
23. $6x + 8y$ 24. $13a + 7$
25a. $d + 8$ b. $2d + 8$ c. There are 16 dogs.
26a. $3x$ b. $4x$ c. Nine are smokers.
27a. $9.8t$ b. 98 feet   c. 15 seconds
28a. $0.08x$ b. $1.08x$ c. $860
29a. $596x - 1355$ b. 30
30. $8b = 5b + 12$; $b = 4$ oz.
31. $3p + 33 = p + 131$; $p = 49$ in
32. $12c + 1.8 = 8c + 5$; $c = 0.8
33. $3g - 70 = 2g + 220$; $g = 290$ mg
34. $9s + 15 + 14 = 10s + 23$; $s = 6$ points

Chapter 2 Review

1. f   3. c   5. b
7. $\frac{5}{3}$ 9. 8   11. $V = 2000 - 200t$

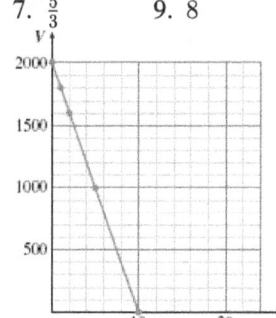

13. $5.75 + 4.50b = 19.25$; 3 bushels
15a. 3   b. -6 c. 6   d. -3
17. $>$ 19. $>$ 21. 20   23. 9
25. 31   27. $\frac{-13}{16}$ 29. -3 31. -1
33. 6°F   35. $460,000   37. -12 yd
39. 6.9, 6, 5   41. 3, 2, 1

43a.
h	-4	-2	0	1	3	5	8
T	30	24	18	15	9	3	-6

b. $T = 18 - 3h$
c. (graph)

d. 24°F   e. 11 p.m.   f. -18°F

45.

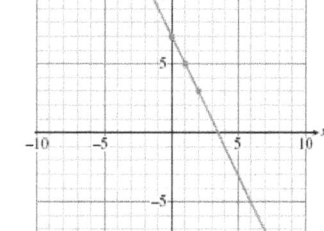

47. $2m + 7n$ 49. $z = -1$

51. $w = -3$ 53. $h = -6$

55. a. -5 b. 2

57. a. $3x$ b. $8x$ c. 6 cm by 18 cm

59. $z \geq 3$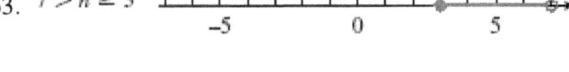

61. $k > -6$

63. $7 > n \geq 3$

Homework 3.1

1. 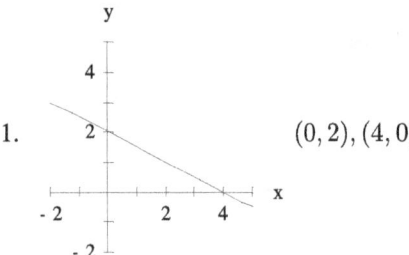 $(0, 2), (4, 0)$

2. 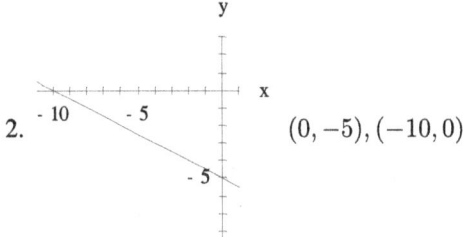 $(0, -5), (-10, 0)$

3. 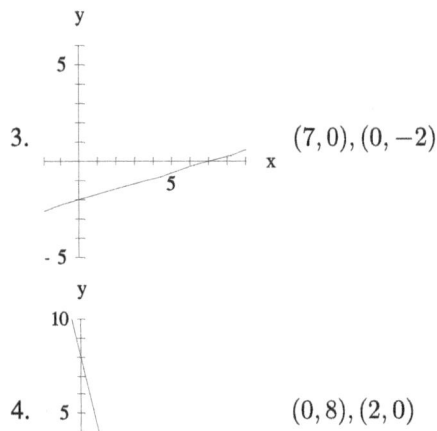 $(7, 0), (0, -2)$

4. $(0, 8), (2, 0)$

5a. $(20, 0), (0, 30)$ b.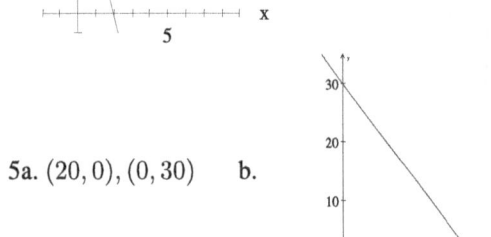

6. 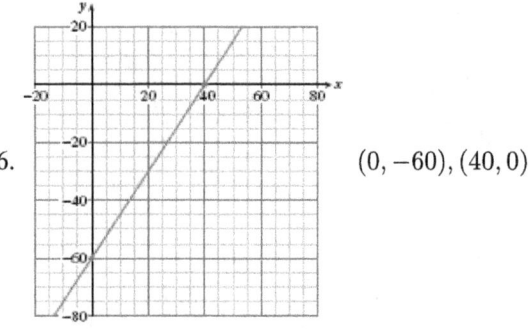 $(0, -60), (40, 0)$

7. d   8. a   9. c   10. b   11. a   12. c

13a. $x = -3$ b. $x = -\frac{k}{2}$

14a. $x = 3$ b. $x = 7 - k$

15a. $x = 14$ b. $x = k + 5$

16a. $x = \dfrac{5}{2}$ b. $x = \dfrac{8 - k}{2}$

17a. $x = 3$ b. $x = \dfrac{k - 15}{-4}$

18a. $x = \dfrac{-10}{3}$ b. $x = \dfrac{-10}{k}$

19a. 360 miles   b. 24 hours   c. 150 miles

d. 60 miles   20a. $(13\frac{1}{3}, 0), (0, 200)$

b.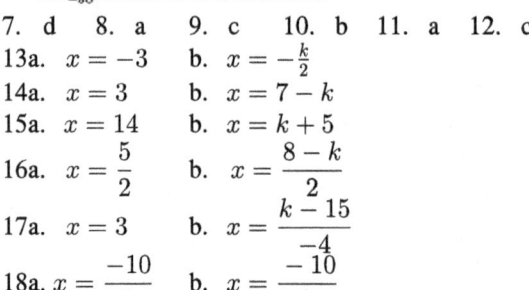

c. The w-intercept at $w = \frac{40}{3}$ shows that the fuel will run out after $\frac{40}{3} \approx 13.3$ weeks. The G-intercept at $G = 200$ shows that there were 200 gallons in the tank when they turned on the furnace.

21a. $(-9, 0), (0, 225)$

b.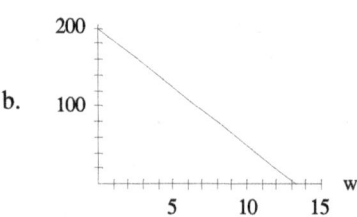

c. The B-intercept at $B = 225$ shows that she has $225 this week. The w-intercept at $w = -9$ shows that she had no balance 9 weeks ago.

22a. $(15, 0), (0, -600)$

b.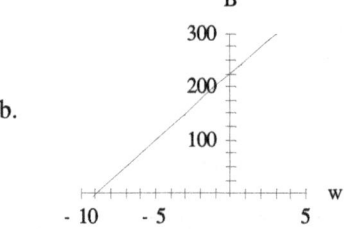

c. The horizontal intercept at $d = 15$ shows that Delbert must groom 15 dogs to break even. The vertical intercept at $P = -600$ shows that Delbert had spent $600 before grooming the first dog.

23a. $(12.5, 0)$ b. $(0, -5)$ 24. Increasing

Homework 3.2

1. $\frac{5}{7}$ 2. 8.72 3. $\frac{11}{6}$ 4. $\frac{30}{7}$

5. a, b, and d are proportions.

6. $x = 6$ 7. $w = \frac{15}{2}$ 8. $b = \frac{63}{5}$ 9. $x = \frac{7}{2}$

10. yes 11. no 12. no 13. no

14. yes 15. yes 16. no 17. 40 lbs

18. 100 liters 19. 80 kilometers

20. 5753 21. about 13.78 in. 22. 2 cups

23a. $w = 7.5h$

b. 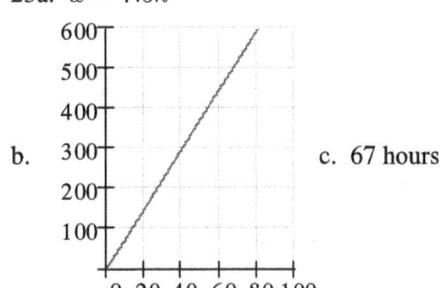 c. 67 hours

24a. $c = 0.03g$

b. 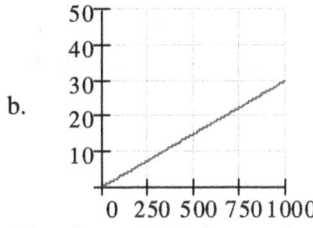 c. 30 grams

25a. $C = 0.04S$ b. \$8000

26a. $d = 8t$

b.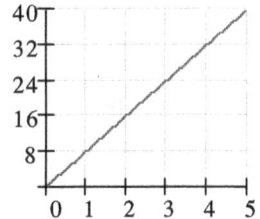

c. The distance doubles also.

27a. $T = 0.06P$

b.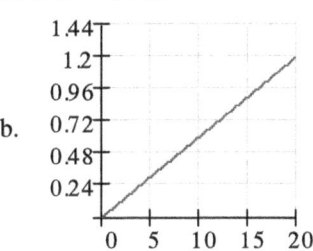

c. The tax is also cut in half.

28a.
Units	Tuition	Units	Tuition
3	620	6	740
5	700	10	900
8	820	16	1140

b. No c. No

29a. $C = 2\pi r$

b.
r	2	5	7	10
C	12.6	31.4	44.0	62.8

c. 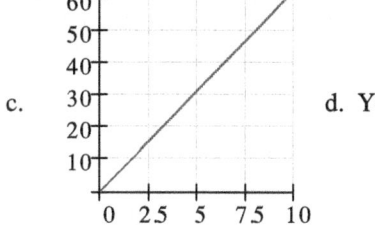 d. Yes

30a. $A = s^2$

b.
s	2	5	6	8
A	4	25	36	64

c. 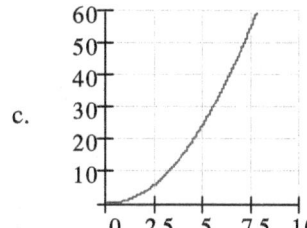 d. No

31a. 3 to 2; $416\frac{2}{3}$ servings; $7\frac{1}{2}$ cans

b. 2.4 oz; 15.5 lb c. 24.75 cans; 90 lb

Homework 3.3

1a. $\frac{3}{2}$ b. -2 c. $\frac{-2}{3}$ d. $\frac{1}{2}$

2a. $\frac{-1}{4}$ b. 3 3. $\frac{1}{15}$

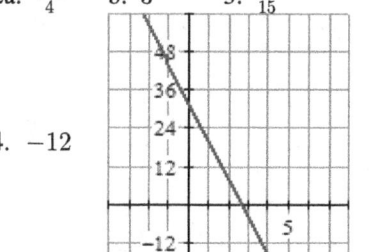

4. -12

5. $\frac{-1}{2}$ 6. $\frac{3}{2}$ 7a. $(0, 4), (6, 0)$

b. 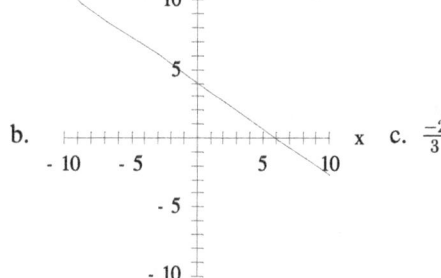 c. $\frac{-2}{3}$

8a. $(0, -5), (2, 0)$

b. 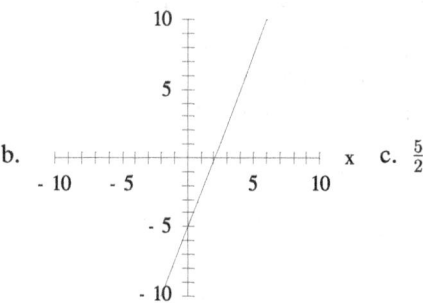 c. $\frac{5}{2}$

9a. $(0, 5), (5, 0)$

Answers-9

b. 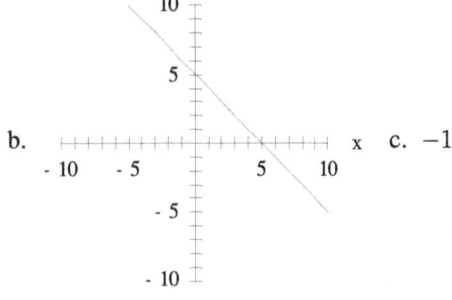 c. -1

10a. $(0, -2), (4, 0)$

b. 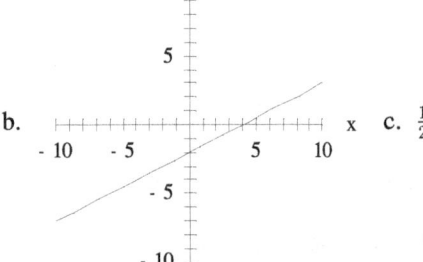 c. $\frac{1}{2}$

11. $\frac{2}{3}$ 12. $\frac{-5}{3}$

13. The line with slope $\frac{5}{3}$ is steeper.

14. The line with slope -2 is decreasing.

15. $\Delta y = 17.5$ 16. $\Delta x = 1.5$

17a. $d = 25g$ b. $\dfrac{25 \text{ miles}}{1 \text{ gallon}}$; the car gets 25 miles

per gallon.

18a. $T = 4p$ b. $\dfrac{4 \text{ cents}}{1 \text{ dollar}}$; the sales tax is 4 cents

per dollar purchased.

19a. $50   b. $(0, 200), (2, 300)$, $\dfrac{\Delta A}{\Delta w} = 50$

c. dollars/week. She is saving $50 per week.

20a. 4 lbs; 3 lbs   b. $\frac{1}{2}$ pound per week

21. 950.4 ft   22. 16 ft

23a. 0   b. All the y-coordinates on the line are
the same, so the change in y is 0.

c. $\dfrac{\Delta y}{\Delta x} = \dfrac{0}{\Delta x} = 0$

24a. undefined   b. All the x-coordinates on the line
are the same, so the change in x is zero.

c. $\dfrac{\Delta y}{\Delta x} = \dfrac{\Delta \text{y}}{0} = 0$

25a. (1) b. (1): $m = 3$; (2): $m = 20$; (2)

26a. $(8, 2.4), (16, 4.8)$ b. 0.3 dollars/oz

Homework 3.4

1. $y = 3x + 4$; slope 3 and y-intercept $(0, 4)$.
2. $y = -2x + \frac{5}{3}$; slope -2 and y-intercept $\left(0, \frac{5}{3}\right)$.
3. $y = \frac{2}{3}x - 2$; slope $\frac{2}{3}$ and y-intercept $(0, -2)$.
4. $y = \frac{5}{4}x$; slope $\frac{5}{4}$ and y-intercept $(0, 0)$
5. $y = 3x - 4$ 6. $y = \frac{2}{3}x - 3$
7. $y = \frac{-8}{5}x + 1$
8a. slope $-\frac{1}{3}$; y-intercept $(0, -1)$ b. $y = -1 - \frac{1}{3}x$

9a.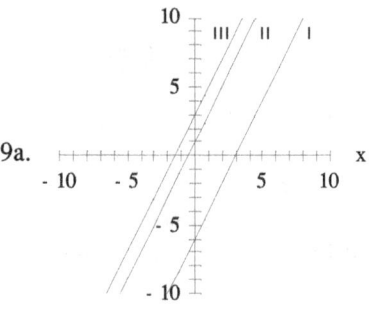

b. The slopes are all 2. c. $(0, -6), (0, 1), (0, 3)$

10a.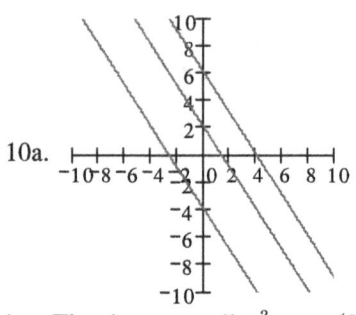

b. The slopes are all $\frac{-3}{2}$. c. $(0, -2), (0, 4), (0, 6)$

11a.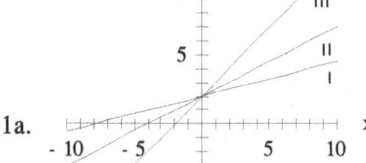

b. I. $\frac{1}{4}$ II. $\frac{1}{2}$ III. 1   c. $(0, 2)$

12a.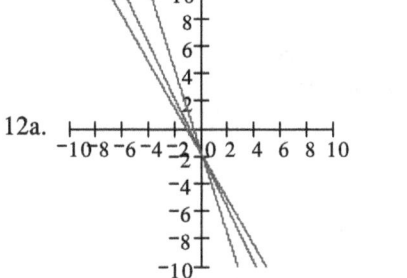

b. I. -3 II. -2 III. $\frac{-5}{3}$ c. $(0, -2)$

13a. $(0, 3), (4, 0)$

b.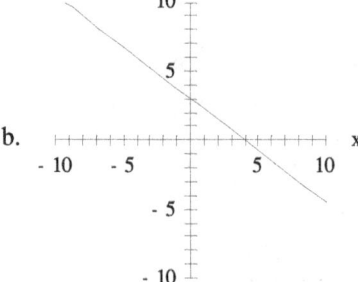

c. $\frac{-3}{4}$ d. $y = \frac{-3}{4}x + 3$

14a. $(0,8), (\frac{8}{3},0)$

b.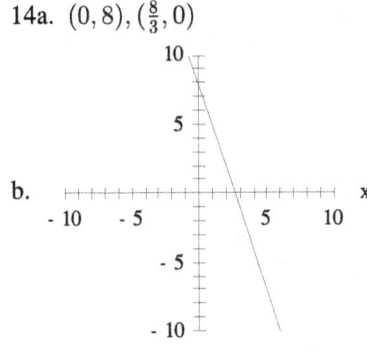

c. -3 d. $y = -3x + 8$

15a. $y = \frac{3}{5}x$ b. $(0,0), \frac{3}{5}$

c.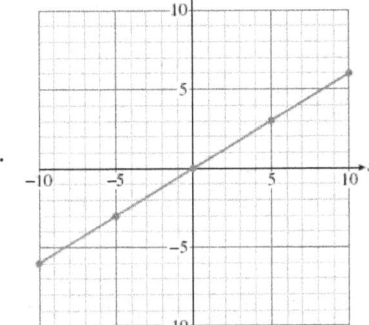

16a. $y = \frac{-5}{4}x$ b. $(0,0), \frac{-5}{4}$

c.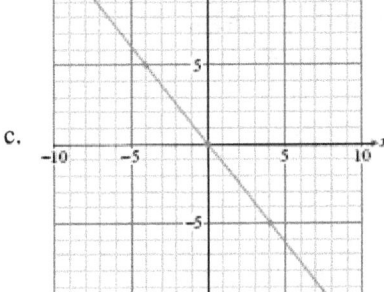

17.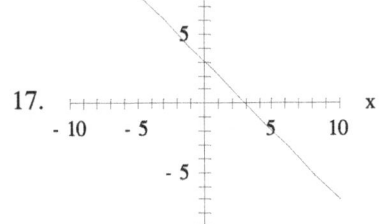

18.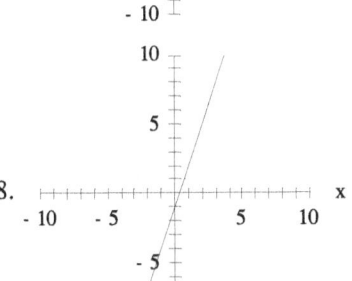

19.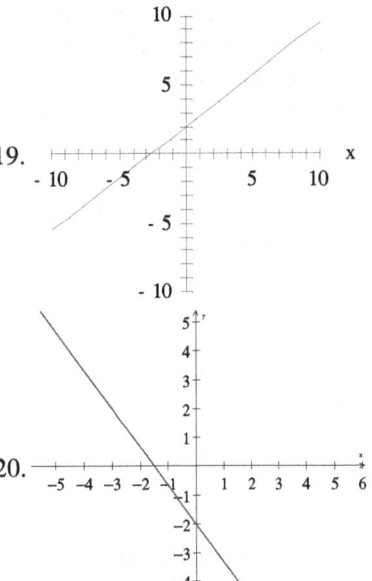

20.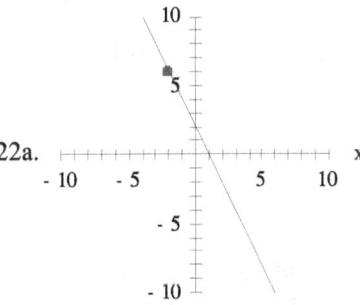

21a. $(0, -240)$ b. 3 c. $P = -240 + 3x$
d. He spent \$240 to open the shop and makes \$3 of profit per smoothie.

22a.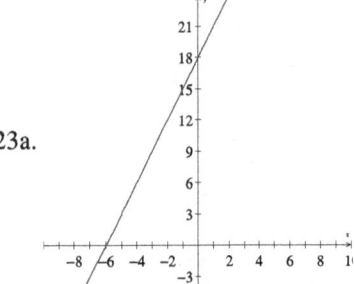

b. The slope means that for eac incorrect answer, she loses 4 points. The S-intercept means that if she has zero wrong answers, her score will be 120 points.

23a.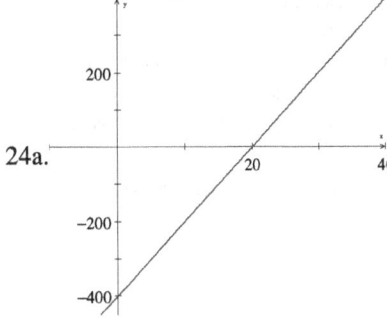

b. The slope gives the weekly rate that the beans are growing: 3 inches per week. The h-intercept, (0,18), means that today, the beans are 18 inces tall.

24a.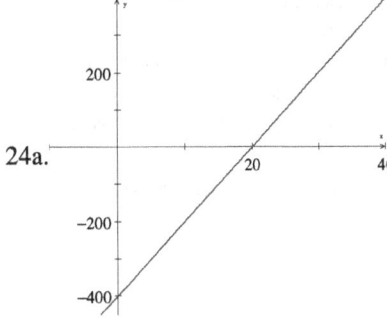

b. The slope, 20 points per question, means that he gets 20 points for each question he gets right. The S-intercept, $(0, -400)$, means that his score at the end of the first round is -400.

Homework 3.5
1. Parallel   2. Neither   3. Neither
4. Parallel: a, g, h; parpendicular: c, f

5a. 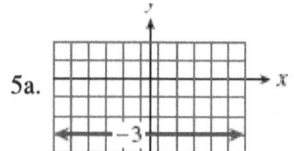   b. $m = 0$

6a. 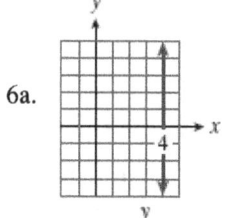   b. m is undefined

7a. 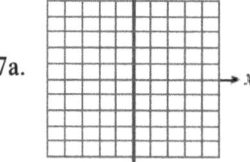   b. m is undefined.

8. $y = -5$ 9. $x = 2$
10. $x = -8$ 11. $y = 0$
12. $x = 4$
13. $m = 0$, $b = -6$ 14. $m = -6$, $b = 0$
15. $m = \frac{-1}{6}$, $b = 0$ 16. $m = -1$, $b = 6$
17. $\frac{5}{3}$ 18. -1 19. $\frac{1}{9}$ 20. 0
21a. $(5, 0)$, $(0, 7)$ b. $\frac{-7}{5}$
22a. $(-2.4, 0)$, $(0, 1.6)$ b. $\frac{2}{3}$
23a. $(\frac{1}{3}, 0)$, $(0, \frac{7}{2})$ b. $\frac{-21}{2}$
24a. $(0, -3)$ b. $(-2, -7)$
25a. $(6, -8)$ b. $(-4, -3)$
26a. \$1200/1 year   b. The college fund grows at te rate of \$1200 per year.
27a. \$0.08/1 page   b. The cost of making copies is 8 cents per page.
28a. -500 feet/1min   b. The elevation decreases at an average rate of 500 feet per minute.
29a. $-3°$/1 hour   b. The temperature dropped at an average rate of $3°$ per hour.
30a. cost per pound of peanuts
b. 1.50 dollars per pound c. \$30
31a. -75 miles/1 hour   b. Roy's distance from home decreases by 75 miles each hour.
c. 800; Roy boarded the train 800 miles away from home.
d. $10\frac{2}{3}$; Roy will arrive home in about ten hours and 40 minutes.

Chapter 3 Review

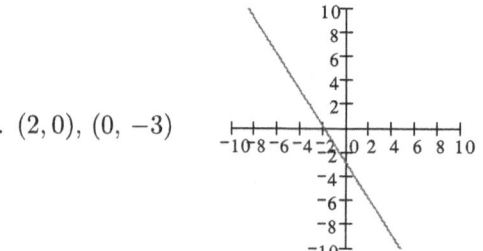

1. $(2, 0)$, $(0, -3)$

3a. $S = 7800 - 600m$

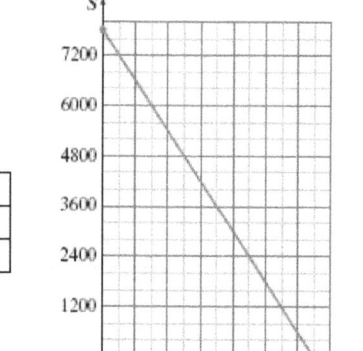

b.

m	S
0	7800
13	0

c. The S-intercept at $S = 7800$ shows that she started with \$7800. The m-intercept at $m = 13$ shows that she will use all the money in 13 months.   d. 9 months
5. $\frac{2}{3}$ 7. 119   9. 50   11. 52 hr
13a. 10 mi by 15 mi b. 50 mi; 30 cm c. $\frac{5}{3}$
d. 150 sq mi; 54 sq cm e. $\frac{25}{9}$
15a. No   b. Yes
17. $\frac{-3}{2}$ 19. 8515 ft
21a. $(0, -6)$, $(3, 0)$

b. 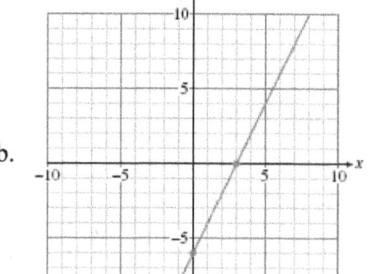   c. $m = 2$

23a. $(2, 0)$, $(0, 3)$

b. 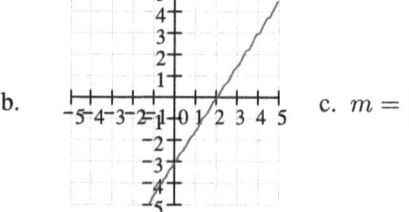   c. $m = \frac{3}{2}$

25a. $m = \frac{-3}{5}$, $b = 3$ b. $y = \frac{-3}{5}x + 3$

27a. $m = \frac{-5}{2}$, $b = -5$

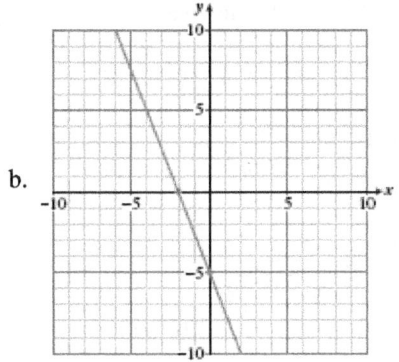

b.

29a. $m = \frac{3}{4}$, $b = 2$

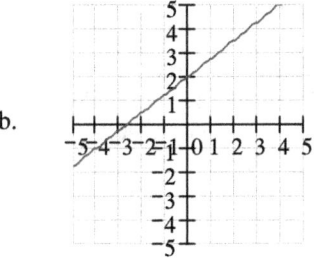

b.

31. $h = 36 + 6t$

33. $b = 500$. At 2 pm, Beryl's altitude was 500 ft.
$m = -15$. Beryl is descending at a rate of 15 ft/sec.

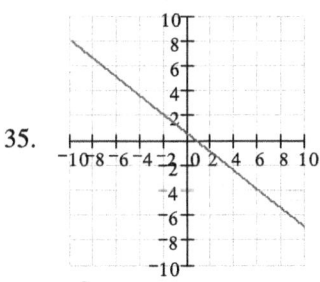

35.

37. $\frac{-3}{2}$ 39. -0.4

41. undefined

43a. 4.2 b. Gas costs $4.20 per gallon.

45. $y = 5x - 1$ 47. $x = -4$

49a. parallel   b. perpendicular

51. 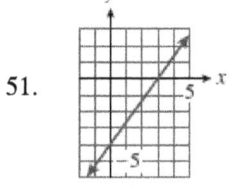   53.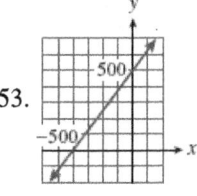

55. 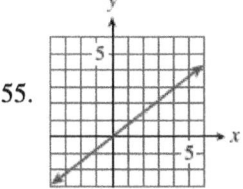   57.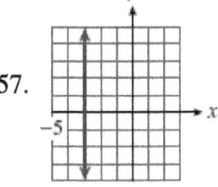

Homework 4.1

1. $10y - 15$ 2. $-8x - 16$ 3. $-5b + 3$
4. $36 - 6t$ 5a. $32c$ b. $32 + 8c$
The distributive law is used in (b).
6a. $-16 - 2t$ b. $-16t$
The distributive law is used in (a).
7. d   8a. $2xy$; 54   b. $2x + 2y$; 24
c. $2 - xy$; -25 d. $-2xy$; -54
9. $-4x - 6$ 10. $x + 23$ 11. $-12z - 15$
12. $-24a + 4$ 13. $y = -2$ 14. $w = 6$
15. $c = 0$ 16. $a \leq -4$ 17. $t < 2$
18. $x = 9$ 19. $a = 4$ 20. $b = 3.25$
21a. $x = 0$: $5(x + 3) = 5(0 + 3) = 15$;
$$5x + 3 = 5(0) + 3 = 3$$
b. $c = 0$ 22a. 48   b. $-12x + 12$ c. 48
23a. $2w - 8$ b. $6w - 16$ c. $2w^2 - 8w$
24a. $108 - c$ b. $216 - 2c$ c. $324 - 2c$
25a. $3s$ b. $3s + 45$ c. 45   d. $6s + 45$
26a. $15 - a$ b. $0.35a$ c. $2.25 - 0.15a$
d. $0.2a + 2.25$ 27a. $8 - x$ b. $0.6x$
c. $2 - 0.25x$ d. $0.35x + 2$ 28a. $10 - x$
b. $0.12x$ c. $5.5 - 0.55x$ d. $5.5 - 0.43x$
29. $5(2x + 3)$; $10x + 15$ 30. $8x(3y + 2)$; $24xy + 16x$
31a. $6w + 12$ b. 5 by 16 yards
32. $13,714,385.71$
33a. $260 - a$ b. $780 - a$ c. 120
34a. $47 - x$ b. $10x$ c. $282 - 6x$ d. $4x + 282$
35. $10x + 6(47 - x) = 330$; she sold 12 reserved tickets.

Homework 4.2

1. No   3. intercept   5. slope-intercept
7. $(-4, -5)$ 9. $(-2, 3)$ 11. inconsistent
12. The graphs have the same slope but different y-intercepts.
13a. Plan A: $y = 0.03x + 20,000$
Plan B: $y = 0.05x + 15,000$

b.

x	Earnings under Plan A	Earnings under Plan B
0	20,000	15,000
50	21,500	17,500
100	23,000	20,000
150	24,500	22,500
200	26,000	25,000
250	27,500	27,500
300	29,000	30,000
350	30,500	32,500
400	32,000	35,000

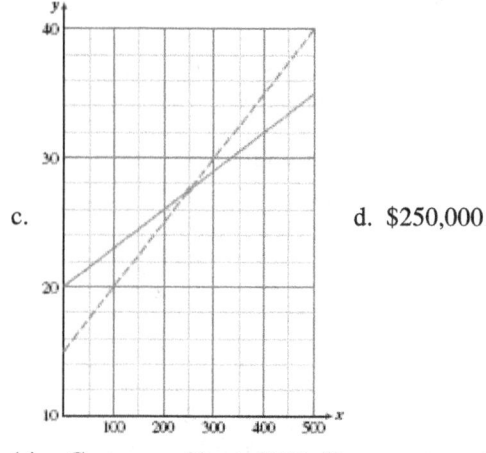

c.

d. $250,000

14a. Cost: $y = 60x + 6000$; Revenue: $y = 80x$

x	Cost	Revenue
0	6,000	0
50	9,000	4,000
100	12,000	8,000
150	15,000	12,000
200	18,000	16,000
250	21,000	20,000
300	24,000	24,000
350	27,000	28,000
400	30,000	32,000

b.

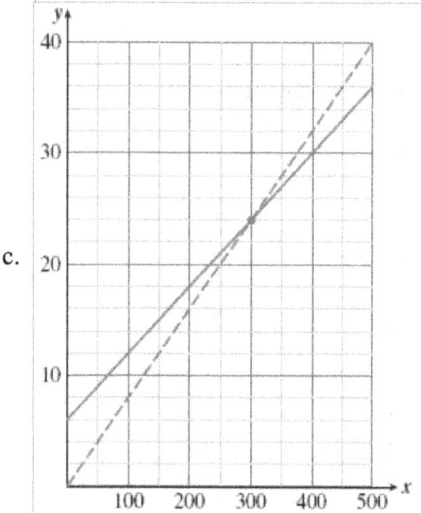

c.

d. 300 clarinets
e. $4000 profit; $-$2000 or a loss of $2000
15. rose: $1.50, carnation: $1.00
16. base angles: 65°, vertex angle: 50°

Homework 4.3

1. $(2, 4)$ 2. $(-1, 2)$
3. $(5, 2)$ 4. $(22, -28)$
5. $(2/3, -4/3)$ 6. $(1/3, -2)$
7. $(3, 2)$ 8. $(1, 1)$
9. $(1, 2)$ 10. $(-3, -2)$
11. $(1, -1)$ 12. $\left(\frac{3}{4}, \frac{-1}{4}\right)$

13. inconsistent   14. inconsistent
15. dependent   16. elimination
17. elimination
18. hamburger: 650 cal; shake: 380 cal
19. 12 tables and 48 chairs
20. 17 meters by 4 meters
21. bacon: $2.20; coffee: $5.60

Homework 4.4

1. bonds: $90,000; mutual fund: $60,000
2. $25,000 at 12%; $5000 at 15%
3. student loan: $5000;   car loan: $10,000
4c. 26.4%   5c. 40%
6. 15 liters of 12% solution;
   30 liters of 30% solution
7. 160 women
8. Delbert: 50 mph; Francine: 45 mph
9. 7 hours; 280 miles   10. 10 hours; 400 miles
11. 40 miles   12a. $g = 0.4t + 43.2$
b. 69.2, 75.2   c,d. 92

Homework 4.5

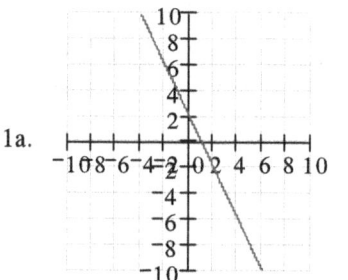

1a.

b. $y = -2x - 2$ c. $(-1, 0)$

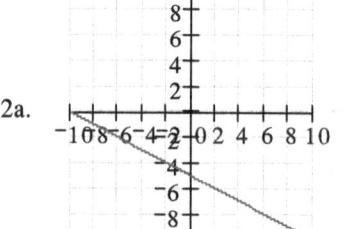

2a.

b. $y = \frac{1}{2}x - 5$ c. $(10, 0)$

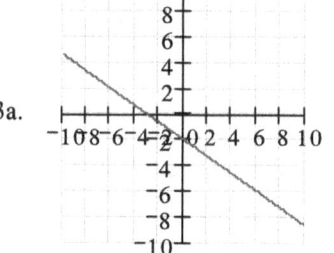

3a.

b. $y = \frac{-2}{3}x - 2$ c. $(-3, 0)$

4a.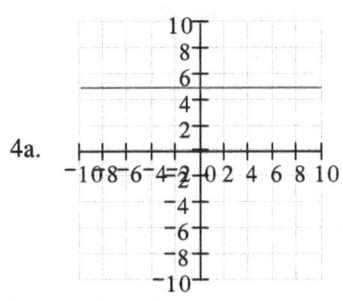

b. $y = 5$ c. no x-intercept

5a.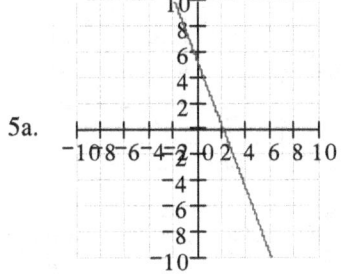

b. $y = -\frac{5}{2}x + 5$ c. $(2, 0)$

6a.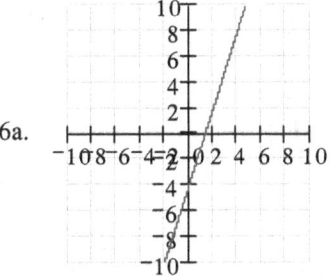

b. $y = 3x - 4$ c. $\left(\frac{4}{3}, 0\right)$

7. $m = \frac{3}{5}, (0, -7)$ 8. $m = 3, (-5, 2)$

9. $m = \frac{4}{3}, (0, 0)$ 10. $y = x + 6$

11. $y = \frac{5}{3}x$ 12. $y = \frac{-1}{8}x + \frac{19}{4}$

13. $y = \frac{-5}{7}x + \frac{11}{7}$ 14. $y = \frac{1}{4}x - \frac{1}{4}$

15. $y = \frac{-5}{3}x - \frac{1}{3}$ 16. $y = \frac{10}{7}x - \frac{8}{7}$

17a. $y = \frac{3}{2}x + \frac{5}{2}$ b. $\frac{3}{2}$

c.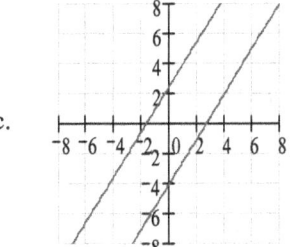

d. $y = \frac{3}{2}x - 4$

18a. $y = \frac{1}{2}x - \frac{5}{2}$ b. -2

c.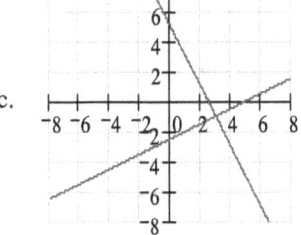

d. $y = -2x + 5$

19a. 2   b. 2   c. $y = 2x + 5$

20a. -3 b. $\frac{1}{3}$ c. $y = \frac{1}{3}x + \frac{8}{3}$

21. $y = -0.0025x + 85$

a. $m = -0.0025$ degrees/ft. The temperature decreases by 0.0025 degrees for every 1 foot increase in altitude.   b. $60°$ c. $85°$

22. $y = 200x + 25,000$

a. $m = 200$ dollars/dryer. It costs \$200 to produce each dryer.   b. 375   c. \$25,000

23. $C = 65h + 125$

a. 65 dollars/hr. The hourly rate for lessons is 65 dollars per hour.   b. \$775   c. \$125

24. $p = 2.2k$

a. 2.2 lbs/kg. One kilogram is equivalent to 2.2 pounds.   b. 22.73 kg

c. Zero kg equals zero lb.

Chapter 4 Review

1a. $30m$ b. Requires distributive law: $-30 + 5m$

3. $2m + 7n$ 5. $-15w + 26$

7. $p = 25$ 9. $a < 1$

11a. $3x$ b. $8x$ c. width: 6 cm; length: 18 cm

13a. $30 - x$ b. $1200x$

c. $800(30 - x)$ d. $400x + 24,000$

e. 12 with speakers and 18 without speakers

15. No   17. $(3, 2)$

19. $(-3, -5)$ 21. $(1, -3)$ 23. $\left(\frac{1}{2}, \frac{7}{2}\right)$

25. $(12, 0)$ 27. Consistent   29. Dependent

31. 12 lb of cereal and 18 lbof dried fruit

33. \$500 in the first account and \$700 in the second

35. 2 lb of the 60% alloy and 6 lb of the 20% alloy

37. Alida: 62 mph, Steve: 31 mph

39. 26   41. \$3181.82 at 8%, \$1818.18 at 13.5%

43a.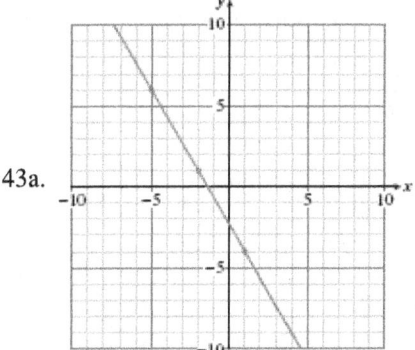

b. $y = \frac{-5}{3}x - \frac{7}{3}$ 45. $y = \frac{-9}{5}x + \frac{2}{5}$

47a. $F = 500 + 0.10C$

b.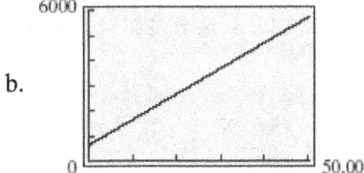

c. $m = 0.10$. The decorator charges 10% of the cost of the job (plus a \$500 flat fee.)

49.

t	0	15
P	4800	6780

b. $P = 4800 + 132t$

c. $m = 132$ people per year gives the rate of population growth

51a. $y = \frac{5}{2}x + b$ b. $y = \frac{-2}{5}x + b$

53. $y = \frac{-2}{3}x + \frac{14}{3}$

Homework 5.1

1a. 64   b. 125   c. 625

2a. $\frac{16}{81}$ b. $\frac{64}{125}$ c. $\frac{121}{81}$

3a. 29.79   b. 45.70   c. 0.41

4a. -25 b. -125 c. 25   d. -125

5a. -4 b. 8   c. -12 d. -36

6a. -40 b. 20   c. 1   d. 13

7a. 192   b. 61   c. 361   d. -84

8a. $3x$ b. x^3 9a. $25a^2$ b. $10a$

10a. $-3q$ b. $-q^3$ 11a. $-6m$ b. $9m^2$

12. $x + x = 2x$ 13. $x \cdot x = x^2$

14. $x^2 + x^2 = 2x^2$ 15. $x \cdot x^2 = x^3$

16. $-2a^2$ 17. $3t - 2t^2$ cannot be simplified.

18. $-2m^2$ 19. $12k^2$ 20. cannot be simplified

21. $7k^2$ 22. $12b^3$ 23. $2y^3 + 2y^2 + y + 1$

24. $3x^3 + 3x^2 + 11$

25. The exponents can't be combined; $14w^3$

26. The terms are not like terms, so they can't be combined; $6 + 3x^2$

27. $7 - (-5) = 12$; $t^2 + 12$

28. The terms are not like terms, so they can't be combined; $5b^2 - 3b$

29. 49   30. $5 + x^3$ 31. $0.25h^3$

32a. $-0.3x^2 + 9x - 50$ b. \$10, \$17.50, \$10

33a. $0.3h$ b. $50 - h$

c. $0.25(50 - h)$ d. $0.3h + 0.25(50 - h)$

34a. For $z = 2$, $3z^2 = 3(2)^2 = 3 \cdot 4 = 12$, but $(3z)^2 = (3 \cdot 2)^2 = 6^2 = 36$

b. $(3z)^2 = (3z)(3z) = 9z^2$, so $(3z)^2 = 9z^2$

35a. $x = 0$, $x = 2$ b. Any values except $x = 0, x = 2$
c. No

36a. 350   b. 3500   c. 35,000

37a. 7.4   b. 74   c. 740

38a. 2400   b. 891,000   c. 30

39a. 39   b. 234   c. 3456

40. $12x$ 41. $60b$

42. area: $63v^2$, perimeter: $32v$

Homework 5.2

1a. -12 b. -2 2a. 9   b. 3

3a. 19/2   b. $-2/3$ 4. 8.660   5. -3.055

6. 1.899   7. -15.544 8. 1.293   9. -1.512

10. b   11. c   12. b   13a. 85

b. 168   c. 4082   14a. 3, 5, 8, 12   b. a

c. 4082   15a. 16   b. 29   c. x

16a. 7   b. 20   c. 4   17a. $\sqrt{6}$

b. $-\sqrt{15}$ c. $2\sqrt{m}$ 18a. $2b$ b. $9a$

c. $8b$ 19. $\frac{5}{4}$, 2, 2.3, $\sqrt{8}$

20. $\sqrt{6}$, 3, $2\sqrt{3}$, $\frac{23}{6}$ 21. $\sqrt{5} \approx 2.236$; 1; 0

22. -1; $2 - \sqrt{7} \approx -0.646$; $10 - \sqrt{103} \approx -0.149$

23. 30; $4 - 4\sqrt{2} \approx -1.657$; $\frac{-7}{8}$

24. ≈ 193.81 miles   25. ≈ 9.61 sec

26. A rational number is one that can be expressed as a ratio of integers, such as 5, $\frac{3}{4}$, and $\frac{-11}{2}$. An irrational number cannot be expressed as a ratio of integers. $\sqrt{3}$, $-\sqrt{5}$, and π are irrational numbers.

27. $\sqrt{4}$, $\sqrt{9}$, $\sqrt{64}$. $\sqrt{2}$, $\sqrt{6}$, $\sqrt{11}$

28. Rational: b, c, e, f

29. No

a	b	$a+b$	$(a+b)^2$	a^2	b^2	a^2+b^2
2	3	5	25	4	9	13
3	4	7	49	9	16	25
1	5	6	36	1	25	26
-2	6	4	16	4	36	40

30. No

a	b	$a+b$	a^2	b^2	a^2+b^2	$\sqrt{a^2+b^2}$
3	4	7	9	16	25	5
2	5	7	4	25	29	$\sqrt{29} \approx 5.4$
1	6	7	1	36	37	$\sqrt{37} \approx 6.1$
-2	-3	-5	4	9	13	$\sqrt{13} \approx 3.6$

31. No

a	b	$a+b$	$\sqrt{a+b}$	$\sqrt{a}$	$\sqrt{b}$	$\sqrt{a}+\sqrt{b}$
2	7	9	3	$\sqrt{2}$	$\sqrt{7}$	$\sqrt{2}+\sqrt{7} \approx 4.1$
4	9	13	$\sqrt{13} \approx 3.6$	2	3	5
1	5	6	$\sqrt{6} \approx 2.4$	1	$\sqrt{5}$	$1+\sqrt{5} \approx 3.2$
9	16	25	5	3	4	7

32. No

a	b	$a+b$	$\sqrt{a}$	$\sqrt{b}$	$\sqrt{a}+\sqrt{b}$	$\left(\sqrt{a}+\sqrt{b}\right)^2$
4	9	13	2	3	5	25
1	4	5	1	2	3	9
3	5	8	$\sqrt{3}$	$\sqrt{5}$	$\sqrt{3}+\sqrt{5}$	≈ 15.7
6	10	16	$\sqrt{6}$	$\sqrt{10}$	$\sqrt{6}+\sqrt{10}$	≈ 31.5

Homework 5.3

1. $16w^2 - 36t^2$ 2. $4\pi w^2 - 25$

3. $\frac{32\pi h^3}{3}$ 4. $42c^3$

5. $22a^3$ 6. $4\sqrt{3}m^2 + 4\sqrt{3}m + 6m$

7. 8   8. 37   9. $\sqrt{68}$ 10. $\sqrt{15}$

11. $k = 5\sqrt{2}$, $3k = 15\sqrt{2}$

12. No   13. Yes   14. No

15. Not a right triangle

16. Sides should be squared

17. The hypotenuse is x, not 11

18. Sides should be squared

19. $w = \dfrac{v}{lh}$ 20. $m = \dfrac{2E}{v^2}$

21. $h = \dfrac{2A}{b+c}$ 22. $C = \dfrac{5}{9}(F-32)$

23. $h = \dfrac{A - 2\pi r^2}{\pi r}$ 24. $x = a - \dfrac{ay}{b}$

25a. $10^2 + 24^2 = x^2$ b. 26 ft

26a. $700^2 + h^2 = 722^2$ b. 177 ft

27a. $90^2 + 90^2 = x^2$ b. 127.3 ft

28a. $5^2 + 7^2 = x^2$ b. 8.6 mi

29. 79.2 m   30. $\sqrt{2}$ in, $\sqrt{3}$ in, 2 in, $\sqrt{5}$ in, $\sqrt{6}$ in, $\sqrt{7}$ in, $\sqrt{8}$ in, 3 in, $\sqrt{10}$ in

31. Area; $49\pi \approx 153.94$ square inches

32. Circumference; 23 inches

33. $\approx 105,904.14$ cubic feet

34. $144\pi \approx 452.40$ square feet

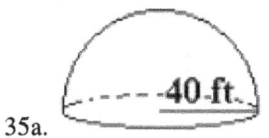

35a.
b. $\frac{128,000\pi}{3}$ cubic feet
c. 3200π sq. ft.
36a. 210π sq. cm.   b. 90 feet
c.

$\neq x^2 - y^2$
43. $x^2 + 7x + 7x + 49 = (x+7)^2$
44. $x^2 - 5x - 5x + 25 = (x-5)^2$
45. No: If $x = 1$, then $(x+4)^2 = (1+4)^2$
   $= 5^2 = 25$, but $x^2 + 4^2 = 1^2 + 4^2 = 1 + 4 = 5$.
56. $x^2 - 4x + 4$ 47. $4x^2 + 4x + 1$
48. $9x^2 - 24xy + 16y^2$ 49. $h^2 + \left(\sqrt{7}\right)^2 = (h+1)^2$
50. $(x-1)^2 + 12^2 = (2x+1)^2$

Homework 5.4

1a. $12n$ b. $12n^2$ c. $64n^3$ 2a. $-30x^3$
b. $30x^3$ c. $25x^2$ 3a. $-16p^2$ b. $-16p^2$
c. $-4p^2$ 4a. $-40xt$ b. $-40x - 8xt$
c. $40x + 8xt$ 5a. $-12n^2$ b. $-n$
c. $3n^2 - 4n$ 6a. $-10x^3$ b. $2x - 10x^2$
c. $2x - 25x^2$ 7. $2a(5a+3) = 10a^2 + 6a$
8. $3xy(2x - 5 + y) = 6x^2y - 15xy + 3xy^2$
9. $-12b^2 + 4b$ 10. $18a^2 - 15a$
11. $15v^2 - 6v^3$ 12. $-8x^3 - 12x^2y$
13. $2y^4 + 6y^2 - 4y$ 14. $-2x^3y + x^2y^2 - 3xy^3$
15. $-ax + 15a$ 16. $4x + 2$
17. $-3ax$ 18. $2ab^2 + 3a^2b^2$
19. $x^2 + 4x + 3x + 12 = (x+3)(x+4)$
20. $4x^2 - 4x - 2x + 2 = (4x-2)(x-1)$

21a.

	a	-3
a	a^2	$-3a$
-5	$-5a$	15

b. $(a-5)(a-3) = a^2 - 8a + 15$

22a.

	$3y$	-2
y	$3y^2$	$-2y$
1	$3y$	-2

b. $(y+1)(3y-2) = 3y^2 + y - 2$

23a.

	$4x$	3
$5x$	$20x^2$	$15x$
-2	$-8x$	-6

b. $(5x-2)(4x+3) = 20x^2 + 7x - 6$

24. $x^2 + xy - 2y^2$ 25. $6s^2 + 11st + 3t^2$
26. $2x^2 - 7ax + 3a^2$ 27. $6x^2 - 20x + 6$
28. $-3x^2 - 9x + 12$ 29. $-4x^2 + 5x + 6$
30. $20x^2 - 5$ 31. 1512
32. $800 + 480 + 20 + 12 = 1312$

33a. $-3x$ b.

	x	-9
x	x^2	$-9x$
6	$6x$	-54

34a. $3x$ b.

	x	4
$2x$	$2x^2$	$8x$
-5	$-5x$	-20

35. $x^2 - 9$ 36. $x^2 - 4a^2$
37. $9x^2 - 1$ 38. $w^2 + 8w + 16$
39. $z^2 - 12z + 36$ 40. $9a^2 - 12ac + 4c^2$
41. No

a	b	$a-b$	$(a-b)^2$	a^2	b^2	a^2-b^2
5	3	2	4	25	9	16
2	6	-4	16	4	36	-32
-4	-3	-1	1	16	9	7

42. $(x-y)^2 = (x-y)(x-y) = x^2 - xy - xy + y^2$

Chapter 5 Review

1a. 147 b. -441 c. -40 d. 16
3a. 1 b. $7/5$ c. $\frac{12}{25}$ d. $\frac{7}{5}$
5a. 4 b. 4 c. 27 d. 27
7a. irrational   b. undefined
c. rational   d. irrational
9a. irrational, 17.32 b. irrational, 7.47
11. -14 13. -2 15. -20.247
17a. -4 b. 4 19a. -4 b. 4
21a. -13 b. 48 23a. 12 b. 18
25a. 6 b. 12 27. 64
29a. 17 b. 40 c. 60
31a. 1 b. $3 + \sqrt{3}$ c. $\dfrac{\sqrt{3}}{2}$
33. $V = 126$; $S = 30\sqrt{7} + 84$
35. $16b^3$ 37. $V = 150p^3$, $S = 190p^2$
39. $v = \dfrac{s}{t} - \dfrac{at}{2}$ 41. $a = \dfrac{2S}{n} - f$
43. 13.9 yd   45a. 2.5 ft   b. 78.54 sq ft
47. 11.40 m   49. 10 m   51. $-20x + 15x^2$
53. $6x^2y - 12xy - 3xy^2$ 55a. $-2t^2$
b. cannot be simplified   c. $-48t^3$
57. $9a^2 - 3a$ 59. $-2a^2 - 9a + 8$ 61. 1
63a. $5x^2 - 2x + 5x - 2$ b. $(x+1)(5x-2)$
65. $x^2 + 2x - 8$ 67. $25a^2 - 1$
69. $4q^2 + 20q + 25$ 71. $u^2 - 7u + 10$
73. $a^2 + 12a + 36$ 75. $6a^2 - 11ac - 10c^2$
77. 18

Homework 6.1

1. a,c
2.

x	-3	-2	-1	0	1	2	3
y	13	18	5	4	5	8	13

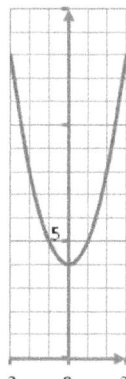

3.

x	-3	-2	-1	0	1	2	3
y	5	0	-3	-4	-3	0	5

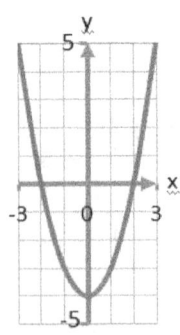

4.

x	-3	-2	-1	0	1	2	3
y	-5	0	3	4	3	0	-5

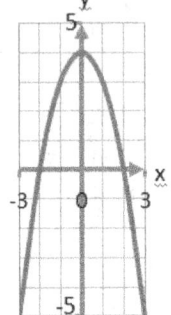

5.

x	-4	-2	-1	0	1	2	4
y	4	1	0.25	0	0.25	1	4

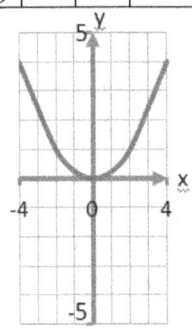

6a. 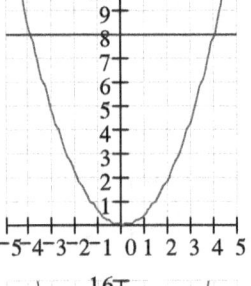   b. $-4, 4$

7a. 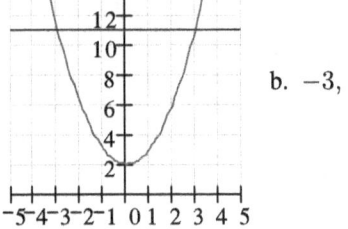   b. $-3, 3$

8a. 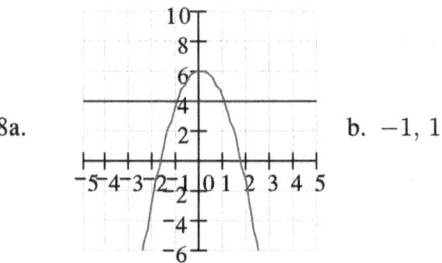   b. $-1, 1$

9. ± 11 10. ± 7 11. $\pm \frac{5}{3}$ 12. $\pm \sqrt{18}$

13. $\pm \sqrt{6}$ 14. ± 30 15. ± 7.25

16. ± 4 17. ± 15 18. No solution

19. ± 5.72 20. ± 5.73 21. $\pm \sqrt{\frac{A}{4\pi}}$

22. $\pm \sqrt{\frac{2F}{m}}$ 23. $\pm \sqrt{\frac{d-6}{k}}$

24. $\pm \sqrt{\frac{2(100-h)}{g}}$

25a.

r	C
0	0
1	0.25
2	1
4	4
6	9
9	20.25
10	25
14	49

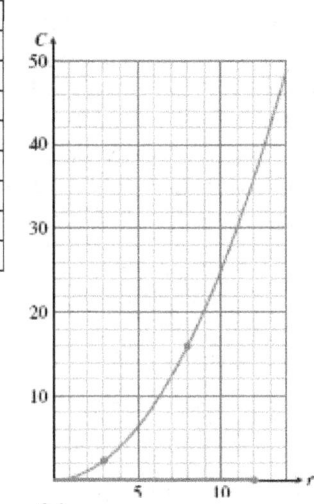

b. $2.25   c. 8 in.

d. The cost of a radius 6" pizza is $9.

26a.

Velocity (kph)	5	10	15	20	40	60
Distance (meters)	0.125	0.5	1.125	2	8	18

b. 8 m   c. 49 kph   d. A distance of 2 m is required to stop a car with a speed of 20 kph.

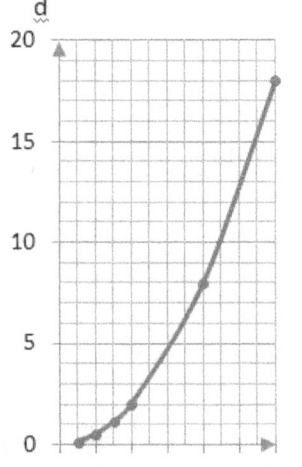

27a. $\sqrt{27}$ ft   b. $3\sqrt{27}$ sq ft

28a. $\sqrt{8}$ cm   b. $\sqrt{14}$ cm   c. $\frac{4\sqrt{14}}{3}$ cm³

29. $\sqrt{x^2 - 4}$

30.

t	h
0	100
0.5	96
1	84
1.25	75
1.5	64
1.75	51
2	36
2.25	19
2.5	0

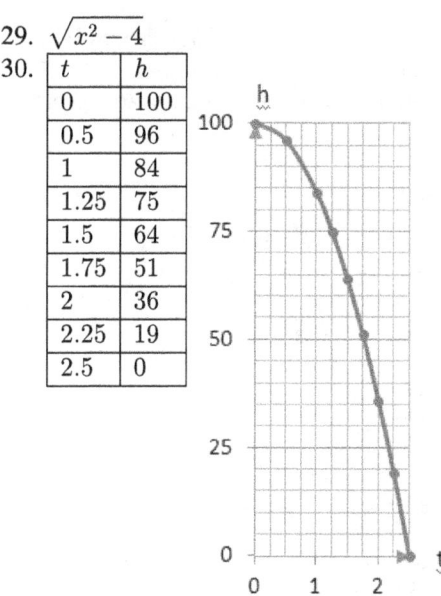

b. 36 ft   c. After 1.25 seconds
d. $75 = 100 - 16t^2$

Homework 6.2

1. $-1, 4$ 2. $0, -7$ 3. $\frac{-3}{2}, \frac{1}{4}$ 4b. $9, -3$
c. $9^2 - 6(9) - 27 = 0$; $(-3)^2 - 6(-3) - 27 = 0$
5b. $9, -9$ c. $9^2 - 81 = 0$; $(-9)^2 - 81 = 0$
6. $4x$ 7. $3ab$ 8. $5w$ 9. -3
10. $2a$ 11. $4 - 3h$ 12. $-3m$ 13. $2a(3a - 4)$
14. $6v(-3v - 1)$ or $-6v(3v + 1)$
15. $h(4 - 9h)$ 16. $d = k(1 - at)$
17. $r(2h - \pi r)$ 18. $V = h(\pi r^2 - \frac{1}{3}s^2)$
19. $0, \frac{3}{2}$ 20. $0, \frac{1}{20}$ 21. $0, -48$
22. $0, \frac{2}{3}$ 23. $0, -1$ 24. $0, 2$

25a.
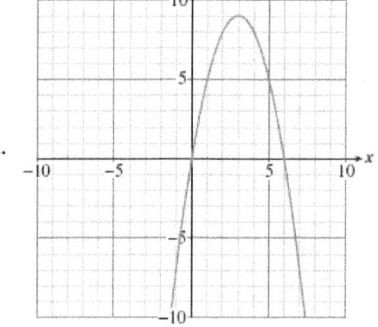

b. $x = 1$, $x = 5$ d. $x = 0$, $x = 6$
f. $-x^2 + 6x$ g. $0, 0$

26a.
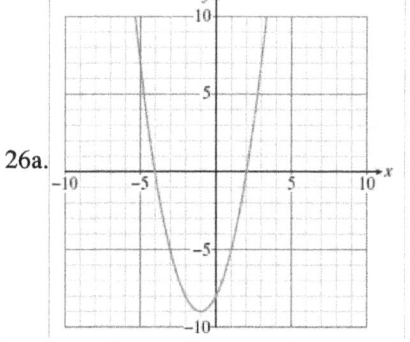

b. $x = -5$, $x = 3$ d. $x = -4$, $x = 2$
f. $x^2 + 2x - 8$ g. $0, 0$

27a. $n(-2n + 80)$; $80 - 2n$ dollars

b.
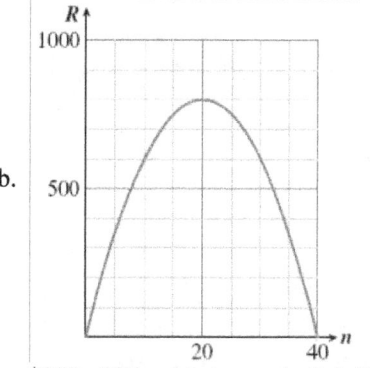

c. $800; 20 bracelets   d. 0 dollars
28a. $-16t(t - 5)$

b.
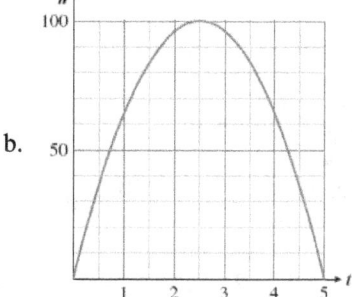

c. 100 ft; 2.5 sec   d. 5 sec
29a. $R = 180p - 3p^2$

b.
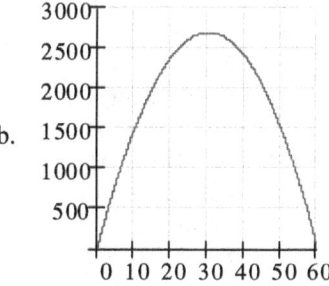

c. $0 and $80   d. $30; $2700
30. $(-2, 0)$, $(\frac{4}{3}, 0)$ 31. $(-4, 0)$, $(0, 0)$
32. $(\frac{-7}{4}, 0)$, $(0, 0)$

33.
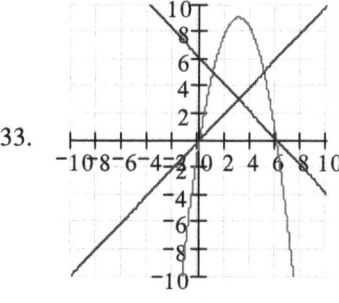

34.
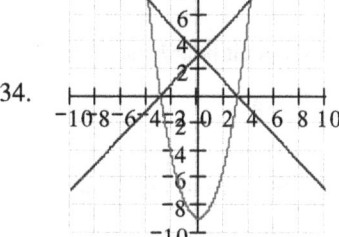

35. $4x^2 = 12x$, $4x^2 - 12x = 0$, $4x(x - 3) = 0$, $x = 0, 3$

36. $9x^2 - 4 = 16$, $9x^2 = 20$, $x^2 = \frac{20}{9}$, $x = \pm\dfrac{\sqrt{20}}{3}$

c. $a = -16$, $a = 5$: The base is 16 inches and the altitude if 5 inches.

37. The solution should be a number, not an algebraic expression.

Homework 6.3

1a.

x	x	6
x	x^2	$6x$
5	$5x$	30

b. $(x + 5)(x + 6) = x^2 + 11x + 30$

2a.

x	x	-3
x	x^2	$-3x$
-9	$-9x$	27

b. $(x - 9)(x - 3) = x^2 - 12x + 27$

3a.

x	x	2
x	x^2	$2x$
8	$8x$	16

b. $(x + 8)(x + 2) = x^2 + 10x + 16$

4. $1 \cdot 24$, $2 \cdot 12$, $3 \cdot 8$, $4 \cdot 6$

5. $1 \cdot 60$, $2 \cdot 30$, $3 \cdot 20$, $4 \cdot 15$, $5 \cdot 12$, $6 \cdot 10$

6. $1 \cdot 40$, $2 \cdot 20$, $4 \cdot 10$, $5 \cdot 8$

7a. $(n + 2)(n + 8)$ b. $-2, -8$

8a. $(h + 2)(h + 24)$ b. $-2, -24$

9a. $(a - 2)(a - 6)$ b. 2, 6

10a. $(t - 3)(t - 12)$ b. 3, 12

11a. $(x + 2)(x - 5)$ b. $-2, 5$

12a. $(a - 2)(a + 10)$ b. 2, -10

13. $(x - 2)(x - 15)$ 14. doesn't factor

15. $(y + 1)(y - 45)$ 16. doesn't factor

17. $(q + 1)(q - 6)$ 18. $(n - 2)(n - 3)$

19. doesn't factor   20. $(b - 4)(b - 8)$

21. $(c + 10)(c - 6)$ 22. $3(b - 3)(b - 8)$

23. $4(x - 9)x + 4)$ 24. $-2(m + 7)(m - 3)$

25. $x = 2$, $x = -5$ 26. $t = 6$, $t = -7$

27. $x = -1$, $x = 6$ 28. $n = 7$

29. $q = 0$, $q = 2$ 30. $x = -3$, $x = 7$

31. $x = -2$, $x = 3$ 32. $x = -6$, $x = -2$

33b. $A = w(w + 3)$

c.

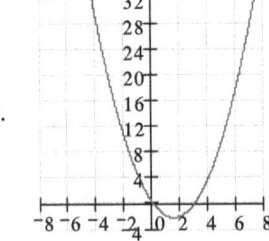

d. $w(w + 3) = 28$ e. 4 yds

34. 6 hr

35a.

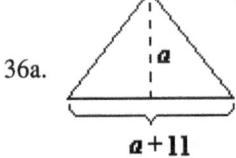   b. $w(2w - 2) = 60$

c. $w = -5$, $w = 6$: The circuit board should be 6 cm by 10 cm.

36a.

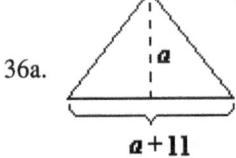   b. $\frac{1}{2}(a + 11)a = 40$

Homework 6.4

1. $(\frac{-5}{2}, 0)$, $(3, 0)$ 2. $(0, 0)$, $(3, 0)$

3. $(0, 0)$, $(\frac{8}{3}, 0)$

4.

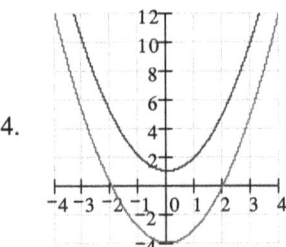

5.

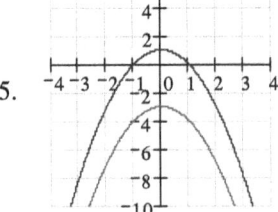

6a. $(0, 1)$, $(0, -4)$, $(0, -3)$, $(0, 1)$

b. $(0, k)$ c. $(0, -k)$

7a.

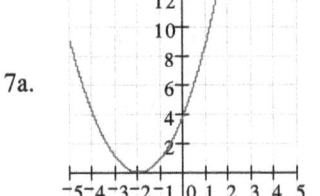   b. $(-2, 0)$

8a.

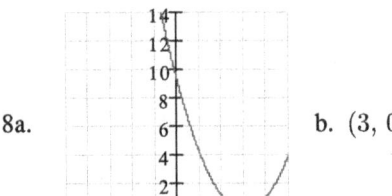   b. $(3, 0)$

9a. $(-2, 0)$, $(0, 0)$ b. $(-1, -1)$

c.

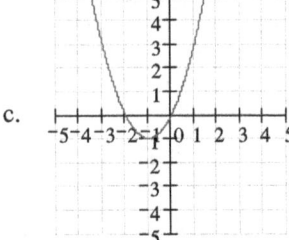

Answers-20

10a. $(-1, 0)$, $(3, 0)$ b. $(1, -4)$

c.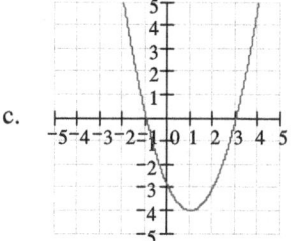

11a. $(-3, 0)$, $(3, 0)$ b. $(0, -9)$

c.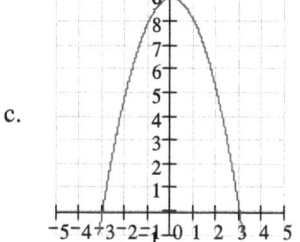

12a. $(1, 0)$, $(5, 0)$ b. $(3, 8)$

c.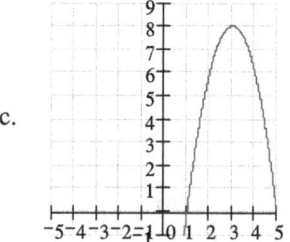

13a. $(-1, 0)$, $(2, 0)$; $(-1, 0)$, $(2, 0)$
b. $(\frac{1}{2}, \frac{-9}{4})$; $(\frac{1}{2}, \frac{9}{4})$

c.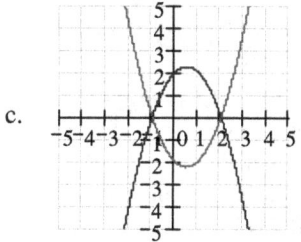

14a. $(-3, 0)$, $(5, 0)$; $(-5, 0)$, $(3, 0)$
b. $(1, -16)$; $(-1, -16)$

c.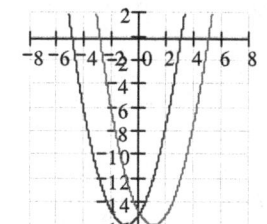

15a.

t	0	0.5	1	1.5	2	2.5	3
h	48	60	64	60	48	28	0

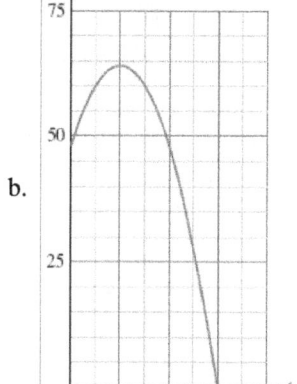

b.   c. 55 ft

16a. 2.2 sec   b. $40 = 48 + 32t - 16t^2$
17. 1 sec   18a. 64 ft   b. 2 sec

Homework 6.5

1a. $\dfrac{8 \pm \sqrt{48}}{2}$ b. 0.54, 7.46

2a. $\dfrac{-2 \pm \sqrt{28}}{6}$ b. -1.22, 0.55

3a. $\dfrac{1 \pm \sqrt{5}}{2}$ b. -0.62, 1.62

4a. $\dfrac{-2 \pm \sqrt{20}}{-8}$ b. -0.31, 0.81

5a. $\dfrac{\pm \sqrt{60}}{6}$ b. ± 1.29

6a. 0, $\frac{1}{3}$ b. 0, $\frac{1}{3}$

7a. $\frac{3}{2}$, 2   b. $\frac{3}{2}$, 2

8. $\dfrac{-6 \pm \sqrt{52}}{2}$ 9. $\dfrac{7}{8}$, 1

10. $\dfrac{9 \pm \sqrt{73}}{4}$ 11. 3, 15

12a. 2.5 sec   b. 5.13 sec
13. width: 15 in, length: 40 in, height: 20 in
14. 9 in by 12 in
15. $x = 8$, $2x - 1 = 15$
16. $k = 6$, $2k - 5 = 7$
17a. $(-0.73, 0)$, $(2.73, 0)$

b.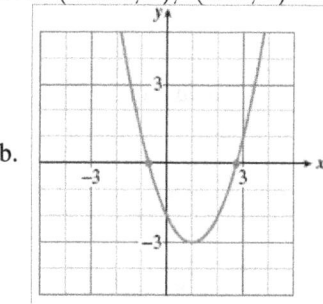

18a. No x-intercepts

b.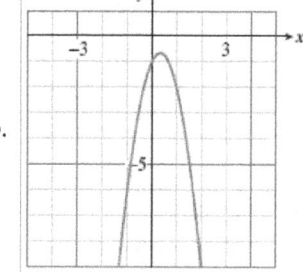

19a. $(-0.24, 0)$, $(4.24, 0)$

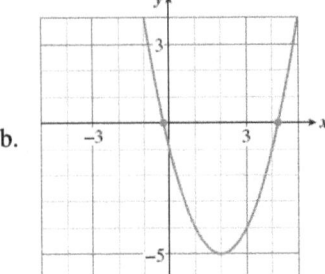

b.

20a. $(-2.4, 0)$, $(3, 0)$

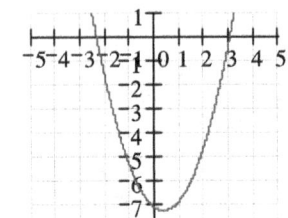

b.

Chapter 6 Review

1. $\pm\frac{2}{3}$　　3. ± 2　　5. $\pm\frac{2}{3}$

7a. 3 sec　　b. 36 ft

9. $6x(4x - 3)$　　11. $(a - 15)(a - 3)$

13. $(2y + 3)(2y + 5)$　　15. $(3w - 1)(w + 5)$

17. $(z + 11)(z - 11)$　　19. $3(2x + 1)(x + 3)$

21. $-5, -5$　　23. $0, 4$　　25. $1, 3$

27. $\dfrac{-6\pm\sqrt{12}}{4}$　　29. $-0.47, 1.07$

31. $s = \pm\sqrt{\dfrac{3V}{h}}$　　32. $d = \pm\sqrt{\dfrac{4A}{\pi}}$

33. $h = \pm\sqrt{\dfrac{C}{br}}$　　34. $r = \pm\sqrt{\dfrac{np}{G}}$

35. 3.81 sec　　37a. $R = 160p - 2p^2$

b. $M = 136p - 2p^2 - 80$

c. 30 cents or 38 cents　　39b. $A = 2w^2$

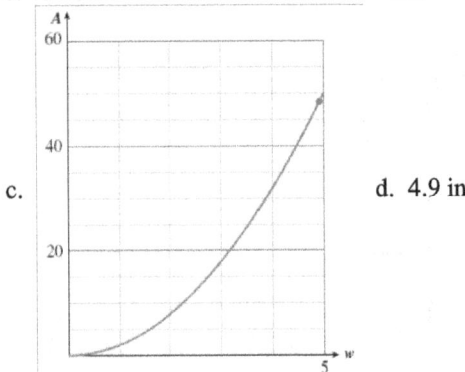

c.　　　　　　　　　　　d. 4.9 in

41a. $(0,0)$; $(0,0)$　　b. $(0,0)$

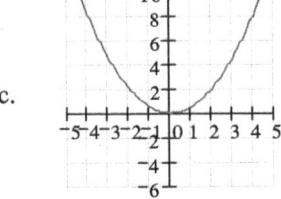

c.

43a. $(0,0), (9,0)$; $(0,0)$　b. $\left(\frac{9}{2}, \frac{-81}{4}\right)$

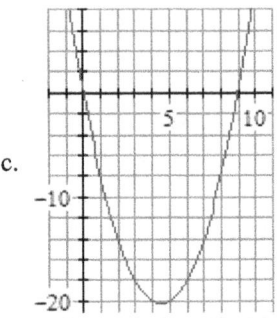

c.

45a. $(-6, 0)$, $(0, 0)$; $(0, 0)$　b. $(-3, -9)$

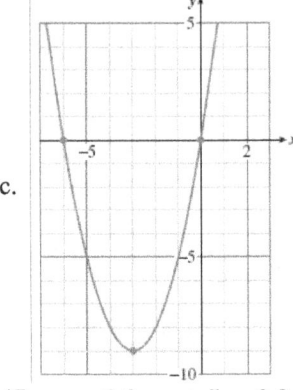

c.

47. $m = 3.9$, $m + 5 = 8.9$

Homework 7.1

1. c　　　2. Possible answers:　a. $3x^4$

b. $2x - 7$　c. $x^2 - 3x - 1$　　d. 4

3a. polynomial　b. not: variable in denominator

c. not: variable in the denominator　　d. polynomial

4. a, b, c　　5a. 2　　b. 7　　c. 4

6. $-1.9x^3 + x + 6.4$　　7. $-2x^2 + 6xy + 2y^3$

8. 14　　9. 15.8972　　10. 24

11. $5 - 3\sqrt{3}$　　12. $12b$　　13. $6x + 6xy - 9y$

14. $14w^3$, exponent does not change

15. $6 + 3x^2$ cannot be simplified

16. $t^2 + 12$, change all signs inside parentheses

17. $2y^3 + 2y^2 + y + 1$　　18. $3x^3 + 3x^2 + 11$

19. $6x^2 + 2x - 3$　　20. $-7x^2 + 7x - 1$

21a. 2000 ft　　　　b. Brenda has landed.

22. $w^3 + 4w^2 - 32$　　23a. 56　　b. $22,100$

c. 1140　　　　　24. 130 ft; No

25a. 567　b. 5133　　c. 1001　　d. $80,080$

26a. $x^3 + 3x^2 + 4x + 1$　b. $x^4 - 7x^3 - 5x^2 + 8x - 3$

27a. 29, before　　b. 6.632

28a. $-x^3 + 10x^2 + 2720x - 180$

b. $27,020$; $50,220$; $35,820$

29a. $28t^2 - 56t + 28$　　b. 28ft; 7 ft　　c. 1 sec

30a. $0.30h$　b. $50 - h$　c. $0.25(50 - h)$

d. $0.05h + 12.5$

31a. $2w$　　b. $w + 30$　c. $2(w + 30)$

d. $4w + 180$

32a, b.

m	T_m	$\frac{1}{3}m^3 + \frac{1}{2}m^2 + \frac{1}{6}m$
1	1	1
2	5	5
3	14	14
4	30	30
5	55	55
6	91	91
7	140	140
8	204	204
9	285	285
10	385	385

c. They are the same.   32. -7.2

Homework 7.2

1a. x^9 b. 5^{14} c. b^9
2a. 5   b. 4   c. 2
3a. $-6x^8$ b. $2x^6$ c. $-x^4y^3$
d. $-9x^5y^5$ 4a. $-x^4$
b, c, d. cannot be simplified   5a. $21z^8$
b. $-10x^4$ 6a. $9c^2d^6$ b. $-10cd^3$
7a. $48w^5$ b. $144w^{10}$ 8a. $34t^6$
b. $52t^{12}$ 9a. $2a(5a+3)$ b. $10a^2+6a$
10a. $3xy(2x-5+y)$ b. $6x^2y-15xy+3xy^2$
11a. $2x^2$ b. x^4 c. 0   d. $-x^4$
12a. $-2x$ b. x^2 c. $-2x^2$ d. x^4
13a. cannot be simplified   b. x^3
c. cannot be simplified   d. $-x^3$
14a. $-x^6$ b. $-x^5$ c. cannot be simplified
d. $-x^5$ 15. $-x^3y - x^2y^2 - xy^3$
16. $-9s^4t^4 + 3s^3t^3 - 18s^2t^2$ 17. $-3ax$
18. $3a^2b^2 + 2ab^2$ 19. $x^2 + xy - 2y^2$
20. $6s^2 + 11st + 3t^2$ 21. $2x^2 - 7ax + 3a^2$
22. $6x^2 - 20x + 6$ 23. $-3x^2 - 9x + 12$
24. $-4x^2 + 5x + 6$ 25. $20x^2 - 5$
26. $4a^3 + 16a^2 - 20a$ 27. $6s^4t^2 + s^3t^3 - s^2t^4$
28. $x^3 - 5x^2 + 8x - 4$ 29. $27x^3 - 18x^2 + 6x - 1$
30. $x^3 + 6x^2 + 11x + 6$ 31a. $x^3 + 7x^2 + 2x - 40$
b. $x = 2, -4, -5$ 32a. $80,000,000$
b. $60,000$ c. $66,000$
33a. $16 - n$ b. $16n - n^2$
34a. Length: $w + 3$, Height: $w - 2$
b. $w^3 + w^2 - 6w$ c. $6w^2 + 4w - 12$
35a. $n + 2$ b. $n + 4$ c. $n^3 + 6n^2 + 8n$
36. $3y^2 + 10y + 8$ 37. $8w^2 - 18w - 18$
38. $6t^2 + 19t + 15$

Homework 7.3

1a. a^3 b. 3^5 c. $\dfrac{1}{z^3}$
2a. $2^2 \cdot 2^3 = 4 \cdot 8 = 32;\ 2^6 = 64$
b. $\dfrac{3^8}{3^2} = \dfrac{6561}{9} = 729;\ 3^4 = 81$
c. $\dfrac{10^3}{10^5} = \dfrac{1000}{100,000} = 0.01;\ 10^2 = 100$
3. $\dfrac{1}{4xy^4}$ 4. $\dfrac{-3x^2}{2}$ 5. $\dfrac{5}{y^2}$

6. $-b(b + c + a)$ 7. $-2k(2k^3 - 2k + 1)$
8. 2   9. $4ab$ 10. $3x^2(3 - 4x^3 + x)$
11. $7xy(2x^2 - 5xy + 3y^2)$ 12. $(x + 6)(2x - 3)$
13. $(2x + 3)(3x^2 - 1)$ 14. $(2x + 1)(x + 5)$
15. $(5t + 2)(t + 1)$ 16. $(3x - 5)(x - 1)$
17. $2(x + 2)(x + 3)$ 18. $b(2a + 7)(2a - 1)$
19. $2z(2z + 3)(z + 1)$ 20. $9b(2a - 3)(a + 1)$
21. $(x - 2y)(x - 3y)$ 22. $(x + 2y)(x + 2y)$
23. $(x + 11a)(x - 7a)$ 24. $4x(x + 2y)(x + y)$
25. $9ab(a + 2b)(a - b)$ 26. $(2t + s)(t - 3s)$
27. $(4by + 1)(by + 1)$ 28. $3a(4b + a)(b + a)$
29. $(2x - 9)(x + 2)$ 30. $(5x - 4)(x + 4)$
31. $(3h + 2)(2h + 1)$ 32. $(9n + 1)(n - 1)$
33. $(2t - 5)(3t + 5)$ 34. $(5x + 6)(x - 4)$
35. $-1, \frac{4}{3}$ 36. $\frac{1}{3}, \frac{3}{2}$ 37. $\frac{1}{2}, \frac{1}{2}$
38. $\frac{-5}{2}, \frac{1}{3}$ 39. $-2, \frac{2}{3}$ 40. $\frac{-1}{2}, 3$
41. $3\frac{1}{3}$ hrs   42. 5 ft by 12 ft
43. $b + 20$ 44. $n + 3$
45. $3u + 1$ 46. $2t - 9$

Homework 7.4

1a. $64t^8$ b. $144a^4$ c. $100h^4k^2$
2a. $4b^8$ b. $11z^{11}$ c. $6p^3q^{12}$
3a. $x^2 + 8x + 8x + 64$ b. $(x + 8)^2$
4a. $x^2 - 9x - 9x + 81$ b. $(x - 9)^2$
5. $n^2 + 2n + 1$ 6. $m^2 - 18m + 81$
7. $4b^2 + 20bc + 25c^2$ 8. $x^2 + 2x + 1$
9. $4x^2 - 12x + 9$ 10. $25x^2 + 20xy + 4y^2$
11. $9a^2 + 6ab + b^2$ 12. $49b^6 - 84b^3 + 36$
13. $4h^2 + 20hk^4 + 25h^8$ 14. $x^2 - 16$
15. $x^2 - 25z^2$ 16. $4x^2 - 9$
17. $9p^2 - 16$ 18. $4x^4 - 1$
19. $h^4 - 49t^2$ 20. $-18a^3 + 60a^2 - 50a$
21. $16x^4 + 96x^3y + 144x^2y^2$ 22. $20m^5p^2 - 5mp^4$
23. $(y + 3)^2$ 24. $(m - 15)^2$
25. $(a^3 - 2b)^2$ 26. $(z - 8)(z + 8)$
27. $(1 - g)(1 + g)$ 28. $(a - 15)(a + 15)$
29. $(x - 3)(x + 3)$ 30. $(6 - ab)(6 + ab)$
31. $(8y - 7x)(8y + 7x)$ 32. $(a^2 + 5)^2$
33. $(6y^4 + 7)(6y^4 - 7)$ 34. $(4x^3 + 3y^2)(4x^3 - 3y^2)$
35. $3(a + 5)(a - 5)$ 36. $2a(a - 3)^2$
37. $9x^3(x^2 + 3)(x^2 - 3)$ 38. $3(4h^2 + k^6)$
39. $(9x^4 + y^2)(3x^2 + y)(3x^2 - y)$
40. $2a^4(9b^4 + a^2)(3b^2 + a)(3b^2 - a)$
41. No. Let $x = 1$, then $(x - 1)^2 = (-2)^2 = 4$,
but $x^2 - 3^2 = 1 - 9 = -8$.
42. The area of the alarge rectangle is $(a + b)^2$.
The sum of the areas of the four smaller rectangles is
$a^2 + 2ab + b^2$.
43. No.   44. $x^2 - 4 = (x + 2)(x - 2)$, but
neither $(x + 2)^2$ nor $(x - 2)^2$ is equivalent to $x^2 + 4$.
45a. $a^3 - 3a^2b + 3ab^2 - b^3$
b. $8x^3 - 36x^2 + 54x - 27$
c. $(a - b)^3 = 27;\ a^3 - b^3 = 117$
46a. $a^3 + b^3$ b. $(a + b)(a^2 - ab + b^2)$
c. $(x + 2)(x^2 - 2x + 4)$

Chapter 7 Review

1a. $\frac{-1}{720}x^6 + \frac{1}{24}x^4 - \frac{1}{2}x^2 + 1$, degree 6

b. $10n^4 + 10^8n^2 + 10^6m$, degree 4

3. 26   5. $\frac{1}{3}$ 7. 0

9. $6x^4 - 12x^3 + 3x^2 - 4x$ 11a. 4845

b. 0, 0, 0   c. 1: There is exactly one way to choose 4 items from a list of 4 — choose all 4 items.

13. $9p^2q - 5pq^2$ 15. $5b^3 + 2b^2 + 6b + 4$

17a. $4x^2$ b. $3x^4$ 19a. cannot be simplified

b. $7a^{10}$ 21a. $\frac{b^6}{3}$ b. $\frac{3}{b^6}$ 23. $-30m^5n^3$

25. $-7x^3$ 27. $-2y^3 + 2y^2 + 24y$

29. $d^3 - 6d^2 + 12d - 8$

31. $144x^5y - 216x^4y^2 + 81x^3y^3$

33. $2a^2 + a - 9$ 35. $\frac{1}{2c}$

37. $56q^2 - 49q^4 - 7q^6$ 39. $81v^2 - 25w^2$

41. $-3p^4 + 9p^3 + 30p^2$ 43. $8x^3 - 27$

45a. $10a^4$ b. $6a^8$ 47a. $8m + 2n$

b. $3m^2 + 5mn - 2n^2$ 49. -32.6

51a. $140p - 2p^2$ b. $528

53. Equation. $x = \frac{-3}{2}, 1$

55. Polynomial. $(2x + 3)(x - 1)$

57. Equation. $x = \frac{-1}{2}, 0$

59. Equation. $x = \frac{-3}{2}$

61. Polynomial. $2x(x + 1)(x - 1)$

63. Equation. $p = -1, 0, 1$

65. $4x^3(3x^2 - 2x + 5)$ 67. $6w^4(5w^5 - 7 + 9w^4)$

69. $-5d^2(2d^2 - 4d + 1)$ 71. $-vw^2(w^3 + w^2 - 1)$

73. $(3r + 2)(-r^3 + 4)$ 75. $3z(z + 2)(z - 2)$

77. $(a^2 + 5)^2$ 79. $2a^2b^2(b^4 + 16a^4)$

81. $-2a^2b(a + 5b)(a - 3b)$

83. $2q^2(q + 8)(q - 5)$ 85. $(x - 2y)(x - y)$

87. $3b(a + 3b)(a + b)$ 89. $(3xy + 2)(3xy - 1)$

91. $(h - 12)^2$ 93. Cannot be factored

95. $(w^4 + 6)^2$ 97. $3(s^2 + 4)(s + 2)(s - 2)$

Homework 8.1

1a. $\frac{-3}{5}, \frac{3}{7}$ b. 3   2a. $\frac{-3}{10}, \frac{-3}{2}$ b. None

3a. $-1, 1$ b. $\frac{-1}{2}$ 4. $\frac{5}{x}$ 5. $\frac{12b}{7}$

6. $\frac{-5}{6}$ 7. $\frac{1}{24v}$ 8. $\frac{-8}{5}$ 9. $\frac{a}{9}$

10. $\frac{-3y^2}{14}$ 11. $\frac{2u^3}{3w}$ 12a. Cannot be reduced

b. $\frac{4}{5}$ 13a. $\frac{m}{2n}$ b. Cannot be reduced

14a. Cannot be reduced   b. $\frac{-1}{3}$

15a. -1 b. Cannot be reduced

16a. Cannot be reduced   b. $\frac{3}{2}$

17a. Cannot be reduced   b. 2   18. b

19. a   20. a   21. b   22. b

23. a   24. b   25. $\frac{x+4}{2}$ 26. $\frac{1}{b-1}$

27. $\frac{n+2}{n-2}$ 28. $\frac{a}{a-3}$ 29. $\frac{3}{4}$ 30. $\frac{1}{x+1}$

31. Cannot be reduced   32. Cannot be reduced

33. -1 34. 1   35. $\frac{-1}{2}$ 36. $\frac{1}{a+b}$

37. $\frac{3x+2y}{2y-3x}$ 38. $\frac{-a}{a+1}$

39a. n : number of immunized first-graders

b. $\frac{n}{84}$

40a. t : total number of employees   b. $\frac{138}{t}$

41a. g : gallons of gas Gladys' car holds   b. $\frac{480}{3g}$

42a. A : area of rectangle; w : width of rectangle

b. $\frac{A}{w}$

43a. $\frac{18}{x-4}$ dollars per gallon   b. 1.8; If the tank holds 14 gallons, gas costs $1.80 per gallon

44a. $\frac{2800}{h-10}$ miles per hour   b. 70; If he drove for 50 hours, his average speed was 70 mph.

45a. $\frac{200}{2x-1}$ square centimeters   b. 8; If $x = 13$, the area of the cross-section is 8cm^2.

46a. 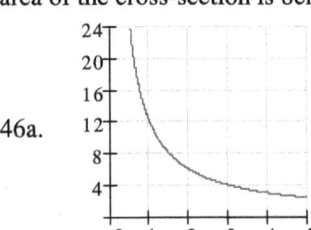   b. 6 mph

c. $\frac{2}{3}$ hr or 40 min

47a. 30 min   b. 50 min   c. 50 min   d. 8 mph

e. The time increases. If the current is 10 mph, the team will not be able to row upstream.

48. b, c, d   49. a, d   50. b, c

51. 3 is not a factor of the numerator

52. $5z$ is not a factor of the numerator

Homework 8.2

1. $\frac{-3c}{4}$ 2. $\frac{5}{3m}$ 3. $\frac{8}{5}$ 4. $\frac{-k}{4}$

5. $\frac{3}{2x}$ 6. $\frac{48}{5}$ 7. $\frac{6}{n^2+2n}$ 8. $\frac{8b}{3b+3}$

9. $\frac{1}{a+1}$ 10. $\frac{p^2}{p+2}$ 11. $\frac{3}{4}$ 12. 5

13a. $\frac{2x}{3}$ b. $\frac{3(a-b)}{4}$ c. $\frac{n+2}{n-2}$

14. $-8t + \frac{1}{4} - \frac{3}{t}$ 15. $\frac{8v^2}{9} - \frac{8}{v} - 1$

16. $\frac{4}{9z^2}$ 17. $\frac{25c^2}{4d^2}$ 18. $\frac{-h^3}{27k^3}$ 19. -3

20. $\frac{1}{8p^2}$ 21. $\frac{-36}{5c^2}$ 22. $\frac{9}{14}$ 23. $\frac{b^2}{6a}$

24. $\frac{2y}{x}$ 25. $\frac{3a}{2}$ 26. $\frac{x}{y-1}$ 27. $\frac{-5}{2a}$

28. $\frac{c^2-1}{c^2+2c-15}$ 29. $\frac{4}{a}$ 30. $\frac{8}{x-1}$

31. $\frac{2x+y}{3x}$ 32. $m-1$ 33. $z-3$ 34. $\frac{4}{b-1}$

35. $3x^3 - 2 - \frac{2}{3x^2}$ 36. $n - 1 + \frac{2}{3n} = \frac{1}{n^2}$

37. $x - 2 + \frac{3}{y}$ 38a. $\frac{7}{x}$ b. $\frac{-3}{x}$

c. $\dfrac{10}{x^2}$ d. $\dfrac{2}{5}$

39a. $\dfrac{p+r-1}{q+2}$

b. $\dfrac{p-r+1}{q+2}$ c. $\dfrac{p(r-1)}{(q+2)^2}$ d. $\dfrac{p}{r-1}$

40a. $\dfrac{1}{2c}$ b. $\dfrac{1}{2c}$ c. $\dfrac{2}{c}$ d. $\dfrac{25}{2c}$

41. $\dfrac{AB^2}{3}$

42. $\dfrac{1}{12k^3T}$

43. $\dfrac{H}{5G^3}$

44. $\dfrac{4LR}{D^2}$

45. $\dfrac{2L}{c}+\dfrac{2LV^2}{c^3}$

46. $\dfrac{3q}{8\pi R}-\dfrac{a^2q}{8\pi R^3}$

47. $\dfrac{4-2m}{m(m+1)}$

48. $-3x+2$

49a. $\dfrac{x}{2}$ b. $2x$ c. $\dfrac{1}{2x}$ d. $\dfrac{1}{a+b}$

e. $\dfrac{3}{4(a+b)}$ f. $\dfrac{4}{3(a+b)}$

50a. $K=\dfrac{-2m}{3}$ b. $H=\dfrac{2t}{5}$

Homework 8.3

1. $\dfrac{3y-4x}{xy}$ 2. $\dfrac{-3b-2a^2}{ab}$ 3. $\dfrac{a-3}{a}$

4. $\dfrac{yz+x^2z+x^2y}{xyz}$ 5. $\dfrac{5-x^2}{xy}$ 6. $\dfrac{3}{2z}$

7. $\dfrac{uv-4v+18}{3v}$ 8. $\dfrac{x+3}{x(x+1)}$

9. $\dfrac{4n^2+15n-9}{(n+3)(n-3)}$ 10. $\dfrac{h^2+2h-3}{h+2}$

11. $\dfrac{v^2+v-1}{v(v-1)}$ 12. $\dfrac{-x^2+6x-4}{x(x-2)}$

13. $\dfrac{-8+5q}{20n}$ 14. $\dfrac{-6s+3}{s^2}$

15. $\dfrac{3z^2-2x}{12x^2z}$ 16. $\dfrac{10x+3}{4x^2}$

17. $\dfrac{10z-17}{24z}$ 18. $\dfrac{13}{4(2a-b)}$

19. $\dfrac{d}{x(d-x)}$ 20. $\dfrac{a^2+N}{2a}$

21. $\dfrac{a^2-2mt}{2at}$ 22. $\dfrac{qr+2aq}{4\pi r^2}$

23. $\dfrac{2Lc}{(c-V)(c+V)}$ 24. $\dfrac{2a^2q}{r(r-a)(r+a)}$

25. $\dfrac{8}{x-1}$ 26. $\dfrac{19}{6(p-2)}$

27. $\dfrac{-m+6}{m(m-3)(m+3)}$ 28. $\dfrac{5k+1}{k(k-3)(k+1)}$

29. $\dfrac{2x^2+2x-3}{x(x+3)^2}$ 30. $\dfrac{-2}{3n}$

31a. $\dfrac{8}{a}$ b. $\dfrac{2b^2-4a}{ab}$ c. $\dfrac{2b^2}{4a}$

d. $\dfrac{2b^2+4a}{ab}$ 32a. $\dfrac{y^2}{9x}$ b. $\dfrac{2y^2+18x}{3x^2y}$

c. $\dfrac{4}{x^3}$ d. $\dfrac{2y^2-18x}{3x^2y}$ 33a. $\dfrac{3a-8}{a^2-4}$

b. $\dfrac{6}{(a+2)^2(a-2)}$ c. $\dfrac{3a-4}{a^2-4}$ d. $\dfrac{3a-6}{2}$

34a. $\dfrac{144}{x(x-2)}$ sq ft   b. $\dfrac{48x-48}{x(x-2)}$ ft

35a. $\dfrac{100}{x-10}$ hrs   b. $\dfrac{100}{x+10}$ hrs

c. $\dfrac{200x}{x^2-100}$ hrs d. $\dfrac{5}{7}$ hrs, $\dfrac{5}{8}$ hrs, $\dfrac{75}{56}$ hrs

36a. $\dfrac{1}{d}$ b. $\dfrac{1}{d-5}$ c. $\dfrac{2d-5}{d(d-5)}$

d. $\dfrac{100}{9}$ or $11\dfrac{1}{9}$ days

Homework 8.4

1. 1   2. 12   3. 4   4. $\dfrac{2}{3},\dfrac{-1}{2}$

5. 1   6. 2   7. $-6, 5$ 8. No solution

9. $\dfrac{1}{2}, 2$ 10. $-4, \dfrac{-3}{2}$ 11. $\dfrac{PV}{h}$

12. $\dfrac{mv^2}{2}$ 13. $\dfrac{y-k+hm}{m}$ 14. $\dfrac{F-aM}{a}$

15. $\dfrac{BR}{B-R}$ 16. $\dfrac{En-Ir}{In}$ 17. $\dfrac{r-dc}{cr}$

18. $\dfrac{Pw}{w+0.622}$ 19. $-4, 5$ 20. $\dfrac{1}{4}$ 21. 50%

22a. 0.207   b. $\dfrac{x+34}{x+164}$ c. 36

23a. $AE=1, ED=x-1, CD=1$ b. $\dfrac{x}{1}=\dfrac{1}{x-1}$

24a. $\dfrac{17x-5}{20}$ b. $\dfrac{25}{17}$

25a. $\dfrac{180}{x+20}=\dfrac{120}{x}$

b. freight train: 40 mph, express train: 60 mph

26a. $\dfrac{360}{x+20}=\dfrac{360}{x}-3$

b. Sam: 40 mph, Reginald: 60 mph

27a. $\dfrac{x}{10}+\dfrac{x+4}{18}=1$ b. 5 hr

28a. $\dfrac{x}{30}-\dfrac{x}{240}=1$ b. $34\dfrac{2}{7}$ min

29. Because $x=1$, dividing by $x-1$ in the fourth steo is dividing by 0.

30. The left side of the equation was multiplied by 6, but the right side was not. $a=6$

31. In the third equation, dividing by m is dividing by 0 because $m=0$ is one of the solutions. $m=6, 0$

32. 2 days

Chapter 8 Review

1a. $\dfrac{-1}{2}$ b. $-5, 2$ 3. Cannot be reduced

5. $\dfrac{5+y}{y}$ 7. $\dfrac{1}{v+2}$ 9. $\dfrac{-1}{2x+3}$

11a. $\dfrac{19}{24}$ b. $\dfrac{5}{32}$ 13a. $\dfrac{10x^2+2x+1}{6x(x+1)}$

b. $\dfrac{1}{4}$ 15a. $\dfrac{2x+1}{x}$ b. $\dfrac{2}{x}$

17. $\dfrac{2c}{9d}$ 19. $m-1$ 21. $\dfrac{8q^2-16q+21}{q(q-3)}$

23a. $\dfrac{5}{x-2}$ b. $\dfrac{5}{x+2}$ c. $\dfrac{10x}{x^2-4}$

25. $-2, 8$ 27. $\dfrac{-1}{2}, 4$ 29. $\dfrac{yz}{y-z}$

31a. $\dfrac{5}{r}=\dfrac{8}{r+2}+1$ b. 1.53 mph

Homework 9.1

1a. t^{15} b. b^8 c. w^{144}

2a. $125x^3$ b. $81w^4z^4$ c. $-a^5b^5$

3a. $\dfrac{w^6}{64}$ b. $\dfrac{625}{v^4}$ c. $\dfrac{-m^3}{p^3}$

4a. x^9 b. x^{18} c. $\dfrac{1}{x^3}$ d. x^3

5. The first law simplifies a product of powers with the same base by adding the exponents. The third law simplifies a power raised to a power by multiplying the exponents.

6a. $2 \cdot 3^2 = 2 \cdot 9 = 36$
b. $-10^2 = -(10 \cdot 10) = -100$
c. $a^4 \cdot a^3 = a^7$

7. $32p^{15}$ 8. $\dfrac{-243}{q^{20}}$ 9. $\dfrac{16h^8}{m^{12}}$

10. x^{13} 11. $4x^{10}y^{14}$ 12. $a^{21}b^{15}$

13. $-a^4$ 14. $-x^3y^4$ 15. $-4p^5q^{10}$

16. $-52y^7$ 17. $2a^9 + 2a^8$

18. $-6v^5 + 3v^4 + 16v^3$ 19a. $4x^2 + 2x^4$

b. $8x^6$ 20a. $-x^7$ b. x^{16}

21a. $324x^{16}$ b. $77x^8$ 22a. $6x^3 - 3x^6$

b. $-18x^9$ 23a. $6x^3 - 3x^6$ b. $3x^6$

24. 6   25. 6   26. 6

27. $x^2 + x^4$ 28. $1 - m^4 + 2m^{12}$

29. <   30. =   31. <

32. =   33. >   34. =

Homework 9.2

1a. $\dfrac{1}{25}$ b. $\dfrac{1}{x^6}$ c. 1   d. $\dfrac{64}{27}$

2a. $\dfrac{81}{b^4}$ b. $\dfrac{1}{32q^5}$ c. $\dfrac{3}{64}$ d. $\dfrac{4}{x^2}$

3a. 2^{-3} b. $3 \cdot 5^{-2}$ c. 3^{-3}

4a. $5^{-4}x$ b. $2z^{-2}$ c. $10^{-5}z^5$

5a. $x^0 = 1$ b. $w^{-3} = \dfrac{1}{w^3}$ c. $2x^{-4} = \dfrac{2}{x^4}$

6a. x^5 b. 5^{-7} c. $15b^{-3}$

7a. c^{-3} b. $2b^4$ c. 6^8

8a. 216   b. 192   c. $8x^3y^5$

9a. 8^{-10} b. w^{18} c. d^{-24}

10a. $p^{-5}q^{-5}$ b. $3^{-2}x^{-2}$ c. $5 \cdot 2^{-3}r^{-3}$

11. $\dfrac{a^{12}}{c^6}$ 12. $\dfrac{1}{8u^{14}}$ 13. $\dfrac{5}{6k^{10}}$

14. $\dfrac{p^{10}}{4}$ 15a. 10^5 b. 10^{-2}

c. 10^2 16a. 10   b. 10^{-8}

17a. $43,000$ b. 0.000008 c. 0.00002

18a. 10^2 b. 10^{-1} c. 10^6

19a. 4.834×10^3 b. 7.2×10^{-2}

c. 7×10^6 20a. 6.85×10^8

b. 5.674×10^5 c. 3.85×10^{-1}

21. 140   22. 0.04   23. 2400

24. 75   25. Positive   26. Negative

27. Positive   28. Negative

29a. 2.9141×10^4 ft   b. 7.6×10^{-5} cm

30a. 0.000005 g   b. $5,000,000,000$ yr

31. 5.88×10^{12} mi   32a. $4,566,666,667$ ft

b. $864,899$ mi   33a. 1.42×10^{18}

b. 4.26×10^{14} g   34. 8.99×10^7 gal

Homework 9.3

1. True   2. False   3. False

4. y^4 5. n^{18} 6. $\pm 4x^2$

7. $-11ab^3$ 8. $3(x + y)$ 9. $\dfrac{-8}{b^3}$

10. $2\sqrt{2}$ 11. $-2\sqrt{5}$ 12. $5\sqrt{5}$

13. $x\sqrt{x}$ 14. $-b^5\sqrt{b}$ 15. $p^{12}\sqrt{p}$

16. $2a\sqrt{2a}$ 17. $\pm 6m^4\sqrt{2m}$ 18. $\dfrac{x^4}{3\sqrt{3}}$

19. $4c^3\sqrt{3d}$ 20. $\dfrac{-3bd\sqrt{5d}}{2}$ 21. $\dfrac{3w\sqrt{w}}{2\sqrt{7z}}$

22. $6x\sqrt{x}$ 23. $-10a^2b\sqrt{2a}$ 24. $-2bk\sqrt{bk}$

25. $3\sqrt{3}$ 26. Cannot be simplified

27. $-7\sqrt{6}$ 28. $-3\sqrt{5}$ 29. $-3\sqrt{3} - 3\sqrt{2}$

30. $-2\sqrt{2a}$ 31. $-3x\sqrt{5x} - 6x\sqrt{5}$

32. $\sqrt{36 + 64} = \sqrt{100} = 10$

33. $\sqrt{3} + \sqrt{3} = 2\sqrt{3}$

34. $\sqrt{9 + x^2}$ cannot be simplified

35. c   36. b   37. a

38. $\sqrt{20}$ 39. $\sqrt{12}$ 40. $\sqrt{54}$

Homework 9.4

1. 4   2. $x\sqrt{6}$ 3. $36a^2$

4. $3\sqrt{2} + \sqrt{6}$ 5. $4\sqrt{5} + 10\sqrt{3}$

6. $2p\sqrt{2} - 2p\sqrt{2p}$ 7. $1 - 2\sqrt{2}$

8. $16 - a$ 9. $7 + 4\sqrt{3}$

10. $2w - 3\sqrt{5w} - 10$ 11. 3   12. $5x$

13. $\dfrac{4}{3}$ 14. $3 - \sqrt{5}$ 15. $\dfrac{-4 + \sqrt{2}}{2}$

16. $\dfrac{2a - \sqrt{2}}{2a}$ 17. $\dfrac{5 + 6\sqrt{2}}{4}$ 18. $\dfrac{9 + \sqrt{3}}{6a}$

19. $\dfrac{3\sqrt{3} + 6}{2}$ 20. $\dfrac{3 - 8\sqrt{y}}{4}$

21. Cannot be simplified 22. $12\sqrt{2}$

23. $1 - 2\sqrt{3}$ 24. Cannot be simplified

25. $\dfrac{5\sqrt{2}}{2}$ 26. $\sqrt{2a}$ 27. $\sqrt{7b}$

28. $\dfrac{\sqrt{21x}}{6}$ 29a. $\dfrac{w\sqrt{3}}{2}$ b. $\dfrac{w^2\sqrt{3}}{4}$

30a. $\dfrac{k\sqrt{14}}{2}$ b. $\dfrac{k^3\sqrt{14}}{6}$

31. $\left(\sqrt{3}\right)^2 - 2\sqrt{3}\sqrt{3} + 3 = 0$

32. $\left(\sqrt{3} + 2\right)^2 + 1 = 4\left(\sqrt{3} + 2\right)$

33. $\left(\dfrac{-3 + \sqrt{5}}{2}\right)^2 + 3\left(\dfrac{-3 + \sqrt{5}}{2}\right) + 1 = 0$

Homework 9.5

1. 21   2. 81   3. No solution

4. 5   5. 5   6a. 16

b. 6.25   c. No solution   d. 28.09

7. $\sqrt{5}, -\sqrt{5}$ 8. 1   9. 5

10. -3 11. $\dfrac{-3}{4}$ 12. 8

13. -125 14. 16   15. $\sqrt[3]{7}$

16. $\dfrac{-1\pm2\sqrt{2}}{2}$ 17. $4\pm\sqrt{5}$ 18. $-3\pm3\sqrt{2}$

19. $\pm\sqrt{b^2-a^2}$ 20. $\pm2\sqrt{1+y^2}$

21a.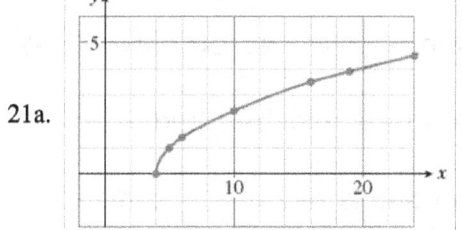

b. 13

22a.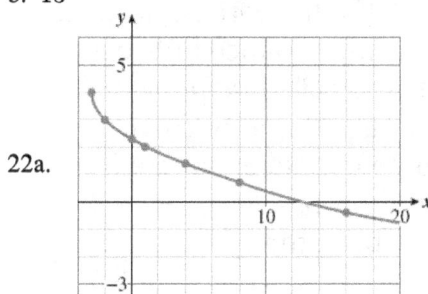

b. 6

23a. 144.5 km   b. It is at least $161\frac{1}{3}$ m

24. At least 24,867 ft   25a. 3 in   b. 2091 in^3

Chapter 9 Review

1a. a^{10} b. a^{24} 3a. $8a^6$ b. $2a^8$

5. $32x^7$ 7. $-32t^8$ 9a. $\dfrac{3}{x^2}$ b. $\dfrac{1}{9x^2}$

11a. $\dfrac{z^2}{25}$ b. $5z^2$ 13. $\dfrac{6}{p^7}$ 15. $\dfrac{8}{k}$

17. 5   19. 5.86×10^5 21. 7×10^{-4}

23. 4.83×10^5 25. $18,240,000,000,000,000$

27. $0.000\,025$ 29. 9.99×10^{-4} kg

31a. $2x^3$ b. Cannot be simplified   c. $(4+x)^3$

33. $-3m^2\sqrt{3m}$ 35. $\dfrac{a\sqrt{ac}}{4}$

37. $\dfrac{4b^2\sqrt{3b}}{3}$ 39. $\sqrt{6}+6\sqrt{2}$

41. $w^3\sqrt{6}$ 43. $2-2\sqrt{3}$

45. $\dfrac{4-3\sqrt{3}}{6}$ 47. $\dfrac{2\pm2\sqrt{6}}{3}$

49. $\dfrac{\pm\sqrt{c^2-2a^2}}{2}$ 51. $\sqrt{6}-3\sqrt{2}$

53. $4-d$ 55. $16+6\sqrt{7}$

57. $\dfrac{2\sqrt{x}}{x}$ 59. $\dfrac{\sqrt{10}}{2}$ 61. $\dfrac{5\sqrt{7}}{7}$

63. $2\left(\dfrac{1+\sqrt{7}}{2}\right)^2-2\left(\dfrac{1+\sqrt{7}}{2}\right)-3=0$

65. $2\sqrt{10a}$ 67. 7   69. 12

71. 41   73. 90 ft

Glossary

absolute value, *n.* the unsigned part of a number; for example, the absolute value of -7 is 7

algebraic expression, *n.* a meaningful combination of numbers, variables, and operation symbols; also called an *expression*

algebraic fraction, *n.* a fraction whose numerator and denominator are polynomials; also called a *rational expression*

altitude, *n.* (1) distance above the ground or above sea level; (2) the vertical distance between the vertex and the base of a geometric object, or the distance between parallel sides of a parallelogram, trapezoid, or rectangle; also called *height*

area, *n.* a measure of the two-dimensional space enclosed by a polygon or curve, typically expressed in terms of square units, such as square meters or square feet

associative law of addition: If a, b, and c are any numbers, then $(a + b) + c = a + (b + c)$

associative law of multiplication: If a, b, and c are any numbers, then $(a \cdot b) \cdot c = a \cdot (b \cdot c)$

average, *n.* a typical or middle value for a set of data

axis (*plural* **axes**), *n.* a line used as a reference for position and/or orientation

axis of symmetry, *n.* a line that cuts a plane figure into two parts, each a mirror image of the other

bar graph, *n.* a picture of numerical information in which the lengths or heights of bars represent the values of variables

base, *n.* (1) a number or algebraic expression that is used as a factor repeated a number of times indicated by the exponent; for example, in 3^5, the base is 3; (2) the bottom side of a polygon; (3) the bottom face of a solid

binomial, *n.* a polynomial consisting of exactly two terms

build a fraction, *v.* to find an equivalent fraction by multiplying numerator and denominator by the same nonzero expression

building factor, *n.* an expression by which both numerator and denominator of a given fraction are multiplied (in order to *build the fraction*)

Cartesian coordinate system, *n.* a grid that associates points in the coordinate plane to ordered pairs of numbers

Cartesian plane, *n.* a plane with a pair of coordinate axes, also called a *coordinate plane*.

circle, *n.* the set of all points in a plane at a fixed distance (the *radius*) from the center

circumference, *n.* the distance around a circle

coefficient, *n.* the numerical factor in a term; for example, in the expression $32a + 7b$, the coefficient of a is 32 and the coefficient of b is 7

combine like terms, *v.* to simplify an expression by adding or subtracting like terms as indicated

common factor, *n.* a quantity that divides evenly into each of two or more given expressions

commutative law of addition: If a and b are any numbers, then $a + b = b + a$

commutative law of multiplication: If a and b are any numbers, then $a \cdot b = b \cdot a$

complementary angles, *n.* two angles whose measures add up to 90°

completing the square, *n.* adding a constant to a quadratic polynomial so that the result is the square of a binomial, a technique used to solve quadratic equations

complex fraction, *n.* a fraction that contains one or more fractions in its numerator and/or in its denominator

complex conjugates, *n.* a pair of complex numbers such as $1 + 2i$ and $1 - 2i$ whose real parts are equal and whose imaginary parts are opposites

complex number, *n.* a number that can be written in the form $a + bi$, where a and b are real numbers and $i^2 = -1$

compound inequality, *n.* a mathematical statement involving two order symbols. For example, the compound inequality $1 < x < 2$ says that   "1 is less than x and x is less than 2."

cone, *n.* a three-dimensional object consisting of a circular base, a vertex, and the points on the line segments joining the circle to the vertex (see page xxx)

conjugate, *n.* see *complex conjugates*

consistent system of equations, *n.* a system of equations having at least one solution

constant, *adj.* unchanging, not variable; for example, the product of two variables is *constant* if the product is always the same number, for any values of the variables

constant, *n.* a number (as opposed to a *variable*)

constant of proportionality, *n.* the quotient of two directly proportional variables, or the product of two inversely proportional variables; also called the *constant of variation*

continuous, *adj.* without holes or gaps; for example, a curve is continuous if it can be drawn without lifting the pencil from the page

coordinate, *n.* a number used with a number line or an axis to designate position

coordinate axis, *n.* one of the two perpendicular number lines used to define the coordinates of points in the plane

coordinate plane, *n.* a plane containing a pair of coordinate axes; also called the *Cartesian plane* or *xy-plane*

corresponding angles, *n.* when two lines ℓ_1 and ℓ_2 are both intersected by a third line ℓ_3 as shown below, angles A and A', B and B', C and C', D and D' are the four pairs of corresponding angles formed.

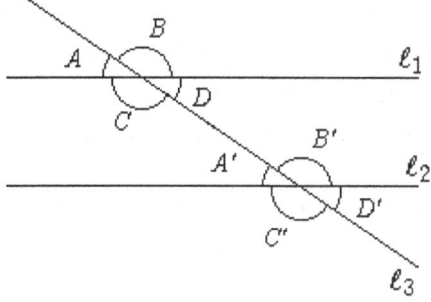

costs, *n.* money that an individual or group must pay out; for example, a company's costs might include payments for wages, supplies, and rent

counting number, *n.* one of the numbers $1, 2, 3, 4, \ldots$; also called a *natural number*

cross-multiply, *v.* to re-write a proportion as an equivalent equation without fractions by equating the two products obtained by multiplying a numerator of one side of the proportion by the denominator of the other side

cube, *n.* (1) a three-dimensional box whose six faces all consist of squares; (2) an expression raised to the power 3

cube, *v.* to raise an expression to the power 3; for example, to cube 2 means to form the product of three 2's, or $2^3 = 2 \times 2 \times 2 = 8$

cube root, *n.* a number which when raised to the power 3 gives a desired value; for example, 2 is the cube root of 8 because $2^3 = 8$.

cylinder, *n.* a three-dimensional figure in the shape of a tin can

decimal fraction, *n.* a decimal number (usually less than 1 in absolute value)

decimal number, *n.* a number such as 2.718 in which the fractional part is indicated by the digits to the right of the decimal point

decimal place, *n.* the relative position of a digit on the right of the decimal point; for example, in the number 3.14159, the digit 4 is in the second decimal place, or hundredths place

decimal point, *n.* the period or dot between the whole number part and the fractional part of a decimal number; for example, the decimal form of $1\frac{3}{10}$ is 1.3

degree, *n.* a measure of angle equal to $\frac{1}{360}$ of a complete revolution

degree of a polynomial (in one variable), *n.* the largest exponent that appears on the variable in the polynomial

denominator, *n.* the expression below the fraction bar in a fraction

dependent system of equations, *n.* a system of equations having infinitely many solutions

dependent variable, *n.* a variable whose value is determined by specifying the value of the *independent variable*; on a coordinate plane, the values of the dependent variable are displayed on the vertical axis

descending powers, *n.* terms in a polynomial arranged with the term of highest degree written first, then the term with the second-highest degree, and so on

diameter, *n.* the distance across a circle through its center, equal to twice the radius

difference, *n.* the result of a subtraction; for example, the expression $a - b$ represents the difference between a and b.

difference of squares, *n.* an expression of the form $a^2 - b^2$

direct variation, *n.* a relation between two variables such that one is a constant multiple of the other (so that the ratio between the two variables is the constant)

directed distance, *n.* the difference between the coordinates of the ending and starting points of a segment of a number line; the directed distance is negative if the ending value is smaller than the starting value; for example, the directed distance from 5 to 2 is $2 - 5 = -3$

directly proportional, *adj.* describing variables related by the equation $y = kx$

discriminant, *n.* the quantity $b^2 - 4ac$, where a, b, and c are the coefficients of the quadratic equation $ax^2 + bx + c = 0$

distributive law: For any numbers $a, b,$ and c, $a(b + c) = ab + ac$

dividend, *n.* a quantity that is divided by another quantity; for example, in the expression $a \div b$, the dividend is a

divisor, *n.* a quantity that is divided into another quantity; for example, in the expression $a \div b$, the divisor is b

elimination method, *n.* a method for solving a system of equations that involves adding together the equations of the system or multiples of the equations of the system

equation, *n.* a mathematical statement that two expressions are equal

equilateral triangle, *n.* a triangle with three sides of equal length

equivalent equations, *n.* equations that have the same solutions

equivalent expressions, *n.* expressions that represent the same value for any value(s) of the variable(s)

evaluate, *v.* to compute the value of an expression when the variable in the expression is replaced by a number

exponent, *n.* the number or expression that indicates how many times a *base* is used as a factor; for example, in 3^5, the exponent is 5, and $3^5 = 3 \times 3 \times 3 \times 3 \times 3$

expression, see *algebraic expression*

extraction of roots, *n.* a method used to solve quadratic equations

extraneous solution, *n.* a value that is not a solution to a given equation but is a solution to an equation derived from the original

extrapolation, *n.* an estimate of the value of a variable beyond the range of the data

factor, *n.* an expression that divides evenly into another expression; for example, 2 is a factor of 6

factor, *v.* to write as a product; for example, to factor 6 we write $6 = 2 \times 3$.

FOIL, *n.* an acronym for a method of computing the product of two binomials: multiply the **F**irst, **O**uter, **I**nner, and **L**ast terms

formula, *n.* an equation involving two or more variables

fraction, *n.* a number in the form $\frac{a}{b}$, where a and b are numbers and $b \neq 0$

fraction bar, *n.* the line segment separating the numerator and denominator of a fraction; it indicates division

fundamental principle of fractions: If a is any number and b and c are nonzero numbers, then $\frac{a \cdot c}{b \cdot c} = \frac{a}{b}$

geometrically similar, *adj.* having the same shape (but possibly different size)

graph, *v.* to draw a graph

graph of an equation (or **inequality**), *n.* a picture of the solutions of an equation (or inequality) using a number line or coordinate plane

greatest common factor (GCF), *n.* the largest factor that divides evenly into two or more expressions

height, *see altitude.*

hemisphere, *n.* half a sphere

histogram, *n.* a type of bar graph in which the height of each bar gives the frequency of a particular outcome or event

horizontal axis, *n.* the horizontal coordinate axis; often called the x-*axis*

horizontal intercept, *n.* the point where the graph meets the horizontal axis; often called x-*intercept.*

hypotenuse, *n.* the longest side of a right triangle (always the side opposite the right angle)

imaginary number, *n.* a complex number of the form bi, where b is a real number and $i^2 = -1$

improper fraction, *n.* a fraction in the form $\frac{a}{b}$, where a and b are natural numbers and $a \geq b$

inconsistent system of equations, *n.* a system of equations having no solution.

independent system of equations, *n.* a system of equations having at most one solution

independent variable, *n.* a variable whose value determines the value of the *dependent variable*, on a coordinate plane, the values of the independent variable are displayed on the horizontal axis

inequality, *n.* a mathematical statement that two quantities are not equal; it takes the form $a < b$, or $a \leq b$, or $a > b$, $a \geq b$, or $a \neq b$

integer, *n.* a whole number or the negative of a whole number

intercept, *n.* a point where a graph meets a coordinate axis

intercept method, *n.* a method for graphing a line by finding its horizontal and vertical intercepts.

interest, *n.* money paid for the use of money; for example, after borrowing money, the borrower must pay the lender not only the original amount of money borrowed (known as the *principal*) but also the interest, which is a certain percentage of the principal

interest rate, *n.* the fraction of the principal that is paid as interest for one year; for example, an interest rate of 10% means that the interest for one year is 10% of the principal

interpolation, *n.* estimating the value of a variable based on data that include both larger and smaller values of the variable

irrational number, *n.* a number that cannot be expressed as a quotient of two integers but does correspond to a point on the number line

isosceles triangle, *n.* a triangle with two sides of equal length

least common denominator (LCD), *n.* the smallest denominator that is a multiple of the denominators of two or more fractions

least common multiple (LCM), *n.* the smallest natural number that two or more natural numbers divide into evenly

leg, *n.* one of the two shorter sides of a right triangle (one side of the right angle.)

like fractions, *n.* fractions with equivalent denominators

like radicals, *n.* square roots with equivalent radicands (or cube roots with equivalent radicands)

like terms, *n.* terms with equivalent variable parts

line graph, *n.* a type of graph in which the points are connected with line segments

linear equation, *n.* an equation such as $2x + 3y = 4$ or $x - 3y = 7$ in which each term has degree zero or one

linear term, *n.* a term that consists of a constant times a variable

mean, *n.* the average of a set of numbers, computed by adding the numbers and dividing by how many are in the set. For example, the mean of 5, 2, and 11 is $\frac{5+2+11}{3} = 6$

median, *n.* the middle number in a set of numbers written in order; for example, the median of $5, 2$, and 11 is 5; if the set has two numbers in the middle when written in order, then the median is the mean of those two: the median of $6, 1, 9$, and 27 is $\dfrac{6 + 9}{2} = 7.5$.

mixed number, *n.* a number such as $33\frac{1}{3}$ written in the form $N\frac{a}{b}$ to mean $N + \frac{a}{b}$, where N is a natural number and $\frac{a}{b}$ is a positive proper fraction

mode, *n.* the number which occurs most frequently in a set of numbers. For example, the mode of $1, 1, 2$, and 3 is 1

model, *n.* a mathematical expression used to represent a situation in the world or a situation described in words; for example, the equation $P = R - C$ is a model for the relationship among the variables profit, revenue, and cost

model, *v.* to create a model

monomial, *n.* an algebraic expression with only one term

natural number, *n.* a counting number

negative number, *n.* a number that is less than 0

negative of, *n.* the opposite of

number line, *n.* a line, such as the one shown below, with coordinates marked on it

numerator, *n.* the expression above the fraction bar in a fraction

numerical coefficient, *n.* see *coefficient*

operation, *n.* addition, subtraction, multiplication, or division, or raising to a power, or taking a root

opposite of a number, *n.* the number with the same absolute value as a given number but the opposite sign; for example, the opposite of -5 is 5

order of operations, *n.* rules that prescribe the order in which to carry out the operations in an expression.

order symbol, *n.* one of the four symbols $<$, $\leq$, $>$, and $\geq$

ordered pair, *n.* a pair of numbers enclosed in parentheses: (x, y); often used to specify a point or a location on the coordinate plane

origin, *n.* the point where the coordinate axes meet, having coordinates $(0, 0)$

parabola, *n.* a curve with the shape of the graph of $y = ax^2$, where $a \neq 0$

parallel lines, *n.* lines that lie in the same plane but do not intersect, even if extended indefinitely

parallelogram, *n.* a four-sided figure in the plane with two pairs of parallel sides

percent, *n.* a fraction with (an understood) denominator of 100; for example, the fraction $\dfrac{51}{100}$ is equal to 51 percent or 51%

perfect square, *n.* the square of an integer; for example, 9 is a perfect square because $9 = 3^2$

perimeter, *n.* the distance around the edge or boundary of a two-dimensional figure

perpendicular lines, *n.* lines that form a right angle

point-slope formula, *n.* an equation for a line in terms of its slope and one point on the line:

$$y - y_1 = m(x - x_1) \text{ or } \frac{y - y_1}{x - x_1} = m$$

polygon, *n.* a closed geometric figure in the plane consisting of line segments that meet only at their endpoints; for example, triangles are polygons with three sides

polynomial, *n.* a sum of terms, each of which is either a constant or a constant times a power of a variable with the exponent a positive integer

positive number, *n.* a number greater than 0

power, *n.* an expression that consists of a base and an exponent

prime number, *n.* an integer greater than 1 whose only whole number factors are itself and 1

principal, *n.* the original amount of money deposited in an account or borrowed from a lender; see also *interest*

principal square root, *n.* the positive square root

product, *n.* the result of a multiplication; for example, the expression $a \cdot b$ represents the product of a and b

profit, *n.* money left after subtracting the costs from the revenue

proper fraction, *n.* a fraction in the form $\frac{a}{b}$, where a and b are integers and $a < b$

proportion, *n.* an equation in which each side is a ratio

proportional, see *directly proportional* and *inversely proportional.*

pyramid, *n.* a three-dimensional object with a polygonal base and triangular faces that meet in a vertex

Pythagorean theorem: If the legs of a right triangle are a and b and the hypotenuse is c, then $a^2 + b^2 = c^2$

quadrant, *n.* one of the four regions into which the coordinate axes divide the plane: the *first quadrant* consists of the points whose two coordinates are positive; the *second quadrant* of points whose first coordinate is negative and second coordinate positive; the *third quadrant* of points whose two coordinates are negative; and the *fourth quadrant* of points whose first coordinate is positive and second coordinate is negative

quadratic, *adj.* relating to the square of a variable or an expression

quadratic formula, *n.* the formula that gives the solutions of a quadratic equation $ax^2 + bx + c = 0$, namely

$$x = \frac{-b \pm \sqrt{b^2 - 4ac}}{2a}$$

quadratic polynomial, *n.* a polynomial whose degree is 2

quadratic term, *n.* a term whose degree is 2

quotient, *n.* the result of a division; for example, the expression $a \div b$ represents the quotient of a and b

radical, *n.* a square root or a cube root

radical equation, *n.* an equation in which the variable appears under a radical

radical sign, *n.* the symbol $\sqrt{}$ used to indicate the principal square root, or the symbol $\sqrt[3]{}$ used to indicate cube root

radicand, *n.* an expression under a radical sign

radius, *n.* the distance from any point on a circle or sphere to its center

raise to a power, *v.* use as a repeated factor, for example, to raise x to the power 2 is the same as multiplying $x \cdot x$

rate, *n.* a ratio that compares two quantities with different units

ratio, *n.* a comparison of two quantities expressed as a fraction

rational, *adj.* having to do with ratios

rational expression, *n.* a ratio of two polynomials; also called an *algebraic fraction*

rational number, *n.* a number that can be expressed as the ratio of two integers $\frac{a}{b}$, $b \neq 0$

rationalize the denominator, *v.* to find an equivalent fraction that contains no radical in the denominator; for example, when we rationalize $\frac{1}{\sqrt{2}}$, we obtain $\frac{\sqrt{2}}{2}$

real line, see *number line.*

real number, *n.* a number that corresponds to a point on a number line

reciprocal, *n.* the result of dividing 1 by a given number; for example, the reciprocal of 2 is $\frac{1}{2}$

rectangle, *n.* a four-sided figure (in the plane) with four right angles and opposite sides that are equal in length.

reduce a fraction, *v.* to find an equivalent fraction whose numerator and denominator share no common factors (other than 1)

regression line, *n.* the line used for linear regression

repeating decimal, *n.* a decimal number with a digit or block of digits that repeats itself endlessly; for example, the repeating decimal 0.16666... (with repeating 6's) is equivalent to $\frac{1}{6}$, and the repeating decimal 0.8414141... (with repeating 41's) is equivalent to $\frac{833}{990}$

revenue, *n.* money that an individual or group receives; for example, a person might have revenues from both a salary and investments

right angle, *n.* an angle of 90°

right triangle, *n.* a triangle that includes one right angle

root, *see cube root, principal square root,* and *square root*

round, *v.* to give an approximate value of a number by choosing the nearest number of a specified accuracy; for example, rounding 3.14159 to two decimal places gives 3.14

satisfy, *v.* to make an equation true (said of a value that is substituted for the variable); for example, the number 5 satisfies the equation $x - 2 = 3$.

scale, *v.* to multiply (measurements) by a fixed number (the *scale factor*)

scientific notation, *n.* a standard method for writing very large or very small numbers using powers of 10; for example, the scientific notation for $12,000$ is 1.2×10^4

semicircle, *n.* half a circle

signed number, *n.* a positive or negative number

similar, see *geometrically similar*

simplify, *v.* to write in an equivalent but simpler or more convenient form; for example, the expression $\sqrt{16}$ can be simplified to 4

slope, *n.* a measure of the steepness of a line or of the rate of change of one variable with respect to another, denoted by m

slope-intercept form, *n.* a standard form for the equation of a nonvertical line: $y = mx + b$

slope-intercept method, *n.* a method for graphing a line that uses the slope and the y-intercept

solution, *n.* a value for a variable that makes an equation or an inequality true; a solution to an equation in two variables is an ordered pair that satisfies the equation

solution to a system, *n.* an ordered pair that satisfies each equation of the system

solve, *v.* (1) to find any and all solutions to an equation, inequality, or system; (2) to write an equation for one variable in terms of any other variables; for example, solving $5x + y = 3$ for y gives $y = -5x + 3$

sphere, *n.* a three-dimensional object in the shape of a ball and consisting of all the points in space at a fixed distance (the radius) from the center of the sphere

square, *n.* (1) the product of any number or expression and itself; (2) a rectangle whose sides are all the same length

square, *v.* to multiply a number or expression by itself, that is, to raise to the power 2

square root, *n.* a number that when squared gives a desired value; for example, 7 is a square root of 49 because $7^2 = 49$

subscript, *n.* a small number written below and to the right of a variable to identify it; for example, in the equation $x_1 = 3$, the variable x has the subscript 1

substitution method, *n.* a method for solving a system of equations that begins by expressing one variable in terms of the other

sum, *n.* the result of an addition; for example, the expression $a + b$ represents the sum of a and b

surface area, *n.* the total area of the faces or surfaces of a three-dimensional object

supplementary angles, *n.* two angles whose measures add up to $180°$

system of (linear) equations, *n.* a pair of (linear) equations involving the same two variables

term, *n.* a quantity that is added to another; for example, in the expression $x - 4$, both x and -4 are terms

trapezoid, *n.* a four-sided figure in the plane with one pair of parallel sides

triangle, *n.* a three-sided figure in the plane

trinomial, *n.* a polynomial with exactly three terms

two-point formula for slope, *n.* a formula that shows how to compute the slope of the line passing through the points (x_1, y_1) and (x_2, y_2):

$$m = \frac{y_2 - y_1}{x_2 - x_1}$$

unlike fractions, *n.* fractions whose denominators are not equivalent

unlike terms, *n.* terms with variable parts that are not equivalent

variable, *adj.* not constant, subject to change

variable, *n.* a numerical quantity that changes over time or in different situations

variation, see *direct variation* and *inverse variation*

vertex, *n.* (1) a point where two sides of a polygon meet; (2) a corner or extreme point of a geometric object; (3) the highest or lowest point on a parabola

vertical angles, *n.* the two pairs of equal angles formed when two lines intersect; for example, in the figure below, angles A and D are one pair of vertical angles, and angles B and C are another pair of vertical angles

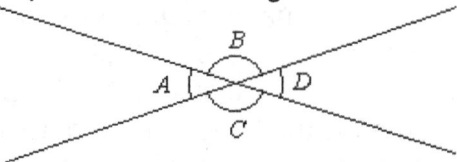

vertical axis, *n.* the vertical coordinate axis; often called the *y-axis*

vertical intercept, *n.* the point where the graph meets the vertical axis; also called the *y-intercept*

volume, *n.* a measure of the three-dimensional space enclosed by a three-dimensional object, typically expressed in terms of cubic units, such as cubic meters or cubic feet

whole number, *n.* one of the numbers 0, 1, 2, 3, …

***x*-axis**, see *horizontal axis*

***x*-intercept**, see *horizontal intercept*

***xy*-plane**, see *coordinate plane*

***y*-axis**, see *vertical axis*

***y*-intercept**, see *vertical intercept*

zero factor principle: If $ab = 0$, then either $a = 0$ or $b = 0$

Notation

.	"decimal point" : 5.3 represents $5\frac{3}{10}$.
$\cdot$	"times" : $5 \cdot 3$ represents 5 times 3, so $5 \cdot 3 = 15$.
$\vert\ \vert$	"absolute value" : $\vert-6\vert$ represents the absolute value of -6, so $\vert-6\vert = 6$.
$<$	"less than" : $2 < 3$ says that 2 is less than 3.
$\leq$	"less than or equal to" : $2 \leq 3$ says that 2 is less than 3 or equal to 3.
$>$	"greater than" : $5 > 1$ says that 5 is greater than 1.
$\geq$	"greater than or equal to" : $5 \geq 1$ says that 5 is greater than 1 or equal to 1.
$\approx$	"approximately equal" : $x \approx 3$ means that x is approximately equal to 3.
$\neq$	"not equal" : $0 \neq 1$ means that 0 is not equal to 1.
x^2	"x-squared" : the 2 is an exponent, so $x^2 = x \cdot x$ and $3^2 = 3 \cdot 3 = 9$.
$\pm$	"plus or minus" : ± 2 represents a positive or negative 2
$\sqrt{}$	"square root" : $\sqrt{4}$ represents the principle square root of 4, so $\sqrt{4} = 2$.
$\sqrt[3]{}$	"cube root" : $\sqrt[3]{8}$ represents the cube root of 8, so $\sqrt[3]{8} = 2$.
$\llcorner$	"perpendicular" : the small square indicates that the two lines are perpendicular
$\circ$	"degrees" : $5°$ represent 5 degrees.
π	"pi" : the ratio of the circumference of a circle to its diameter. $\pi \approx 3.14159$
%	"per cent" : the numerator of a fraction whose denominator is 100
Δ	"change in" : Δx (read "delta x") represents the change in x

Associative Laws
Addition: If a, b, and c are any numbers, then $(a+b)+c = a+(b+c)$.
Multiplication: If a, b, and c are any numbers, then $(a \cdot b) \cdot c = a \cdot (b \cdot c)$.

Commutative Laws
Addition: If a and b are any numbers, then $a+b = b+a$.
Multiplication: If a and b are any numbers, then $a \cdot b = b \cdot a$.

Distributive Law: $a(b+c) = ab + ac$ for any numbers a, b, and c.

Properties of Equality
Addition: If $a = b$ and c is any number, then $a+c = b+c$.
Subtraction: If $a = b$ and c is any number, then $a-c = b-c$.
Multiplication: If $a = b$ and c is any number, then $a \cdot c = b \cdot c$.
Division: If $a = b$ and c is any nonzero number, then $\dfrac{a}{c} = \dfrac{b}{c}$.

Fundamental Principle of Fractions
If a is any number, and b and c are nonzero numbers, then $\dfrac{a \cdot c}{b \cdot c} = \dfrac{a}{b}$.

Laws of Exponents:
First: $a^m \cdot a^n = a^{m+n}$

Second: $\dfrac{a^m}{a^n} = a^{m-n} \quad (n < m)$

$\dfrac{a^m}{a^n} = \dfrac{1}{a^{n-m}} \quad (n > m)$

Third: $(a^m)^n = a^{mn}$

Fourth: $(ab)^n = a^n b^n$

Fifth: $\left(\dfrac{a}{b}\right)^n = \dfrac{a^n}{b^n}$

Product Rule for Radicals
If a and b are both nonnegative, then $\sqrt{ab} = \sqrt{a}\,\sqrt{b}$.

Quotient Rule for Radicals
If $a \geq 0$ and $b > 0$, then $\sqrt{\dfrac{a}{b}} = \dfrac{\sqrt{a}}{\sqrt{b}}$.

Zero Factor Principle
If $ab = 0$, then either $a = 0$ or $b = 0$.

Geometric Formulas

Rectangle
Area $= lw$
Perimeter $= 2l + 2w$

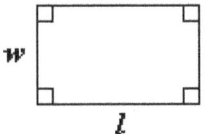

Box
Volume $= lwh$
Surface Area $= 2lw + 2lh + 2wh$

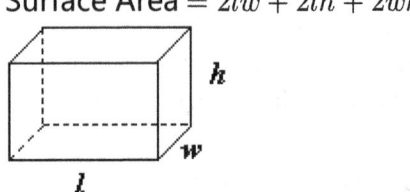

Circle
Area $= \pi r^2$
Circumference $= 2\pi r$

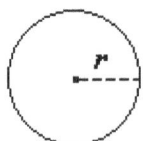

Sphere
Volume $= \frac{4}{3}\pi r^3$
Surface Area $= 4\pi r^2$

Triangle
Area $= \frac{1}{2}bh$
Perimeter $= a + b + c$

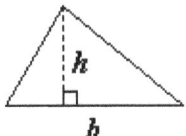

Cone
Volume $= \frac{1}{3}\pi r^2 h$
Surface Area $= \pi r^2 + \pi rs$

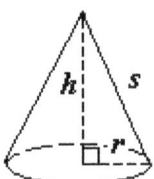

Parallelogram
Area $= bh$

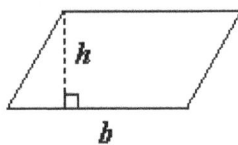

Cylinder
Volume $= \pi r^2 h$
Surface Area $= 2\pi rh + 2\pi r^2$

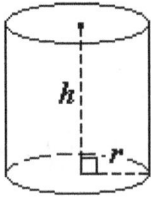

Trapezoid
Area $= \frac{1}{2}h(a + b)$

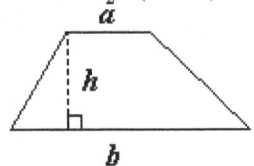

Common Units of Measure

English Units

Length

    12 inches = 1 foot
    3 feet = 1 yard
    5280 feet = 1 mile

Weight

    16 ounces = 1 pound
    2000 pounds = 1 ton

Fluid Capacity

    16 fluid ounces = 1 pint
    2 pints = 1 quart
    4 quarts = 1 gallon

Metric Units

    100 centimeters = 1 meter
    1000 meters = 1 kilometer

    1000 milligrams = 1 gram
    1000 grams = 1 kilogram

    1 cubic centimeter = 1 milliliter
    1000 milliliters = 1 liter

Unit Conversion

English to Metric

    1 inch = 2.54 centimeters
    1 foot = 0.305 meters
    1 mile = 1.609 kilometers

    1 ounce = 28.350 grams
    1 pound = 0.454 kilograms

    1 quart = 0.946 liter

Metric to English

    1 centimeter = 0.394 inch
    1 meter = 3.28 feet
    1 kilometer = 0.621

    1 gram = 0.035 ounces
    1 kilogram = 2.205 pounds

    1 liter = 1.057 quarts